POLYNOMIAL OPERATOR EQUATIONS

in Abstract Spaces and Applications

IOANNIS K. ARGYROS

Department of Mathematics
Cameron University
Lawton, Oklahoma

CRC Press

Taylor & Francis Group
Boca Raton London New York

CRC Press is an imprint of the
Taylor & Francis Group, an **informa** business

CRC Press
Taylor & Francis Group
6000 Broken Sound Parkway NW, Suite 300
Boca Raton, FL 33487-2742

© 1998 by Taylor & Francis Group, LLC
CRC Press is an imprint of Taylor & Francis Group, an Informa business

First issued in paperback 2019

No claim to original U.S. Government works

ISBN-13: 978-0-367-44787-8 (pbk)
ISBN-13: 978-0-8493-8702-9 (hbk)

Visit the Taylor & Francis Web site at
http://www.taylorandfrancis.com

and the CRC Press Web site at
http://www.crcpress.com

Library of Congress Cataloging-in-Publication Data

Catalog record is available from the Library of Congress.

Cover Design: *Dawn Boyd*

Contents

List of Tables

Introduction

My goal in the text is to present new and important old results about polynomial equations as well as an analysis of general new and efficient iterative methods for their numerical solution in various space settings. To achieve this goal, we made the text as self-contained as possible by proving all the results in great detail. Exercises have been added at the end of each chapter, which complement the material. Some of the exercises are considered really to be results (theorems, propositions, etc.) that we decided not to include in the main body of each chapter. Several applications of our results are given for the solution of integral as well as differential equations throughout every chapter.

We have provided material that can be used on the one hand as a required text for senior undergraduate, graduate students, and others in the following study areas: Advanced Numerical Analysis, Numerical Functional Analysis, Functional Analysis, and Approximation Theory. On the other hand, the text can be recommended for an Integral or Differential Equations course. Moreover, to make the work useful as a reference source, literature citations will be supplied at the end of each chapter with possible extensions of the facts contained here or open problems. We will use graphics and exercises designed to allow students to apply the latest technology. In addition, the text will end with a very updated and comprehensive bibliography in the field. The main prerequisite for the reader is the material covered in: Advanced Calculus, second course in Numerical-Functional Analysis, and a first course in Algebra and Integral-Differential Equations. A comprehensive modern presentation of the subject to be described here appears to be needed due to the rapid growth in this field and should benefit not only those working in the field, but also those interested in, or in need of, information about specific results or techniques.

Abstract polynomial equations are evidently systems of algebraic polynomial equations. Polynomial systems can arise directly in applications, or be approximations to equations involving operators having a power series expansion at a certain point. Another source of polynomial systems is the discretization of polynomial equations taking place when a differential or an integral equation is solved. Finite polynomial systems can be obtained by taking a segment of an infinite system, or by other approximation techniques applied to equations in infinite dimensional space.

Chapters 1, 2, and 3 cover special cases of nonlinear operator equations. In particular, the solution of polynomial operator equations of positive integer degree n is discussed. The so-called polynomial operators are a natural generalization of linear operators. Equations in such operators are the linear space analog of ordinary polynomials in one or several variables over the fields of real or complex numbers. Such equations encompass a broad spectrum of applied problems including all linear equations. Often the polynomial nature of many nonlinear problems goes unrecognized by researchers. This is most likely due to the fact that unlike polynomials in a single variable, polynomial operators have received little attention. It must certainly be mentioned that existence theory is far from complete and what little is there is confined to local small solutions in neighborhoods that are often of very small radius. Here an attempt is made to partially fill this space by doing the following:

(a) Numerical methods for approximating distinct solutions of quadratic ($n = 2$) (in Chapters 1 and 2) and polynomial equations ($n \geq 2$) (in Chapter 3) are given.

(b) Results on global existence theorems not related with contractions are provided.

(c) Moreover, for those of a qualitative rather than computational frame of mind, it has been suggested that polynomial operators should carry a Galois theory. In an attempt to inform and contribute in this area, we have provided our results at the end of each chapter.

Chapter 4 deals with polynomial integral as well as polynomial differential equations appearing in radiative transfer, heat transfer, neutron transport, electromechanical networks, elasticity, and other areas. In particular, results on the various Chandrasekhar equations (Nobel Prize of Physics, 1983) are given using Chapters 1 through 3. These results are demonstrated through the examination of different cases.

In Chapter 5 we study the Weierstrass theorem, Matrix representations, Lagrange and Hermite interpolation, completely continuous multilinear operators, and the bounds of polynomial equations in the following settings: Banach space, Banach algebra, and Hilbert space.

Finally in Chapter 6 we provide general methods for solving operator equations. In particular we use inexact Newton-like methods to approximate solutions of nonlinear operator equations in Banach space. We also show how to use these general methods to solve polynomial equations.

Chapter 1

Quadratic Equations and Perturbation Theory

This chapter introduces existence and uniqueness theorems to find a solution "close" to zero as well as a solution "away" from zero for the quadratic equation in a Banach space. We make use of the well known contraction mapping principle mainly in order to achieve this goal [15, 17, 18, 33, 60, 114, 119, 187].

1.1 Algebraic Theory of Quadratic Operators

In this section we give a very brief introduction to the theory of bilinear and quadratic operators needed for what follows.

Let X, Y, Z denote linear spaces over the field R ($R = \mathbf{R}$ or $R = \mathbf{C}$).

DEFINITION 1.1
Let B be a bilinear operator on $X \times X$. Define the mean $\bar{B}$ of B on $X \times X$ by

$$\bar{B}(x, y) = \frac{1}{2}(B(x, y) + B(y, x)) \quad \text{for all } x, y \in X .\qquad (1.1)$$

Note that $\bar{B}$ is symmetric.

DEFINITION 1.2 An operator $Q: X \to Z$ is said to be quadratic if there exists a bilinear operator $B: X \times X \to Z$ such that

$$Q(x) = B(x, x)\qquad (1.2)$$

for all $x \in X$; Q can be regarded as the restriction of B to the diagonal axis of

1

$X \times X$, *denoted also as* X^2 *wherever convenient.*

REMARK 1.1   Denote by $Q(X; Z)$ the set of all quadratic operators from X to Z. This set becomes a vector space when addition and scalar multiplication are defined by

$$(Q_1 + Q_2)(x) = Q_1(x) + Q_2(x)$$

$$(cQ)(x) = cQ(x),$$

for all $Q_1, Q_2, Q \in Q(X; Z)$, $x \in X$, $c \in S$.   ∎

THEOREM 1.1

An operator $Q: X \rightarrow Z$ *is a quadratic operator if and only if there exists a symmetric bilinear operator* $\bar{B}: X \times X \rightarrow Z$ *satisfying*

$$Q(c_1 x + c_2 y) = c_1^2 Q(x) + 2c_1 c_2 \bar{B}(x, y) + c_2^2 Q(y) \qquad (1.3)$$

for all $x, y \in X$, $c_1, c_2 \in S$. *Such a symmetric* $\bar{B}$ *is unique.*

PROOF   To show the sufficiency of the condition, set $c_1 = c_2 = \frac{1}{2}$ and $x = y$. Then,

$$Q(x) = \frac{1}{4}Q(x) + \frac{1}{2}\bar{B}(x, x) + \frac{1}{4}Q(x) \quad \text{for all } x \in X \qquad (1.4)$$

i.e., $Q(x) = \bar{B}(x, x)$, showing that Q is quadratic. Conversely, suppose Q is quadratic. Then we have a bilinear operator $B: X \times X \rightarrow Z$ such that

$$Q(x) = B(x, x) \quad \text{for all } x \in X. \qquad (1.5)$$

Define a bilinear operator $\bar{B}: X \times X \rightarrow Z$ by

$$\bar{B}(x, y) = \frac{1}{2}(B(x, y) + B(y, x)) \quad \text{for all } x, y \in X. \qquad (1.6)$$

Then $\bar{B}(x, y) = \bar{B}(y, x)$ and therefore $\bar{B}$ is symmetric.
    Suppose there exists a symmetric bilinear operator $\bar{B}_1: X \times X \rightarrow Z$ such that

$$Q(x) = \bar{B}(x, x) = \bar{B}_1(x, x) \quad \text{for all } x \in X.$$

Replacing x by $x + y$, we obtain

$$Q(x + y) = \bar{B}(x + y, x + y) = \bar{B}_1(x + y, x + y),$$

so that

$$\bar{B}(x, x) + \bar{B}(x, y) + \bar{B}(y, x) + \bar{B}(y, y) = \bar{B}_1(x, x) + \bar{B}_1(x, y) + \bar{B}_1(y, x) + \bar{B}_1(y, y)$$

which implies

$$\bar{B}(x, y) + \bar{B}(y, x) = \bar{B}_1(x, y) + \bar{B}_1(y, x)$$

or $\bar{B}(x, y) = \bar{B}_1(x, y)$. Hence, $\bar{B}$ is unique.

Finally,

$$Q(c_1 x + c_2 y) = \bar{B}(c_1 x + c_2 y, c_1 x + c_2 y)$$

$$= c_1^2 \bar{B}(x, x) + 2 c_1 c_2 \bar{B}(x, y) + c_2^2 \bar{B}(y, y)$$

$$= c_1^2 Q(x) + 2 c_1 c_2 \bar{B}(x, y) + c_2^2 Q(y).$$

Thus, the theorem follows. ∎

We now introduce the following concept.

DEFINITION 1.3 *Let $B : X \times Y \to Z$ be bilinear. Then for each $x \in X$ let*

$$B_X(x) : Y \to Z, \quad \text{sending } y \in Y \text{ to } B(x, y). \tag{1.7}$$

Then $B_X(x)$ is obviously a linear operator, and so $B_X(x) \in L(Y; Z)$. The operator B_X, defined by

$$B_X : X \to L(Y; Z), \quad \text{sending } x \in X \text{ to } B_X(x), \tag{1.8}$$

is called the associated left linear operator for B. Similarly we define the right linear operator for B by

$$B_Y : Y \to L(X; Z), \quad \text{sending } y \in Y \text{ to } B_Y(y). \tag{1.9}$$

Note that for all $x \in X$, $y \in Y$, we have

$$(B_X(x)) y = B(x, y) = (B_Y(y)) x. \tag{1.10}$$

DEFINITION 1.4   *Let m, n, r be integers such that $m, n, r \geq 1$. A family $c^* = \{c_i^{jk}/c_i^{jk} \in S,\ j = 1, \ldots, m;\ k = 1, \ldots, n;\ i = 1, \ldots, r\}$ is called a 2-matrix of size $(m, n; r)$.*

DEFINITION 1.5   *Let $B: X \times Y \to Z$ be bilinear. The left radical of B is defined as the intersection set $\bigcap_{y \in Y} ker(B_Y(y))$ in X. The right radical of B, defined similarly, is the subset $\bigcap_{x \in X} ker(B_X(x))$ in Y.*

DEFINITION 1.6   *Let $Q: X \to Z$ be quadratic and let $Q^*: X \times X \to Z$ be the symmetric bilinear operator associated by Theorem 1.1. The radical of Q is defined to be the set*

$$rad(Q) = rad\left(Q^*\right) = ker\left(Q_X^*\right) . \tag{1.11}$$

DEFINITION 1.7   *The null set of Q is defined to be the set $null(Q) = \{x \in X \mid Q(x) = O_Z\}$.*

THEOREM 1.2
$x \in rad(Q)$ if and only if $Q(x + y) = Q(y)$, $y \in X$.

PROOF   Suppose $x \in rad(Q)$; then $Q^*(x, y) = O_Z$, $y \in X$. Now,

$$Q(x + y) = Q^*(x + y, x + y) = Q^*(x, x) + Q^*(y, x) + Q^*(x, y) + Q^*(y, y)$$

$$= Q^*(x, x) + Q^*(x, y) + Q^*(x, y) + Q^*(y, y) ;$$

since Q^* is symmetric, this implies

$$Q(x + y) = Q(y) .$$

Conversely, suppose $Q(x + y) = Q(y)$, $y \in X$. Then

$$Q^*(x, x) + 2Q^*(x, y) = O_Z, \quad y \in X .$$

In particular $Q^*(x, x) + 2Q^*(x, x) = O_Z$, i.e., $x \in ker(Q_X^*) = rad(Q)$. ∎

We now close this section by giving examples of the operators defined above.

Example 1.1

Let $X = Y = Z = \mathbf{R}^2$. A 2-matrix c^* over $\mathbf{R}$ of size $(2, 2; 2)$ can be displayed in a unique way by an array,

$$c^* \equiv \begin{pmatrix} c_1^{11} & c_1^{12} \\ c_1^{21} & c_1^{22} \\ \hline c_2^{11} & c_2^{12} \\ c_2^{21} & c_2^{22} \end{pmatrix}, \quad \text{where } c_i^{jk} \in \mathbf{R}. \tag{1.12}$$

Let the bases of the $\mathbf{R}^2$s be $\{(1, 0), (0, 1)\}$. Let $x, y \in \mathbf{R}^2$ and $x = \begin{pmatrix} x_1 \\ x_2 \end{pmatrix}$, $y = \begin{pmatrix} y_1 \\ y_2 \end{pmatrix}$. Define an operator $B: \mathbf{R}^2 \times \mathbf{R}^2 \to \mathbf{R}^2$ by the following scheme:

$$B(x, y) \equiv \left[(x_1, x_2) \begin{pmatrix} c_1^{11} & c_1^{12} \\ c_1^{21} & c_1^{22} \\ \hline c_2^{11} & c_2^{12} \\ c_2^{21} & c_2^{22} \end{pmatrix} \begin{pmatrix} y_1 \\ y_2 \end{pmatrix} \right]$$

$$\equiv \left[\begin{pmatrix} (x_1, x_2) \begin{pmatrix} c_1^{11} & c_1^{12} \\ c_1^{21} & c_1^{22} \end{pmatrix} \begin{pmatrix} y_1 \\ y_2 \end{pmatrix} \\ (x_1, x_2) \begin{pmatrix} c_2^{11} & c_2^{12} \\ c_2^{21} & c_2^{22} \end{pmatrix} \begin{pmatrix} y_1 \\ y_2 \end{pmatrix} \end{pmatrix} \right]$$

$$\equiv \begin{pmatrix} c_1^{11} x_1 y_1 + c_1^{21} x_2 y_2 + c_1^{12} x_1 y_2 + c_1^{22} x_2 y_2 \\ c_2^{11} x_1 y_1 + c_2^{21} x_2 y_1 + c_2^{12} x_1 y_2 + c_2^{22} x_2 y_2 \end{pmatrix}. \tag{1.13}$$

Choosing $x_1 = y_1$, $x_2 = y_2$ we obtain a quadratic operator $Q: \mathbf{R}^2 \to \mathbf{R}^2$ defined by

$$Q(x) \equiv \begin{pmatrix} c_1^{11} x_1^2 + (c_1^{12} + c_1^{21}) x_1 x_2 + c_1^{22} x_2^2 \\ c_2^{11} x_1^2 + (c_2^{12} + c_2^{21}) x_1 x_2 + c_2^{22} x_2^2 \end{pmatrix}.$$

Now the ordinary 2×2 matrix representation of the linear operator $B_X(x): Y \to Z$ is

$$\begin{pmatrix} c_1^{11} x_1 + c_1^{21} x_2 & c_1^{12} x_1 + c_1^{22} x_2 \\ c_2^{11} x_1 + c_2^{21} x_2 & c_2^{12} x_1 + c_2^{22} x_2 \end{pmatrix}.$$

Similarly for the linear operator $B_Y(y): X \to Z$, we obtain

$$\begin{pmatrix} c_1^{11}y_1 + c_1^{12}y_2 & c_2^{11}y_1 + c_2^{12}y_2 \\ c_1^{21}y_1 + c_1^{22}y_2 & c_2^{21}y_1 + c_2^{22}y_2 \end{pmatrix}.$$

DEFINITION 1.8   *Let T_1, T_2 be linear operators from X to Y; then T_1 is called the square root of T_2 if $T_1(T_1x) = T_2X$ for all $x \in X$. We also write*

$$T_1^2 = T_2 . \tag{1.14}$$

DEFINITION 1.9   *Let B denote the unique symmetric bilinear operator associated with the quadratic operator Q. The subset $F\{B\} = \{x \in X \mid (B_X(x))^2 = B_X(B(x,x))\}$ of X is called the factor set of B.*

Example 1.2
Using Example 1.1, with $X = Y = Z = \mathbf{R}^2$, we obtain $(B_X(x))^2 = B_X(B(x,x))$, where $x = \begin{pmatrix} x_1 \\ x_2 \end{pmatrix}$, $x_1, x_2 \in \mathbf{R}$; i.e., $B_X(x)B_X(x)(y) = B_X(B(x,x))(y)$ for all $y \in \mathbf{R}^2$, where $y = \begin{pmatrix} y_1 \\ y_2 \end{pmatrix}$, $y_1, y_2 \in \mathbf{R}$. Therefore,

$$B(x, B(x, y)) = B(B(x, x), y) . \tag{1.15}$$

Let us choose the array

$$\begin{pmatrix} 1 & 0 \\ 0 & 0 \\ \overline{} & \overline{} \\ 0 & 0 \\ 0 & 1 \end{pmatrix},$$

using (1.14) above.   Then we obtain $B(x, y) = \begin{pmatrix} x_1 y_1 \\ x_2 y_2 \end{pmatrix}$ and $B(y, x) = \begin{pmatrix} y_1 & x_1 \\ x_2 & y_2 \end{pmatrix}$, so $B(x, y) = B(y, x)$ for all $(x, y) \in R^2$ and therefore B is a symmetric bilinear operator from $\mathbf{R}^2 \times \mathbf{R}^2$ to $\mathbf{R}^2$.
   Let us apply (1.14) from Example 1.1 in (1.15) above. Then we get

$$B\left(\begin{pmatrix} x_1 \\ y_1 \end{pmatrix}, \begin{pmatrix} x_1 y_1 \\ x_2 y_2 \end{pmatrix} \right) = B\left(\begin{pmatrix} x_1^2 \\ x_2^2 \end{pmatrix}, \begin{pmatrix} y_1 \\ y_2 \end{pmatrix} \right)$$

i.e.,

$$\begin{pmatrix} x_1^2 y_1 \\ x_2^2 y_2 \end{pmatrix} = \begin{pmatrix} x_1^2 y_1 \\ x_2^2 y_2 \end{pmatrix},$$

which gives

$$x_1^2 y_1 = x_1^2 y_1$$

$$x_2^2 y_2 = x_2^2 y_2 .$$

But the above are identities and therefore x_1, x_2 can be chosen arbitrarily. Since x_1, x_2 can be chosen arbitrarily, $x = \begin{pmatrix} x_1 \\ x_2 \end{pmatrix}$ can be chosen arbitrarily and hence $F\{B\} = \mathbf{R}^2$.

REMARK 1.2   Example 1.2 shows that if $z = \begin{pmatrix} z_1 \\ z_2 \end{pmatrix}$ where $z_1, z_2 \in \mathbf{R}^+$, then the equation $B(x, x) = z$, where B is the bilinear operator defined in Example 1.2, has a solution

$$u = \begin{pmatrix} \sqrt{z_1} \\ \sqrt{z_2} \end{pmatrix} ,$$

because

$$B(u, u) = \begin{pmatrix} (\sqrt{z_1})^2 \\ (\sqrt{z_2})^2 \end{pmatrix} = \begin{pmatrix} z_1 \\ z_2 \end{pmatrix} . \quad \blacksquare$$

DEFINITION 1.10   *Let $\{(x_n, y_n)\}$, $n = 1, 2, \ldots$ be a sequence in $X \times Y$. We say the sequence converges to (x, y), symbolized by $(x_n, y_n) \to (x, y)$, $n \to \infty$, if*

$$\|(x_n, y_n) - (x, y)\| = \|(x_n - x), (y_n - y)\| \to 0, \quad as \; n \to \infty . \qquad (1.16)$$

It follows that $(x_n, y_n) \to (x, y)$ if and only if $\|x_n - x\| \to 0$ and $\|y_n - y\| \to 0$ as $n \to \infty$.

DEFINITION 1.11   *Let B be a bilinear operator from $X \times Y$ to Z. The operator B is said to be continuous at the point (x_0, y_0) in $X \times Y$ if for any $\varepsilon > 0$, there exists $\delta > 0$ such that $\|B(x, y) - B(x_0, y_0)\| < \varepsilon$, whenever $\|(x, y) - (x_0, y_0)\| < \delta$. A bilinear operator is said to be continuous on $X \times Y$ if it is continuous at every point in $X \times Y$.*

DEFINITION 1.12   *The quadratic operator from X to Z defined previously is said to be bounded if there exists $c > 0$ such that*

$$\|Q(x)\| \le c\|x\|^2, \quad for \; all \; x \in X . \qquad (1.17)$$

The quantity $\|Q\| = \sup_{\|x\|\leq 1} \|Q(x)\|$ is called the *norm* of Q. Obviously Q is bounded if and only if $\|Q\| < \infty$. Now we can easily obtain the following:

$$\|Q\| = \sup_{\|x\|\neq 0} \frac{\|Q(x)\|}{\|x\|^2}$$

$$= \inf\{c > 0 \mid \|Q(x)\| \leq c\|x\|^2, \text{ for all } x \in X\}. \tag{1.18}$$

Using Remark 1.1 and the above inequalities, we can easily prove that the set $Q^*(X; Z)$, denoting the set of bounded quadratic operators, becomes a normed space.

DEFINITION 1.13   *The operator Q defined previously is said to be continuous at $x_0 \in X$ if for any $\varepsilon > 0$, there exists $\delta > 0$ such that*

$$\|Q(x) - Q(x_0)\| < \varepsilon, \quad \text{whenever } \|x - x_0\| < \delta.$$

The operator Q is said to be continuous on X if it is continuous at every point $x \in X$.

1.2 Perturbation Theory

Let X denote a Banach space over the field S of real or complex numbers and consider the equations

$$x = y + B(x, x) \tag{1.19}$$

and

$$x = y + L(x) + B(x, x) \tag{1.20}$$

where $L: X \to X$ is a continuous linear operator, $B: X \times X \to X$ is a continuous bilinear operator and y is fixed in X.

From now on we denote $U(z, 2) = \{x \in X : \|x - z\| < 2\}$ and $\bar{U}(z, 2) = \{x \in X : \|x - z\| \leq 2\}$.

In this section we prove a consequence of the contraction mapping principle which can be used to prove existence and uniqueness for the solutions of (1.19) and (1.20).

More precisely we prove existence and uniqueness for a solution x^* of (1.19) in a specific ball $\bar{U}(z, r)$.

The principal new idea in our general theorem is the introduction of a second quadratic equation

$$z = y + F(z, z) \tag{1.21}$$

for comparison with (1.19) [similarly for (1.20)]. The estimates on (1.19) are then obtained under suitable choices of F. We also prove a similar theorem for (1.20).

Our theorem is applied to the famous Chandrasekhar equation, which is of the form

$$x(s) = 1 + \lambda x(s) \int_0^1 \frac{s}{s+t} x(t)dt , \tag{1.22}$$

which arises in the theory of radiative transfer [77]; $x(s)$ is the unknown function which is sought in $C[0, 1]$.

The physical background of this equation is fairly elaborate. It was developed by Ambartsumian [7] and Chandrasekhar (Nobel prize of Physics, 1983) [77] to solve the problem of determination of the angular distribution of the radiant flux emerging from a plane radiation field.

This radiation field must be isotropic at a point, that is, the distribution is independent of direction at that point. The positive parameter 2λ is called the albedo for scattering. It represents the fraction of the radiation lost due to scattering on a plane radiation field. Explicit definitions of these terms may be found in the literature [77]. It is considered to be the prototype of the equation

$$x(s) = 1 + \lambda s \, x(s) \int_0^1 \frac{p(s)}{s+t} x(t)dt$$

for more general laws of scattering, where $p(s)$ is an even polynomial in s with

$$\int_0^1 p(x)ds \leq \frac{1}{2} .$$

Integral equations of the above form also arise in other studies [115, 116, 143].

Rall [167] proved existence and uniqueness for a solution x^* of (1.22) in a ball centered at 0 provided that

$$\lambda < .36067 \ldots . \tag{1.23}$$

Here we prove existence and uniqueness for a solution x^* of (1.22) in a smaller ball centered at a specific $z \in X$ provided that

$$\lambda < .424059 \ldots . \tag{1.24}$$

Our theorem is also applied to the Anselone–Moore system [9]

$$x_j(s) = y_j(s) + \gamma \int_0^1 L_{j1}(s,t)x_1(t)x_2(t)dt$$

$$+ \gamma \int_0^1 L_{j2}(s,t)\frac{1}{2}x_1^2(t)dt\,, \tag{1.25}$$

$$0 \leq s \leq 1, \qquad j = 1,2\,,$$

where $X = C[0,1]$, γ is a nonnegative parameter and $L_{jk}(s,t)$, $j,k = 1,2$ are given. This problem is from nonlinear elasticity theory. It pertains to the buckling of a thin shallow spherical shell clamped at the edge and under uniform external pressure. The functions x_1 and x_2 represent the coordinates of the function describing the displacement. The positive parameter γ is proportional to the applied pressure. The overall discussion of the physical problem, derivations of equations, detailed numerical results, and their physical significance can be found in [9].

Anselone and Moore [9] proved existence and uniqueness for a solution x^* of (1.25) in a ball centered at 0 provided that

$$\gamma < \left[4\|y\| \max_{j=1,2}(\|L_{j1}\| + \|L_{j2}\|)\right]^{-1}\,.$$

Here we prove existence and uniqueness for a solution x^* of (1.25) in a smaller ball centered at a specific $z \in X$ provided that

$$\gamma < \left[4\|y\| \max_{j=1,2}\left(\|L_{j1}\| + \frac{1}{2}\|L_{j2}\|\right)\right]^{-1}\,.$$

Similar techniques could be applied to the Riccati equation [89, 168] and to the two-dimensional Navier–Stokes equation [89].

We provide a lower bound for the solutions of (1.19) and (1.20). Assuming that a solution of (1.19) or (1.20) is known, we provide conditions for a second solution.

We also show how we can use the solutions of finite rank equations to approximate solutions of (1.19) when B is the uniform limit of finite rank operators.

All the above results help us approximate a solution "close" to zero, i.e., $\|x^*\| \to 0$ as $\|y\| \to 0$. However, we develop techniques which guarantee that the solution found is "away" from zero. This is of extreme importance because most techniques developed thus far can only assist us in finding "small" solutions (see [15, 17, 18, 129, 167, 168] and the references therein).

We will need the following results, which compare the norm of a bilinear operator with the norm of the corresponding quadratic operator.

PROPOSITION 1.1

Let $B: X \times X \to Z$ be a bounded bilinear operator. Then the quadratic operator $Q(x) = B(x, x)$ is a bounded operator. Moreover

$$\|Q\| \leq \|B\| .$$

PROOF

$$\|Q(x)\| = \|B(x, x)\|$$

$$\leq \|B\| \|x\| \|x\|$$

$$= \|B\| \|x\|^2$$

so Q is bounded and the result follows. ∎

PROPOSITION 1.2

Let $Q: X \to Z$ be a bounded quadratic operator. Then the unique symmetric bilinear operator B associated with Q is bounded and

$$\|B\| \leq 2\|Q\| . \tag{1.26}$$

Moreover if the parallelogram law holds in X then

$$\|Q\| = \|B\| . \tag{1.27}$$

PROOF   We know that

$$4B(x, y) = Q(x + y) - Q(x, y)$$

so

$$4\|B(x, y)\| \leq \|Q(x + y) - Q(x - y)\|$$

$$\leq \|Q(x + y)\| + \|Q(x - y)\|$$

$$\leq \|Q\| \|x + y\|^2 + \|Q\| \|x - y\|^2$$

$$= \|Q\|(\|x + y\|^2 + \|x - y\|^2)$$

$$\leq \|Q\|(\|x\|^2 + \|y\|^2 + 2\|x\|\|y\| + \|x\|^2 + \|y\|^2 + 2\|x\|\|y\|)$$

so

$$4 \sup_{\substack{\|x\|\leq 1 \\ \|y\|\leq 1}} \|B(x,y)\| \leq 8\|Q\|$$

i.e.,

$$\|B\| \leq 2\|Q\| .$$

Therefore B is bounded. Now, if the parallelogram law holds then

$$\|x+y\|^2 + \|x-y\|^2 = 2\left(\|x\|^2 + \|y\|^2\right)$$

and the inequality becomes

$$\|B\| \leq \|Q\| .$$

The result now follows from Proposition 1.1.   ∎

PROPOSITION 1.3
Let $B: X \to Y \to Z$ be a bilinear operator. Then

(i) If B is continuous at $(0,0)$, then B is bounded.

(ii) If B is bounded, then B is continuous.

PROOF   (i) Since B is continuous at $(0,0)$, there exists $\delta > 0$ such that

$$\|B(x,y) - B(0,0)\| \leq 1 \quad \text{when} \quad \|(x,y) - (0,0)\| \leq \delta$$

so

$$\|B(x,y)\| \leq 1 \quad \text{when} \quad \|(x,y)\| \leq \delta .$$

Now let $(x,y) \in X \times Y$ be such that $x, y \neq 0$. Set

$$x_1 = \frac{\delta x}{\|x\|}, \quad y_1 = \frac{\delta y}{\|y\|} .$$

Then $\|x_1\| = \|y_1\| = \delta$ and since $\|(x_1, y_1)\| \leq \delta$ holds we obtain

$$\|B(x_1, y_1)\| \leq 1$$

or

$$\left\| B\left(\frac{\delta x}{\|x\|}, \frac{\delta y}{\|y\|} \right) \right\| \leq 1,$$

so

$$\|B(x, y)\| \leq c\|x\|\,\|y\|$$

where $c = \frac{1}{\delta^2}$. This inequality is true even if $x = 0$ or $y = 0$. So B is bounded.
(ii) Since B is bounded, there exists $c > 0$ such that

$$\|B(x, y)\| \leq c\|x\|\,\|y\| \quad \text{for all } (x, y) \in X \times Y.$$

Let $(x_1, y_1) \in X \times Y$ and $\varepsilon > 0$. Pick any value δ such that

$$0 < \delta < 1 \quad \text{and} \quad \delta < \frac{\varepsilon}{c\,(\|x_1\| + \|y_1\| + 1)}.$$

Now if

$$\|(x, y) - (x_1, y_1)\| < \delta \quad \text{then} \quad \|x - x_1\| < \delta \text{ and } \|y - y_1\| < \delta.$$

Also
$$\|y\| \leq \|y_1\| + \delta < \|y_1\| + 1$$

so

$$\|B(x, y) - B\,(x_1, y_1)\| = \|B\,(x - x_1, y) + B\,(x_1, y - y_1)\|$$

$$\leq \|B\,(x - x_1, y)\| + \|B\,(x_1, y - y_1)\|$$

$$\leq c\|x - x_1\|\,\|y\| + c\|x_1\|\,\|y - y_1\|$$

$$\leq c\delta\,(\|y_1\| + 1) + c\,\|x_1\|\,\delta$$

$$= c\delta\,(\|x_1\| + \|y_1\| + 1) < \varepsilon.$$

Therefore, B is continuous at (x_1, y_1).   ∎

PROPOSITION 1.4
The quadratic operator $Q: X \to Z$ is continuous if and only if it is bounded.

PROOF   Immediate from Proposition 1.3.   ∎

In the remainder of this section (unless otherwise stated) we assume that B in (1.19) and (1.20) is symmetric. This does not involve any loss of generality since B agrees with the mean $\bar{B}$ on the diagonal of $X \times X$.

DEFINITION 1.14   *Let* $T : U \to U$ *be an operator with domain* $U \subset X$, *satisfying a Lipschitz condition*

$$\|T(x) - T(y)\| \le q\|x - y\| \quad \textit{for all } x, y \in U . \tag{1.28}$$

If $0 < q < 1$ *then* T *is called a contraction operator.*

We now state without proof the contraction mapping principle. The proof can be found in [153].

THEOREM 1.3
Let U *be a closed set and assume that the contraction operator* T *maps* U *into itself. Then* T *has a unique fixed point* $x^* \in U$. *Moreover the sequence*

$$x_n = Tx_{n-1}, \quad n = 1, 2, \ldots \tag{1.29}$$

converges to x^* *for any* $x_0 \in U$ *in such a way that*

$$\left\| x_n - x^* \right\| \le \frac{q^n}{1 - q} \, \|x_0 - T(x_0)\| . \tag{1.30}$$

If the desired tolerance of the error is ε then the number k of iterations to obtain the inequality $\|x_n - x^*\| \le \varepsilon$ is given by

$$k = [p] + 1$$

where $[p]$ is the integer part of p and

$$p = \frac{1}{\ln q} \ln \frac{\varepsilon(1 - q)}{\|x_0 - T(x_0)\|} . \tag{1.31}$$

A direct application of Theorem 1.3 yields the following main theorem.

THEOREM 1.4
Let B *be a bilinear operator on* $X \times X$ *and suppose* y *and* z *belong to* X. *Define* $T : X \to X$ *by*

$$T(x) = y + B(x, x) . \tag{1.32}$$

Set

$$a = \frac{1}{2\|B\|} - \|z\| , \tag{1.33}$$

$$b = a - \left(a^2 - \frac{\|T(z) - z\|}{\|B\|} \right)^{1/2} , \tag{1.34}$$

and assume b is nonnegative and $a \neq 0$. Then

(i) *T has a unique fixed point in $U(z, a)$;*

(ii) *this fixed point actually lies in $\bar{U}(z, b)$.*

PROOF   The hypotheses $b \geq 0$ and $a \neq 0$ imply that $a > 0$ and

$$a^2 - \frac{\|T(z) - z\|}{\|B\|} \geq 0 .$$

Fix r such that $b \leq r < a$.

CLAIM 1.1
T is a contraction operator on $\bar{U}(z, r)$.

If $x_1, x_2 \in \bar{U}(z, r)$, then

$$\|T(x_1) - T(x_2)\| = \|B(x_1, x_1) - B(x_2, x_2)\|$$

$$= \|B(x_1, x_1) - B(x_1, x_2) + B(x_2, x_1) - B(x_2, x_2)\|$$

$$= \|B(x_1, x_1 - x_2) + B(x_2, x_1 - x_2)\|$$

$$\leq \|B\| \, \|x_1\| \, \|x_1 - x_2\| + \|B\| \, \|x_2\| \, \|x_1 - x_2\|$$

$$= \|B\| \, (\|x_1\| + \|x_2\|) \, \|x_1 - x_2\|$$

$$= \|B\| \, (\|x_1 - z + z\| + \|x_2 - z + z\|) \, \|x_1 - x_2\|$$

$$\leq \|B\| \, (\|x_1 - z\| + \|z\| + \|x_2 - z\| + \|z\|) \, \|x_1 - x_2\|$$

$$\leq 2(r + \|z\|)\|B\| \ \|x_1 - x_2\| \ .$$

Set $q = 2(r + \|z\|)\|B\|$. By hypothesis,

$$r < \frac{1}{2\|B\|} - \|z\|$$

so, $0 < q < 1$ and the claim is proved.

CLAIM 1.2
T maps $\bar{U}(z, r)$ into $\bar{U}(z, r)$. We have

$$\|T(x) - z\| = \|T(x) - T(z) + T(z) - z\|$$

$$\leq \|B(x, x) - B(z, z)\| + \|T(z) - z\|$$

$$= \|B(x, x - z) + B(x - z, z)\| + \|T(z) - z\|$$

$$\leq \|B\| \ \|x\| \ \|x - z\| + \|B\| \ \|x - z\| \ \|z\| + \|T(z) - z\|$$

$$\leq \|B\| \ \|x - z + z\| \ \|x - z\| + \|B\| \ \|x - z\| \ \|z\| + \|T(z) - z\|$$

$$\leq \|B\|(\|x - z\| + \|z\|)\|x - z\| + \|B\| \ \|x - z\| \ \|z\| + \|T(z) - z\|$$

$$\leq \|B\|r^2 + 2\|B\| \ \|z\|r + \|T(z) - z\| \ .$$

Define the real quadratic polynomial $g(r)$ by,

$$g(r) = \|B\|r^2 + (2\|B\| \ \|z\| - 1)r + \|T(z) - z\| \ . \tag{1.35}$$

To establish the claim we must show that $g(r) \leq 0$ for all r, $b \leq r < a$. Now the quadratic function $g(r)$ is convex, with smallest root at b and minimum occurring at a. So for $b \leq r < a$

$$\|B\|r^2 + 2\|B\| \ \|z\|r + \|T(z) - z\| \leq r \ .$$

The theorem now follows from Theorem 1.3.   ■

THEOREM 1.5
Assume a) $\|L\| < 1$, b) $\|y\| < \frac{(1-\|L\|)^2}{4\|B\|}$. Then

(i) *Equation (1.20) has a unique solution* $x^* \in U(0, r_2')$, *where*

$$r_2' = \frac{1 - \|L\|}{2\|B\|} ;$$

(ii) *moreover* $x^* \in \bar{U}(0, r_1')$, *where*

$$r_1' = \frac{(1 - \|L\|) - \left[(1 - \|L\|)^2 - 4\|B\| \, \|y\|\right]^{1/2}}{2\|B\|} .$$

PROOF Similar to Theorem 1.4. (Here we take $z = 0$ for simplicity.) ∎

COROLLARY 1.1
If

$$4\|B\| \, \|y\| < 1 \tag{1.36}$$

then

(i) *the equation*

$$x = y + B(x, x)$$

has a unique solution x^* *in the open ball* $U(0, r_2)$, *where*

$$r_2 = \frac{1}{2\|B\|} ;$$

(ii) *moreover* $x^* \in \bar{U}(0, r_1)$, *where*

$$r_1 = \left[1 - (1 - 4\|B\| \, \|y\|)^{1/2}\right] (2\|B\|)^{-1} .$$

PROOF Take $z = 0$ in Theorem 1.4. ∎

We now state Rall's theorem for comparison. The proof can be found in [168].

THEOREM 1.6
If

$$4\|B\| \, \|y\| < 1$$

then

(i)   *Equation (1.19) has a solution $x^* \in X$ satisfying*

$$\|x^*\| \le \frac{1 - \sqrt{1 - 4\|B\|\,\|y\|}}{2\|B\|} \; ;$$

(ii)   *moreover x^* is unique in $U(x, R)$, where*

$$R = \frac{\sqrt{1 - 4\|B\|\,\|y\|}}{2\|B\|} .$$

Rall was seeking a solution of the equation

$$x = y + \lambda B(x, x) \tag{1.37}$$

expressed as a series of the form

$$x = x_0 + \lambda x_1 + \ldots + \lambda^n x_n + \ldots \tag{1.38}$$

where

$$x_0 = y$$

$$x_1 = B(x_0, x_0)$$

and

$$x_n = \sum_{j=0}^{n-1} B\left(x_j, x_{n-j-1}\right), \quad n = 1, 2, \ldots . \tag{1.39}$$

He proved the convergence of this series by dominating it with the convergent series

$$\frac{1 - \sqrt{1 - 4\lambda fg}}{2\lambda g} = e_0 + \lambda e_1 + \ldots + \lambda^n e_n + \ldots$$

where

$$f = \|y\|, \quad g = \|B\|$$

and

$$e_0 = f$$
$$e_1 = g e_0 e_0$$
$$\vdots \tag{1.40}$$
$$e_n = \sum_{j=0}^{n-1} g e_j e_{n-j-1}.$$

Comparing Corollary 1.1 and Rall's theorem, we observe that Corollary 1.1 guarantees existence and uniqueness for a solution x^* of (1.19) in a ball centered at 0, but Rall's theorem guarantees uniqueness in a smaller ball centered at the solution x^*.

Corollary 1.1 is a crude application of Theorem 1.4. Sometimes, it is possible to introduce an auxiliary quadratic equation which is "close to" (1.19), but easier to handle. In the next section, we will use Theorem 1.4 in a more subtle way to exploit this idea. In particular, we will learn how to solve (1.19) in cases not covered by Rall's theorem.

We will need the auxiliary quadratic equations.

THEOREM 1.7

Consider the equation

$$z = y + F(z, z) \tag{1.41}$$

where $F: X \times X \to X$ is a bounded symmetric bilinear operator and y is fixed in X. Suppose that there exists a solution z satisfying (1.41) and

$$\|z\| < \left[2\sqrt{\|B\|} \left(\sqrt{\|B - F\|} + \sqrt{\|B\|} \right) \right]^{-1}. \tag{1.42}$$

Then

(i) *Equation (1.19) has a unique solution $x^* \in U(z, a)$;*

(ii) *moreover $x^* \in \bar{U}(z, b)$, where*

$$b = \left\{ 1 - 2\|B\| \, \|z\| - \left[(2\|B\| \, \|z\| - 1)^2 - 4\|B - F\| \, \|B\| \, \|z\|^2 \right]^{1/2} \right\}$$

$$(2\|B\|)^{-1}.$$

PROOF   We have

$$\|T(z) - z\| = \|(B - F)(z, z) + (F(z, z) + y - z)\|$$

$$\leq \|(B - F)(z, z)\| + \|F(z, z) + y - z\|$$

$$\leq \|B - F\| \, \|z\|^2.$$

So

$$\|T(z) - z\| \leq \|B - F\| \, \|z\|^2. \tag{1.43}$$

Now (1.42) implies the hypothesis of Theorem 1.4 since

$$\|z\| < (2\|B\|)^{-1} \Rightarrow a > 0$$

while by (1.42) and (1.43) we have

$$\frac{1}{2\|B\|} - \|z\| > \sqrt{\frac{\|B - F\|}{\|B\|}} \, \|z\|$$

or

$$a > \sqrt{\frac{\|T(z) - z\|}{\|B\|}} \Rightarrow b \geq 0 . \quad \blacksquare$$

In practice, an exact solution of the auxiliary Equation (1.41) can seldom be obtained. The following theorem, whose proof is similar to that of Theorem 1.7, guarantees that the original Equation (1.19) has a solution even when we can only find an approximate solution of (1.41).

THEOREM 1.8
Let B and F be bounded bilinear operators on X × X and suppose y and z belong to X. Define T: X → X by

$$T(x) = y + B(x, x) ,$$

and set

$$a = \frac{1}{2\|B\|} - \|z\|$$

$$\varepsilon = \|F(z, z) + y - z\| \, \|z\|^{-2}$$

$$b = a - \left[a^2 - \frac{\|B - F\| \, \|z\|^2 + \varepsilon \|z\|^2}{\|B\|} \right]^{1/2}$$

Assume that

$$\|z\| < \left[2\sqrt{\|B\|} \left(\sqrt{\|B\|} + \sqrt{\|B - F\| + \varepsilon} \right) \right]^{-1}$$

Then

(i)   *T has a unique fixed point in $U(z, a)$;*

(ii) *this fixed point actually lies in $\bar{U}(z, b)$.*

Because Theorem 1.8 relies on the contraction mapping principle, it actually provides an iteration procedure for solving (1.19), namely set $x_0 = z$ and

$$x_{n+1} = y + B(x_n, x_n), \quad n = 1, 2, \dots. \tag{1.44}$$

PROPOSITION 1.5
The iteration (1.44) converges for any $x_0 \in \bar{U}(z, b)$ to the solution x^* of (1.19) at the rate of a geometric progression with quotient

$$q = 1 - \left[(2\|B\|\,\|z\| - 1)^2 - 4\|B - F\|\,\|B\|\,\|z\|^2 \right]^{1/2}.$$

PROOF   By Theorem 1.7 we have

$$q = 2(b + \|z\|)\|B\|$$

$$= 1 - \left[(2\|B\|\,\|z\| - 1)^2 - 4\|B\|\,\|B - F\|\,\|z\|^2 \right]^{1/2}. \quad \blacksquare$$

COROLLARY 1.2
Under the hypotheses of Theorem 1.7, the solution x^* obtained in Theorem 1.7 satisfies

$$\|x^*\| < \frac{1}{2\|B\|}. \tag{1.45}$$

PROOF   By Theorem 1.7

$$\|x^* - z\| < a,$$

so that

$$\|x^*\| \leq \|z\| + a,$$

i.e.,

$$\|x^*\| < \frac{1}{2\|B\|}. \quad \blacksquare$$

COROLLARY 1.3
For any $y \in X$ such that $\|y\| < \frac{1}{4\|B\|}$

(i)   *Equation (1.19) has a unique solution $x^* \in U(y, a)$, where*

$$a = \frac{1 - 2\|B\|\,\|y\|}{2\|B\|} \; ; \tag{1.46}$$

(ii) *moreover $x^* \in \bar{U}(y, b)$, where*

$$b = \frac{1 - 2\|B\|\,\|y\| - \sqrt{1 - 4\|B\|\,\|y\|}}{2\|B\|} \; .$$

PROOF   Apply Theorem 1.7 with $F = 0$ and $z = y$.   ∎

Example 1.3
This example shows that Theorem 1.7 can be applied even if $\|y\| \geq \frac{1}{4\|B\|}$. We take $X = \mathbf{R}$,

$$x = -.251 + x^2 \, ,$$

$$z = -.251 + .8z^2 \, .$$

Here $|y| = .251 > \frac{1}{4 \cdot 1} = .25$,

$$z_1 = -.214270516 \quad \text{and} \quad z_2 = 1.46270517 \, .$$

Then

$$\|z_1\| = .214270516 < \frac{1}{2\sqrt{1}(\sqrt{1 - .8} + \sqrt{1})} = .345491502$$

so according to Theorem 1.7

(i)   there exists a unique solution x^* of (1.19) in $U(z_1, a)$;

(ii) moreover $x^* \in \bar{U}(z_1, b)$,

where

$$b = .0165474506$$

and

$$a = .285729484 \, .$$

The solutions of (1.19) are

$$x_1 = -.207813534, \quad x_2 = 1.207813535$$

and

$$x_1 \in \bar{U}(z_1, b) \quad \text{but,} \quad x_2 \notin U(z_1, a) .$$

PROPOSITION 1.6
Assume

    (a)   the hypotheses of Theorems 1.6 and 1.7 and Corollary 1.1, are satisfied;

    (b) $(\|B\| - \|B - F\|)\|z\|^2 - \|z\| + \|y\| > 0.$

Then Theorem 1.7 provides a sharper estimate on $\|x^*\|$ than Theorem 1.6 and Corollary 1.1.

PROOF   By Theorem 1.7

$$\|x^* - z\| \le b \quad \text{so} \quad \|x^*\| \le b + \|z\| .$$

By Theorem 1.6 and Corollary 1.1

$$\|x^*\| \le \frac{1 - \sqrt{1 - 4\|B\|\,\|y\|}}{2\|B\|}$$

so it is enough to show

$$\frac{1 - \left[(2\|B\|\,\|z\| - 1)^2 - 4\|B\|\,\|B - F\|\,\|z\|^2\right]^{1/2}}{2\|B\|} < \frac{1 - \sqrt{1 - 4\|B\|\,\|y\|}}{2\|B\|}$$

or

$$(\|B\| - \|B - F\|)\|z\|^2 - \|z\| + \|y\| > 0$$

and the result follows from (b).   ■

REMARK 1.3   If the evaluation of $\|B - F\|$ in Theorem 1.7 is difficult, then,

    a)   we can look for a z such that:

$$\|z\| < \left[2\sqrt{\|B\|} \left(\sqrt{\|B\| + \|F\|} + \sqrt{\|B\|}\right)\right]^{-1}$$

$$\le \left[2\sqrt{\|B\|} \left(\sqrt{\|B - F\|} + \sqrt{\|B\|}\right)\right]^{-1}$$

and start the iteration with $x_0 = z$.

b) We can apply the theorem in the ball $\bar{U}(z, a')$, where $b \leq a' < a$ and

$$a' = \frac{1 - 2\|B\|\,\|z\| - \left[(2\|B\|\,\|z\| - 1)^2 - 4\|B\|\,(\|B\| + \|F\|)\|z\|^2\right]^{1/2}}{2\|B\|},$$

provided that the quantity under the radical is nonnegative. Also note that since $b \leq a' < a$, we have

$$\bar{U}(z, b) \subset \bar{U}(z, a') \subset U(z, a) . \quad \blacksquare$$

1.3 Chandrasekhar's Integral Equation

Example 1.4
For the Equation (1.22), choose $X = C[0, 1]$ with the sup-norm. The operator $Q: X \rightarrow X$ defined by

$$Q(x) = x(s) \int_0^1 \frac{s}{s+t} x(t) dt$$

is quadratic since the symmetric bilinear operator $G: X \times X \rightarrow X$ defined by

$$G(x, y) = \frac{1}{2}\left[x(s) \int_0^1 \frac{s}{s+t} y(t) dt + y(s) \int_0^1 \frac{s}{s+t} x(t) dt\right]$$

satisfies

$$G(x, x) = Q(x) \quad \text{for all } x \in X .$$

We will prove that the norm $\|G\| = \ln 2$. Now

$$\|Q\| = \max_s \int_0^1 \left|\frac{s}{s+t}\right| dt = \ln 2$$

and since always

$$\|Q\| \leq \|G\|$$

we obtain

$$\ln 2 \leq \|G\| .$$

The proof will be completed if we prove that

$$\|G\| \le \ln 2 .$$

But by the definition of G,

$$\|G\| \le \frac{1}{2} \max_s \left(2 \int_0^1 \left| \frac{s}{s+t} \right| dt \right) = \ln 2$$

so

$$\|G\| = \ln 2 .$$

We now apply Corollary 1.1, Theorem 1.7, and Corollary 1.3 to (1.22) with $B = \lambda G$. According to Corollary 1.1

(i) Equation (1.22) has a unique solution $x^* \in U(0, r_2)$, where

$$r_2 = \frac{1}{2\lambda \ln 2} ;$$

(ii) moreover $x^* \in \bar{U}(0, r_1)$, where

$$r_1 = \frac{1 - \sqrt{1 - 4\lambda \ln 2}}{2\lambda \ln 2} ,$$

provided that $1 - 4\lambda \ln 2 > 0$, i.e., $\lambda < .36067 \ldots$.

According to Theorem 1.6, Equation (1.22) has a unique solution x^* in $U(x^*, R)$, where

$$R = \frac{\sqrt{1 - 4\lambda \ln 2}}{2\lambda \ln 2} ,$$

provided that $1 - 4\lambda \ln 2 > 0$, i.e., $\lambda < .36067 \ldots$.
Finally, according to Corollary 1.3,

(i) Equation (1.22) has a unique solution $x^* \in U(1, b)$, where

$$b = \frac{1 - 2\lambda \ln 2}{2\lambda \ln 2} ;$$

(ii) moreover $x^* \in \bar{U}(1, a)$, where

$$a = \frac{1 - 2\lambda \ln 2 - \sqrt{1 - 4\lambda \ln 2}}{2\lambda \ln 2} ,$$

provided that $1 - 4\lambda \ln 2 > 0$, i.e., $\lambda < .36067 \ldots$.

Our next goal is to use Theorem 1.8 to obtain solutions of (1.22) for a wider range of λ. It is not necessary to assume B has any connection with Chandrasekhar's equation in Propositions 1.7 or 1.8.

PROPOSITION 1.7
If $z \in X$ is a solution of the equation

$$z = y + \lambda B(z, z) ,$$

satisfying

$$2\lambda \|B\| \, \|z\| < 1 ,$$

then for

$$\lambda \leq \lambda_1 < c_1 ,$$

where

$$c_1 = [4\|B\| \, \|z\|(I - \lambda \|B\| \, \|z\|)]^{-1} ,$$

the conclusions of Theorem 1.7 for the equation

$$x = y + \lambda_1 B(x, x)$$

hold.

PROOF   To apply Theorem 1.7 we need

$$\|z\| < \left[2\sqrt{\lambda_1 \|B\|} \left(\sqrt{|\lambda - \lambda_1| \, \|B\|} + \sqrt{\lambda_1 \|B\|} \right) \right]^{-1} .$$

Since

$$\lambda_1 < c_1 = [4\|B\| \, \|z\|(1 - \lambda \|B\| \, \|z\|)]^{-1} ,$$

we have

$$\lambda_1^2 - \lambda_1 \lambda < \left(\frac{1}{2\|B\| \, \|z\|} \right)^2 + \lambda_1^2 - \frac{\lambda_1}{\|B\| \, \|z\|}$$

or by taking the square root of both sides of the last inequality and using

$$\lambda_1 < (2\|B\| \, \|z\|)^{-1}$$

we get

$$\sqrt{\lambda_1(\lambda_1 - \lambda)} < \frac{1}{2\|B\| \|z\|} - \lambda_1 .$$

The result now follows by solving the last inequality for $\|z\|$.   ∎

If z is not an exact solution of the quadratic equation

$$z = y + f(z, z) ,$$

then we can use the following generalization of Proposition 1.7.

PROPOSITION 1.8
Let B be a bilinear operator on $X \times X$, suppose y and z belong to X and λ is a positive parameter.
  Set

$$\varepsilon = \|\lambda B(z, z) + y - z\| \|B\| \|z\|^{-2}$$

and

$$c_I = \{4\|z\| \|B\| [I - (\lambda - \varepsilon)\|B\| \|z\|]\}^{-1} .$$

Then for any λ_1 satisfying $\lambda \leq \lambda_1 < c_1$, the equation

$$x = y + \lambda_1 B(x, x)$$

has a unique solution in $U(z, a)$ and in fact this solution lies in $\bar{U}(z, b)$. Here

$$a = \frac{1}{2\lambda_1 \|B\|} - \|z\|$$

$$b = a - \left[a^2 - \left(1 - \frac{\lambda}{\lambda_1}\right) \|z\|^2 - \frac{\varepsilon}{\lambda_1} \|z\|^2 \right]^{1/2} .$$

PROOF   Similar to Proposition 1.7.   ∎

REMARK 1.4   According to Corollary 1.1 or Theorem 1.6 and the discussion following Example 1.4, Chandrasekhar's equation (1.22) has a solution z provided that $\lambda < .36067376$. But now using Proposition 1.8, we can extend the range of λ until .424059379 as follows. We first construct an arithmetic model of (1.22) and then we use the solution of the arithmetic problem to generate an initial approximation z_0 to the solution of (1.22). We assume now that $\lambda = .35$ and introduce a numerical integration rule of the form

$$\int_0^1 f(s)ds \cong \sum_{i=1}^n w_i f(s_i)$$

where the points s_i, $0 \le s_i \le 1$, $i = 1, 2, \ldots, n$ are called the nodes and the numbers w_i, $i = 1, 2, \ldots, n$, are called the weights of the given rule. The number n is called the order of the rule. A Gaussian integration rule will be used. This rule has the property that it will integrate any polynomial on [0, 1] exactly if the degree of the polynomial does not exceed 17 and the exact values of the nodes and weights are used.

The integral in (1.22) can be approximated as

$$\int_0^1 \frac{x(t)}{s+t}dt \cong \sum_{j=1}^n \frac{w_j}{s+s_j}x(s_j), \quad 0 \le s \le 1.$$

Thus, for

$$d_{ij} = \frac{s_i w_j}{s_j + s_i}, \quad i, j = 1, 2, \ldots, n,$$

the Equation (1.22) may now be approximated by the arithmetic fixed point problem

$$\xi_i = 1 + .35\xi_i \sum_{j=1}^n d_{ij}\xi_j, \quad i = 1, 2, \ldots, n$$

where

$$\xi_i \cong x(s_i), \quad i = 1, 2, \ldots, n$$

or

$$\underline{x} = 1 + .35\underline{x} \odot D\underline{x},$$

where $\underline{x} = (\xi_1, \ldots, \xi_n)$, 1 denotes the vector $(1, 1, \ldots, 1)$, D the matrix defined by the d_{ij}s, $i, j = 1, 2, \ldots, n$, and $\odot$ represents inner product. The iteration formula now takes the form

$$x_{m+1} = 1 + .35x_m \odot Dx_m$$

or

$$\xi_i^{(m+1)} = 1 + .35\xi_i^{(m)} \sum_{j=1}^n d_{ij}\xi_j^{(m)}, \quad i = 1, 2, \ldots, n$$

for some

$$x_0 = \left(\xi_1^{(0)}, \xi_2^{(0)}, \ldots, \xi_n^{(0)}\right).$$

The norm of the matrix D was found to be in $\mathbf{R}^9_\infty$

$$\|D\| = .69314701 . \quad \blacksquare$$

We can now use (the discrete version of) Corollary 1.3 with $B = \lambda D = .35D$ and $X = \mathbf{R}^9$ and the above estimate to justify the initial choice of the approximation $x_0 = (1, 1, \ldots, 1)$ to the solution $\underline{x}$ of the arithmetic fixed point problem

$$\underline{x} = 1 + .35\underline{x} \odot D\underline{x}$$

which is guaranteed by Corollary 1.3. The number of iterations required to obtain convergence to nine decimal places is 21.

Once the solutions ξ_i, $i = 1, 2, \ldots, n$ have been obtained, the question of their relationship to the corresponding solution z of the Equation (1.22) arises. We must construct a function by some interpolating procedure to use the values obtained from the solution of the arithmetic problem. This is done because the nodes of the Gaussian numerical integration rule are not the values of s at which approximate solutions are desired. We can make use of the fact

$$z(0) = 1 \qquad \text{in (1.22)}$$

and extend the straight-line segment joining (s_8, ξ_8) and (s_9, ξ_9) to $s = 1$. This is equivalent to using the basis

$$p_0(s) = \begin{cases} 1 - \frac{s}{s_1}, & 0 \le s \le s_1 \\ 0, & s_1 \le s \le 1, \end{cases}$$

$$p(s) = 0, \qquad 0 \le s \le s_{i-1} ,$$

$$= \frac{s - s_{i-1}}{s_i - s_{i-1}}, \qquad s_{i-1} \le s_i$$

$$= \frac{s_{i+1} - s}{s_{i+1} - s_i}, \qquad s_i \le s \le s_{i+1}$$

$$= 0 \qquad s_{i+1} \le s \le 1, \quad i = 1, 2, \ldots, 8 ,$$

$$p_9(s) = \begin{cases} 0, & 0 \le s \le s_8 \\ \frac{s - s_8}{s_9 - s_8}, & s_8 \le s \le 1 \end{cases}$$

for a 10-dimensional subspace of $C[0, 1]$. The resulting function is

$$\tilde{z}(s) = p_0(s) + \sum_{i=1}^{9} \xi_i \, p_i(s), \quad 0 \le s \le 1 \, .$$

Since the integral transforms

$$P_i(s) = \int_0^1 \frac{s}{s+t} p_i(t) dt, \quad 0 \le s \le 1, \quad i = 0, 1, \dots, 9$$

can be calculated explicitly, we could set

$$z_0(s) = \tilde{z}(s), \quad 0 \le s \le 1,$$

and obtain $z_n(s)$ by the iterative formula suggested in Theorem 1.3

$$z_{n+1}(s) = 1 + .35s \, z_n(s) \int_0^1 \frac{z_n(t)}{s+t} dt, \quad 0 \le s \le 1, \quad n = 0, 1, 2, \dots \, .$$

We have Table 1.1.

Table 1.1 Numerical solution of Chandrasekhar's equation I

i	s_i	ξ_i	s	$z_{295}(s)$
1	.015919912	1.027289881	0	1
3	.019331432	1.178454651	.2	1.182516314
4	.337873312	1.254112032	.4	1.280619257
6	.662126712	1.367934132	.6	1.350079131
7	.806685713	1.404454533	.8	1.402900122
9	.984080101	1.441732192	1	1.444745320

We need 295 iterations to achieve an accuracy of 9 decimal places. Note that $\|z_{295}\| = 1.44745321$. We can now evaluate c_1 using Proposition 1.8:

$$c_1 = .384363732 \, .$$

Now we choose $\lambda_1 = .38$ say and proceed as follows. We use as the initial approximation to the arithmetic problem

$$\underline{x} = 1 + .38\underline{x} \odot D\underline{x}$$

the solution $\underline{\xi} = (\xi_1, \ldots, \xi_9)$ of this equation obtained for $\lambda = .35$ (the components $\xi_1, \xi_3, \xi_4, \xi_6, \xi_7, \xi_9$ for example are given in Table 1.1).

This process stops when $\lambda_1 \cong .42405937$. Denote by N the number of iterations needed to get

$$\| z - z_N \| \le 10^{-9}$$

where z is the exact solution of Equation (1.22) for a fixed λ. Here are some characteristic values for λ the norm of the corresponding approximation solution z_N, c_1, and N.

Table 1.2 Numerical solution of Chandrasekhar's equation II

λ	$\| z_N \|$	c_1	N
.35	1.44474532	.384363732	295
$\vdots$	$\vdots$	$\vdots$	$\vdots$
.38	1.534201867	.394512252	320
.39	1.558263525	.399942101	329
$\vdots$	$\vdots$	$\vdots$	$\vdots$
.4	1.59821923	.405244331	343
$\vdots$	$\vdots$	$\vdots$	$\vdots$
.42	1.68363661	.420163281	360
$\vdots$	$\vdots$	$\vdots$	$\vdots$
.423	1.69644924	.423011429	371
$\vdots$	$\vdots$	$\vdots$	$\vdots$
.424	1.70085561	.424070047	384
$\vdots$	$\vdots$	$\vdots$	$\vdots$
.424059378	1.700973716	.424059379	397
.424059379	1.700973721	.424059379	410

1.4 Anselone and Moore's Equation

DEFINITION 1.15 *Consider the following system in X*

$$x_i = y_i + B_i(x, x), \quad i = 1, 2, \ldots, n \tag{1.47}$$

where

$$x = (x_1, x_2, \ldots, x_n) \in X^n$$

and

$$B_i : X^n \times X^n \to X$$

are bounded symmetric bilinear operators, $i = 1, 2, \ldots, n$. The norm of a vector $x \in X^n$ is given by

$$\|x\| = \max\{\|x_1\|, \ldots, \|x_n\|\} .$$

Define the operators $\tilde{Q} : X^n \to X^n$ by

$$\tilde{Q}(x) = (B_1(x, x), \ldots, B_n(x, x))$$

and $\tilde{B} : X^n \times X^n \to X^n$ by

$$\tilde{B}(u, v) = \frac{1}{4} \left(\tilde{Q}(u + v) - \tilde{Q}(u - v) \right) .$$

Then (1.47) can be written as

$$x = y + \tilde{B}(x, x) .$$

PROPOSITION 1.9
If $\tilde{B}, \tilde{Q}$ are as in Definition 1.15, then $\tilde{B}$ is a bounded symmetric bilinear operator and $\tilde{Q}$ is a bounded quadratic operator. Moreover

$$\left\| \tilde{B} \right\| = \max\{\|B_1\|, \ldots, \|B_n\|\} .$$

PROOF   The fact that $\tilde{B}$ is symmetric and bilinear on $X^n \times X^n$ is a direct consequence of the fact that each B_i, $i = 1, 2, \ldots, n$ is a bounded symmetric bilinear operator on $X^n \times X^n$. The boundedness of $\tilde{B}$ follows from the computation

$$\left\| \tilde{B}(x, x) \right\| = \max\{\|B_1(x, x)\|, \ldots, \|B_n(x, x)\|\}$$

$$\leq \max\{\|B_1\|, \ldots, \|B_n\|\} \|x\|^2 ,$$

valid for each $x \in X^n$. On the other hand, consideration of vectors in X^n with a single nonzero component shows $\|\tilde{B}\| \geq \|B_i\|$ for each $i = 1, \ldots, n$, which establishes the last statement of the proposition. Finally, Definition 1.3 and Proposition 1.2 show that $\tilde{Q}$ is bounded and quadratic. ∎

REMARK 1.5   Equation (1.25) is of the form (1.47). Here, $X = C[0, 1]$. Define the linear operators $L_{jk}: X \to X$, $j, k = 1, 2$ by

$$(L_{jk}x)(s) = \int_0^1 L_{jk}(s, t)x(t)dt, \quad 0 \le s \le 1$$

with

$$\|L_{jk}\| = \sup_s \int_0^1 |L_{jk}(s, t)| dt, \quad j, k = 1, 2$$

and the symmetric bilinear operator $\tilde{B}: X \times X \times X \times X \to X \times X$ by

$$\tilde{B}(x, y) = \frac{1}{4}(E(x + y) - E(x - y)), \quad x = (x_1, x_2), \quad y = (y_1, y_2)$$

where $E: X \times X \to X \times X$ is the quadratic operator given by

$$E(x) = \begin{pmatrix} L_{11}x_1x_2 + \frac{1}{2}L_{12}x_1^2 \\ L_{21}x_1x_2 + \frac{1}{2}L_{22}x_1^2 \end{pmatrix} . \quad \blacksquare$$

Finally, according to Proposition 1.8 and the definition of $\tilde{B}$ we have

$$\|\tilde{B}\| \le \max_{j=1,2} \left(\|L_{j1}\| + \frac{1}{2} \|L_{j2}\| \right) .$$

We now state for comparison Anselone–Moore's theorem which was proved using Theorem 1.3, concerning the solution of the Equation (1.25).

THEOREM 1.9
If [9]

$$4\gamma \|y\| \|E\| < 1$$

where

$$\|E\| \le \max_{j=1,2} (\|L_{j1}\| + \|L_{j2}\|)$$

then

    (i)   *Equation (1.25) has a unique solution $x^* \in \bar{U}(0, r_2)$, where*

$$r_2 = \frac{1}{2\gamma \|E\|} ;$$

(ii)   *moreover $x^* \in \bar{U}(0, r_1)$, where*

$$r_1 \left[1 - (1 - 4\gamma \|E\| \|y\|)^{1/2} \right] (2\gamma \|E\|)^{-1} .$$

REMARK 1.6   By Remark 1.5, since

$$\max_{j=1,2} \left(\|L_{j1}\| + \frac{1}{2} \|L_{j2}\| \right) \le \max_{j=1,2} \left(\|L_{j1}\| + \|L_{j2}\| \right)$$

Corollary 1.3 gives on the one hand a wider range for γ and on the other hand better information about where the solution lies than Theorem 1.9. Finally we can use Proposition 1.7 (or 1.8) to extend the range of γ even more, but we omit the details here. ∎

1.5   Other Perturbation Theorems

Here we list a number of variants of Theorem 1.7. Their proofs, being similar to that of Theorem 1.7, are omitted.

THEOREM 1.10
Consider the equation

$$z = y_1 + B(z, z) . \tag{1.48}$$

Suppose that there exists a solution z satisfying (1.48) and

$$\|z\| < \frac{1 - 2\sqrt{\|B\| \|y - y_1\|}}{2\|B\|}, \quad \text{with } 1 - 4\|B\| \|y - y_1\| > 0 .$$

Then

(i)   *Equation (1.19) has a unique solution $x^* \in U(z, a)$, where*

$$a = \frac{1 - 2\|B\| \|z\|}{2\|B\|} ;$$

(ii)   *moreover $x^* \in \bar{U}(z, b)$, where*

$$b = \frac{1 - 2\|B\| \|z\| - \left[(1 - 2\|B\| \|z\|)^2 - 4\|B\| \|y - y_1\| \right]^{1/2}}{2\|B\|} .$$

THEOREM 1.11

Consider the equation

$$z = y + L_1(z) + B_1(z, z) \qquad (1.49)$$

where $L_1: X \to X$ is a bounded linear operator $B_1: X \times X \to X$ is a bounded symmetric bilinear operator and $y \in X$ is fixed. If
  (a) $\|L\| < 1$;
  (b) *there exists $z \in X$ satisfying (1.49) and*

$$\|z\| < c$$

where

$$c = \frac{c_1}{c_1 + c_2}$$

with $c_1 = (1 - \|L\|)^2$, $c_2 = 2\|B\|(1 - \|L\| + \|L - L_1\|)$ and

$$c_3 = 2[\|B\|^2(1 - \|L\| + \|L - L_1\|)^2 - \|B\|(\|B\| - \|B - B_1\|)(1 - \|L\|)^2]^{1/2} .$$

Then

  (i)   *Equation (1.20) has a unique solution $x^* \in U(z, a)$, where*

$$a = \frac{1 - \|L\| - 2\|B\| \, \|z\|}{2\|B\|} \; ;$$

  (ii) *moreover $x^* \in \bar{U}(z, b)$, where*

$$b = \frac{1 - \|L\| - 2\|B\| \, \|z\| - \sqrt{D}}{2\|B\|}$$

  and

$$D = 4\|B\| (\|B\| - \|B - B_1\|) \, \|z\|^2$$

$$- 4\|B\|(1 - \|L\| + \|L - L_1\|)\|z\| + (1 - \|L\|)^2 .$$

The following theorem shows how we can use the solutions of finite rank equations to approximate solutions of (1.19) when B is the uniform limit of finite rank operators.

THEOREM 1.12
Consider the equations

$$z = y + F_n(z, z) \tag{1.50}$$

where $F_n: X \times X \to X$, $n = 1, 2, \ldots$ are bounded symmetric bilinear operators and y is fixed in X. If

(a) *the sequence, $\{F_n\}$, $n = 1, 2, \ldots$ converges to B uniformly as $n \to \infty$, and*

(b) *for each n, $n = 1, 2, \ldots$ there exists z_n, satisfying (1.50) and $\sup \|z_n\| < \frac{1}{2\|B\|}$, then the sequence $\{z_n\}$, $n = 1, 2, \ldots$ converge to a solution z of (1.19).*

(Note that B and the F_ns are not necessarily of finite rank here.)

PROOF   We have

$$z_m - z_n = F_m(z_m, z_m) - F_n(z_n, z_n)$$

$$= F_m(z_m, z_m) - F_n(z_m, z_m) + F_n(z_m, z_m) + F_m(z_n, z_n)$$

$$- F_m(z_n, z_n) - F_n(z_n, z_n)$$

$$= (F_m - F_n)(z_m, z_m) + (F_m - F_n)(z_n, z_n) + F_n(z_m, z_m) - F_m(z_n, z_n)$$

$$- F_n(z_m, z_n) + F_n(z_m, z_n) + F_m(z_m, z_n) - F_m(z_m, z_n)$$

$$= (F_m - F_n)(z_m, z_m) + (F_m - F_n)(z_n, z_n) + F_n(z_m, z_m - z_n)$$

$$+ F_m(z_m - z_n, z_n) + (F_n - F_m)(z_m, z_n) .$$

Now, since $\sup \|z_n\| < \frac{1}{2\|B\|}$, there exists $c > 0$ such that

$$\|z_n\| \le c < \frac{1}{2\|B\|}, \quad n = 1, 2, \ldots .$$

Moreover,

$$\|z_m - z_n\| \le \|F_m - F_n\| \, \|z_m\|^2 + \|F_m - F_n\| \, \|z_n\|^2 + \|F_n\| \, \|z_m\| \, \|z_m - z_n\|$$

$$+ \|B\| \, \|z_m - z_n\| \, \|z_n\| + \|F_m - F_n\| \, \|z_m\| \, \|z_n\|$$

$$\leq 2 \|B\| c \, \|z_m - z_n\| \quad \text{as } n, m \to \infty$$

so

$$\|z_m - z_n\| \, (2\|B\|c - 1) \geq 0, \quad \text{since } c < \frac{1}{2\|B\|}$$

$2\|B\|c - 1 < 0$, so we must have $\|z_m - z_n\| \to 0$ as $n, m \to \infty$, therefore the sequence $\{z_n\}$, $n = 1, 2, \ldots$ is a Cauchy sequence in a Banach space and as such it converges to a $z \in X$. The element $z \in X$ is a solution of (1.19) since

$$z = \lim_{n \to \infty} z_n = \lim_{n \to \infty} (y + F_n (z_n, z_n))$$

$$= y + \lim_{n \to \infty} F_n (z_n, z_n) = y + B(z, z) . \quad \blacksquare$$

Example 1.5
Take $X = \mathbf{R}$

$$x = 0 + x^2 \quad \text{for } x = y + B(x, x)$$

$$z = 0 + \frac{n - 1}{n} z^2 \quad \text{for } z = y + f_n(z, z) .$$

Here $F_n \to B$ since $\frac{n-1}{n} \to 1$ as $n \to \infty$

$$z_n^- = 0 \to 0 \quad \text{as } n \to \infty$$

$$z_n^+ = \frac{n}{n - 1} \to 1 \quad \text{as } n \to \infty$$

so,

$$|z_n^-| = |0| = 0 < \frac{1}{2} .$$

Now, $z_n^- \to 0$ as $n \to \infty$ which is a solution of (1.19).

We now give some conditions for faster convergence of iteration (1.44) to a solution x^* of (1.19). We find it convenient to restate Corollary 1.1 in the form:

THEOREM 1.13
If condition (1.36) holds, then

(i)   the equation

$$x = y + B(x, x)$$

has a unique solution x^* in the open ball $U(0, r_2)$, where

$$r_2 = \frac{1}{2\|B\|} \; ; \tag{1.51}$$

(ii)   moreover $x^* \in \bar{U}(0, r_1)$, where

$$r_1 = \frac{\sqrt{1 - 4\|B\| \cdot \|y\|}}{2\|B\|} \, . \tag{1.52}$$

The iteration

$$x_{n+1} = y + B\,(x_n, x_n), \quad n = 0, 1, 2, \ldots$$

converges to x^* for any $x_0 \in U(0, r)$ for some r such that

$$r_1 \leq r < r_2 .$$

The following estimate is also true

$$\|x^* - x_n\| \leq \frac{q^n}{1 - q} \, \|x_0 - x_1\|$$

where, the rate of convergence is given by

$$q = q(r) = 2\|B\| \cdot r . \tag{1.53}$$

THEOREM 1.14
If x^* is a solution of (1.19) such that

$$\|x^*\| \leq \frac{\sqrt{2} - 1}{2\|B\|} , \tag{1.54}$$

then

(i)   the hypotheses and conclusions of Theorem 1.13 are satisfied;

(ii)   *moreover, if*

$$4\sqrt{2} - 5 < 4\|B\| \cdot \|y\| < 1 \tag{1.55}$$

the iteration (1.44) converges faster if $x_0 \in \bar{U}(x^*, R)$ *than if* $x_0 \in \bar{U}(0, r_1) - \bar{U}(x^*, R)$, *where*

$$0 < R < \frac{2 - \sqrt{2} - \sqrt{1 - 4\|B\| \cdot \|y\|}}{2\|B\|}. \tag{1.56}$$

PROOF   (i) If x^* is a solution of (1.19) then

$$\|y\| = \|x^* - B(x^*, x^*)\| \le \|x^*\| \cdot (1 + \|B\| \cdot \|x^*\|) .$$

Now, (1.36) is satisfied if

$$4\|B\| \, \|x^*\| \cdot (1 + \|B\| \cdot \|x^*\|) < 1$$

which is true by (1.54).
   (ii) *Claim 1.* The operator T on X given by

$$T(x) = y + B(x, x)$$

is a contraction operator on $\bar{U}(x^*, R)$. If x_1, x_2 are chosen in $\bar{U}(x^*, R)$ then

$$\|T(x_1) - T(x_2)\| = \|B(x_1 + x_2, x_1 - x_2)\|$$

$$= \|B(x_1 - x^* + x_2 - x^* + 2x^*, x_1 - x_2)\|$$

$$\le 2\|B\| (R + \|x^*\|) \|x_1 - x_2\|$$

$$= \tilde{q}(R) \cdot \|x_1 - x_2\|, \quad \tilde{q}(r) = 2\|B\| (R + \|x^*\|)$$

but

$$0 < \tilde{q}(R) < 1$$

by the choice of x^* and R.
   Claim 2. T maps $\bar{U}(x^*, R)$ into $\bar{U}(x^*, R)$. If $x \in \bar{U}(x^*, R)$

$$\|T(x) - x^*\| = \|B(x + x^*, x - x^*)\|$$

$$= \|B(x - x^* + 2x^*, x - x^*)\|$$

$$\leq \|B\| R \left(R + 2 \|x^*\| \right)$$

so,

$$\|T(x) - x^*\| \leq R \quad \text{if} \quad \|B\| R \left(R + 2 \|x^*\| \right) \leq R$$

which is true by the choice of x^*, R and (1.55).

If $x \in \bar{U}(x^*, R)$, then $x \in \bar{U}(0, r_1)$; that is, $\bar{U}(x^*, R) \subset \bar{U}(0, r_1)$. This is true because

$$\|x\| \leq \|x^*\| + R \leq \frac{\sqrt{2} - 1}{2\|B\|} + R < r_1 . \qquad (1.57)$$

The proof will now be complete if we show

$$\tilde{q}(R) < q\,(r_1)$$

or

$$2\|B\| \left(R + \|x^*\| \right) < 2\|B\| r_1$$

which is true by (1.57). $\quad\blacksquare$

If we use Theorem 1.13, then we see that x^* lies in a "ring" included in the ball $\bar{U}(0, r_1)$, that is, we have a better information on the location of the solution.

We now provide a lower bound for the solution of (1.19) and (1.20).

Equations (1.19) and (1.20) are special cases of the equation $P(x) = 0$, where P is given by $P(x) = y + L'(x) + B(x, x)$. In fact, Equations (1.19) and (1.20) correspond to

$$L' = -I \quad \text{and} \quad L' = L - I ,$$

respectively.

PROPOSITION 1.10
Let $x^* \in X$ be a solution of $P(x) = 0$. Then

$$\|x^*\| \geq \frac{-\|L'\| + \left[\|L'\|^2 + 4\|B\| \|y\|\right]^{1/2}}{2\|B\|}, \quad \text{if } \|B\| \neq 0 .$$

PROOF   We have $y + L'(x^*) + B(x^*, x^*) = 0$ so $-y = L'(x^*) + B(x^*, x^*)$ so

$$\|y\| = \|L'(x^*) + B(x^*, x^*)\|$$

$$\leq \|L'\| \|x^*\| + \|B\| \|x^*\|^2$$

so

$$\|B\| \ \|x^*\|^2 + \|L'\| \ \|x^*\| - \|y\| \geq 0$$

and the result follows. ∎

Note that the same inequality holds if we replace B by Q.

Example 1.6

Let $X = C[0, 1]$ with $\|x\| = \max_s |x(s)|$. Consider the equation

$$x(s) = s + \lambda \int_0^1 (s - t)x^2(t)dt \ .$$

Here we define

$$B(x, y)(s) = \int_0^1 (s - t)x(t)y(t)dt \ .$$

Then we get

$$\|B(x, y)\| \leq \left(\sup_s \int_0^1 |s - t|dt \right) \|x\| \ \|y\|$$

or

$$\|B\| \leq \frac{1}{2} \ .$$

Also, let $y(t) = x(t) = 1$ for all $t \in [0, 1]$ so $\|x\| = \|y\| = 1$ and

$$\|B(x, y)\| = \sup_s \left| s \int_0^1 dt - \int_0^1 tdt \right| = \sup_s \left| s - \frac{1}{2} \right| = \frac{1}{2}$$

so

$$\|B\| \geq \frac{1}{2}$$

and therefore $\|B\| = \frac{1}{2}$.

Let $\lambda = \frac{1}{4}$. Then, according to Corollary 1.1, since

$$4|\lambda| \ \|B\| \ \|y\| = 4\frac{1}{2}\frac{1}{4}1 < 1 \ ,$$

there exists a solution x^* of (1.19), $r_1 = 4 - 2\sqrt{2}$ and $r_2 = 2$. According to Theorem 1.4, the iteration

$$x_{n+1}(s) = s + \frac{1}{4} \int_0^1 (s - t) x_n^2(t) dt$$

converges to a solution of (1.19) for any $x_0(s) \in U(0, r_2)$. We can now find a good initial guess because by Proposition 1.10, $\|x^*\| \geq .88$. Also

$$\|x^*\| \leq \frac{1 - \sqrt{I - 4|\lambda| \|B\| \|y\|}}{2|\lambda| \|B\|} = 1.16$$

so

$$.88 \leq \|x^*\| \leq 1.16$$

but, since

$$.88 \leq \left| 1 + \frac{1}{4} \xi_1 \right| + \frac{1}{4} |\xi_2| \leq 1.16 ,$$

for $\xi_1 = .3$, $\xi_2 = .2$ and $s \in [0, 1]$, we make the initial guess

$$x_0(s) = \left[1 + \frac{1}{4}(.3) \right] s + \frac{1}{4}(.2)$$

$$= 1.075s + .05 .$$

The following theorem perturbs both y and B. It can be used to cover cases when condition (1.36) is violated.

THEOREM 1.15
Consider the equation

$$z = y_1 + F(z, z) \tag{1.58}$$

where $F: X \times X \rightarrow Z$ is a bounded symmetric bilinear operator and y_1 is fixed in X. Suppose that there exists a solution z of (1.58) satisfying

$$\|z\| < \frac{1 - \left[4 \|y - y_1\| (\|B\| - \|B - F\|) + \frac{\|B-F\|}{\|B\|} \right]^{1/2}}{2[\|B\| - \|B - F\|]}$$

$$\textit{if } \|B\| \neq \|B - F\| ; \tag{1.59}$$

or

$$\|z\| < \frac{1 - 4\|B\| \ \|y - y_1\|}{4\|B\|} \qquad \text{if } \|B\| = \|B - F\|$$

provided that

$$1 - 4\|B\| \cdot \|y - y_1\| > 0 , \tag{1.60}$$

in either case.
   Then

(i)   *Equation (1.19) has a unique solution* $x^* \in U(z, a)$;

(ii) *moreover,* $x^* \in \bar{U}(z, b)$, *where*

$$a = \frac{1 - 2\|B\| \cdot \|z\|}{2\|B\|} \tag{1.61}$$

   and

$$b = \frac{1 - 2\|B\| \cdot \|z\| - \left[4\|B\|(\|B\| - \|B - F\|)\|z\|^2 \right.}{2\|B\|} \left. {-4\|B\| \cdot \|z\| + 1 - 4\|B\| \cdot \|y - y_1\|\right]^{1/2}} . \tag{1.62}$$

PROOF   Note that by the choice of z and (1.60) the quantities under the radicals in (1.59) and (1.62) are nonnegative. Let us define the operator T on $U(z, r)$ for some r such that

$$b \le r < a , \tag{1.63}$$

by

$$T(x) = y + B(x, x) . \tag{1.64}$$

The result will be established by proving the following two claims:

CLAIM 1.3
T is a contraction on $U(z, r)$.

   Let $w, v \in \bar{U}(z, r)$. Then, using (1.64) we have

$$\|T(w) - T(v)\| = \|B(w, w) - B(v, v)\| = \|B[(w - z) + (v - z) + 2z, w - v]\|$$

$$\le 2\|B\|(r + \|z\|)\|w - v\|$$

$$= q\|w - v\|$$

with $0 < q < 1$, by the choice of r.

CLAIM 1.4
T maps (U, z) into $U(z, r)$.

Let $x \in U(z, r)$. Then, using (1.64) we have

$$\|T(x) - z\| \leq \|y - y_1\| + \|B - F\| \cdot \|z\|^2 + \|B\|(r + 2\|z\|)r .$$

To show

$$\|T(x) - z\| \leq r$$

it is enough to show

$$\|B\| \cdot r^2 + (2\|B\| \cdot \|z\| - 1)r + \|y - y_1\| + \|B - F\| \cdot \|z\|^2 \leq 0$$

which is true by the choice of r, z, and (1.60). That completes the proof of the theorem. ∎

REMARK 1.7   The hypotheses (1.59) and (1.60) for the application of Theorem 1.14 for some interesting special choices of F, y_1, and z become:

(a)   for $y_1 = y$ and $F = B$, we have

$$\|z\| < \frac{1}{2\|B\|} ;$$

(b)   for $y_1 = y$, we have

$$\|z\| < \frac{1}{2\sqrt{\|B\|} \left(\sqrt{\|B - F\|} + \sqrt{\|B\|}\right)} ;$$

(c)   for $F = 0$ and $z = y_1 = y$, we have

$$4\|B\| \cdot \|y\| < 1 ;$$

and

(d)   for $F = B$, we have

$$\|z\| < \frac{1 - \sqrt{4\|y - y_1\| \cdot \|B\|}}{2\|B\|} \tag{1.65}$$

provided that $1 - 4\|B\| \, \|y - y_1\| > 0$. ∎

It is easy to see by referring to case (c) of Remark 1.7 that Theorem 1.6 (Rall's Theorem) is a special case of Theorem 1.15.

PROPOSITION 1.11
Under the hypotheses of Theorem 1.15 the following are true:

(i) *the iteration (1.44) converges to the solution $x^* \in \bar{U}(z, b)$ of Equation (1.19) at the rate of a geometric progression with quotient*

$$q = 1 - [4\|B\|(\|B\| - \|B - F\|)\|z\|^2$$

$$- 4\|B\| \cdot \|z\| + 1 - 4\|B\| \, \|y - y_1\|]^{1/2} \, ;$$

(ii) *The solution x^* is such that*

$$\|x^*\| \leq \frac{1}{2\|B\|} \, ;$$

(iii) *If the following is true*

$$(\|B\| - \|B - F\|)\|z\|^2 - \|z\| + \|y\| - \|y - y_1\| > 0 \, ,$$

then under the hypotheses of Theorem 1.15 and Theorem 1.6, Theorem 1.15 provides a sharper estimate on $\|x^*\|$ than the latter.

PROOF   (i) By the proof of Theorem 1.15

$$q = 2(b + \|z\|)\|B\|$$

$$= 1 - \left[4\|B\|(\|B\| - \|B - F\|)\|z\|^2 - 4\|B\| \, \|z\| + 1 - 4\|B\| \cdot \|y - y_1\| \right]^{1/2} .$$

(ii) By Corollary 1.1 we have

$$\|x^* - z\| < a \, ,$$

so that

$$\|x^*\| \leq \|z\| + a < \frac{1}{2\|B\|} \, .$$

(iii) We have

$$\|x^* - z\| \le b \quad \text{so} \quad \|x^*\| \le b + \|z\| .$$

By Theorem 1.15

$$\|x^*\| \le \frac{1 - (1 - 4\|B\| \cdot \|y\|)^{1/2}}{2\|B\|}$$

so it is enough to show

$$1 - \left[4\|B\|(\|B\| - \|B - F\|)\|z\|^2 - 4\|B\| \cdot \|z\| + 1 - 4\|B\| \cdot \|y - y_1\| \right]^{1/2}$$

$$< 1 - (1 - 4\|B\| \cdot \|y\|)^{1/2}$$

which is true if (c) above holds. ∎

REMARK 1.8   We have for $y = 1$:

(a) according to Corollary 1.1, Equation (1.22) has a unique solution in $U(x^*, R)$,

$$R = \frac{(1 - 4\lambda \ln 2)^{1/2}}{2\lambda \ln 2} .$$

(b) By (c) in Remark 1.7, Equation (1.22) has a unique solution $x^* \in U(1, r)$, $b \le r_1 < a$, where

$$a = \frac{1 - 2\lambda \ln 2}{2\lambda \ln 2} ,$$

$$b = \frac{1 - 2\lambda \ln 2 - (1 - 4\lambda \ln 2)^{1/2}}{2\lambda \ln 2} . \quad ∎$$

Note that under the same hypothesis Remark 1.7 (c) provides better information on the location of the solution than Theorem 1.4.

For $y = 1$, Equation (1.22) becomes

$$z(s) = 1 + \lambda K(z(s), z(s)) = 1 + \lambda z(s) \int_0^1 \frac{s}{s + t} z(t) dt . \tag{1.66}$$

The iteration (1.44) for $y_1 = 1$ and $\lambda K = F$ becomes

$$z_{n+1} = 1 + \lambda K (z_n(s), z_n(s)) .$$

Take $z_0(s) = 1 \in U(1, r)$. Then for $\lambda = .35$ say we can find the solution $z(s)$ of (1.66). Simpson's rule was used for the numerical quadratures over s in the range $0(0.05)(1.0)$, to eventually find that

$$\|z\| = 1.44474532 .$$

This result agrees with the one in [77] or [89] at least to six decimal places.

Consider now Chandrasekhar's integral equation (1.22) corresponding to (1.19) for $c = 1.08$ and $B = \lambda K$, namely

$$x(s) = 1.08 + (.35)x(s) \int_0^1 \frac{s}{s+t} x(t)dt . \tag{1.67}$$

The condition

$$4\|B\| \, \|y\| < 1$$

is violated since

$$4\|B\| \cdot \|y\| = 4(.35)(\ln 2)(1.08) = 1.048038537 > 1 .$$

Therefore, Rall's Theorem or (c) in Remark 1.7 cannot be applied.

However, we can apply (d) in Remark 1.7. Our conditions are satisfied because they become

$$1.44474532 < 1.48674657$$

$$1 - 4(.35)(\ln 2)|1.08 - 1| = .92236757 > 0 ,$$

respectively.

Moreover,

$$a = .61624759$$

and

$$b = .39263584.$$

Therefore, (1.67) has a unique solution $x^*(s) \in U(z, r)$ for

$$.39263584 \leq r < .61624759 .$$

Setting $x_0(s) = \|z\| = 1.44474532$, the iteration

$$x_{n+1} = 1.08 + (.35)x_n(s) \int_0^1 \frac{s}{s+t} x_n(t)dt$$

converges to the solution $x^*(s)$, which can be calculated as the z above to obtain

$$\|x^*\| = 1.63791832 .$$

Exercises

1. Show that the space $C[0, 1]$ of continuous real functions $x = x(s)$, $0 \leq s \leq 1$, is a Banach space for the norm

$$\|x\| = \max_{[0,1]} |x(s)| .$$

2. Derive an upper bound for $\|B\|$, where

$$B = \begin{pmatrix} b_{111} & b_{112} & b_{121} & b_{122} \\ b_{211} & b_{212} & b_{221} & b_{222} \end{pmatrix}$$

   is a linear operator from $\mathbf{R}^2_\infty$ into $L(\mathbf{R}^2_\infty, \mathbf{R}^2_\infty)$, with

$$\|L\| = \max \{|\lambda_{11}| + |\lambda_{12}|, |\lambda_{21}| + |\lambda_{22}|\}$$

   for

$$L = \begin{pmatrix} \lambda_{11} & \lambda_{12} \\ \lambda_{21} & \lambda_{22} \end{pmatrix} \quad \text{in } L\left(\mathbf{R}^2_\infty, \mathbf{R}^2_\infty\right) .$$

3. Construct arithmetic fixed point problems for each of the following numbers: (a) $\sqrt[3]{5}$; (b) $1 + \sqrt[4]{3} + \sqrt{7}$; (c) $3 + \sqrt[6]{1 + \sqrt{4}}$. In each case choose an x_0 by estimating the number sought and calculate a few terms of the fixed point iteration. If the iteration apparently diverges, try to reformulate the fixed point problem to obtain a sequence which appears to converge.

4. Let $f(x)$ be a continuous function on $[0, 1]$ such that $0 \leq f(x) \leq 1$ for $x \in [0, 1]$. Define iteration

$$x_n = x_{n-1} + \frac{1}{n} \left[f(x_{n-1}) - x_{n-1} \right] .$$

   Show that this iteration converges to a fixed point of f for any $x_0 \in [0, 1]$.

5. Let $X = \{x \in \mathbf{R} \mid x \geq 1\} \subseteq \mathbf{R}$ and let the mapping $T: X \longrightarrow X$ be defined by $Tx = \frac{x}{2} + \frac{1}{x}$. Show that T is a contraction and find the smallest q [see (1.28)].

6. It is important that in Banach's Theorem 1.3 the condition (1.28) cannot be replaced by $\|T(x) - T(y)\| < |x - y|$ when $x \neq y$. To see this, consider $X = \{x \mid 1 \leq x < +\infty\}$, taken with the usual metric of the real line, and $T: X \longrightarrow X$ defined by $x \longmapsto x + x^{-1}$. Show that $|Tx - Ty| < |x - y|$ when $x \neq y$, but the mapping has no fixed points.

7. If $T: X \longrightarrow X$ satisfies $d(Tx, Ty) < d(x, y)$ when $x \neq y$ and T has a fixed point, show that the fixed point is unique; here, (X, d) is a metric space.

8. If T is a contraction, show that T^n ($n \in \mathbf{N}$) is a contraction. If T^n is a contraction for an $n > 1$, show that T need not be a contraction.

9. In analysis, a usual sufficient condition for the convergence of an iteration $x_n = g(x_{n-1})$ is that g be continuously differentiable and

$$\left| g'(x) \right| \leq \alpha < 1 .$$

Verify this by the use of Banach's fixed point theorem.

10. To find approximate numerical solutions of a given equation $f(x) = 0$, we may convert the equation to the form $x = g(x)$, choose an initial value x_0 and compute

$$x_n = g(x_{n-1}) \qquad n = 1, 2, \cdots .$$

Suppose that g is continuously differentiable on some interval $J = [x_0 - r, x_0 + r]$ and satisfies $|g'(x)| \leq \alpha < 1$ on J as well as

$$|g(x_0) - x_0| < (1 - \alpha)r .$$

Show that then $x = g(x)$ has a unique solution x on J, the iterative sequence (x_m) converges to that solution, and one has the error estimates

$$|x - x_m| < \alpha^m r, \qquad |x - x_m| \leq \frac{\alpha}{1 - \alpha} |x_m - x_{m-1}| .$$

11. Using Banach's Theorem 1.3, set up an iteration process for solving $f(x) = 0$ if f is continuously differentiable on an interval $J = [a, b]$, $f(a) < 0$, $f(b) > 0$ and $0 < k_1 \leq f'(x) \leq k_2$ ($x \in J$); use $g(x) = x - \lambda f(x)$ with a suitable λ.

12. **(Newton's method)** Let f be real-valued and twice continuously differentiable on an interval $[a, b]$, and let x^* be a simple zero of f in (a, b). Show that *Newton's method* defined by

$$x_{n+1} = g\,(x_n)\,, \qquad g\,(x_n) = x_n - \frac{f\,(x_n)}{f'\,(x_n)}$$

is a contraction in some neighborhood of x^* (so that the iterative sequence converges to x^* for any x_0 sufficiently close to x^*).

13. **(Square root)** Show that an iteration for calculating the square root of a given positive number c is

$$x_{n+1} = g\,(x_n) = \frac{1}{2}\left(x_n + \frac{c}{x_n}\right)$$

where $n = 0, 1, \cdots$. What condition do we get from Problem 9? Starting from $x_0 = 1$, calculate approximations $x_1, \cdots, x_4$ for $\sqrt{2}$.

14. Let $T: X \longrightarrow X$ be a contraction on a complete metric space, so that (1.28) holds. Because of rounding errors and for other reasons, instead of T one often has to take a mapping $S: X \longrightarrow X$ such that for all $x \in X$,

$$d(Tx, Sx) \le \eta \qquad (\eta > 0, \ \text{suitable})\,.$$

Using induction, show that then for any $x \in X$,

$$d\left(T^m x, S^m x\right) \le \eta\,\frac{1 - \alpha^m}{1 - \alpha} \qquad (m = 1, 2, \cdots)\,.$$

15. The mapping S in Problem 14 may not have a fixed point; but in practice, S^n often has a fixed point y for some n. Using Problem 14, show that then for the distance from y to the fixed point x of T we have

$$d(x, y) \le \frac{\eta}{1 - \alpha}\,.$$

16. In Problem 14, let $x = Tx$ and $y_m = S^m y_0$. Using Problem 14, show that

$$d\,(x, y_m) \le \frac{1}{1 - \alpha}\left[\eta + \alpha^m d\,(y_0, Sy_0)\right]\,.$$

What is the significance of this formula in applications?

17. **(Lipschitz condition)** A mapping $T:[a, b] \longrightarrow [a, b]$ is said to satisfy a *Lipschitz condition* with a *Lipschitz constant k* on $[a, b]$ if there is a constant k such that for all $x, y \in [a, b]$,

$$|Tx - Ty| \leq k|x - y| .$$

(a) Is T a contraction? (b) If T is continuously differentiable, show that T satisfies a Lipschitz condition. (c) Does the converse of (b) hold?

18. Show that $x = \frac{1}{2} \cos x$ has a solution x^*. Find an interval $[a, b]$ containing x^* such that for every $x_0 \in [a, b]$, the iteration

$$x_{n+1} = \frac{1}{2} \cos (x_n) \qquad (n \geq 0)$$

will converge to x^*. Calculate the first few iterates and estimate the rate of convergence.

19. To find a root for $f(x) = 0$ by iteration, rewrite the equation as

$$x = x + af(x) = g(x)$$

for some constant $a \neq 0$. If x^* is a root of $f(x)$ and if $f'(x^*) \neq 0$, how should a be chosen in order that the sequence $x_{n+1} = g(x_n)$ $(n \geq 0)$ converge to x^*?

20. Let A be an operator mapping a closed set D in a Banach space X into itself, and let some power A^k be a contraction operator. Prove that the successive approximations (1.29) converge to the unique fixed point of A in D.

21. Let A be an operator mapping a compact set $D \subseteq X$ into itself, such that

$$\|A(x) - A(y)\| < \|x - y\| \qquad \text{for all } x, y \in D$$

with $x \neq y$. Show that the successive approximations (1.29) converge to a fixed point of A.

22. Give an example of a linear operator $T \in L(\mathbf{R}^2)$ and of two norms on $\mathbf{R}^2$ such that T is a contraction in one norm, but not in the other.

23. Define $f:[0, 1] \subseteq R \to R$ by $f(x) = \frac{1}{2}x + 2$, $x \in [0, 1]$. Show that f is a contraction on $[0, 1]$, but has no fixed point.

24. Show that the set $Q^*(X; Z)$ (Definition 1.12) is a normed space.

25. Prove Proposition 1.4.

26. Prove Theorem 1.3.

27. Prove Theorem 1.5.

28. Prove Theorem 1.6.

29. Prove Proposition 1.8.

30. Prove Theorem 1.9.

31. Examine the details of Remark 1.6.

32. Prove Theorem 1.10.

33. Prove Theorem 1.11.

34. Prove Theorem 1.13.

35. Show part (a) of Remark 1.8.

36. Let $X = C[0, 1]$ be the Banach space of real continuous functions on $[0, 1]$ with the uniform norm $\|f\| = \max_{0 \le t \le 1} |f(t)|$. Define the linear operators $L_{jk}: X \to X$ by

$$\left(L_{jk} f \right) (s) = \int_0^1 L_{jk}(s, t) f(t) dt, \quad 0 \le s \le 1, \; j, k = 1, 2,$$

where the kernels $L_{jk}(s, t)$ are given. Show:

(i) The operators L_{jk} are bounded and compact with

$$\|L_{jk}\| = \sup_{0 \le s \le 1} \int_0^1 |L_{jk}(s, t)| \, dt, \quad j, k = 1, 2.$$

Let $Y = X \times X$ be the Banach space with $\|x\| = \max(\|x_1\|, \|x_2\|)$ where $x = (x_1, x_2)$ or $x = \begin{pmatrix} x_1 \\ x_2 \end{pmatrix}$, as convenience dictates. Define the linear operator $L: Y \to Y$ by

$$L(x) = \begin{pmatrix} L_{11} & L_{12} \\ L_{21} & L_{22} \end{pmatrix} \begin{pmatrix} x_1 \\ x_2 \end{pmatrix} = \begin{pmatrix} L_{11} x_1 + L_{12} x_2 \\ L_{21} + L_{22} x_2 \end{pmatrix}.$$

Show:

(ii) L is bounded and compact with

$$\|L\| \le \max_{j=1,2} \left(\|L_{j1}\| + \|L_{j2}\| \right).$$

Moreover, define $F: Y \to Y$ by

$$F(x) = \left(x_1, x_2, \frac{1}{2}x_1^2\right) .$$

Show:

(iii) F is bounded and compact with

$$\|F(x)\| \le \|x\|^2$$

and

$$\|F(x) - F(y)\| \le (\|x\| + \|y\|)\|x - y\| .$$

Let $Z = \mathbf{R} \times Y$ and consider this set as a Banach space with elements (γ, x) and $\|(\gamma x)\| = \max(|\gamma|, \|x\|)$. Define the operator $P: Z \to Y$ by

$$P(\gamma, x) = (1 - \gamma L F)(x) - g$$

where $g = (g_1, g_2)$ is known. (See also the Anselone and Moore equation [9] in Section 1.4 that can be formulated this way.)

Now consider an operator T defined on an open subset of Z with values $T(\alpha, x)$ in a Banach space X_1. Show:

(iv)   Suppose $T(\alpha_0, x_0) = 0$, T is continuously differentiable with respect to x at (α_0, x_0), and $[T_x(\alpha_0, x_0)]^{-1}$ exists. Then there exist constants $\varepsilon > 0$ and $\delta > 0$ such that for $|\alpha - \alpha_0| < \varepsilon$ the equation $T(\alpha, x) = 0$ has a unique solution $x = g(\alpha)$ with $\|g(\alpha) - x_0\| < \varepsilon$. Moreover, $g(\alpha)$ is continuous.

(v)   Suppose that

(v$_1$) T is twice differentiable and $\|T_{xx}(x)\| \le k$ on $U(x_0, r_0)$, where r_0 is defined below;

(v$_2$) $\Gamma_0 = [T_x(x_0)]^{-1}$ (α fixed) exists and $\|\Gamma_0\| \le b_0$;

(v$_3$) $\|x_1 - x_0\| \le \|\Gamma_0 T(x_0)\| \le \eta_0$;

(v$_4$) $\tau_0 = b_0 \eta_0 k \le \frac{1}{2}$; where

$$r_0 = N(\tau_0)\eta_0, \quad N(\tau) = \frac{1 - \sqrt{1 - 2\tau}}{\tau} .$$

Then the equation $T(x) = 0$ has a solution $x^* \in \bar{U}(x_0, r_0)$ and the Newton method generates a sequence $\{x_n\}$ ($n \ge 0$) which converges to x^*, and

$$\|x_n - x^*\| \le \frac{1}{2^{n-1}} (2h_0)^{2^n - 1} \eta_0 .$$

Moreover, if (v_1) holds in $U(x_0, g_0)$, where

$$g_0 = M(\tau_0)\, \eta_0, \quad M(\tau) = \frac{1 + \sqrt{1 - 2\tau}}{\tau}$$

then the solution x^* is unique in $U(x_0, g_0)$.

(vi) Set $T = P$ above and compare the results obtained here with ones obtained in Theorem 1.8.

37. Consider the quadratic equation [129]

$$y = B(x, x) + L(x) \tag{1.68}$$

in a Banach space X, where B is a bounded symmetric bilinear operator in X, L is a bounded linear operator in X and $y \in X$ is fixed. Define the iteration

$$x_0 = z, \quad x_{n+1} = (L + Bx_n)^{-1}(y) \ (n \geq 0) \text{ with } z \in X. \tag{1.69}$$

Show:

(i) If the (1.69) iteration is defined and if there exists $x^* \in X$ such that $x_n \to x^*$ as $n \to \infty$, then x^* is a solution of Equation (1.68).

(ii) If
L^{-1} exists
$0 < \|L^{-1}\| \cdot \|B\| \cdot \|w\| \leq \frac{1}{4}, w = L^{-1}(y)$
and
$\dfrac{1 - (1 - 4\|L^{-1}\| \cdot \|B\| \cdot \|w\|)^{1/2}}{2} \leq \|L^{-1}B(z)\|$

$\leq \dfrac{1 + (1 - 4\|L^{-1}\| \cdot \|B\| \cdot \|w\|)^{1/2}}{2}$

then iteration (1.69) is well defined.

(iii) If
L^{-1} exists,
$0 < \|L^{-1}\| \cdot \|B\| \cdot \|w\| \leq \delta < \frac{1}{4}$,
and
$\dfrac{1 - (1 - 4\delta)^{1/2}}{2} \leq \|L^{-1}B(z)\| < \dfrac{1}{2}$,
then there exists an $x^* \in X$ such that $x_n \to x^*$ as $n \to \infty$, and

$$\|x^* - x_n\| \leq \frac{b^n}{1 - b}\, \|(L + B(z))^{-1}y - z\|$$

where

$$b = \frac{\|L^{-1}B(z)\|}{1 - \|L^{-1}B(z)\|} < 1.$$

38. Consider the quadratic equation as above given by (1.69). Show [167]:

(i) If $P(u) = P(v) = 0$ then $P'\left(\frac{u+v}{2}\right)(u - v) = 0$.

(ii) If $P(x^*) = 0$ and $P'(x^*)^{-1}$ exists, the solution x^* of equation $P(x^*) = 0$ is unique in $U(x^*, r)$ where

$$r = \frac{1}{\|B\| \cdot \|P'(x^*)^{-1}\|}.$$

Let $P'(x_0) = 0$ for some $x_0 \in X$, then (1.69) can be expressed as

$$B(x, x) = z, \tag{1.70}$$

where $x = u - x_0$, $B = \frac{1}{2}P''(x_0)$ and $z = -P(z_0)$.
A quadratic equation such that $P'(x_0) = 0$ for some $x_0 \in X$ is said to be of first kind. The corresponding Equation (1.70) is called a normal form of (1.69).

(iii) If $u \in X$ is a solution of (1.69) of first kind then null $P(w + u) = 0$ for all $w \in \text{null}(B)$.

(iv) The roots of (1.69) of first kind appear in pairs: $P(u) = P(x_0 + x) = 0 \Leftrightarrow P(u) = P(x_0 - x)$, where $P'(x_0) = 0$. If $P'(x_0)^{-1}$ exists then (1.69) becomes

$$B_1(x, x) + I(x) + z = 0 \tag{1.71}$$

with $x = u - x_0$, $B_1 = \frac{1}{2}[P'(x_0)^{-1}]^{-1}P''(x_0)$, $z = P'(x_0)^{-1}(y)$.
The equation is then called of second kind.

(v) If $P(u) = 0$ is a quadratic equation of the second kind

$$P'(x_1) \neq -P'(x_2) \quad \text{for all } x_1, x_2 \in X.$$

(vi) If $P(u) = 0$ for some $u \in X$ and $w \neq 0$ and in null (B_1), then $P(u + w) \neq 0$.

(vii) If $u_1 \neq u_2$ and $P(u_1) = P(u_2)$ with P of second kind, then

$$P'(u_1) \neq P'(u_2).$$

If P is not of first or second kind then it is called of the third kind.

(viii) The quadratic equation (1.70) has a solution $x \in F(B)$ (factor set of B) if and only if $B(z)^{1/2}$ exists such that

$$B(x) = B(z)^{1/2} \quad \text{and} \quad B(z)^{1/2}(x) = z \ .$$

(ix) The quadratic equation (1.71) has a solution $x \in F(B_1)$ if and only if $(I - 4B_1(z))^{1/2}$ exists such that

$$B_1(x) = \frac{1}{2}\left[-I + (I - 4B_1(z))^{1/2}\right]$$

and

$$\frac{1}{2}\left[I + (I - 4B_1(z))^{1/2}\right]x + z = 0 \ .$$

(x) If $\text{null}(B) = \{0\}$, Equation (1.70) has at most one solution $x \in F(B)$ for each distinct square root of $B(z)$.

(xi) The quadratic equation (1.71) has at most one solution $x \in F(B_1)$ for each distinct pair $(I - 4B_1(z))^{1/2}$, $-(I - 4B_1(z))^{1/2}$ of square roots of $I - 4B_1(z)$.

(xii) Equation (1.70) has a solution $x \in \text{null}(B)$ if and only if $z = 0$, in which case all $x \in \text{null}(B)$ satisfy (1.70).

(xiii) Equation (1.71) has a solution $x \in \text{null}(B_1)$ if and only if $z \in \text{null}(B_1)$, in which case $x = -z$ is the unique solution of (1.71) in $\text{null}(B_1)$.

39. Consider the quadratic equation (1.19). If $v \in X$ is any solution, then any other solution x^* is given by $x^* = v + h$, where h is a nonzero solution of the equation

$$(I - 2B(v))(h) = B(h, h) \ . \tag{1.72}$$

Assume that $I - 2B(v)$ is invertible and set $B_1 = (I - 2B(v))^{-1}B$. Then (1.72) becomes

$$h = B_1(h, h) \ . \tag{1.73}$$

Introduce the iteration

$$h_{n+1} = B_1 (h_n)^{-1} (h_n) \quad (n \geq 0)$$

to find nonzero solutions of (1.73) based on the contraction mapping principle.

40. Consider the quadratic equation in the form

$$B(x, x) = y \ . \tag{1.74}$$

Case 1. Let $y = 0$ and set $x = v - h$ for some v such that $2B(v)$ is invertible (if it exists!). Then Equation (1.74) becomes

$$B_1(h, h) = h - y_0 \qquad (1.75)$$

where
$$B_1 = 2B(v)^{-1}B, \quad y_0 = 2B(v)^{-1}B(v, v) .$$

Introduce the iteration

$$h_{n+1} = B_1(h_n)^{-1}(h_n - y_0) \quad \text{for some } h_0 \in X$$

to find a solution h^* of (1.75) such that $h^* \neq v$.
Case 2. Let $y \neq 0$, and introduce the iteration

$$x_{n+1} = B(x_n)^{-1}(y) \quad \text{for some } x_0 \in X$$

to find solutions x^* of (1.74).
Produce theorems based on the construction mapping principle for both cases.

41. Consider Equation (1.19), and define the numbers

$$p = \frac{-1 + \sqrt{1 + 4\|B\| \cdot \|y\|}}{2\|B\|}$$

$$s_1 = \frac{1 - \sqrt{1 - 4\|B\| \cdot \|y\|}}{2\|B\|},$$

$$s_2 = \frac{1 + \sqrt{1 - 4\|B\| \cdot \|y\|}}{2\|B\|},$$

$$r = \frac{1 - 4\|B\| \cdot \|y\|}{2\|B\|}$$

Assume that $1 - 4\|B\| \cdot \|y\| > 0$. Then show:

(i) Equation (1.19) has a solution x^* which is unique in $U(x^*, r)$.
(ii) Any solution x^* of (1.19) is such that

$$\|x^*\| \geq p$$

(condition $1 - 4\|B\| \cdot \|y\| > 0$ is not required here).

(iii)   The following are true:

$$p \leq \|x^*\| \leq s_1 \quad \text{or} \quad \|x^*\| \geq s_2 .$$

42.   (A) Let P be a nonlinear operator defined on a convex subset D of a Banach space X with values in a Banach space Y, which is twice Fréchet differentiable. Assume:

(a) $\Gamma_0 = P'(z)^{-1}$ exists and is bounded for some fixed $z \in D$;

(b) $\|P(z)\| \leq v$;

(c) $\|P''(x)\| \leq b$ if $x \in U(z, r) \subseteq D$;

(d) $h = \|\Gamma_0\|^2 vb \leq \frac{1}{2}$; and

(e) $r_0 = \frac{(1-\sqrt{1-2h})v\|\Gamma_0\|}{h} < r$.

Then show that there exists $x^* \in U(z, r_0)$ such that $P(x^*) = 0$, which is the only solution of P contained in $U(z, r) \cap U(z, r_1)$, where $r_1 = \frac{(1+\sqrt{1-2h})\|\Gamma_0\|v}{h}$.

(B) Let $z \in X$ be such that $z = y + f(z, z)$ for some auxiliary bounded symmetric bilinear operator F defined on D. Define P on D by

$$P(x) = x - z + F(z, z) - B(x, x) .$$

Note that every solution x^* of the above equation is a solution of (1.19). Assume:

(f) $(I - 2B(z))^{-1}$ exists and is bounded;

(g) $\|P(z)\| = \|(F - B)(z, z)\| \leq \|F - B\| \cdot \|z\|^2 = v$;

(h) $\|P''(x)\| \leq z\|B\| = b$ if $x \in U(z, r) \subseteq D$);

(i) $h_0 = \|(I - 2B(z))^{-1}\|^2 vb \leq \frac{1}{2}$; and

(j) $r_0 = \frac{(1-\sqrt{1-2h_0})v\|(I-2B(z))^{-1}\|}{h_0} < r$.

Then show that there exists a solution x^* of (1.19) which is unique in $U(z, r) \cap U(z, r_1)$, where

$$r_1 = \frac{\left(1 + \sqrt{1 - 2h_0}\right) v \,\|(I - 2B(z))^{-1}\|}{h_0} .$$

(C) Assume:

(k) $(I - 2F(z))^{-1}$ exists and is bounded above by some $K > 0$;

(l) $4\|F(z)B(z) - B(z)B(z)\| \leq \frac{1}{\|(I-2F(z))-1\|}$;

(m) $\|P(z)\| \le v$;

(n) $2\|B\| \le b$ if $x \in U(z, r) \subseteq D$;

(o) $h_1 = k_0^2 vb \le \frac{1}{2}$,

$$K_0 = \frac{1 + 2\|(I - 2F(z))^{-1}\| \cdot \|B(z)\|}{1 - 4\|(I - 2F(z))^{-1}\| \cdot \|F(z)B(z) - B(z)F(z)\|};$$

and

(p) $r_0 = \frac{(1 - \sqrt{1 - 2h_1})K_0 v}{h} < r$.

Then show that there exists a solution x^* of (1.19) which is unique in $U(z, r) \cap U(z, r_1)$, where

$$r_1 = \frac{\left(1 + \sqrt{1 - 2h_1}\right) K_0 v}{h_1}.$$

43. Let Q, Q_n $(n \ge 0)$ be bounded quadratic operators such that $Q_n(x^*) \to Q(x^*)$ as $n \to \infty$, where x^* is a solution of Equation (1.19). Let B_n be the unique symmetric bounded bilinear operators associated with Q_n $(n \ge 0)$. Define the real functions $f_n(r)$ and $g_n(r)$ by

$$f_n(r) = 2\|B_n\| r + 2\|x_0\| \cdot \|B_n\| - 1,$$

$$g_n(r) = \|B_n\| r^2 + (2\|x^*\| \cdot \|B_n\| - 1) r + \|Q_n(x^*) - Q(x^*)\|$$

$$(n \ge 0)$$

and assume: $f_n(r) < 0$ and $g_n(r)$ $(n \ge 0)$.

Show:

(i) There exists $r > 0$ with $r_n \le r < R$ such that equation $x = y + Q_n(x)$ $(n \ge 0)$ has a unique solution $x_n \in \bar{U}(x_0, r)$ where r_n denotes the small root of the equation $g_n(r) = 0$ $(n \ge 0)$.

(ii) $\lim_{n \to \infty} x_n = x^*$.

(iii) Conditions of the functions f_n and g_n are satisfied provided that

$$\|x^*\| < \frac{1 - \sqrt{\|B_n\| \|Q_n(x^*) - Q(x^*)\|}}{2\|B_n\|}$$

and

$$r_n \le r < \frac{1 - 2\|x^*\| \|B_n\|}{2\|B_n\|} \quad (n \ge 0),$$

respectively.

44.   With the notation introduced in Theorem 1.4, assume:

   (a)   hypotheses of Theorem 1.4 are true for $x_0 = z$;

   (b)   there exists c, with $0 < c < 1$ such that

$$rc^3 + \left(\|B\| \cdot \|y\|^2 - r \right) c^2 + (r_0 - r)\,(2\|B\|$$

$$\|y\| + 1)c + \|B\| \cdot (r_0 - r)^2 \le 0$$

   for any $r \in [r_1, r_0]$ and $r_1 \le r_0 < r_2$.

Show:

   (i)   Iteration (1.29) generates a sequence $\{x_n\}$ $(n \ge 0)$ which converges to a solution x^* of Equation (1.19) such that

$$\|x_n - x_0\| \le \sum_{k=1}^{n} \omega^{(k)}\,(r_0) \quad (n \ge 0)$$

and
$$\left\|x_n - x^*\right\| \le \sigma\left(\omega^{(n)}\,(r_0) \right) \quad (n \ge 0)$$

where

$$\omega^{(n)}(r) = r(1 - c)^n \quad \text{and} \quad \sigma\left(\omega^{(n)}(r) \right) = \frac{r}{c}(1 - c)^n \quad (n \ge 0)\,.$$

   (ii)   The second estimate above is better than the corresponding one in Theorem 1.4 with q given in Proposition 1.5 if (c) satisfies (b) above and

$$0 < 1 - 2(r + \|z\|)\|B\| < c < 1 \quad \text{for } r \in [r_1, r_2]\,.$$

45.   Consider the Chandrasekhar equation in the form

$$x(s) = 1 + \int_0^1 \frac{sx(s)x(t)}{s+t}\psi(t)dt\,, \qquad (1.76)$$

where ψ is a known function. Use any n-point quadrature formula of the form

$$\int_0^1 f(t)dt = \sum_{j=1}^{n} c_j f(t_j) + r$$

where $0 < t_1 < t_2 < \cdots < t_n < 1$ are the grid points of the formula, $c_1, \ldots, c_n$ are the weights, and r is the remainder or error term. Find a discrete analog of Equation (1.76). In particular show that the resulting system of equations has the solution

$$x_i = \left(\prod_{j=1}^{n} t_j \right)^{-1} \prod_{j=1}^{n} (t_j + t_i) \Big/ \prod_{j=1}^{n} (1 + d_j t_i) \quad i = 1, 2, \ldots, n \, ,$$

where $d_1, \ldots, d_n$ are the nonnegative solutions of the equation

$$1 = 2 \sum_{j=1}^{n} \left[c_j \psi\left(t_j\right) \Big/ \left(1 - d_j^2 t_j^2\right) \right] \, .$$

46.   For the scalar equation $x = F(x)$, a fixed point x^* in the real case will be the abscissa of an intersection point of the straight line $y = x$, with the curve $y = F(x)$. Give a geometric procedure for constructing the points $x_{m+1} = F(x_m)$, $m = 0, 1, 2, \ldots$. Sketch examples of convergent and divergent sequences.

47.   Suppose that the operator F in R_∞^n defined by

$$f_i = f_i\left(c_1, c_2, \ldots, c_n\right), \quad i = 1, 2, \ldots, n \, ,$$

is a contraction mapping on the closed ball $u(x_0, 10)$, $\|x_1 - x_0\| = 1$, and $q = .75$. Find the minimum integer k such that

$$\left\| x_k - x^* \right\| \leq 2^{-25}$$

using Theorem 1.3.

48.   Give sufficient conditions for the solvability of the quadratic scalar equation

$$x = a + bx^2 \, ,$$

using Theorem 1.3 with $x_0 = 0$. Derive explicit formulas for x_1, x_2, x_3, using iteration (1.19).

49.   Apply Theorem 1.3 to the ordinary differential equation

$$\frac{dy}{dt} = 1 + y^2, \quad y(0) = 0, \ 0 \leq t \leq a \, .$$

Determine a value of a for which existence and uniqueness of a solution are guaranteed.

Chapter 2

More Methods for Solving Quadratic Equations

This chapter introduces more existence and uniqueness results for quadratic equations on various space settings. In the first section we work with Banach algebras [33]. Section 2.2 uses the majorant method [22]. Finite rank equations are examined in Section 2.3 [15]. Section 2.4 deals with noncontractive solutions [16]. Quadratic integral equations with perturbation are finally discussed in Section 2.5 [22, 23].

2.1 Banach Algebras

In this section we consider a special case of Equation (1.19) given by

$$x = y + x\tilde{K}(x) \tag{2.1}$$

where $\tilde{K}$ is a linear operator on a Banach algebra X_A and $y \in X_A$ is fixed [109, 118, 128]. Note that Equation (1.22) can be brought in the form (2.1).

Here we suggest the iteration

$$x_{n+1} = (L(x_n))(K(x_n)) \tag{2.2}$$

for solving (2.1), where

$$L(x) = x - y$$

and

$$K(x) = \frac{1}{\tilde{K}(x)}$$

provided that $K(x)$ is well defined and $L(x) \neq 0$ on $U(z, r)$ for some $z \in X_A$ and $r > 0$.

Using a theorem of Darbo [87], we provide conditions for the convergence of (2.2) to a solution x of (2.1).

We denote by $C[0, 1]$ the Banach space of all real continuous functions on $[0, 1]$ with the maximum norm,

$$\|x\|_C = \max_{0 \le s \le 1} |x(s)| . \tag{2.3}$$

We let $C^p[0, 1]$, $0 < p < 1$, denote the Banach space of all real continuous functions on $[0, 1]$ such that

$$\sup_{0 \le t, s \le 1} \frac{|x(t) - x(s)|}{|t - s|^p} < \infty ,$$

with the norm

$$\|x\|_{C^p} = \max_{0 \le s \le 1} |x(s)| + \sup_{0 \le t, s \le 1} \frac{|x(t) - x(s)|}{|t - s|^p} . \tag{2.4}$$

We note that the spaces $X_A^1 = C[0, 1]$ and $X_A^p = C^p[0, 1]$ with norms given by (2.3) and (2.4), respectively, are Banach algebras.

Following Kuratowski [123] we define the measure of noncompactness $M_M(D)$ of a bounded set D of a metric space M to be

$$M_M(D) = \inf\{\varepsilon > 0 \mid D \text{ can be covered by a finite number}$$

of sets of diameter less than or equal to $\varepsilon\} .$

Suppose g maps M continuously into a metric space N, and suppose g takes bounded sets to bounded sets. If for some $e \in [0, \infty)$ $M_M(g(D)) \le eM_M(D)$ for every bounded subset D of M, we say that g is an e-set contraction. If $e < 1$, then g is a strict set contraction.

We will use the following theorem [87].

THEOREM 2.1

Let D be a subset of a Banach algebra X_A and suppose that $T: D \to X_A$ is of the form

$$T(x) = x_0 + L(x)K(x) ,$$

where

(a) $x_0 \in X_A$;

(b) $L: D \to X_A$ *satisfies* $\|L(x) - L(y)\| \le b\|x - y\|$ *for some $b \ge 0$ and all* $x, y \in D$;

(c) $K: D \rightarrow X_A$ is compact;

(d) $a = \sup_{x \in D} \|K(x)\| < \infty$.

If $ab < 1$, then T is a strict set contraction.

Moreover, if T leaves a closed bounded convex subset of D of a Banach algebra invariant, then T has a fixed point in D.

From now on except for the norm of $\tilde{f}$ and related operators, $\|\cdot\|$ will denote that norm.

We will prove the following:

THEOREM 2.2

Let $f(s, t)$ be a continuous function on $[0, 1] \times [0, 1]$ for which

$$\sup_{0 \leq s_1, s_2 \leq 1} \frac{1}{|s_1 - s_2|^p} \int_0^1 |(f(s_1, t) - f(s_2, t))| \, dt \leq M . \qquad (2.5)$$

for some p, $0 < p < 1$, and some $M < \infty$. Define a linear operator $\tilde{f}$ on X_A^1 by

$$\left(\tilde{f} x \right)(s) = \int_0^1 f(s, t) x(t) dt ,$$

and assume there exists $z \in X_A^1$ such that $\tilde{f}(z)$ never vanishes on $[0, 1]$. Let $\|\tilde{f}\|$ be defined by $\|\tilde{f}\| = \sup_{0 \leq s \leq 1} \int_0^1 |f(s, t)| dt$ and fix $r_1 \in (0, r_0)$ where

$$r_0 = \frac{1}{\left\| \tilde{f} \right\| \cdot \left\| \tilde{f}(z)^{-1} \right\|} .$$

Then the operator K on $U(z, r_1)$ given by

$$K(x) = \frac{1}{\tilde{f}(x)} \qquad (2.6)$$

is well defined, satisfies

$$a = \sup_{x \in U(z, r_1)} \|K(x)\| \leq \frac{\left\| \tilde{f}(z)^{-1} \right\|}{1 - \left\| \tilde{f} \right\| \left\| \tilde{f}(z)^{-1} \right\| \cdot r_1} , \qquad (2.7)$$

and is compact on $U(z, r_1)$.

PROOF   The result follows immediately from the Banach lemma for invertible operators the choice of r_2 and the identity

$$\tilde{f}(x) = \tilde{f}(z)\left[1 + \tilde{f}(z)^{-1}\tilde{f}(x - z)\right] .$$

We will use the result, namely, if for every uniformly bounded sequence $\{x_n\}$ in a subset of X_A^1 there is a p, $0 < p < 1$, such that $K(x_n) \in X_A^p$ for every n and $\{K(x_n)\}$ is bounded in the norm of X_A^p, then K is a compact operator.

Now let x_n belong to $U(z, r_1)$ such that $\|x_n\| \le M_1$ for some $M_1 > 0$. Consider

$$h_n(s) = \int_0^1 f(s, t)x_n(t)dt .$$

Then for p as in (2.5)

$$\sup_{0 \le s_1, s_1 \le 1} \frac{|h_n(s_1) - h_n(s_2)|}{|s_1 - s_2|^p} = \sup_{0 \le s_1, s_2 \le 1} \frac{1}{|s_1 - s_2|^p}\left|\int_0^1 (f(s_1, t)\right.$$

$$\left. - f(s_2, t))\, x_n(t)dt\right| \le MM_1 . \qquad (2.8)$$

By (2.7), there exists $q > 0$ such that

$$\left|\int_0^1 f(s, t)x_n(t)dt\right| \ge q . \qquad (2.9)$$

From (2.6), (2.8), and (2.9)

$$\|K(x_n)\|_{X_A^p} \le \frac{1}{q} + \frac{MM_1}{q^2} . \qquad (2.10)$$

From (2.10) we obtain that K is a compact operator on $U(z, r_1)$.
    That completes the proof.   ∎

Let z, f, $\tilde{f}$, K and r_0 be as in Theorem 2.1. Next, fix $y \ne z$ and denote by $P(y, z)$ the constant $y + z\tilde{f}(z) - z$. Assume the constants c_1, c_2 and Δ given by

$$c_1 = 1 - \left\|\tilde{f}(z)^{-1}\left(I - z\tilde{f}\right)\right\|$$

$$c_2 = 1 - \left\|\tilde{f}(z)^{-1}\right\| ,$$

and

$$\Delta = \left(\left\| \tilde{f}(z)^{-1} \left(I - z\tilde{f} \right) \right\| - 1 \right)^2 - 4 \left\| \tilde{f}(z)^{-1} \right\| \cdot \left\| \tilde{f} \right\| \cdot \left\| \tilde{f}(z)^{-1} P(z) \right\|,$$

are positive. Define the constants $r_2 - r_5$ by

$$r_2 = \frac{1 - \left\| \tilde{f}(z)^{-1} \right\|}{\left\| \tilde{f} \right\| \cdot \left\| \tilde{f}(z)^{-1} \right\|},$$

$$r_3 = \frac{1 - \left\| \tilde{f}(z)^{-1} \left(I - z\tilde{f} \right) \right\| - \sqrt{\Delta}}{2 \left\| \tilde{f} \right\| \cdot \left\| \tilde{f}(z)^{-1} \right\|},$$

$$r_4 = \frac{1 - \left\| \tilde{f}(z)^{-1} \left(I - z\tilde{f} \right) \right\| + \sqrt{\Delta}}{2 \left\| \tilde{f} \right\| \cdot \left\| \tilde{f}(z)^{-1} \right\|},$$

and

$$r_5 = \| z - y \|.$$

THEOREM 2.3
Let z, f, $\tilde{f}$, K, y and the constants c_1, c_2, Δ, r_0 and $r_2 - r_5$ be as above. Assume the following conditions are satisfied:

$$\begin{cases} \text{The } z, f, \tilde{f}, K \text{ are as in Theorem 2.2 and } y \in X_A^1 \text{ is fixed;} \\ \text{the following are true:} \\ c_1 < 0, \\ c_2 > 0, \\ \Delta > 0, \\ r_3 < \min(r_2, r_4, r_5). \end{cases}$$

and choose $r_6 \in (r_3, \min(r_2, r_4, r_5))$.
  Define the operator T on $\bar{U}(z, r_6)$ by

$$T(x) = (L(x))(K(x))$$

where L, K are operators given by

$$L(x) = x - y$$

$$K(x) = \frac{1}{\tilde{f}(x)} \cdot$$

Then the operator T has a fixed point in $\bar{U}(z, r_6)$.

PROOF   One only needs to show that the hypotheses of Theorem 2.1 are satisfied. Note that the hypotheses (a), with $b = 1$, and (b) are obvious and that (c) follows from Theorem 2.2 since $r_6 < r_2 < r_0$. We now prove the claims:

CLAIM 2.1
T is a strict set contraction on $\bar{U}(z, r_6)$.

Since, $b = 1$ it is enough to show that

$$a = \sup_{x \in U(z, r_6)} \|K(x)\| < 1$$

or by (2.8)

$$\frac{\left\|\tilde{f}(z)^{-1}\right\|}{1 - \left\|\tilde{f}\right\| \cdot \left\|\tilde{f}(z)^{-1}\right\| \cdot r_6} < 1$$

which is true by the choice of r_6.

CLAIM 2.2
T maps $\bar{U}(z, r_6)$ into $\bar{U}(z, r_6)$.

Let $x \in \bar{U}(z, r_6)$. Then $T(x) \in \bar{U}(z, r_6)$ if

$$\|T(x) - z\| = \left\| \tilde{f}(x)^{-1} \left[\left(I - z\tilde{f} \right)(x - z) - P(z) \right] \right\|$$

$$\leq \frac{\left\| \tilde{f}(z)^{-1} \left(I - z\tilde{f} \right) \right\| \cdot r_6 + \left\| \tilde{f}(z)^{-1} P(z) \right\|}{1 - \left\| \tilde{f}(z)^{-1} \right\| \cdot \left\| \tilde{f} \right\| \cdot r_6} \leq r_6$$

which is also true because $r_3 \leq r_6 \leq r_4$.
   That completes the proof.   ∎

Note that results similar to those in Theorems 2.2 and 2.3 can easily be proved if we work in the space X_A^p.

2.2 The Majorant Method

We introduce the iteration

$$x_{n+1} = B(x_n)^{-1}(x_n - y), \quad n = 0, 1, 2, \dots \tag{2.11}$$

for approximating solutions x^* of Equation (1.19).

Special cases of (1.19) appear in many interesting problems arising in astrophysics, in the kinetic theory of gases as well as the theory of ordinary and partial differential equations [89, 100]. Equation (1.19) has been studied extensively.

A common hypothesis for the above techniques is the estimate (1.36).

It turns out that under this hypothesis the previously mentioned techniques approximate a small solution v^* of Equation (1.19) for any starting point x_0 close enough to the solution. The obtained solution v^* is such that $v^* = v^*(y) \to 0$ as $y \to 0$. We make use of the "theory of majorants" and under assumptions similar to the ones introduced in the above-mentioned techniques, iteration (2.11) can be used to approximate a second solution x^* of (1.19) with $x^* \neq v^*$ and $x^* = x^*(y) \to 0$ as $y \to 0$. Moreover, under the same assumptions we show that the Newton–Kantorovich method [113, 114] can be used to obtain a solution $z_N^* = x^*$ also. This result is not known even for quadratic systems in $\mathbf{R}^n$, $n > 1$. Some sufficient conditions are also given for the existence of more than one distinct solution of (1.19). Our results are illustrated with the solution of a quadratic system in $E = \mathbf{R}^2$ as well as the solution of a Riccati differential equation.

THEOREM 2.4

Let B be a bounded symmetric bilinear operator on $X \times X$ and suppose that $x_0, y \in X$ with $x_0 \neq 0$ and $x_0 \neq y$. Assume:

(a) *The inverse of the linear operator $B(x_0): X \to X$ with $B(x_0)(x) = B(x_0, x)$ for all $x \in X$ exists and is bounded.*

(b) *The estimates*

$$0 \leq c < 1 \tag{2.12}$$

and

$$0 \leq d < \frac{(1-c)^2}{4ab} \tag{2.13}$$

are true where we have denoted

$$a \geq \left\| B(x_0)^{-1} \right\|, \tag{2.14}$$

$$b \geq \|B\|, \tag{2.15}$$

$$c \geq \left\| B(x_0)^{-1} (I - B(x_0)) \right\| \tag{2.16}$$

and

$$d \geq \left\| B(x_0)^{-1} (B(x_0, x_0) + y - x_0) \right\|. \tag{2.17}$$

Then:

(i)   *The real sequence $\{t_n\}$, $n = 0, 1, 2, \ldots$ given by*

$$t_{n+2} = t_{n+1} - \frac{1+c-2abt_{n+1}}{1+c+2abt_{n+1}} (t_n - t_{n+1}), \quad n = 0, 1, 2, \ldots,$$

$$t_0 = \frac{1-c}{2ab}, \quad t_1 = \left[\frac{1+c}{2} \right] t_0, \tag{2.18}$$

is positive, decreasing and converges to zero.

(ii) *The sequence $\{x_n\}$, $n = 0, 1, 2, \ldots$ generated by (2.11) is well defined, remains in $U(x_0, r_0) = \{x \in X / \|x - x_0\| < r_0 = \frac{1-c}{2ab}\}$ and converges to a unique solution $x^* \in \bar{U}(x_0, r_0)$ of Equation (1.19).*

Moreover, the following estimates are true for all $n = 0, 1, 2, \ldots$,

$$\|x_{n+1} - x_n\| \leq t_n - t_{n+1}$$

and

$$\|x_n - x^*\| \leq t_n \leq \left[\frac{1+c}{2} \right]^n t_0.$$

PROOF   (i) It can easily be seen by (2.18) that the sequence $\{t_n\}$, $n = 0, 1, 2, \ldots$ is certainly nonnegative if

$$(1 + c)t_{k+1} + 2abt_k t_{k+1} - (1 + c)t_k \geq 0 \quad \text{for all } k = 0, 1, 2, \ldots. \tag{2.19}$$

Inequality (2.19) is true as equality for $k = 0$. Let us assume that it is true for $k = 0, 1, 2, \ldots, n$. We shall show that it is true for $k = n + 1$. Using (2.18), the left-hand side of inequality (2.19) for $k = n + 1$ becomes

$$\frac{2ab(1 + c + 2abt_k) t_{k+1}^2 + (1 + c)^2 t_{k+1} - (1 + c)^2 t_k}{1 + c + 2abt_{k+1}}$$

which is nonnegative if $t_{k+1} \geq \frac{(1+c)t_k}{1+c+2abt_k}$ and that is true by our assumption.
By the choice of t_0, t_1, and (2.17), $t_0 - t_1 > 0$.
Let us assume that

$$t_k - t_{k+1} > 0, \quad k = 0, 1, 2, \ldots, n . \tag{2.20}$$

Using (2.18), we see that (2.20) is true for $k = n + 1$ if

$$1 + c - 2abt_{k+1} > 0 \quad \text{for} \quad k = 0, 1, 2, \ldots, n . \tag{2.21}$$

Inequality (2.21) is true for $k = 0$ by the choice of t_1. Let us assume that (2.21) is true for $k = 0, 1, 2, \ldots, n$. To show (2.21) for $k = n + 1$ it suffices to show $t_{k+2} < \frac{1+c}{2ab}$ or by (2.18)

$$2ab \left[2(1 + c)t_{k+1} + 2abt_k t_{k+1} - (1 + c)t_k \right] \leq (1 + c)^2 + 2ab(1 + c)t_{k+1}$$

or $t_{k+1} \leq \frac{1+c}{2ab}$ which is true by hypothesis.

We have now shown that the real sequence $\{t_n\}$, $n = 0, 1, 2, \ldots$ is positive and decreasing and as such it converges to some $t^* \geq 0$. But using simple induction and (2.18) we can easily show that $t_{n+1} \leq \left[\frac{1+c}{2}\right] t_n \leq \left[\frac{1+c}{2}\right]^{n+1} t_0$. That is $t^* = 0$.

(ii) Let us observe that the linear operator $B(x)$ is invertible for all $x \in U(x_0, r_0)$. Indeed we have

$$\left\| B (x_0)^{-1} B (x - x_0) \right\| \leq \left\| B (x_0)^{-1} \right\| \cdot \| B \| \cdot \| x - x_0 \| \leq ab \| x - x_0 \| < 1$$

so that according to Banach's lemma on invertible operators

$$\left\| B(x)^{-1} \right\| = \left\| \left[I + B (x_0)^{-1} B (x - x_0) \right]^{-1} B (x_0)^{-1} \right\|$$

$$\leq \frac{a}{1 - ab \| x - x_0 \|} . \tag{2.22}$$

We shall prove that

$$\| x_n - x_{n+1} \| \leq t_n - t_{n+1} \quad \text{for} \quad n = 0, 1, 2, \ldots . \tag{2.23}$$

By (i) it follows that if (2.11) is well defined for $n = 0, 1, 2, \ldots, k$ and if (2.23) holds for $n \geq k$ then $\| x_0 - x_n \| \leq t_0 - t_n < t_0 - t^*$ for $n \leq k$. This shows that (2.22) is satisfied for $x = x_i$, $i \leq k$. Thus (2.11) will be defined for $n = k+1$, too. By (2.11) and (2.17) $\|x_1 - x_0\| \leq d \leq t_0 - t_1$. That is, (2.23) is true for $n = 0$. Suppose (2.23) holds for $n = 0, 1, 2, \ldots, k$. Observing that

$$B(x_{k+1})(x_{k+1} - x_{k+2}) = B(x_{k+1}, x_{k+1}) + y - x_{k+1}$$

$$- B(x_k, x_k) - y + x_k + B(x_k)(x_k - x_{k+1})$$

$$= B(x_{k+1} - x_k, x_{k+1} + x_k) + x_k - x_{k+1}$$

$$+ B(x_k)(x_k - x_{k+1}) \qquad (2.24)$$

$$= B(x_{k+1} - x_k, x_{k+1} - x_k)$$

$$+ B(x_k - x_{k+1}, x_{k+1} - x_k)$$

$$+ B(x_{k+1}, x_{k+1} - x_k) - (x_{k+1} - x_k)$$

$$= (B(x_{k+1}) - I)(x_{k+1} - x_k) \, ,$$

we get

$$x_{k+1} - x_{k+2} = B(x_{k+1})^{-1} \left[B(x_{k+1} - x_0) + B(x_0) - I \right]$$

$$(x_{k+1} - x_k) \, . \qquad (2.25)$$

By taking norms in (2.25) and using (2.22) we obtain

$$\|x_{k+1} - x_{k+2}\| \leq \frac{\left[c + ab(t_0 - t_{k+1}) \right](t_k - t_{k+1})}{1 - ab(t_0 - t_{k+1})} = t_{k+1} - t_{k+2}$$

by choice of t_0. Inequality (2.24) shows that $\{x_n\}$, $n = 0, 1, 2, \ldots$ is a Cauchy sequence in a Banach space X and as such it converges to some $x^* \in X$. By taking the limit as $n \to \infty$ in (2.11) we get $x^* = y + B(x^*, x^*)$. That is x^* is a solution of Equation (1.19). Fix n and let $p = 0, 1, 2, \ldots$. Then

$$\|x_n - x^*\| \leq \|x_n - x_{n+p}\| + \|x_{n+p} - x^*\|$$

$$\leq t_n - t_{n+p} + \|x_{n+p} - x^*\| \, . \qquad (2.26)$$

By letting $p \to \infty$ we obtain

$$\|x_n - x^*\| \le t_n - t^*, \quad n = 0, 1, 2, \dots . \tag{2.27}$$

By (2.27) for $n = 0$ we get

$$\|x_0 - x^*\| \le t_0 - t^* = \frac{1 - c}{2ab} - t^* = \frac{1 - c}{2ab} .$$

That is, $x^* \in \bar{U}(x_0, r_0)$.

Finally, let us assume that there exists a second solution $z^* \in U(x_0, r_0)$ of Equation (1.19). By (2.11) we have

$$x_{n+1} - z^* = x_n - B(x_n)^{-1}(y + B(x_n, x_n) - x_n) - z^*$$

$$= B(x_n)^{-1}\left[B(x_n)(x_n - z^*) + x_n - y - B(x_n, x_n)\right.$$

$$\left. + y + B(z^*, z^*) - z^*\right]$$

$$= -B(x_n)^{-1}\left[B(z^* - z_0) + B(x_0) - I\right](x_n - z^*) .$$

By taking the norms in the above identity and using (2.22) we obtain

$$\|x_{n+1} - z^*\| \le 2\left[\frac{c + ab\,\|z^* - x_0\|}{1 + c + 2abt_{n+1}}\right]\|x_n - z^*\| .$$

By the choice of r_0 the factor of $\|x_n - z^*\|$ is less than 1 so that $\|x_n - z^*\|$ goes to zero as $n \to \infty$; hence $z^* = \lim_{n \to \infty} x_n = x^*$.

That completes the proof of the theorem. ∎

Moreover, we can show the following theorem:

THEOREM 2.5

Let B be a bounded symmetric bilinear operator on $X \times X$ and suppose that $x_0, y \in X$ with $x_0 \ne 0$, $x_0 \ne y$. Assume:

(a) *The following estimate is true:*

$$4be < 1 \tag{2.28}$$

where
$$e \geq \|y\| . \tag{2.29}$$

(b)   *The hypotheses of Theorem 2.4 are satisfied for some $x_0 \in X$ such that*

$$\|x_0\| > p \quad \text{with a certain } p \in (p_1, p_2) ,$$

where p_1 and p_2 are the two positive solutions of the scalar quadratic equation
$$bz^2 - z + e = 0 . \tag{2.30}$$

Then:

(i)   *The iteration*
$$v_{n+1} = y + B(v_n, v_n) \tag{2.31}$$

remains in $U(0, p_1)$ and converges to a unique solution v^* of Equation (1.19) in $U(0, \frac{1}{2\|B\|})$ for any $v_0 \in \bar{U}(0, p_1)$. Moreover, for all $n = 0, 1, 2, \ldots$

$$\|v_n - v^*\| \leq p_1 - e \sum_{j=0}^{n} \frac{(2j)!}{j!(j+1)!} (eb)^j .$$

(ii)   *The solution x^* of Equation (1.19) obtained via iteration (2.11) is such that $x^* \neq v^*$.*

PROOF   (i) The first part of the result in (i) follows from Corollary 1.3, whereas the second part follows from (1.30).

(ii) We shall show that $\|x_n\| > p$ for a certain $p \in (p_1, p_2)$. By (2.11) we obtain $\|x_n - y\| = \|B(x_n, x_{n+1})\| \leq \|B\| \cdot \|x_n\| \|x_{n+1}\|$ or

$$\|x_{n+1}\| \geq \frac{\|x_n - y\|}{\|B\| \cdot \|x_n\|} .$$

Assume that $\|x_k\| > p$ for all $k = 0, 1, 2, \ldots, n$. Since

$$\|x_n\| > p > e , \tag{2.32}$$

it is enough to show

$$\frac{\|x_n\| - e}{b \|x_n\|} > p \tag{2.33}$$

or

$$\|x_n\| > \frac{e}{1 - pb} .$$

By (2.32) it finally suffices to show $p > \frac{e}{1-pb}$, which is true for $p \in (p_1, p_2)$. By taking the limit as $n \to \infty$ in (2.32) we get $\|x^*\| \geq p$. Therefore, we obtain $x^* \neq v^*$.

That completes the proof of the theorem. ∎

Furthermore, we can prove the following theorem concerning the number of solutions of Equation (1.19).

THEOREM 2.6
Let B be a bounded symmetric bilinear operator on $X \times X$ and suppose that $x_0, y \in X$ with $x_0 \neq y$ and $y \neq 0$.
Assume:

(a) *point $x_0 \in X$ is such that*
$$B (x_0) = I ; \qquad (2.34)$$

(b) *inequality (2.28) is true.*

Then elements v^*, x^*, $x_0 - x^*$ and $x_0 - v^*$ are solutions of Equation (1.19) with

$$x^* \neq x_0 - x^* \qquad (2.35)$$

and

$$v^* \neq x_0 - v^* . \qquad (2.36)$$

PROOF   It follows by (a) that the hypotheses of Theorem 2.4 are satisfied. That is x^* is a solution of Equation (1.19). By (b) v^* is a solution of Equation (1.19). For $z = x_0 - x^*$ we have

$$y + B(z, z) = y + B \left(x_0 - x^*, x_0 - x^*\right)$$

$$= y + B (x_0, x_0) - 2B \left(x_0, x^*\right) + B \left(x^*, x^*\right)$$

$$= x^* + x_0 - 2x^* = x_0 - x^* .$$

Similarly we show that $x_0 - v^*$ is a solution of Equation (1.19).
Let us assume now that

$$x_0 - x^* = x^* . \qquad (2.37)$$

Then by (2.37) and (1.19) we have

$$x_0 = 2\left(y + B\left(x^*, x^*\right)\right) = 2\left(y + \frac{1}{4}B(x_0, x_0)\right)$$

which implies

$$x_0 = 4y .$$

That is,

$$x^* = 2y . \tag{2.38}$$

But then by (2.38) and (1.19) $2y = y + B(2y, 2y)$ or $4\|B\| \|y\| \geq 1$ since $y \neq 0$ contradicting (b). This shows (2.35). Similarly we show (2.36) and that completes the proof of the theorem. ∎

We can show the following.

THEOREM 2.7

Let B be a bounded symmetric bilinear operator on $X \times X$ and suppose that $x_0, y \in X$ with $x_0 \neq 0$, $x_0 \neq y$. Assume:

(a) hypotheses of Theorem 2.4 are satisfied;

(b) inequality (1.19) is true;

(c) inequality

$$\|x_0 - y\| > \frac{1 - 2b\left(e - r_0\right) - \sqrt{1 - 4eb}}{2b} = R \tag{2.39}$$

is true.

Then solutions x^* and v^* obtained via Theorem 2.4 and 2.5 are distinct.

PROOF　Assume that $x^* = v^*$. The solution x^* is such that

$$\left\|B\left(x^*, x^*\right)\right\| = \left\|x^* - y\right\| \tag{2.40}$$

and since $R > r_0$, (2.40) gives

$$\frac{1 - \sqrt{1 - 4eb}}{2b} \geq \|x^*\| \geq \sqrt{\frac{\|x_0 - y\| - r_0}{b}} . \tag{2.41}$$

But then from (2.41) we deduce

$$\|x_0 - y\| \leq R$$

contradicting (2.39).

That completes the proof of the theorem. ∎

Note that under the hypotheses of Theorems 2.6 and 2.7 it follows immediately that

$$x_0 - x^* \neq x_0 - v^* .$$

REMARK 2.1 (a) It can easily be seen that (2.14) and (2.15) can be replaced by the weaker condition

$$\left\| B(x_0)^{-1} B \right\| \leq q . \qquad (2.42)$$

(b) If we know the constants a, b, c, d then we may compute the sequence $\{t_n\}$, $n = 0, 1, 2, \ldots$ before obtaining the sequence $\{x_n\}$, $n = 0, 1, 2, \ldots$ via the iterative algorithm (2.11). Therefore, the estimates on the distances $\|x_n - x^*\|$ and $\|x_{n+1} - x_n\|$ obtained in Theorem 2.4 may be called *a priori* error estimates. Moreover, the convergence of iteration to a solution x^* of Equation (1.19) is only linear. Let us assume that the linear operator

$$\Gamma_0 = (I - 2B(x_0))^{-1} \qquad (2.43)$$

exists for some $x_0 \in X$ and

$$\|\Gamma_0\| \leq b_0, \ \|\Gamma_0(x_0 - y - B(x_0, x_0))\| \leq \eta_0, \ h_0 = 2b_0\|B\|\eta_0 \leq \frac{1}{2}. \qquad (2.44)$$

Then the Newton–Kantorovich iteration

$$z_{n+1} = z_n - (I - 2B(z_n))^{-1}(z_n - y - B(z_n, z_n)) ,$$

$$n = 0, 1, 2, \ldots, z_0 = x_0 \qquad (2.45)$$

for solving (1.19) converges to a unique solution z_N^* of Equation (1.19) in $U(x_0, r_N)$ with

$$r_N = \frac{1 - \sqrt{1 - 4bb_0\eta_0}}{2bb_0} . \qquad (2.46)$$

Moreover, the order of convergence is quadratic. However, we do not know if $\|z_n\| > p$ for a certain $p \in (p_1, p_2)$ whenever $\|z_0\| > p$. That is, we do not know if $\|z_N^*\| \not> p$ or if $z_N^* \neq v^*$. ∎

It will be shown later that whenever the hypotheses of Theorem 2.4 are satisfied then the Newton–Kantorovich hypotheses (2.44) are satisfied also and $x^* = z_N^*$.

That is, if we choose $x_0 = z_0$ with $\|x_0\| \geq p$, then

$$\|z_N^*\| \not\geq p \quad \text{and} \quad z_N^* \neq v^* \qquad (2.47)$$

even if $z_n \geq p$ for some n, $n = 0, 1, 2, \ldots$. Therefore, in practice we will prefer to use iteration (2.45) instead of (2.11) to find bounded away from zero solution x^* of Equation (1.19), since (2.45) converges faster than (2.11). However, our main concern is that the property (2.46) could only be proved through iteration (2.11) as the following theorem indicates.

THEOREM 2.8

Under the hypotheses of Theorem 2.5 the Newton–Kantorovich iteration (2.45) for $z_0 = x_0$ converges to a unique solution z_N^* of Equation (1.19) in $U(x_0, R_N)$ and $z_N^* = x^*$. Moreover, if $\|z_0\| > p$ for a certain $p \in (p_1, p_2)$ then

$$\|z_N^*\| \geq p, \qquad (2.48)$$

and

$$\|z_n - z_N^*\| \leq \frac{1}{2^n} (2h_0)^{2^n - 1} \eta_0, \quad n = 0, 1, 2, \ldots .$$

Furthermore, the solution z_N^* can be written as $z_N^* = x_0 + h$ where h is a solution of the quadratic equation

$$h = y_1 + B_1(h, h) \qquad (2.49)$$

with

$$y_1 = (I - 2B(x_0))^{-1} (B(x_0, x_0) + y - x_0) \quad \text{and} \quad B_1 = (I - 2B(x_0))^{-1} B .$$

PROOF   By the Banach lemma on invertible operators [114, 119] the linear operator

$$B(x_0)^{-1} - 2I = \left(B(x_0)^{-1} - I \right) - I$$

is invertible because $\|I\| \cdot \|I - B(x_0)^{-1}\| \leq c < 1$ and

$$\left\| \left(B(x_0)^{-1} - 2I \right)^{-1} \right\| \leq \frac{1}{1 - c} .$$

Equation (2.48) has a solution h if

$$4 \|y_1\| \|B_1\| \leq 4 \left[\frac{1}{1 - c} \cdot d \right] \left[\frac{1}{1 - c} \cdot ab \right] < 1 \qquad (2.50)$$

which is true by (2.13). It can easily be seen now that $w^* = x_0 + h$ is a solution of Equation (1.19) if and only if h is a solution of Equation (2.49). The linear operator $(I - 2B(x_0))^{-1}$ exists since

$$(I - 2B(x_0))^{-1} = \left(B(x_0)^{-1} - 2I\right)^{-1} B(x_0)^{-1} . \tag{2.51}$$

The Newton–Kantorovich hypotheses (2.44) are now satisfied and by the definition of r_N, (2.28) and (2.29) we deduce that $z_N^* = w^*$. By the uniqueness of the solutions x^* and z_N^* in the balls $U(x_0, r_0)$ and $U(x_0, r_N)$ it follows that $z_N^* = x^*$ (the balls have the same center).

The rest of the theorem follows from part (b) of Theorem 2.5. ∎

All the results obtained here can apply to iteration (2.45). Note that the result (2.48) is not known even for quadratic systems in $\mathbf{R}^n$, $n > 1$.

To cover the cases when B is not symmetric, we can state the following theorem whose proof is identical to that of Theorem 2.4 which is omitted.

THEOREM 2.9
Let B be a bounded bilinear operator on $X \times X$ and suppose that $x_0, y \in X$ with $x_0 \neq 0$, $x_0 \neq y$. Further, let

$$\bar{a} \geq \left\| B(x_0)^{-1} B \right\| , \bar{b} \geq \left\| B(x_0)^{-1} (2\bar{B} - B) \right\| ,$$

$$\bar{c} \geq \left\| B(x_0)^{-1} (2\bar{B} - B)(x_0) - I \right\| ,$$

and let A, B be defined as

$$A = \frac{-[\bar{b}(\bar{a} + \bar{b})\bar{t}_0 + 2(\bar{a} \cdot \bar{c} + \bar{b})] + \left\{ [\bar{b}(\bar{a} + \bar{b})\bar{t}_0 + 2(\bar{a}\bar{c} + \bar{b})]^2 + 4(\bar{a}\bar{c} + \bar{b})\bar{t}_0(\bar{a}^2 - \bar{b}^2) \right\}^{1/2}}{2(\bar{a}^2 - \bar{b}^2)} ,$$

$$\bar{t}_0 = \frac{1 - \bar{c}}{\bar{a} + \bar{b}} ,$$

and

$$B = \frac{(1 + \bar{c})(1 - \bar{c})}{2\bar{a}(\bar{c} + 3)} .$$

Assume:

(a) *The inverse of the linear operator* $B(x_0): X \rightarrow X$ *with* $B(x_0)(x) = B(x_0, x)$ *for all* $x \in X$ *exists and is bounded;*

(b) *The following estimates are true:*

$$\bar{a} \geq \bar{b},$$

$$0 \leq c < 1,$$

$$0 \leq d < \bar{t}_0 - A \quad \text{if } \bar{a} > \bar{b}$$

and

$$0 \leq d < \bar{t}_0 - B \quad \text{if } \bar{a} = \bar{b}.$$

Then

(i) *the real sequence* $\{\bar{t}_n\}$, $n = 0, 1, 2, \ldots$ *given by*

$$\bar{t}_{n+2} = \bar{t}_{n+1} - \frac{\bar{c} + \bar{b}\bar{t}_0 - \bar{b}\bar{t}_{n+1}}{1 - \bar{a}\bar{t}_0 + \bar{a}\bar{t}_{n+1}} \left(\bar{t}_n - \bar{t}_{n+1} \right), \quad n = 0, 1, 2, \ldots$$

$$\bar{t}_1 = B \quad \text{if } \bar{a} > \bar{b}$$

and

$$\bar{t}_1 = B \quad \text{if } \bar{a} = \bar{b}$$

is positive, decreasing and converges to zero.

(ii) *The sequence* $\{x_n\}$, $n = 0, 1, 2, \ldots$ *generated by (2.11) is well defined, remains in* $U(x_0, \bar{r}_0)$ *and converges to a unique solution* $x^* \in \bar{U}(x_0, \bar{r}_0)$ *of Equation (1.19) with* $\bar{r}_0 = \bar{a}^{-1}$.

Moreover, the following estimates are true for all $n = 0, 1, 2, \ldots$

$$\|x_{n+1} - x_n\| \leq \bar{t}_n - \bar{t}_{n+1} \quad \text{and} \quad \|x_n - x^*\| \leq \bar{t}_n.$$

Remarks similar to the ones made after Theorem 2.7 can now easily follow for Theorem 2.9.

The results obtained in the next three examples can also be obtained through the use of iteration (2.45). However, we will only use iteration (2.11) for demonstrational purposes.

Example 2.1

Let $X = \mathbf{R}^2$ and define a bilinear operator on X by (1.13).

Consider the quadratic equation on X given by

$$w = y + B(w, w) \qquad (2.52)$$

or equivalently

$$w_1 = \frac{1}{48} - 3w_1^2 + 2w_1 w_2 - w_2^2$$

$$w_2 = -\frac{1}{48} + w_1^2 - 2w_1 w_2 - w_2^2 \qquad (2.53)$$

where

$$c_1^{11} = -3, \quad c_2^{21} = -1,$$

$$c_1^{12} = 1, \; c_2^{22} = -1, \; c_1^{21} = 1, \; c_1^{22} = -1, \; c_2^{11} = 1, \; c_2^{12} = -1,$$

$$y = \begin{bmatrix} y_1 \\ y_2 \end{bmatrix}, \quad w = \begin{bmatrix} w_1 \\ w_2 \end{bmatrix}$$

$$y_1 = \frac{1}{48}$$

and

$$y_2 = -\frac{1}{48}.$$

For $x \in X$, let $\|x\| = \max_{(i)} |x_i|, i = 1, 2$. Using the norm on $L(X, X)$ one can define the norm of B on X by

$$\|B\| = \sup_{\|x\|=1} \max_{(i)} \sum_{j=1}^{2} \sum_{k=1}^{2} \left| \sum_{k=1}^{2} b_{ijk} x_k \right|,$$

(see also Section 1.1) from which it follows at once that

$$\|B\| \le \max_{(i)} \sum_{i=1}^{2} \sum_{k=1}^{2} |b_{ijk}|.$$

Let $x_0 = \begin{bmatrix} -.5 \\ -.5 \end{bmatrix}$. With the above values it can easily be seen that B is a bounded, symmetric operator on X and

$$B(x_0) = I, \ e = d = \|y\| = \tfrac{1}{48}, \ b = 6, \ a = 1, \ c = 0,$$

$$r_0 = \tfrac{1}{12}, \ R = .08690776$$

and

$$\|x_0 - y\| = .520833333 .$$

According to Theorem 2.5 (b), Equation (2.53) has a small solution $v^* \in \bar{U}(0, p_1)$ which can be found to be $v^* = \begin{bmatrix} .0200308 \\ .0200308 \end{bmatrix}$ using the iteration (2.31) for v_0 for $v_0 = y$. We took $v_8 = v^*$. According to Theorem 2.4, Equation (2.53) has a solution in $U(x_0, r_0)$ which can be found to be $x^* = \begin{bmatrix} -.5200308 \\ -.5200308 \end{bmatrix}$ using the iteration (2.11) for $x_0 = \begin{bmatrix} -.5 \\ -.5 \end{bmatrix}$. We took $x_9 = x^*$. Since $\|x_0 = y_0\| > R$, it was known before actually computing v^* and x^* that $x^* \neq v^*$. Note, however, that $x^* - x_0 \neq v^*$.

It can easily be seen that $v_1^* = \begin{bmatrix} -.25 \\ .1318813 \end{bmatrix}$ is the third solution of (2.53).

Finally, the fourth solution x_1^* of Equation (2.53) is given by $x_1^* = x_0 - v_1^*$. We have now found all four solutions of Equation (2.53).

A more interesting example is given by the following.

Example 2.2
Consider the Riccati differential equation

$$x^2(t) + 2z(t)x(t) + y_1(t) - \frac{dx}{dt} = 0, \quad 0 \leq t < T < 1, \ x(0) = 0 . \quad (2.54)$$

As X takes $C_0^1[0, T]$, the space of all continuously differentiable functions $x = x(t)$, such that $x(0) = 0$, and as Y takes the space $C[0, T]$ of all continuous real functions. Let us equip the above spaces with the usual sup-norm. That is,

$$\|x\| = \sup_{0 \leq t \leq T} |x(t)| \quad \text{for } x \in E \ (\text{or } Y) . \quad (2.55)$$

Equation (2.54) is the quadratic equation of the form (1.19) with $B(x_1, x_2) =$

$B(x_1)(x_2)$ where $B(x_1)$ is a linear operator for fixed x_1 given by

$$B(x_1)(w)(t) = \left[\left[\frac{d}{dt} - 2z\right]^{-1} x_1 w\right](t) = \left[\exp\left[\int_0^t 2z(q)dq\right]\right] \cdot$$

$$\cdot \int_0^t \exp\left[-\int_0^s 2z(q)dq\right] x_1(s)w(s)ds, \qquad (2.56)$$

for all $w \in E$ and $0 \le t \le T$, and

$$y = \left[\frac{d}{dt} - 2z\right]^{-1} y_1. \qquad (2.57)$$

The linear operator $\frac{d}{dt} - 2z$ is indeed invertible for all $x \in E$; in fact, the inverse transformation $u = \left[\frac{d}{dt} - 2z\right]^{-1} v$ has the explicit representation

$$u(t) = \left[\exp\left[\int_0^t 2z(q)dq\right]\right]\int_0^t$$

$$\exp\left[-\int_0^s 2z(q)dq\right] v(s)ds, \quad 0 \le t \le T, \qquad (2.58)$$

where $u \in X$ for $v \in Y$. It can easily be seen that the bilinear operator B defined above is bounded and symmetric. We can easily deduce for $T = \frac{1}{2}$,

$$\|B\| = \frac{1}{2} \sup_{0 \le t \le T} \left|\left(1 - t^2\right)\ln\left(1 - t^2\right)\right| \le .375 \text{ for } z(t) = -\frac{1}{1 - t^2}.$$

Take $y_1(t) = -.14\frac{1+t^2}{1-t^2}$. Then easily, $y(t) = -.14t$ for all $0 \le t \le T$ and $\|y\| = .07$.

The condition (a) in Theorem 2.5 is now satisfied. Moreover, if the condition (b) in Theorem 2.5 is satisfied for some x_0, then using iterations (2.1) and (2.31) we can obtain the solutions x^* and v^*, respectively, with $x^* \ne v^*$.

Example 2.3

There are examples of interesting linear operators satisfying condition (2.34). Indeed, with the notation of the previous example, let us define a linear operator $B(\cdot)$ by $B(v) = \left[\frac{d}{dt} - 2z\right]^{-1}(v)$. Choose z as before and $v(t) = x_0(t) = \frac{1+t^2}{1-t^2}$. It can then easily be seen that $B(x_0)(t) = I(t) = t$ for all $0 \le t \le T$, that is,

$B(x_0) = I$. Therefore, the differential equation $\frac{du}{dt} - 2z(t)u(t) = v(t)$, $u(0) = 0$, has the unique solution u given by $u(t) = t$, $0 \le t \le T$.

Example 2.4

Consider the scalar equation $x = \delta + \beta x^2$ with $\delta, b > 0$ and $1 - 4\delta\beta > 0$.

Let us choose $\frac{1}{2\beta} < x_0 < \frac{1+\sqrt{1-4\delta\beta}}{2\beta}$. The conditions (2.12), (2.13), and (2.30) become, respectively,

$$x_0 \ge \tfrac{1}{2\beta} \, ,$$

$$\frac{2\beta + \sqrt{1(1 - 4\delta\beta)}}{2\beta} < x_0,$$

$$x_0 > p \quad \text{for} \quad p \in (p_1, p_2) \, ,$$

$$p_1 = \frac{1 - \sqrt{1 - 4\delta\beta}}{2\beta}, \quad p_2 = \frac{1 + \sqrt{1 - 4\delta\beta}}{2\beta} \, .$$

That is, x_0 must be chosen such that

$$\frac{2\beta + \sqrt{2(1 - 4\delta\beta)}}{4\beta} < x_0 < \frac{1 + \sqrt{1 - 4\delta\beta}}{2\beta} \, .$$

The large solution of the scalar quadratic equation can now be obtained using iteration (2.11) for the above choice of x_0.

2.3   Compact Quadratic Equations

In this section we approximate an isolated solution of a compact operator equation using the solutions of a family of collectively compact operator equations [10, 62, 63].

We study the quadratic equation

$$x = y + Q(x) \tag{2.59}$$

in a Banach space X, where $y \in X$ is fixed and Q is a compact quadratic operator from a subset K of X into X. We assume in the first part that solutions of the family of quadratic equations

$$x = y + Q_n(x), \quad n = 1, 2, \ldots \tag{2.60}$$

where $\{Q_n\}$, $n = 1, 2, \ldots$ are collectively compact and the Q_n converge pointwise to Q are known. We then use these solutions to approximate a solution of (2.59).

For linear equations, if (2.59) is nonsingular, (2.60) is uniquely solvable for sufficiently large n and the approximate solutions converge to the true solution [63].

For nonlinear equations, under certain differentiability conditions on Q and Q_n, $n = 1, 2, \ldots$, (2.60) has a unique solution in a neighborhood of an isolated solution of (2.59) if n is sufficiently large and these solutions converge to the isolated solution [10].

In [63] use of the Shauder–Leray Theory has been made to extend the above results.

Here we give a simplified proof based on the contraction mapping principle of the existence, uniqueness, and convergence of solutions of (2.60) in a neighborhood of an isolated solution of (2.59).

Let Q, Q_n be as before and let B, B_n, $n = 1, 2, \ldots$ be the symmetric bilinear operators associated with Q and Q_n, respectively. Let $x_0 \in K$ be a solution of (2.59); then it can easily be checked that the first and second Fréchet derivatives of Q_n, $n = 1, 2, \ldots$ at x_0 are

$$Q_n'(x_0) = 2B_n(x_0),$$

$$Q_n''(x_0) = 2B_n, \quad n = 1, 2, \ldots.$$

Let $I - 2B(x_0) = I - Q'(x_0)$ be nonsingular. The fact that $Q'(x_0)$ exists implies that $Q'(x_0)$ is compact. Let $c > 0$ be a uniform bound on the norm of the Q_ns, $n = 1, 2, \ldots$ that is

$$\|Q_n\| \le c, \quad n = 1, 2, \ldots.$$

Then the hypotheses on $\{Q_n\}$, $n = 1, 2, \ldots$ imply that $\{Q_n'(x_0)\}$, $n = 1, 2, \ldots$ is a sequence of collectively compact linear operators on X and $Q_n'(x_0)$ converges pointwise to $Q'(x_0)$. Moreover, by applying the collectively compact approximation theory to $Q'(x_0)$, $\{Q_n'(x_0)\}$, $n = 1, 2, \ldots$ we obtain that the linear operators $[I - Q_n'(x_0)]$ are nonsingular for sufficiently large n, say $n \ge N$, and

$$\left\|[I - Q_n'(x_0)]^{-1}\right\| \le d < \infty, \ n \ge N \quad \text{for some } d > 0.$$

We have now proved the following theorem.

PROPOSITION 2.1

Assume that $[I - Q'(x_0)]$ is nonsingular and that Q, Q_n are as in the introduction. Then the linear operators $[I - Q'_n(x_0)]$ are nonsingular with bounded inverse for sufficiently large n.

Let Q, Q_n, B, B_n be as before. Assume that the hypotheses of Proposition 2.1 are true. Let c, d denote the uniform bounds on the norms of the B_n and $(I - Q'_n(x_0))^{-1}$, $n = 1, 2, \ldots$, respectively. Define the real functions $f(r)$ and $g(r)$ by

$$f(r) = 4dcr - 1$$

and

$$g(r) = dcr^2 - r + dc \|x_0\|^2 + d\|y\| .$$

Note that

$$f(r) < 0$$

and

$$g(r) \leq 0$$

$$\text{if} \quad \left. \begin{array}{l} 1 - 4dc\|y\| > 0 \\ 1 - 4d^2c^2 \|x_0\|^2 - 4d^2c\|y\| > 0 \\ r_- \leq r < \frac{1}{4dc} \equiv R \end{array} \right\} \qquad (2.61)$$

where r_- is the small root of $g(r) = 0$.

We can now prove the main theorem:

THEOREM 2.10

Let $x_0 \in X$ be an isolated solution of (2.59) satisfying (2.61). Moreover, assume that the hypotheses of Theorem 2.9 are true. Then,

(i)   *there exists $r > 0$, with*

$$r_- \leq r < R$$

such that for $n \geq N$, (2.60) has a unique solution

$$x_n \in \bar{U}(x_0, r)$$

and

(ii) $\lim_{n \to \infty} x_n = x_0$.

PROOF (i) Define the operator T on X by

$$T(x) \equiv x = \left[I - Q_n'(x_0)\right]^{-1} \left[x - y - Q_n(x)\right] .$$

Note that the fixed points of T are solutions of (2.60).

CLAIM 2.3
T is a contraction operator on $\bar{U}(x_0, r)$.

Let $w, v \in \bar{U}(x_0, r)$. Then

$$\|T(w) - T(v)\| = \left\| w - v - \left[I - Q_n'(x_0)\right]^{-1} \left[w - v - Q_n(w) + Q_n(v)\right]\right\|$$

$$= \left\| \left[I - Q_n'(x_0)\right]^{-1} \left[B_n(x - x_0) + B_n(v - x_0)\right](w - v)\right\|$$

$$\leq 4cdr \|w - v\| .$$

So, T is a contraction if $f(r) < 0$ which is true by hypothesis (2.61).

CLAIM 2.4
T maps $\bar{U}(z, r)$ into $\bar{U}(z, r)$.

Let $w \in \bar{U}(x_0, r)$. Then,

$$\|T(w) - x_0\| \leq r$$

if

$$\|T(w) - x_0\| = \left\| w - \left[I - Q_n'(x_0)\right]^{-1} \left[w - y - Q_n(w)\right] - x_0\right\|$$

$$= \left\| \left[I - Q_n'(x_0)\right]^{-1} \left[Q_n(w - x_0) - Q_n(x_0) + y\right]\right\|$$

$$\leq cdr^2 + cd\|x_0\|^2 + d\|y\| \leq r, \quad \text{or } g(r) \leq 0$$

which is true if (2.61) holds. Part (i) now follows from the contraction mapping principle (see Theorem 1.8).

(ii) We have

$$\left(I - Q'_n\left(x_0\right)\right)\left(x_n - x_0\right) = \left(I - 2B_n\left(x_0\right)\right)\left(x_n - x_0\right)$$

$$= B_n\left(x_n, x_n\right) - B\left(x_0, x_0\right) - 2B_n\left(x_0\right)\left(x_n - x_0\right)$$

$$= B_n\left(x_0, x_0\right) - B\left(x_0, x_0\right) + B_n\left(x_n - x_0, x_n - x_0\right)$$

$$= Q_n\left(x_0\right) - Q\left(x_0\right) + Q_n\left(x_n - x_0\right) .$$

Therefore,

$$\|x_n - x_0\| \le d\left\|Q_n\left(x_0\right) - Q\left(x_0\right)\right\| + dc\left\|x_n - x_0\right\|^2$$

or,

$$\|x_n - x_0\| \le \frac{d}{1 - dc\left\|x_n - x_0\right\|}\left\|Q_n\left(x_0\right) - Q\left(x_0\right)\right\|$$

$$\le \frac{d}{1 - dcr}\left\|Q_n\left(x_0\right) - Q\left(x_0\right)\right\| \to 0 \quad \text{as } n \to \infty .$$

That is $\lim_{n\to\infty} x_n = x_0$ and the theorem is proved. ∎

In practice, many examples of interest are such that the operators Q_n in (2.60) are bounded and of finite rank; and, therefore, compact [10]. The evaluation of the parameter c is of considerable difficulty even then.

Here is an example that illustrates how the Q_ns, $n = 1, 2, \ldots$ may look like and how we evaluate c.

Example 2.5
Consider the Banach space $X = C[0, 1]$, the space of continuous functions on $[0, 1]$ equipped with the max-norm. Define the finite rank operator Q on X by

$$Q(x)(s) = \int_0^1 (s - t)x^2(t)dt, \quad 0 \le s \le 1, \ \text{rank}(Q) = 2 .$$

Now,

$$|Q(x)(s)| = \left| s\int_0^1 x^2(t)dt - \int_0^1 tx^2(t)dt \right| \le \|x\|^2\left(1 + \frac{1}{2}\right), \ \text{for all } x ,$$

$$\max_s \|Q(x)\| \le \tfrac{3}{2}\|x\|^2 \,,$$

so

$$\|Q\| \le \frac{3}{2} \,.$$

We can do better here: take a fixed x, $\|x\| \le 1$, then

$$\|Q(x)\| = \max_s |sa - b| \,,$$

$$a = \int_0^1 x^2(t)dt \ge 0 \quad \text{and} \quad b = \int_0^1 tx^2(t)dt \ge 0 \,.$$

Note that $ab = \int_0^1 (1-t)x^2(t)dt \ge 0$, so

$$\|Q(x)\| = \max\{b, a - b\} \,.$$

But, $tx^2(t) \le t$ for all $x^2(t) \le 1$, that is

$$b \le \int_0^1 tdt = \frac{1}{2}$$

and

$$(1-t)x^2(t) \le 1 - t \quad \text{for all } x^2(t) \le 1$$

so,

$$a - b \le \int_0^1 (1-t)dt = \frac{1}{2}$$

and

$$\|Q\| = \sup_{\|x\| \le 1} \|Q(x)\| \le \max\left\{\frac{1}{2}, \frac{1}{2}\right\} = \frac{1}{2} \,.$$

Finally, take $x(t) = 1$ for all $t, 0 \le t \le 1$, so $\|x\| = 1$ and

$$\|Q(x)\| = \sup_s \left| s\int_0^1 dt - \int_0^1 tdt \right| = \sup_s \left| s - \frac{1}{2} \right| = \frac{1}{2}$$

that is

$$\|Q\| = \frac{1}{2} \,.$$

The problem of evaluating the x_ns, $n = 1, 2, \ldots$ whose existence in $\bar{U}(x_0, z)$ is guaranteed by Theorem 2.10 may be as difficult as that of evaluating solutions of (2.59). In the next section we show, however, how to find all the solutions of (2.60) when the Q_ns, $n = 1, 2, \ldots$ are of finite rank. In particular we show that the problem of solving (2.60) is equivalent to solving a quadratic system in $\mathbf{R}^n$ or $(\mathbf{C}^n)$. Assuming then that the algebraic system can be solved, we can select for each n, $n = N, N + 1, \ldots$, a solution x_n such that $x_n \in \bar{U}(x_0, r)$ and then apply Theorem 2.10.

2.4 Finite Rank Equations

DEFINITION 2.1   *An operator $P : X \to Z$ has finite rank n, rank $(P) = n$, if span $(Rang(P))$ has dimension n.*

REMARK 2.2   Denote by $Q_F^*(X)$ the set of all bounded quadratic operators Q in X such that Q has finite rank. Obviously $Q_F^*(X)$ is a subspace of $Q^*(X)$.   ∎

REMARK 2.3   Denote by X^{2^*} the set of all bounded quadratic functionals $f : X \to S$.   ∎

REMARK 2.4   Let $f \in X^{2^*}$, $b \in X$; the operator $f \otimes b : X \to X$ sending $x \in X$ to $f(x)b \in X$ is a bounded quadratic operator of rank one. Thus,

$$Q = \sum_{i=1}^{n} f_i \otimes b_i \in Q_F^*(X)$$

for any $f_i \in X^{2^*}$, $i = 1, 2, \ldots, n$, $b_i \in X$, $i = 1, 2, \ldots, n$.   ∎

LEMMA 2.1
If $Q : X \to Y$ is a quadratic operator and $L : Y \to Z$ is a linear operator, then $L \circ Q : X \to Z$ is quadratic.
   Moreover, if Q and L are bounded so is $L \circ Q$. (Q and L need not be of finite rank).

PROOF   This follows easily from the definition of linear and quadratic operators (see Section 1.1).   ∎

REMARK 2.5   Denote by $X^{2^*} \otimes X$ the vector subspace generated in $Q^*(X)$ by the set $\{Q \in Q_F^*(X) \mid Q = f \otimes b, \ f \in X^{2^*}, b \in X\}$ so $Q \in X^{2^*} \otimes X$ if and only if

$$Q = \sum_{i=1}^{n} f_i \otimes b_i . \quad \blacksquare$$

THEOREM 2.11
$Q_F^*(X) = X^{2^*} \otimes X.$

PROOF   Let $\{b_1, \ldots, b_n\}$ be a basis for $\mathrm{rang}(Q)$ and choose $g_i \in X^{2^*}$ such that $g_i(b_j) = \delta_{ij}, i, j = 1, 2, \ldots, n$. Since $\mathrm{rang}(Q)$ is finite dimensional, the $\{g_i\}$, $i = 1, 2, \ldots, n$ are bounded and by the Hahn–Banach theorem [153, 154] they can be extended to bounded linear functionals on X without increasing their norms. Let

$$f_i = g_i \circ Q, \quad i = 1, 2, \ldots, n .$$

Then the $f_i, i = 1, 2, \ldots, n$ are bounded quadratic functionals and

$$Q = \sum_{i=1}^{n} f_i \otimes b_i . \quad \blacksquare$$

DEFINITION 2.2   *Let $f_i^*: X \times X \to S, i = 1, 2, \ldots, n$ denote the symmetric bilinear functionals associated with the $f_i, i = 1, 2, \ldots, n$, by*

$$f_i^*(x, y) = \frac{1}{4} \left(f_i(x + y) - f_i(x - y) \right) .$$

Denote by A' the matrix of the linear transformation $2B_X(y)$ restricted to $\mathrm{rang}(Q)$ relative to the basis $b_1, \ldots, b_n$. Define the $n \times n$ matrix A, by $A = I - A'$,

$$\underline{1} = \begin{bmatrix} 1_1 \\ \vdots \\ 1_n \end{bmatrix}, \quad \text{by } 1_i = f_i(y), \ i = 1, 2, \ldots, n ,$$

the 2-matrix $\underline{C}, \underline{C} = \begin{bmatrix} C_1 \\ \vdots \\ C_n \end{bmatrix}$ by $C_i = \{c_i^{jk}\}$ where $c_i^{jk} = f_i^*(b_j, b_k), i, j, k = 1, 2, \ldots, n.$

Define $\underline{v}$ by $\underline{v} = A^{-1}\underline{1}$ and the 2-matrix $\underline{M} = \begin{bmatrix} M_1 \\ \vdots \\ M_n \end{bmatrix}$ with $M_k = |A|^{-1} M_k'$ where each $M_k', k = 1, 2, \ldots, n$ is the $n \times n$ matrix which results from the deter-

minant of the matrix A if we replace the kth column by $\begin{bmatrix} C_1 \\ \vdots \\ C_n \end{bmatrix}$. *Define A$\underline{M}$ by*

$\begin{bmatrix} AM_1 \\ \vdots \\ AM_n \end{bmatrix}$.

Note that $M'_k, k = 1, 2, \ldots, n$ is indeed an $n \times n$ matrix: For the $n = 2$ case

$$M'_1 = \begin{vmatrix} C_1 & a_{12} \\ C_2 & a_{22} \end{vmatrix} = a_{22}C_1 - a_{12}C_2 \, ,$$

$$M'_2 = \begin{vmatrix} a_{11} & C_1 \\ a_{21} & C_2 \end{vmatrix} = a_{11}C_2 - a_{21}C_1 \, .$$

REMARK 2.6   The problem of solving (1.19) can be translated to a finite dimensional one by making the substitution $z = x - y$ to obtain

$$z = Q(z + y) \, , \tag{2.62}$$

which shows that it is a finite dimensional problem since z must lie in rang(Q). More precisely, we have the following theorem.   ∎

THEOREM 2.12
The point $w \in X$ is a solution of (1.19) if and only if

$$w = y + \sum_{i=1}^{n} \xi_i b_i$$

where the vector $\underline{\xi} = \begin{bmatrix} \xi_1 \\ \vdots \\ \xi_n \end{bmatrix} \in S^n$ is a solution of

$$\underline{x} = \underline{1} + A'\underline{x} + \underline{x}^{tr}\underline{C}\,\underline{x} \quad in \; S^n \, . \tag{2.63}$$

Moreover if $|A| = |I - A'| \neq 0$, then Cramer's rule transforms (2.63) to

$$\underline{x} = \underline{v} + \underline{x}^{tr}\underline{M}\,\underline{x} \quad in \; S^n \, . \tag{2.64}$$

PROOF   Assume that (1.19) has a solution $w \in X$. Then

$$w = y + Q(w)$$

$$w = y + \sum_{i=1}^{n} f_i(w) b_i .$$

Apply $f_1, f_2, \ldots, f_n$ in turn to this vector identity to obtain for $p = 1, 2, \ldots, n$

$$f_p(w) = f_p \left(y + \sum_{i=1}^{n} f_i(w) b_i \right)$$

$$= f_p(y) + \sum_{k=1}^{n} f_k^2(w) f_p(b_k) + 2 \sum_{k=1}^{n} f_k(w) f_p^*(y, b_k)$$

$$+ 2 \sum_{i \neq j}^{n} f_i(w) f_j(w) f_p^*(b_i, b_j) .$$

Letting $f_i(w) = x_i$, $i = 1, 2, \ldots, n$, and writing these equations in vector form we obtain

$$\underline{x} = \underline{1} + A' \underline{x} + \underline{x}^{tr} C \underline{x}$$

or

$$A \underline{x} = \underline{1} + \underline{x}^{tr} C \underline{x} .$$

Since $|A| \neq 0$, we obtain (2.64) by multiplying both sides of the above by A^{-1}.

Conversely, given (2.64) assume (2.63) has a solution vector $\underline{\xi} = \begin{bmatrix} \xi_1 \\ \vdots \\ \xi_n \end{bmatrix} \in S^n$.

Let $w \in X$ be defined as

$$w = y + \sum_{i=1}^{n} \xi_i b_i .$$

Apply $f_1, f_2, \ldots, f_n$ in turn to this vector identity to obtain for $p = 1, 2, \ldots, n$

$$f_p(w) = f_p(y) + \sum_{k=1}^{n} \xi_k^2 f_p(b_k) + 2 \sum_{k=1}^{n} \xi_k f_p^*(y, b_k)$$

$$+ 2 \sum_{i \neq j}^{n} \xi_i \xi_j f_p^*(b_i, b_j)$$

or in matrix notation

$$\underline{f}(w) = \underline{1} + A'\underline{\xi} + \underline{\xi}^{tr}\underline{C}\underline{\xi} \,.$$

Now since $\underline{\xi}$ satisfies (2.63) we have

$$\underline{\xi} = \underline{1} + A'\underline{\xi} + \underline{\xi}^{tr}\underline{C}\underline{\xi} \,.$$

Comparing the last two equations we get

$$\xi_i = f_i(w), \quad i = 1, 2, \ldots, n$$

so

$$w = y + \sum_{i=1}^{n} f_i(w)b_i$$

or

$$w = y + Q(w)$$

Therefore, w is a solution of (1.19) and the theorem is proved. ∎

We now give an example of Theorem 2.12.

Example 2.6
Let $X = C[0, 1]$ and consider the equation

$$x(s) = s + s \int_0^1 x^2(t)dt$$

where $s \in [0, 1]$. This equation is of the form (1.19), with rank$(Q) = 1$,

$$y(s) = s,$$

$$b = s, \quad \text{and}$$

$$f(x) = \int_0^1 x^2(t)dt \,.$$

Using the formula

$$f^*(v, w) = \frac{1}{4}(f(v + w) - f(v - w)) \,,$$

we have

$$A = 1 - 2f^*(y, b) = 1 - 2\frac{1}{4}\int_0^1 4s^2 ds = \frac{1}{3}$$

$$\underline{1} = f(y) = f(s) = \int_0^1 s^2 ds = \frac{1}{3}$$

$$\underline{C} = f(b) = f(s) = \int_0^1 s^2 ds = \frac{1}{3}$$

$$\underline{v} = 3\frac{1}{3} = 1$$

$$\underline{M} = 3\frac{1}{3} = 1 .$$

Therefore, (2.64) becomes $\xi = 1 + \xi^2$ in $\mathbf{R}$, which obviously has the solutions $\frac{1 \pm i\sqrt{3}}{2}$; since $x = y + \xi b$, we finally have that the solutions of (1.19) in this case are

$$x(s) = s + \left(\frac{1 \pm i\sqrt{3}}{2}\right) s$$

or

$$x(s) = \left(\frac{3 \pm i\sqrt{3}}{2}\right) s .$$

REMARK 2.7   It is clear that the map

$$\begin{bmatrix} \xi_1 \\ \vdots \\ \xi_n \end{bmatrix} \leftrightarrow y + \sum_{i=1}^n \xi_i b_i$$

is a bijection between the solutions of (2.64) and those of (1.19). Also note that (2.64) has either infinitely many solutions or at most 2^n. ∎

THEOREM 2.13
If $Q: X \to X$ is a quadratic operator written as $Q = \sum_{i=1}^n f_i \otimes b_i$, then

$$null(Q) = \bigcap_{i=1}^n null(f_i) .$$

PROOF   We have

$$Q(x) = 0 \Leftrightarrow \sum_{i=1}^{n} f_i(x)b_i = 0 \Leftrightarrow f_i(x) = 0 \, ,$$

$$i = 1, 2, \ldots, n \Leftrightarrow x \in \bigcap_{i=1}^{n} \text{null}\,(f_i) \, . \quad \blacksquare$$

All the matrices in the following theorem are defined similarly to the matrices in Definition 2.2.

THEOREM 2.14
Consider the equation

$$0 = y + Q(x) \, . \tag{2.65}$$

Then if the point $w \in X$ is a solution of (2.65) we obtain the algebraic equation

$$0 = \underline{1}' + A'\underline{x} + \underline{x}^{tr}\underline{C}'\underline{x} \quad \text{in } S^n \, ,$$

or, if $|A'| \neq 0$,

$$\underline{x} = \underline{v}' + \underline{x}^{tr}\underline{M}'\underline{x} \, . \tag{2.66}$$

Moreover, if $w \in X$ satisfies

$$w = y + \sum_{i=1}^{n} \xi_i b_i$$

where the vector $\underline{\xi} = \begin{bmatrix} \xi_1 \\ \vdots \\ \xi_n \end{bmatrix}$ *is a solution of (2.66), then $w \in \text{null}(Q)$.*

PROOF   Let $w \in X$ be the solution of (2.65). Then exactly as in Theorem 2.11 we obtain

$$0 = \underline{1}' + A'\underline{x} + \underline{x}^{tr}\underline{C}'\underline{x} \, ,$$

or

$$-A'\underline{x} = \underline{1}' + \underline{x}^{tr}\underline{C}'\underline{x} \, ,$$

and if $|A'| \neq 0$, then

$$\underline{x} = \underline{v}' + \underline{x}^{tr}\underline{M}'\underline{x} \, .$$

Set

$$w = y + \sum_{i=1}^{n} \xi_i b_i ,$$

and apply $f_1, \ldots, f_n$ in turn to w to get as in Definition 2.2

$$\underline{f(w)} = \underline{1}' + A'\underline{\xi} + \underline{\xi}^{tr} \underline{C}' \underline{\xi} .$$

But $\underline{\xi}$ satisfies the equation

$$0 = \underline{1}' + A'\underline{\xi} + \underline{\xi}^{tr} \underline{C}' \underline{\xi} .$$

By comparing the last two equations, we obtain $f_i(w) = 0$, $i = 1, 2, \ldots, n$, so $w \in \bigcap_{i=1}^{n} \text{null}(f_i) = \text{null}(Q)$. ∎

THEOREM 2.15

Consider the problem (2.65) and assume

(a) *$\text{null}(Q) = \{0\}$, and*

(b) *there exists $s \in X$ such that*

$$f_i(s) = \xi_i, \quad i = 1, 2, \ldots, n .$$

Then if $w = y + \sum_{i=1}^{n} \xi_i b_i$

(i) *$w = 0$, and*

(ii) *the point $s \in X$ is a solution of (2.65).*

PROOF   We have $w \in \text{null}(Q) = \{0\}$, so $w = 0$. Also

$$y + Q(s) = y + \sum_{i=1}^{n} f_i(s) b_i$$

$$= y + \sum_{i=1}^{n} \xi_i b_i$$

$$= w$$

$$= 0 .$$

Thus, $s \in X$ is a solution of (2.14).   ■

Example 2.7

In $X = C[0, 1]$, consider the equation

$$0 = s + \int_0^1 sx^2(t)dt$$

which is of the form (2.65) with rank(Q) = 1. Computing as in Example 2.6 we obtain the equivalent of (2.66). In this case,

$$\xi = -\frac{1}{2} - \frac{1}{2}\xi^2$$

or

$$(\xi + 1)^2 = 0, \quad \text{so } \xi = -1 .$$

According to Theorems 2.14 and 2.15,

$$x = y + f(x)b = s - s = 0 \quad \text{so null}(Q) = \{0\} .$$

Now we want to find $w \in X$ such that

$$f(w) = -1$$

and

$$\int_0^1 w^2(t)dt = -1 ,$$

obviously

$$w(s) = \pm i \quad \text{and} \quad w(s) = \pm i\sqrt{3}\, s$$

satisfy the above and (2.65). Therefore, we have an example in which (2.65) has four solutions. This also shows that Remark 2.6 does not apply to (2.65) and (2.66) of Theorem 2.14.

REMARK 2.8   Equation (1.20) can be reduced to an equation without "L" by the following device. Set

$$\bar{X} = X \oplus S$$

$$\bar{y} = y \oplus \left(\frac{1}{4}\right)$$

$$\bar{Q}(x \oplus c) = (Q(x) + 2cL(x)) \oplus c^2$$

then $x \oplus c$ is a solution of

$$\bar{x} = \bar{y} + \bar{Q}(\bar{x})$$

if and only if

$$x \oplus c = y \oplus \frac{1}{4} + (Q(x) + 2cL(x)) \oplus c^2$$

if and only if

$$c = \frac{1}{4} + c^2$$

i.e., $c = \frac{1}{2}$ and

$$x = y + L(x) + Q(x),$$

i.e., x satisfies (1.20). ∎

DEFINITION 2.3   Let $L: X \to X$ *be a bounded linear operator. Then L is called of nearly finite rank if*

$$L = L_1 + L_2$$

where $L_1: X \to X$ is a bounded linear operator of finite rank and $L_2: X \to X$ is a linear operator such that $\|L_2\| < 1$.

Recall that if $\|L_2\| < 1$, then $R = [I - L_2]^{-1}$ exists and RL_1 is of finite rank whenever L_1 is.

THEOREM 2.16
Consider the quadratic equation (1.20) and let $L: X \to X$ be a linear operator of nearly finite rank. Then the point $w \in X$ is a solution of (1.20) if and only if w is a solution of the finite rank quadratic equation

$$x = R(y) + RL_1(x) + RQ(x) \tag{2.67}$$

where R and L_1 are as in Definition 2.3.

PROOF   Let w be a solution of (1.20). Then

$$w = y + L(w) + Q(w)$$

$$= y + L_1(w) + L_2(w) + Q(w),$$

so

$$(I - L_2)(w) = y + L_1(w) + Q(w),$$

or

$$w = R(y) + RL_1(w) + RQ(w).$$

Conversely, if $v \in X$ is a solution of (2.67), then

$$v = R(y) + RL_1(v) + RQ(v)$$

so

$$(I - L_2)(v) = y + L_1(v) + Q(v),$$

or

$$v = y + L_1(v) + L_2(v) + Q(v)$$

$$= y + (L_1 + L_2)(v) + Q(v)$$

$$= y + L(v) + Q(v). \quad \blacksquare$$

2.5   Noncontractive Solutions

In this section we develop techniques for approximating a solution x^* of Equation (1.19) [or (1.20)] that do not make use of contractions. This way it is hoped to obtain a solution "away" from zero. We consider several cases.

CASE 2.1
A solution by series of the quadratic equation (1.19) is presented in the next theorem. The new idea is based on some factoring properties of a certain class of bounded linear operators.

We will consider the quadratic equation (1.19) in the form

$$x = y + \lambda B(x, x) \tag{2.68}$$

for a real or complex number, and study the convergence of the iteration

$$x_{n+1} = y + \lambda B(x_n, x_n), \quad n \geq 0 \tag{2.69}$$

for various $x_0 \in X$ to a solution x^* of (2.68).

(A) For $x_0 = y$ in (2.69), we show that if the sequence of linear operators $\{(B(y))^k\}$, $k = 0, 1, 2, \ldots$, $(B(y)^0 = I)$, belong to a certain subspace of $L(X)$ and the following estimate holds

$$4\lambda \| B(y) \| < 1 \quad (\lambda > 0) , \tag{2.70}$$

then there exists a solution x^* of (2.68) given by

$$x^* = \sum_{k=0}^{\infty} 2^k \frac{1 \cdot 3 \cdots (2k-1)}{1 \cdot 2 \cdots (k+1)} \lambda^k (B(y))^k (y) . \tag{2.71}$$

(B) For $x_0 \neq y$, if the inverse of the linear operator $I - 2B(x_0)$ exists, set

$$\tilde{y} = (I - 2B(x_0))^{-1} (y + \lambda B(x_0, x_0) - x_0)$$

and

$$\tilde{B} = (I - 2B(x_0))^{-1} B .$$

If the rest of the hypotheses of (A) are satisfied for $\tilde{B}$ and $\tilde{y}$, we obtain a solution x^* of (2.41) given by

$$x^* = x_0 + \sum_{k=0}^{\infty} 2^k \frac{1 \cdot 3 \cdots (2k-1)}{1 \cdot 2 \cdots (k+1)} \lambda^k (\tilde{B}(\tilde{y}))^k (\tilde{y}) . \tag{2.72}$$

PROPOSITION 2.2

Let A denote the set defined by

$$A = \{L \in L(X) B(L(y)) = B(y)L, \quad \text{with } B, y \text{ as in } (2.68)\} .$$

Then A is a vector subspace of $L(X)$.

PROOF   Obviously $A \neq \phi$ since the identity operator $I \in A$. Let c_1, c_2 be arbitrary numbers in the field of X and assume that L_1 and $L_2 \in A$. Then

$$B[(c_1 L_2 + c_2 L_2)(y)] = B(c_1 L_1(y) + c_2 L_2(y))$$

$$= c_1 B(L_1(y)) + c_2 B(L_2(y))$$

$$= c_2 B(y)L_1 + c_2 B(y)L_2$$

$$= B(y)(c_1 L_1 + c_2 L_2) ,$$

so, $c_1 L_1 + c_2 L_2 \in A$ and the proof is complete. ∎

We seek a solution expressed as

$$x = z_0 + \lambda z_1 + \cdots + \lambda^n z_n + \cdots .\tag{2.73}$$

Formal substitution of (2.73) into (2.68) and equation of like powers of λ gives

$$z_0 = y$$

$$z_1 = B\,(z_0, z_0)$$

$$z_2 = B\,(z_0, z_1) + B\,(z_1, z_0)\tag{2.74}$$

$$z_n = \sum_{j=0}^{n-1} B\left(z_j, z_{n-j-1}\right)$$

We now state the result:

THEOREM 2.17
Assume that the sequence $\{(B(y))^k\} \in A$, $k = 0, 1, 2, \dots$ and (2.70) holds. Then there exists a solution x^* of (2.68) given by (2.71).

PROOF   For $x_0 = y$, (2.69) becomes

$$x_n = z_0 + \lambda z_1 + \lambda^2 z_2 + \cdots + \lambda^n z_n\tag{2.75}$$

where,

$$z_n = \sum_{j=0}^{n-1} B\left(z_j, z_{n-j-1}\right)$$

$$= \sum_{j=0}^{n-1} B\left[2^j \frac{1 \cdot 3 \cdots (2j-1)}{1 \cdot 2 \cdots (j+1)}(B(y))^j(y)\,,\right.$$

$$\left. 2^{n-j-1} \frac{1 \cdot 3 \cdots [2(n-j-1)-1]}{1 \cdot 2 \cdots (n-j)}(B(y))^{n-j-1}(y)\right]$$

$$= 2^{n-1}(B(y))^n(y) \sum_{j=0}^{n} \frac{1 \cdot 3 \cdot 5 \cdots (2j-1)}{1 \cdot 2 \cdot 3 \cdots (j+1)}$$

$$\frac{1 \cdot 3 \cdots [2(n-j-1)-1]}{1 \cdot 2 \cdot 3 \cdots (n-j)}$$

$$= 2^{n-1} B(y)^n(y) 2 \cdot \frac{1 \cdot 3 \cdots (2n-1)}{1 \cdot 2 \cdots (n+1)}$$

$$= 2^n \frac{1 \cdot 3 \cdots (2n-1)}{1 \cdot 2 \cdots (n+1)} (B(y))^n(y) .$$

Now,

$$\lim_{n \to \infty} x_n = \lim_{n \to \infty} \left(\sum_{k=0}^{n} \lambda^k z_k \right) = \sum_{n=0}^{\infty} \lambda^n z_n . \tag{2.76}$$

The above series converges if the series (dominating)

$$\sum_{n=0}^{\infty} 2^n \lambda^n \frac{1 \cdot 3 \cdots (2n-1)}{1 \cdot 2 \cdots (n+1)} \|B(y)\|^n \|y\| \tag{2.77}$$

converges. Applying the ratio test, we can easily see that the series given by (2.77) converges if (2.70) holds which is true by hypothesis. The proof is now completed if we set

$$x^* = \lim_{n \to \infty} x_n . \quad \blacksquare$$

REMARK 2.9   Under the hypotheses of Theorem 2.17, since

$$\|B(y)\| \le \|B\| \cdot \|y\|$$

we see that:

(i) If both (2.70) and (2.74) hold, then (2.70) allows a wider range for λ than (2.74).

(ii) If (2.74) holds, then (2.70) holds also, but the converse is not necessarily true.

$\blacksquare$

The evaluation of the z_ks, $k = 0, 1, 2, \ldots$ in (2.74) is difficult in practice. However, the same evaluation under the hypotheses of Theorem 2.17 becomes much easier.

Note that it will be easy, but pointless, to construct a simple example to show that Theorem 2.17 succeeds where Corollaries 1.1 or 1.3 fail.

Finally for $x_0 \neq y$, one can set $x = x_0 + h$ in (2.68) to obtain

$$h = \tilde{y} + \tilde{B}(h, h) ,$$

provided that the linear operator $(I - 2B(x_0))^{-1}$ exists. If the rest of the hypotheses of Theorem 2.17 are satisfied for $\tilde{y}$ and $\tilde{B}$ a solution x^* of (2.68) given by (2.72) is easily obtained.

Note that the hypotheses on x_0, $\tilde{y}$ and $\tilde{B}$ are similar (but not the same) to the hypotheses of Newton's Kantorovich theorem for the solution of (2.68).

CASE 2.2

Set $\lambda = 1$ in (2.68) for simplicity. Then we can reason as follows. Newton's method

$$\bar{x}_{n+1} = \bar{x}_n - (2B(\bar{x}_n) - I)^{-1} P(\bar{x}_n), \quad \text{with } P(x) = B(x, x) + y - x \quad (2.78)$$

can be used to approximate a solution x^* of (1.19). Moreover, the sequences y_n, z_n, x_n, B_n, $n = 0, 1, 2, \ldots$ given by

$$y_0 = y ,$$

$$y_n = (I - 2B(y_{n-1}))^{-1} B_{n-1}(y_{n-1}, y_{n-1}) ,$$

$$B_0 = B ,$$

$$B_n = (I - 2B_{n-1}(y_{n-1}))^{-1} B_{n-1} , \tag{2.79}$$

$$z_0 = x^* ,$$

$$z_{n+1} = B_n(z_n, z_n) ,$$

and

$$x_n = \sum_{k=0}^{n} y_k$$

are well defined for all $n \geq 0$ provided that (1.36) holds.

We show that x_n is an approximate solution of (1.19) such that

$$\|x^* - x_n\| \leq \frac{1}{4\|B\|} \cdot \frac{1}{2^n} \cdot R_0^{2^n} = E_n \tag{2.80}$$

where

$$R_0 = 1 - \sqrt{1 - 4\|B\| \cdot \|y\|} . \tag{2.81}$$

The corresponding estimate for (2.78) is given by

$$\|x^* - \bar{x}_n\| \leq \frac{1}{2^n} (2h_0)^{2^n - 1} \cdot \eta_0 = \bar{E}_n \tag{2.82}$$

where

$$2h_0 = \left(\frac{2\|B\| \cdot \|y\|}{1 - 2\|B\| \cdot \|y\|} \right)^2 \tag{2.83}$$

and

$$\eta_0 = \frac{\|B\| \cdot \|y\|^2}{1 - 2\|B\| \cdot \|y\|} . \tag{2.84}$$

Set $p = 4\|B\| \cdot \|y\|$. Then for $p \in (p_1, 1)$, $p_1 \approx .916$, we show

$$E_k < \bar{E}_k, \quad \text{for some } N_1 \text{ and any } k \text{ such that } k \geq N_1 \tag{2.85}$$

whereas for $p \in (0, p_1]$,

$$\bar{E}_m \leq E_m, \quad \text{for some } N_2 \text{ and any } m \text{ such that } m \geq N_2 . \tag{2.86}$$

In fact we can show:

THEOREM 2.18
Assume that (2.71) is satisfied, and

$$\left\| (2B(y) - I)^{-1} \right\| \leq b_0 ,$$

$$\left\| (2B(y) - I)^{-1} P(y) \right\| \leq \eta_0 ,$$

$$h_0 = b_0 L \eta_0 < \frac{1}{2} , \tag{2.87}$$

where

$$L = 2\|B\| .$$

If

$$R \geq r_0 = \frac{1 - \sqrt{1 - 2h_0}}{h_0} \eta_0 ,$$

then iteration (2.78) converges to a unique solution x^* of (1.19) in the ball

$$U(y, r_0) .$$

PROOF   Since by (2.71)

$$2\|B(y)\| \leq 2\|B\| \cdot \|y\| < 1$$

the inverse of the linear operator $2B(y) - I$ exists and

$$\left\|(2B(y) - I)^{-1}\right\| \leq \frac{1}{1 - 2\|B\| \cdot \|y\|} .$$

Inequality (2.87) is now easily verified for

$$b_0 = \frac{1}{1 - 2\|B\| \cdot \|y\|},$$

$$\eta_0 = \frac{\|B\| \cdot \|y\|^2}{1 - 2\|B\| \cdot \|y\|}$$

and

$$L = 2\|B\|$$

provided that (1.36) holds.

The result now follows from the Kantorovich Theorem 11.3 in [119] for $\bar{x}_0 = y$ (or see also Section 2.2). ∎

Denote by x^* the solution of Equation (1.19) guaranteed by Theorem 2.18 and set $x^* = y + z_1$ in (1.19) to obtain

$$z_1 = y_1 + B_1(z_1, z_1) .$$

Since

$$4\|B_1\| \cdot \|y_1\| \leq 4\left\|(I - 2B(y))^{-1}\right\|^2 \|B\|^2\|y\|^2$$

$$\leq \left(\frac{2\|B\| \cdot \|y\|}{1 - 2\|B\| \cdot \|y\|}\right)^2 \leq 1 ,$$

the solution z_1 can be obtained using Theorem 2.18. This procedure can easily be continued inductively for any n to finally obtain the sequences given by (2.79).

The solution x^* of (1.19) can then be represented as

$$x^* = y_0 + y_1 + \cdots + y_n + z_{n+1} + B_n (z_n, z_n) \ .$$

Moreover,

$$\|x^* - x_n\| = \|B_n (z_n, z_n)\| \le \|B_n\| \cdot \|z_n\|^2 \le \|B_n\| \left[\|B_{n-1}\| \cdot \|z_{n-1}\|^2 \right]^2$$

$$\le \cdots \le \|B_n\| \cdot \|B_{n-1}\|^2 \|B_{n-2}\|^{2^2} \cdots \|B_0\|^{2^n} \|x^*\|^{2^{n+1}}$$

$$\le (2^n \|B_0\|) \left(2^{n-1} \|B_0\|\right)^2 \cdots \|B_0\|^{2^n} \|x^*\|^{2^{n+1}}$$

(since $\|x^*\| \le \|y\| + r_0 = \frac{1-\sqrt{1-4\|B\|\cdot\|y\|}}{2\|B\|}$)

$$\le 2^{n(2^n-1)-(2+2\cdot2^2+\cdots+(n-1)2^{n-1})} \|B\|^{(2^{n+1}-1)} \frac{R_0^{2^{n+1}}}{(2\|B\|)^{2^{n+1}}}$$

$$\le \frac{1}{4\|B\|} \cdot \frac{1}{2^n} \cdot R_0^{2^n} \ .$$

That is, $\lim_{n\to\infty} \|x^* - x_n\| = 0$.

We have now proved the following theorem:

THEOREM 2.19

Assume that (1.36) holds and denote by x^* the solution of (1.19) guaranteed to exist by Corollary 1.1 under the above hypothesis.

Then

(i) *the solution x^* can be represented as*

$$x^* = y_0 + y_1 + \cdots + y_n + B_n (z_n, z_n) = x_{n-1} + z_n \qquad (2.88)$$

and

$$\|x^* - x_n\| \le E_n, \quad n = 0, 1, 2, \dots \ .$$

(ii)   Moreover, there exist positive integers N_1 and N_2 such that

$$E_k < \bar{E}_k \quad \text{for } p \in (p_1, 1) \text{ and } k \geq N_1 \tag{2.89}$$

where

$$p \approx .916$$

and

$$\bar{E}_m \leq E_m, \quad \text{for } p \in (0, p_1] \text{ and } m \geq N_2 .$$

Note that part (ii) above can easily be proved by considering the difference $2h_0 - R_0$.

Finally, we remark that under the same hypotheses the approximate solution x_n is preferred to be constructed instead of $\bar{x}_m$ if (2.89) holds, since we then have

$$n < m$$

$$E_n \leq \varepsilon$$

$$E_m \geq \varepsilon$$

for some given $\varepsilon > 0$.

That is to say, the required number n of iterates x_n to be computed to achieve the same accuracy $\varepsilon > 0$ is smaller than the corresponding number m of iterates $\bar{x}_m$.

The results can be extended to the case when $\bar{x}_0 \neq y$.

CASE 2.3

We now return back to Equation (2.68). We showed in Corollary 1.3 and elsewhere that if (1.36) holds (with B replaced by λB), then a small solution x^* of (2.68) exists with

$$\|x^*\| \leq \frac{1 - \sqrt{1 - 4|\lambda| \, \|B\| \, \|y\|}}{2\lambda \|B\|} . \tag{2.90}$$

A number of authors have raised the question of the existence of a solution x^* of (2.68) under the hypothesis that

$$|\lambda| > \frac{1}{4\|B\| \cdot \|y\|} . \tag{2.91}$$

In this case we show a positive result in this direction. In fact, we show that if

$$2\|2\lambda B(y) - I\| < 1 \tag{2.92}$$

which implies (2.91), then Equation (2.68) has a solution x^* belonging to a new space $\bar{X}$ constructing from X in a way similar to the way that the complex numbers are constructed from the reals.

A simple example when $X = \mathbf{R}^2$ is provided.

Example 2.8
Consider the real quadratic equation

$$\zeta = a + \lambda b \zeta^2 . \tag{2.93}$$

Then the following results for the solution ζ_1 and ζ_2 of (2.93) can easily be proved.

(i) if $4|\lambda ab| \leq 1$, then

$$\zeta_1 = \frac{1 - \sqrt{1 - 4\lambda ab}}{2\lambda b} = \sum_{n=0}^{\infty} 2^n a^{n+1} b^n \frac{1 \cdot 3 \cdots (2n-1)}{1 \cdot 2 \cdots (n+1)} \lambda^n ,$$

$$\zeta_2 = \frac{1}{\lambda b} - \zeta_1 ;$$

(ii) if $2|2\lambda ab - 1| < 1$

$$\zeta_1 = \frac{1 - i\sqrt{4\lambda ab - 1}}{2\lambda b} = \frac{1}{2\lambda b} - \frac{i}{2\lambda b}$$

$$\left[a + (2\lambda ab - 1) - \sum_{n=2}^{\infty} (-1)^n \frac{1 \cdot 3 \cdot 5 \cdots (2n-3)}{n!} \cdot (2\lambda ab - 1)^n \right]$$

and $\zeta_2 = \frac{1}{\lambda b} - \zeta_1$.

Note that the solution in case (i) does not belong to the real number system R but in the set $R(i)$ the set of complex numbers, where each $z \in R(i)$ is such that

$$z = c_1 + i c_2, \quad c_1, c_2 \in R$$

and

$$|z| = \sqrt{|c_1|^2 + |c_1|^2} .$$

These ideas can generalize as follows: Let $\bar{X} = X(i)$ denote the elements

$$w = v_i + iv_2, \quad v_1, v_2 \in X .$$

The set $\bar{X}$ becomes a linear space with the usual addition and multiplication by a scalar. Define a norm on $\bar{X}$ by

$$\|w\| = \sqrt{\|v_1\|^2 + \|v_2\|^2} .$$

The above condition is trivially a norm on $\bar{X}$.

Finally, one can check that $(\bar{X}, \| \cdot \|)$ is a Banach space.

Let $y \in X$ be fixed. Assume that the linear operator $B(y)$ is invertible. Then in parallel to (i) and (ii) above we have:

(i′)   if $4|\lambda| \cdot \|B(y)\| < 1$, then let $x_1, x_2 \in X$ be

$$x_1 = \frac{1}{2\lambda} B(y)^{-1} \left[I - \sqrt{I - 4\lambda B(y)} \right] (y)$$

$$= \sum_{n=0}^{\infty} 2^n B(y)^n (y) \frac{1 \cdot 3 \cdots (2n - 1)}{1 \cdot 2 \cdots (n + 1)} \lambda^n \qquad (2.94)$$

and

$$x_2 = \frac{1}{\lambda} B(y)^{-1}(y) - x_1 ;$$

(ii′)   if $2\|2\lambda B(y) - I\| < 1$, then let $\bar{x}_1, \bar{x}_2 \in \bar{X}$ be

$$\bar{x}_1 = \frac{1}{2\lambda} B(y)^{-1} \left[I - i\sqrt{4\lambda B(y) - I} \right] (y)$$

$$= \frac{1}{2\lambda} B(y)^{-1}(y) - \frac{i}{2\lambda} B(y)^{-1}(y)$$

$$\left[y + 2(\lambda B(y) - I) - \sum_{n=2}^{\infty} (-1)^n \frac{1 \cdot 3 \cdots (2n - 3)}{n!} (2\lambda B(y) - I)^n \right]$$

and $\bar{x}_2 = \frac{1}{\lambda} B(y)^{-1}(y) - x_1$.

Let w be the sum of the series in $\bar{x}_1$ (or $\bar{x}_2$), since

$$\left\| B(y)^{-1}(y) \right\| \leq \left\| B(y)^{-1} \right\| \cdot \|y\| \leq \frac{\|y\|}{\|B\| \cdot \|y\|} = \frac{1}{\|B\|}$$

then

$$\|\|\bar{x}_1\|\| = \|\|\bar{x}_2\|\| \leq \sqrt{\left(\frac{1}{\|B\|}\right)^2 + \left(\frac{\|w\|}{\|B\|}\right)^2} \cdot \frac{1}{2|\lambda|} = \frac{1}{2|\lambda| \, \|B\|} \sqrt{1 + \|w\|^2} \,.$$

We will now show that $x_1, x_2 \in X$ are under certain additional conditions solutions of (2.68).

We first prove the proposition:

PROPOSITION 2.3

Let $v \in X$ be fixed and let $L(X)$ denote the set of all bounded linear operators on X.

Define the set $A(v)$ by $A(v) = \{L \in L(x)/B(L(y)(v) = LB(y)(v), \ v \in X$ fixed and y as in (2.68)\}. Then, $A(v)$ is a vector subspace of $L(X)$.

PROOF Note that $A(v) \neq \emptyset$ since the identity operator $I \in A(v)$. Let $c_1 L_1, c_2 \cdot L_2 \in A(v)$ for $L_1, L_2 \in L(X)$ and $c_1, c_2 \in \mathbf{R}$. We have

$$B\left((c_1 L_1 + c_2 L_2)(y)\right)(v) = B\left(c_1 L_1(y) + c_2 L_2(y)\right)(v)$$

$$= c_1 B\left(L_1(y)(v) + c_2 B(L_2(y))\right)(v)$$

$$= c_1 L_1 B(y)(v) + c_2 L_2 B(y)(v) = (c_1 L_1 + c_2 L_2) B(y)(v) \,,$$

so $c_1 L_1 + c_2 L_2 \in A(v)$ and the Proposition is proved. ∎

We now state and prove the result:

THEOREM 2.20

Let B and y in (2.68) be such that the linear operator $B(y)$ is invertible and assume that:

(a) $4|\lambda| \|B(y)\| < 1;$

(b) $B(y)^{-1} \in A(B(y)^{-1}(y))$,

$$B(y)^{-1}, B(y)^{-1}\sqrt{I - 4\lambda B(y)} \in A\left(B(y)^{-1}\sqrt{I - 4\lambda B(y)}\,(y)\right)$$

then x_1, x_2 are solutions of (2.68) having a series representation given by (2.94).

PROOF   We have

$$\lambda B \left[\frac{B(y)^{-1}(y) \pm B(y)^{-1}\sqrt{I - 4\lambda B(y)}(y)}{2\lambda}, \frac{B(y)^{-1}(y) \pm B(y)^{-1}\sqrt{I - 4\lambda B(y)}(y)}{2\lambda} \right]$$

$$= \frac{1}{4\lambda}\left[B\left(B(y)^{-1}(y)\right), B(y)^{-1}(y) \pm 2B\left(B(y)^{-1}(y), B(y)^{-1}\sqrt{I - 4\lambda B(y)}(y)\right) \right]$$

$$= \frac{1}{4\lambda}\left[B(y)^{-1}(y) \pm 2B(y)^{-1}\sqrt{I - 4\lambda B(y)}(y) + B(y)^{-1}(I - 4\lambda B(y))(y) \right]$$

$$= \frac{B(y)^{-1}\left[I \pm \sqrt{I - 4\lambda B(y)}\right](y)}{2\lambda} - y = x - y$$

so,

$$x = y + \lambda B(x, x)$$

for $x = x_1$ or x_2, i.e., x_1 or x_2 are solutions of (2.68).   ▮

Note that (i) above is not used to show that x_1, x_2 are solutions of (2.88) but it is needed to guarantee the convergence of the series in (2.94). The proof of the following theorem is similar to Theorem 2.20 and is omitted.

THEOREM 2.21
Let B and y in (2.68) be such that the linear operator $B(y)$ is invertible and assume that:

(a) $2\|2\lambda B(y) - I\| < 1$;

(b) $B(y)^{-1} \in A(B(y)^{-1}(y))$,

$$B(y)^{-1}, B(y)^{-1}\sqrt{4\lambda B(y) - I} \in A\left(B(y)^{-1}\sqrt{4\lambda B(y) - I}(y)\right)$$

then $\bar{x}_1, \bar{x}_2$ are solutions of (2.68).

We now give an example for Theorem 2.21.

Example 2.9

Let $X = \mathbf{R}^2$ with the ordinary bases $\{(1, 0), (0, 1)\}$.

Consider the quadratic equation in X given by (2.68), where

$$x = \begin{pmatrix} x_1 \\ x_2 \end{pmatrix}, \quad B \sim \left[\begin{array}{cc} 3 & 1 \\ 1 & 2 \end{array} \middle/ \begin{array}{cc} -\frac{1}{2} & -1 \\ -1 & -7 \end{array} \right],$$

$$y = \begin{bmatrix} \frac{4}{5} \\ \frac{2}{5} \end{bmatrix} \quad \text{and} \quad \lambda = \frac{5}{8}.$$

Then (2.68) can be written as [using (1.13)]

$$x_1 = \frac{4}{5} + \frac{15}{8}x_1^2 + \frac{5}{4}x_1 x_2 + \frac{5}{4}x_2^2$$

$$x_2 = -\frac{2}{5} - \frac{5}{16}x_1^2 - \frac{5}{4}x_1 x_2 - \frac{35}{8}x_2^2.$$

Moreover, we have

$$B(y) = 2I, \quad B(y)^{-1} = \frac{1}{2}I \quad \text{and} \quad \sqrt{4\lambda B(y) - I} = 2I.$$

Therefore,

$$x_1 = \frac{8}{25} \pm \frac{16}{25}i \quad \text{and} \quad x_2 = -\frac{4}{25} \mp \frac{8}{25}i.$$

In practice if $\sqrt{I - 4\lambda B(y)}$ or $\sqrt{4\lambda B(y) - I}$ are too difficult to evaluate, then one can use the power series representation of x_1, x_2 or $\bar{x}_1$ and $\bar{x}_2$. Finally, note that since

$$| \|2\lambda B(y)\| - \|I\| | \leq \|2\lambda B(y) - I\|,$$

condition (2.92) implies

$$2|\lambda| \cdot \|B(y)\| > 1,$$

or (2.91). However, (2.91) cannot replace (2.92) in Theorem 2.20.

CASE 2.4

Here motivated by the solution of the real quadratic equation we seek a solution *x* of (2.68) expressed as*

$$x^* = \frac{1}{\lambda}v + \sum_{n=0}^{\infty} \lambda^n x_n , \qquad (2.95)$$

where v, $x_n \in X$, $n = 0, 1, 2, \ldots$ are to be specified. Under certain assumptions on v we show that if (1.36) holds (for B belonging to λB) the solution x^* of (2.68) given by (2.95) is such that

$$\|x^*\| \geq \frac{1 + \sqrt{1 - 4\lambda \|B\| \cdot \|y\|}}{2\lambda \|B\|} . \qquad (2.96)$$

We now state the result:

THEOREM 2.22
Assume:

(a) *there exists $v \in X$ satisfying*

$$B(v, v) = v, \qquad v \neq 0 \qquad (2.97)$$

 and such that the linear operator $(I - 2B(v))^{-1}$ exists on X.

(b) *Let k denote the norm of $(I - 2B(v))^{-1}$ and set*

$$x_0 = (I - 2B(v))^{-1}(y)$$

$$x_n = \sum_{j=0}^{n-1} (I - 2B(v))^{-1} B\left(x_j, x_{n-j-1}\right), \qquad n = 1, 2, \ldots$$

 with
$$4\lambda k^2 \|B\| \cdot \|y\| \leq 1 \quad and \quad \|B\| \neq 0 . \qquad (2.98)$$

Then there exists a solution x^* of (2.68) given by (2.95) and satisfying

$$\|x^*\| \leq \frac{1}{\lambda}\|v\| + \frac{1 - \sqrt{1 - 4\lambda k^2 \|B\| \cdot \|y\|}}{2\lambda \|B\| \cdot k} . \qquad (2.99)$$

PROOF   Formal substitution of (2.95) into (2.68) (as before) and equation of like powers of λ shows that if x^* is a solution then $v, x_n, n = 0, 1, 2, \ldots$ must be given in (2.97) and (2.98).

The real series

$$\frac{1}{\lambda}\|v\| + \sum_{n=0}^{\infty} \lambda^n z_n \, ,$$

where

$$z_0 = k\|y\|$$

$$z_n = \sum_{j=0}^{n-1} k\|B\|z_j z_{n-j-1} \, ,$$

obviously dominates the series given by (2.95). Moreover, by (2.99), we have

$$\sum_{n=0}^{\infty} \lambda^n z_n = \frac{1 - \sqrt{1 - 4\lambda k^2 \|B\| \cdot \|y\|}}{2\lambda \|B\| \cdot k} \, .$$

Therefore, the series given by (2.95) converges to a solution x of (2.68) satisfying (2.99) and the proof is completed. ∎

We now prove the existence of a "not small" solution. For simplicity we take $\lambda = 1$.

PROPOSITION 2.4

If the hypotheses of Theorem 2.20 are satisfied, and k is such that

$$0 < k \leq 1 \quad and \quad 1 - 4\|B\| \cdot \|y\| > 0 \, ,$$

then there exists a solution x^* of (2.68) given by (2.95) and satisfying

$$\|x^*\| \geq \frac{1 + \sqrt{1 - 4\|B\| \cdot \|y\|}}{2\|B\|} \, . \tag{2.100}$$

PROOF   The solution x^* of (2.68) given by (2.95) is guaranteed by Theorem 2.20. Hence, it is enough to show (2.100). By (2.95) we have

$$\|x^*\| \geq \|v\| - \left\| \sum_{n=0}^{\infty} x_n \right\| \geq \|v\| - \sum_{n=0}^{\infty} \|x_n\|$$

$$\geq \|v\| - \frac{1 - \sqrt{1 - 4k^2 \|B\| \cdot \|y\|}}{2\|B\|k} \, . \tag{2.101}$$

If v is a nonzero solution of (2.97), then

$$\|v\| = \|B(v, v)\| \leq \|B\| \cdot \|v\|^2 .$$

Therefore, we get

$$\|v\| \geq \frac{1}{\|B\|} . \tag{2.102}$$

Now, (2.101), because of (2.102), becomes

$$\|x^*\| \geq \frac{1}{\|B\|} - \frac{1 - \sqrt{1 - 4k^2\|B\| \cdot \|y\|}}{2\|B\|k} . \tag{2.103}$$

By (2.103), to show (2.100), it is enough to show

$$\frac{(2k - 1) + \sqrt{1 - 4k^2\|B\| \cdot \|y\|}}{2\|B\| \cdot k} \geq \frac{1 + \sqrt{1 - 4\|B\| \cdot \|y\|}}{2\|B\|} . \tag{2.104}$$

After the simplification showing (2.104) becomes easily equivalent to showing

$$0 < k \leq 1 ,$$

which is true by hypothesis and the proof is completed.   ∎

Example 2.10
Note that in the case of the real quadratic equation

$$r = \alpha + \lambda\beta r^2 ,$$

where $\alpha = \|y\|$ and $\beta = \|B\|$ equality is achieved in (2.100) and (2.96). The solutions are then given by

$$r^- = \frac{1 - \sqrt{1 - 4\lambda\alpha\beta}}{2\lambda\beta}$$

$$= \sum_{n=0}^{\infty} \lambda^n z_n$$

$$= \sum_{n=0}^{\infty} 2^n \lambda^n \alpha^{n+1} \beta^n \frac{1 \cdot 3 \cdots (2n - 1)}{1 \cdot 2 \cdots (n + 1)} ,$$

and

$$r^+ = \frac{1}{\lambda\beta} - r^- .$$

Finally, note that for $v = 0$ in Theorem 2.22, we obtain the results in Corollaries 1.1 and 1.3.

CASE 2.5
The following result is also useful:

THEOREM 2.23
If (1.36) is true, then

(i) *Equation (1.19) has a unique solution x_0 satisfying*

$$\|x_0\| \leq \frac{1 - \sqrt{1 - 4\|B\|\,\|y\|}}{2\|B\|} ;$$

(ii) *any other solution x of (1.19) is given by $x = x_0 + h$, where h satisfies*

$$h = B_1(h, h) \quad \text{with} \quad h \neq 0$$

where

$$B_1(h, h) \equiv (1 - 2B(x_0))^{-1} B(h, h) .$$

PROOF (i) True by Corollary 1.1.
(ii)

$$x_0 + h = y + B(x_0 + h, x_0 + h)$$

$$= y + B(x_0, x_0) + 2B(x_0, h) + B(h, h) ,$$

so

$$h = 2B(x_0, h) + B(h, h) = 2B(x_0)(h) + B(h, h) .$$

But we have

$$\|2B(x_0)\| \leq 2\|B\|\,\|x_0\| \leq 2\|B\| \frac{1 - \sqrt{1 - 4\|B\|\,\|y\|}}{2\|B\|} < 1$$

so, $(I - 2B(x_0))^{-1}$ exists and

$$(I - 2B(x_0))(h) = B(h, h) ,$$

i.e., $h = B_1(h, h)$. The proof is now complete. $\blacksquare$

Example 2.11
Take $X = \mathbf{R}$,

$$x = y + Bx^2, \quad x_0 = \frac{1 - \sqrt{1 - 4By}}{2B}$$

provided that $4By < 1$ then,

$$x_0 + h = y + B(x_0 + h)^2$$

$$= y + Bx_0^2 + Bh^2 + 2Bx_0h \, ,$$

so $h(1 - 2Bx_0) = Bh^2$,

$$h = \frac{1 - 2B_0}{B} = \frac{2\sqrt{1 - 4By}}{2B} \, ,$$

provided that $B \neq 0$, and

$$x = x_0 + h = \frac{1 - \sqrt{1 - 4By}}{2B} + \frac{2\sqrt{1 - 4By}}{2B} \, ,$$

so we obtain

$$x = \frac{1 + \sqrt{1 - 4By}}{2B} \, .$$

PROPOSITION 2.5
If there exists $x_0 \in X$ such that $B(x_0) = I$, then x_0 is a solution of $B(x, x) = x$.

PROOF $B(x_0) = I$ so $B(x_0)(x_0) = I(x_0) = x_0$. $\blacksquare$

PROPOSITION 2.6
If there exists $x_0 \in X$ such that $B(x_0) = I$ then $Ker(B_X) = \{0\}$.

PROOF $Ker(B) = \bigcap_{x \in X} Ker(B(x)) \subseteq Ker(B(x_0)) = Ker(I) = \{0\}$. $\blacksquare$

Example 2.12
Take $X = \mathbf{R}^2$. The converse of Proposition 2.6 is not always true. Let $B: X \times X \to$

X be a symmetric bilinear operator given by the 2-matrix

$$\left[\begin{array}{cc} 1 & 1 \\ -1 & 1 \\ \hline 1 & 1 \\ 1 & 1 \end{array}\right],$$

with respect to the basis $\{(1, 0), (0, 1)\}$ in X. Then if

$$B(y) = \left[\begin{array}{cc} y_1 - y_2 & -y_1 + y_2 \\ y_1 + y_2 & y_1 + y_2 \end{array}\right] = \left[\begin{array}{cc} 0 & 0 \\ 0 & 0 \end{array}\right],$$

where

$$y = \left[\begin{array}{c} y_1 \\ y_2 \end{array}\right],$$

we obtain $y_1 = y_2 = 0$, so $\text{Ker}(B) = \left[\begin{array}{c} 0 \\ 0 \end{array}\right] = \text{Ker}(I)$, where

$$I = \left[\begin{array}{cc} 1 & 0 \\ 0 & 1 \end{array}\right].$$

But the equation $B(x) = I$, with $x = \left[\begin{array}{c} x_1 \\ x_2 \end{array}\right]$, i.e.,

$$\left[\begin{array}{cc} x_1 - x_2 & -x_1 + x_2 \\ x_1 + x_2 & x_1 + x_2 \end{array}\right] = \left[\begin{array}{cc} 1 & 0 \\ 0 & 1 \end{array}\right]$$

obviously has no solution in X.

PROPOSITION 2.7

If there exists $x_0 \in X$ such that $\text{Ker}(B(x_0)) = \{0\}$, then $B(x_0) = I$ has at most one solution.

PROOF

$$\text{Ker}(B) = \bigcap_{x \in X} \text{Ker}(B(x_0))$$

$$\subseteq \text{Ker}(B(x_0)) = \{0\}$$

so B is injective, hence the result follows.   ■

CASE 2.6
It is well known that the real quadratic equation

$$x = y + bx^2$$

has at most two solutions x_1, x_2 *which satisfy the Vietta relations*

$$b(x_1 + x_2) = 1$$

$$b(x_1 x_2) = y.$$

Here we study how the above results can be extended in a Banach space. Assume there exists a $z \in X$ *such that the linear operator* $B(z)$ *satisfies the condition*

$$B(z) = I, \quad \text{where } I \text{ is the identity operator on } X$$

and

$$x_1 + x_2 = z.$$

Then if x_1 *and* x_2 *are solutions of (1.19), the following are true:*

$$B(x_1 + x_2) = I$$

$$B(x_1, x_2) = y. \tag{2.105}$$

We can now prove the validity of relations (2.105).

THEOREM 2.24
Let $z \in X$ *be such that* $B(z) = I$ *and* $x_1, x_2 \in X$ *be solutions of (1.19) with*

$$x_1 + x_1 = z.$$

Then

$$B(x_1 + x_1) = I$$

$$B(x_1, x_2) = y.$$

PROOF We have
$$B\,(x_1 + x_2) = B(z) = I\;.$$

Also,

$$x_1 + x_2 = 2y + B\,(x_1, x_1) + B\,(x_2, x_2)$$

$$= 2y + B\,(x_1 + x_2, x_1 + x_2) - 2B\,(x_1, x_2)$$

$$= 2y + x_1 + x_2 = 2B\,(x_1, x_2)\;.$$

That is
$$B\,(x_1, x_2) = y\;,$$

which completes the proof of the theorem. ∎

Moreover we can show:

THEOREM 2.25
Let $z \in X$ be such that $B(z) = I$ and let x^* be a solution of (1.19), then

$$\bar{x} = z - x$$

is a solution of (1.19) also.

PROOF We have

$$y + B\,(z - x^*, z - x^*) = y + B(z, z) - 2B\,(x^*, z) + B\,(x^*, x^*)$$

$$= y + z - 2x^* + B\,(x^*, x^*)$$

$$= y - y + z - x^*$$

$$= z - x^*$$

$$= \bar{x}\;,$$

which shows that $\bar{x}$ is a solution of (1.19). ∎

Furthermore we can show:

PROPOSITION 2.8

Let x_1, x_2, x_3, x_4 be solutions of (1.19) with

$$x_1 + x_2 = z$$

$$x_3 + x_4 = \bar{z}$$

where $B(z) = B(\bar{z}) = I$. Then

$$z = \bar{z} \ .$$

PROOF   We have

$$x_1 = y + B\,(x_1, x_1) = y + B\,(x_3 + x_4 - x_2 + z - \bar{z}, x_3 + x_4 - x_2 + z - \bar{z})$$

$$= \cdots$$

$$= x_3 + x_4 + x_2 - 2B\,(x_2, x_3 + x_4)$$

$$= x_3 + x_4 - x_2$$

(by expanding and rearranging the right-hand side of the first equation). So, $x_1 + x_2 = x_3 + x_4$; that is,

$$z = \bar{z} \ . \quad \blacksquare$$

PROPOSITION 2.9

Assume:

  (a) *there exist $r, s \in X$ such that*

$$B(r, s) = 0 \tag{2.106}$$

   and $x_1, x_2 \in X$ solutions of (1.19) satisfying (2.105).

  (b) *the following estimate is true:*

$$r + x_1 = s + x_2 \ . \tag{2.107}$$

Then the element $x \in X$ defined by

$$x = r + x_1$$

or

$$x = s + x_2$$

is a solution of (1.19).

PROOF By (2.106) and (2.107) we have,

$$B(x, x) - x + y = B(x, x) - B(x, x_1 + x_2) + B(x_1, x_2)$$

$$= B(x - x_1, x - x_2)$$

$$= B(r, s) = 0 ,$$

if $x = r + x_1$ or $x = s + x_2$. That completes the proof of the proposition. ∎

We now provide the example.

Example 2.13
Let $X = \mathbf{R}^2$ and define a bilinear operator on B on X by (1.13). Consider the quadratic equation on X (1.19)

$$w = y + B(w, w)$$

where

$$y = \begin{bmatrix} y_1 \\ y_2 \end{bmatrix} = \begin{bmatrix} \frac{1}{48} \\ -\frac{1}{48} \end{bmatrix} ,$$

$b_1^{11} = -3$, $b_1^{12} = 1$, $b_1^{21} = 1$, $b_1^{22} = -1$, $b_2^{11} = 1$, $b_2^{12} = -1$, $b_2^{21} = -1$ and $b_2^{22} = -1$ or

$$w_1 = \frac{1}{48} - 3w_1^2 + 2w_1 w_2 - w_2^2$$

$$w_2 = -\frac{1}{48} + w_1^2 - 2w_1 w_2 - w_2^2 . \tag{2.108}$$

The solutions w^1, w^2, w^3, w^4 of (2.108) are now given by

$$w_1^1 = .0200308, \quad w_2^1 = .0200308 ,$$

$$w_1^2 = -.5200308, \quad w_2^2 = -.5200308 ,$$

$$w_1^3 = -.25 \,,$$

$$w_2^3 = .1318813 \,,$$

$$w_1^4 = -.25 \,,$$

$$w_2^4 = -.6318812 \,.$$

It can easily be seen that $B(z) = I$ if $z = \begin{bmatrix} -.5 \\ -.5 \end{bmatrix}$. Note that

$$w^1 + w^2 = w^3 + w^4 = z$$

$$B\left(w^1, w^2\right) = B\left(w^3, w^4\right) = y$$

and

$$B\left(w^1 + w^2\right) = B\left(w^3 + w^4\right) = I \,.$$

CASE 2.7

We now end this section with a result for (1.19) related with the null space of a certain linear operator.

In particular, under the assumption that for all $z_1, z_2, \dots, z_n, x \in X$

$$\lim_{n \to \infty} B\,(z_1)\, B\,(z_2) \cdots B\,(z_n)\,(x) \tag{2.109}$$

exists, we give more insight into the behavior of Equation (1.19).

Let $z_1, z_2, \dots, z_n$ be fixed in X. If the limit in (2.109) exists as $n \to \infty$, then (2.109) defines a linear operator L, depending on the choice of $z_1, \dots, z_n$, on X, given by

$$L(x) = \lim_{n \to \infty} B\,(z_1)\, B\,(z_2) \cdots B\,(z_n)\,(x) \,. \tag{2.110}$$

We denote the domain of L by $D(L)$ and the null space of L by $N(L)$. By $x \in D(L)$ or $x \in N(L)$ we mean

$$L(x) \quad \text{exists or} \quad L(x) = 0$$

where L is given by (2.110). Note that $N(L) \neq \emptyset$, since $0 \in N(L)$.

THEOREM 2.26
If the linear operator $B(z)$ has no nonzero fixed point for all $z \in X$, then (1.19) has at most one solution.

PROOF   If (1.19) has no solution, there is nothing to prove. Let $\bar{x}$ be a solution of (1.19). Then any other solution x can be given by

$$x = \bar{x} + h, \quad \text{for some } h \in X .$$

We now have by (1.19),

$$\bar{x} + h = y + B(\bar{x} + h, \bar{x} + h)$$

or

$$B(z)(h) = h, \quad z = 2\bar{x} + h .$$

By hypothesis $h = 0$. Therefore, the solution $\bar{x}$ is unique.   ∎

If Equation (1.19) has a solution $\bar{x}$, then the totality of the solutions is given by $x = \bar{x} + h$ where h is a fixed point of the linear operator $B(z)$ with $z = 2\bar{x} + h$. The solution $\bar{x}$ is unique if $h = 0$. Every fixed point of the operator $B(z)$ is a fixed point of the operator

$$L(x) = \lim_{n \to \infty} (B(z))^n(x), \quad \text{with } z = 2\bar{x} + h .$$

The equation

$$L(h) = h \quad \text{shows that} \quad h = 0 \quad \text{if } X = N(L) .$$

Thus, the solution $\bar{x}$ is unique in this case. Therefore, for this particular choice of L the condition $X = N(L)$ is a sufficient condition for the uniqueness of $\bar{x}$.

The assumption that X is indeed a Banach space was not used in the derivation of the above results. It will be used, however, in the following results.

THEOREM 2.27
Let w be a solution of (1.19). Define the sequence $\{x_n\}$, $n = 1, 2, \ldots$ by

$$x_n = y + B(x_{n-1}, x_{n-1}) \quad \text{for some } x_0 \in X . \tag{2.111}$$

Then $\{x_n\}$, $n = 1, 2, \ldots$ converges to some point $v \in X$ if and only if

$$x_0 - w \in D(L) ,$$

where L is given by (2.110) for $z_n = x_{n-1} + w$, $n = 1, 2, 3, \ldots$.

PROOF   We have

$$x_n - w = (y + B (x_{n-1}, x_{n-1})) - (y + B(w, w))$$

$$= B (x_{n-1}, x_{n-1}) - B(w, w)$$

$$= B (x_{n-1} + w, x_{n-1} - w)$$

$$= B (x_{n-1} + w) B (x_{n-2} + w, x_{n-2} - w)$$

$$\cdots$$

$$= B (z_n) B (z_{n-1}) \cdots B (z_1) (x_0 - w) \ ,$$

where $z_k = x_{k-1} + w$, $k = 1, 2, \ldots, n$. Now if $x_0 - w \in D(L)$ the limit on the right-hand side of the above exists, therefore $\{x_n\}$, $n = 1, 2, \ldots$ converges to some point $v \in X$.

Conversely, if $\{x_n\}$, $n = 1, 2, \ldots$ converges to some $v \in X$ as $n \to \infty$ then

$$x_0 - w \in D(L) . \quad \blacksquare$$

Sometimes this weaker condition suffices, as indicated next.

THEOREM 2.28

Let x_0 be an approximate solution of Equation (1.19) in the sense that $x_0 - B(x_0, x_0) - y \in N(L)$, where L is given by (2.110) with $z_k = x_{k-1} + x_k$, $k = 1, 2, \ldots$, and the x_ns are given by (2.111).

Then every limit point of the sequence $\{x_n\}$, $n = 1, 2, \ldots$ satisfies Equation (1.19).

PROOF   As in Theorem 2.27,

$$x_{n-1} - x_n = B (z_n) B (z_{n-1}) \cdots B (z_1) (x_0 - x_1)$$

where $z_k = x_{k-1} + x_k$, $k = 1, 2, \ldots, n$.

Now if $x_n \to x$ for $n = n_i \to \infty$, then $x_{n+1} \to x$ for $n = n_i \to \infty$, and the result follows if we let $n_i \to \infty$ in $x_{n+1} = y + B(x_n, x_n)$. $\quad \blacksquare$

Furthermore we can show:

THEOREM 2.29

If $x \in D(L)$ is a solution of (1.19), then every other solution w is in $D(L)$ where L is given by (2.110) with $z_n = x + w$, $n = 0, 1, 2, \ldots$.

PROOF Since $x = y + B(x, x)$ and $w = y + B(w, w)$, then $w - x$ is a fixed point of $B(x + w)$ and hence $w - x \in D(L)$. Therefore, $w = (w - x) + x \in D(L)$. ∎

The following result complements Theorem 2.27.

THEOREM 2.30

Equation (1.19) has a solution $x \in D(L)$ where L is as in Theorem 2.27 if and only if the sequence $\{t_n\} = \{s_0 + s_1 + \cdots + s_n\}$ converges, where

$$s_0 = y \,,$$

$$s_k = \sum_{j=0}^{k-1} B\left(s_j, s_{k-j-1}\right) \,, \quad k = 1, 2, \ldots \,.$$

PROOF The given sequence $\{t_n\}$, $n = 1, 2, \ldots$ corresponds to $\{x_{n+1}\}$ with $x_0 = 0$. If it converges to some $x \in X$ then x is a solution of (1.19). By Theorem 2.27, $x \in D(L)$ (for $x_0 = 0$, $w = x$). Now since $x_0 = 0 \in D(L)$, the converse follows from Theorem 2.27. ∎

Example 2.14

We now complete this section with an example. Let $X = \mathbf{R}^2$ and consider the bilinear operator B on X given by

$$B(x, y) = \left\{ (x_1, x_2) \begin{bmatrix} 1 & 0 \\ 0 & 1 \\ - & - \\ 1 & 0 \\ 0 & -1 \end{bmatrix} \begin{bmatrix} y_1 \\ y_2 \end{bmatrix} \right\}$$

$$x = \begin{bmatrix} x_1 \\ x_2 \end{bmatrix}, \ y = \begin{bmatrix} y_1 \\ y_2 \end{bmatrix}$$

$$= \begin{bmatrix} x_1 & x_2 \\ x_1 & -x_2 \end{bmatrix} \begin{bmatrix} y_1 \\ y_2 \end{bmatrix}$$

$$= \begin{bmatrix} x_1 y_1 + x_2 y_2 \\ x_1 y_1 - x_2 y_2 \end{bmatrix} .$$

Note that B is a symmetric bilinear operator on X. We consider the equation

$$x = 0 + B(x, x)$$

or

$$x_1 = x_1^2 + x_2^2$$

$$x_2 = x_1^2 - x_2^2 . \tag{2.112}$$

Define a norm on $\mathbf{R}^2$ by

$$\|x\| = \max_{1 \le i \le z} \|x_i\| .$$

Let $z = \begin{bmatrix} z_1 \\ z_2 \end{bmatrix}$. Then

$$B(z) B(x)(y) = \begin{bmatrix} z_1 & z_2 \\ z_1 & -z_2 \end{bmatrix} \begin{bmatrix} x_1 y_1 + x_2 y_2 \\ x_1 y_1 - x_2 y_2 \end{bmatrix}$$

$$= \begin{bmatrix} z_1 x_1 y_1 + z_1 x_2 y_2 + z_2 x_1 y_1 - z_2 x_2 y_2 \\ z_1 x_1 y_1 + z_1 x_2 y_2 - z_2 x_1 y_1 + z_2 x_2 y_2 \end{bmatrix} .$$

If we apply the above for $z_i = \begin{bmatrix} z_i^i \\ z_i^i \end{bmatrix}$, $i = 1, 2, \ldots, n$, then $B(z_1) B(z_2) \cdots B(z_n)(x)$ will be of the form

$$v_n = \begin{bmatrix} \displaystyle\sum_{k=1}^{2^n} c_1^k c_2^k \cdots c_{2^{n+1}}^k \\ \displaystyle\sum_{k=1}^{2^n} d_1^k d_2^k \cdots d_{2^{n+1}}^k \end{bmatrix}$$

where $d_m^k, c_m^k \in \{x, z_i\}$, $i = 1, 2, \ldots, n$, $k = 1, 2, \ldots, 2^n$, $m = 1, 2, \ldots, 2^{n+1}$.

Now let us restrict X to $[0, \frac{1}{4}] \times [0, \frac{1}{4}]$, then

$$\|v_n\| \le 2^n \cdot \left(\frac{1}{4}\right)^{n+1} = \frac{1}{2^{n+1}} \to 0 \quad \text{as } n \to \infty,$$

therefore

$$L(x) = 0 \quad \text{for all } x \in \left[0, \frac{1}{4}\right] \times \left[0, \frac{1}{4}\right] \equiv \bar{X}$$

so $\bar{X} = N(L)$ and Equation (2.112) has a unique solution in $\bar{X}$, namely $x = \begin{bmatrix} 0 \\ 0 \end{bmatrix}$.

2.6 On a Class of Quadratic Integral Equations with Perturbation

Here we consider the equation

$$H(x) = 1 + x H(x) \int_0^1 k(x, t) \psi(t) H(t) dt$$

$$+ \int_0^1 F(x, t, H(x), H(t)) dt, \tag{2.113}$$

which is a generalization of (1.22) for

$$k(x, t) = \frac{x}{x + t}, \quad k(0, 0) = 1$$

and

$$F \equiv 0$$

for all $x, t \in [0, 1]$.

The kernel function $k(x, t)$ is known on $[0, 1] \times [0, 1]$ and satisfies the conditions
(a)

$$0 < k(x, t) < 1, \quad k(0, 0) = 1 \tag{2.114}$$

and
(b)

$$k(x, t) + k(t, x) = 1 \tag{2.115}$$

for all $x, t \in [0, 1]$.

The operator F is a perturbation of the Chandrasekhar H-equation (1.22).

We prove the existence of two positive distinct solutions of (2.113) in the spaces $X = C[0, 1]$ and $X^p = C^p[0, 1]$, $(0 < p < 1)$, respectively, under certain assumptions on F and ψ. See also Section 2.1.

We equip the Banach space X of all real continuous functions on $[0, 1]$ with the usual maximum norm

$$\|w\|_X = \max_{0 \leq t \leq 1} |w(t)| .$$

Denote by X_+ the cone of nonnegative functions in X. Let $X^p, 0 < p < 1$, denote the Banach space of all real continuous functions on $[0, 1]$ such that

$$\sup_{x,t \in [0,1]} \frac{|w(x) - w(t)|}{|x - t|^p} < \infty ,$$

with the norm

$$\|x\|_{X^p} = \max_{0 \leq x \leq 1} |w(x)| + \sup_{0 \leq x, t} \frac{|w(x) - w(t)|}{|x - t|^p}$$

and X_+^p the cone of nonnegative functions in X^p.

For $r > 0$ and $R > 0$, let

$$U_r = \left\{ H(x) \in X_+ \mid 1 - \int_0^1 k(x, t)\psi(t)H(t)dt \geq r \right\} ,$$

$$U_r^R = \{ H(x) \in U_r \mid \|H\|_X \leq R \},$$

$$U_r^p = \left\{ H(x) \in X_+^p \mid 1 - \int_0^1 k(x, t)\psi(t)H(t)dt \geq r \right\} ,$$

and

$$U_r^{pR} = \left\{ H(x) \in U_r^p \mid \|H\|_{X^p} \leq R \right\} .$$

As in Section 2.1 we will use Theorem 2.1 due to Darbo to show:

THEOREM 2.31

Suppose that the functions ψ, F, and $k(x, t)$ satisfy the following conditions:

(a) $|F(x, t, w_1, v_1) - F(x, t, w_2, v_2)| \leq l_1|w_1 - w_2| + l_2|v_1 - v_2|$ *for all $x, t \in [0, 1]$ and all $w_i, v_i \in \mathbf{R}^+$, $i = 1, 2$;*

(b) $|F(x, t, w, v)| \le \varepsilon;$

(c) $c = \sup_{0 \le x \le 1} \left| \int_0^1 k(x, t) \psi(t) dt \right| \le \frac{1}{4(1+\varepsilon)};$

(d)   there exist $p, q_3 > 0$ with $0 < p < 1$ such that

$$\frac{1}{|x_1 - x_2|} \left| \int_0^1 (k(x_1, t) - k(x_2, t)) \psi(t) dt \right| \le q_3$$

for all $x_1, x_2 \in U_r$, where r is as defined in (e) and (f).

Then for all $r > 0$ such that

(e) $\frac{1}{r}(l_1 + l_2) < 1$ and

(f) $r_1 \le r \le r_2$, where

$$r_1 = \frac{1 - \sqrt{1 - 4c(1+\varepsilon)}}{2}, \quad r_2 = \frac{1 + \sqrt{1 - 4c(1+\varepsilon)}}{2}$$

and for all $R > 0$ such that $R_1 \le R \le R_2$, where

$$R_1 = \frac{1 - \sqrt{1 - 4c(1+\varepsilon)}}{2c}, \quad R_2 = \frac{1 + \sqrt{1 - 4c(1+\varepsilon)}}{2c}$$

there exists a solution H to Equation (2.113) such that $H \in U_r^R$.

PROOF   Equation (2.113) can equivalently be written in the form

$$H(x) = \frac{1 + \int_0^1 F(x, t, H(x), H(t)) dt}{1 - \int_0^1 k(x, t) \psi(t) H(t) dt} \tag{2.116}$$

or

$$T(H) = L(H) P(H) \tag{2.117}$$

where

$$L(H)(x) = 1 + \int_0^1 F(x, t, H(x), H(t)) dt \tag{2.118}$$

and

$$P(H)(x) = \left[1 - \int_0^1 k(x, t) \psi(t) H(t) dt \right]^{-1}. \tag{2.119}$$

The space X with the maximum norm in a Banach algebra.

We will first show that P is a compact operator on U_r. We know that if for every uniformly bounded sequence $\{H_n\}$ in a subset of X, there is a $p, 0 < p < 1$, such that $P(H_n) \in X^p$ for every n and $\{K(H_n)\}$ is bounded in the norm of X^p, then P is a compact operator.

Let $H_n \in U_r$, where r satisfies (f) and such that $\|H_n\|_X \leq q$, for some $q > 0$. Consider the sequence h_n defined by

$$h_n(x) = \int_0^1 k(x, t)\psi(t)H_n(t)dt , \qquad (2.120)$$

then

$$\frac{|h_n(x_1) - h_n(x_2)|}{|x_1 - x_2|^p} = \frac{1}{|x_1 - x_2|^p} \left| \int_0^1 (k(x_1, t) - k(x_2, t)) \psi(t)H_n(t)dt \right| .$$

It can easily be seen using (d) that there exists $q_1 > 0$ such that

$$\frac{|h_n(x_1) - h_n(x_2)|}{|x_1 - x_2|^p} \leq q_1 \quad \text{for } x_1, x_2 \in [0, 1] . \qquad (2.121)$$

Therefore, by (2.119), $h_n \in X^p$ and

$$\|P(H_n)\|_{X^p} \leq \frac{1}{r} + \frac{q_1}{r^2} = q_2 . \qquad (2.122)$$

From (2.122) we get that P is a compact operator on U_r and

$$\sup_{H \in U_r} \|P(H)\|_X \leq \frac{1}{r} . \qquad (2.123)$$

Using condition (a), we obtain the estimate

$$\|L(H_1) - L(H_2)\|_X = \max_{0 \leq x \leq 1} \left| \int_0^1 (F(x, t, H_1(x), H_1(t))) \right.$$

$$\left. -F(x, t, H_2(x), H_2(t)) dt \right| < (l_1 + l_2) \|H_1 - H_2\|_X . \qquad (2.124)$$

Let $H \in U_r^R$. Then $T(H) \in U_r^R$ if

$$1 - \int_0^1 k(x, t)\psi(t)\left(1 + \int_0^1 P(x, t, H(x), H(t))dt\right)$$

$$\cdot \left[1 - \int_0^1 k(x, t)\psi(t)H(t)dt\right]^{-1} \geq r \qquad (2.125)$$

and

$$\|T(H)\|_X = \|L(H)P(H)\|_X \leq (1 + \varepsilon)\left[1 - \int_0^1 k(x, t)\psi(t)H(t)dt\right]^{-1}$$

$$\leq R. \qquad (2.126)$$

The inequalities (2.125) and (2.126) are now satisfied if

$$1 - \frac{1}{r}(1 + \varepsilon)c \geq r \qquad (2.127)$$

and

$$\frac{1}{1 - cR}(1 + \varepsilon) \leq R \qquad (2.128)$$

hold, respectively. However, inequalities (2.127) and (2.128) are satisfied by the choice of r and R in condition (f).

Hence, U_r^R is an invariant bounded convex set under T. Moreover from (2.123), (2.124), condition (e), and Theorem 2.1, T is a strict set-contraction operator.

Therefore, T has a fixed point $H \in U_r^R$; that is, H satisfies (2.116) and (2.113). That completes the proof of the theorem. ∎

THEOREM 2.32
Assume:

(g) *the conditions (a) through (d) and (f) in Theorem 2.31 are satisfied;*

(h) *the partial derivatives $\frac{\partial^2 F}{\partial x \partial w}$, $\frac{\partial^2 F}{\partial x \partial v}$, $\frac{\partial^2 F}{\partial w \partial v}$, $\frac{\partial^2 F}{\partial v^2}$ exist and*

$$l_3 = \sup\left|\frac{\partial^2 F}{\partial x \partial w}\right|, \quad l_4 = \sup\left|\frac{\partial^2 F}{\partial x \partial v}\right|,$$

$$l_5 = \sup \left| \frac{\partial^2 F}{\partial x \partial v} \right| , \quad l_6 = \sup \left| \frac{\partial^2 F}{\partial v^2} \right| ,$$

are finite, and that

(i) $\sup \| P(H) \|_{X^p} \cdot D < 1$, *where*

$$D = 2l_1 + l_2 + l_3 + l_4 + R(l_5 + l_6) .$$

Then there exists a solution $H \in U_r^{pR}$ of (2.113).
Here r is as defined in (f) and

$$R_1 \leq R \leq R_2 \qquad\qquad (2.129)$$

where

$$R_1 = \frac{1 - (ac + db) - \sqrt{\Delta}}{2ad} ,$$

$$R_2 = \frac{1 - ac - bd + \sqrt{\Delta}}{2ad}$$

$$\Delta = 1 + (ac + bd)^2 - 2ac - 2bd ,$$

$$a = \frac{m}{r^2}$$

$$m = \sup_{x_1, x_2 \in (0,1)} \left| \int_0^1 (k(x_1, t) - k(x_2, t)) \psi(t) dt \right| ,$$

(assuming that it is finite)

$$b = \frac{1}{r} ,$$

$$c = 1 + \varepsilon + l_0 ,$$

$$d = l_1 ,$$

and

$$l_0 = \sup \left| \frac{\partial F}{\partial x} \right| .$$

For m and l_1 sufficiently small, the following is true $0 < R_1 < R_2$.

PROOF   As in Theorem 2.31 we show that if $H_n \in X^{p_1}$ and $\|H_n\|_{X^{p_1}} \le q$ then $\|P(H)\|_{X^p} \le q_1$ for all $0 < p_1 < p < 1$. Then, P is a compact operator on U_r^p.
Let $H_1, H_2 \in U_r^{pR}$ and set

$$I = \int_0^1 F\left(x_2, t, H_2\left(x_2\right), H_2(t)\right) dt - \int_0^1 F\left(x_2, t, H_1\left(x_2\right), H_1(t)\right) dt$$

$$- \int_0^1 F\left(x_1, t, H_2\left(x_1\right), H_2(t)\right) dt + \int_0^1 F\left(x_1, t, H_1\left(x_1\right), H_1(t)\right) dt \ ,$$

$$I_1 = \int_0^1 F\left(x_2, t, H_2\left(x_2\right), H_2(t)\right) dt - \int_0^1 F\left(x_1, t, H_2\left(x_2\right), H_2(t)\right) dt$$

$$- \int_0^1 F\left(x_2, t, H_1\left(x_2\right), H_1(t)\right) dt + \int_0^1 F\left(x_1, t, H_1\left(x_2\right), H_1(t)\right) dt \ ,$$

$$I_2 = \int_0^1 F\left(x_1, t, H_2\left(x_2\right), H_2(t)\right) dt - \int_0^1 F\left(x_1, t, H_2\left(x_2\right), H_1(t)\right) dt$$

$$- \int_0^1 F\left(x_1, t, H_2\left(x_1\right), H_2(t)\right) t + \int_0^1 F\left(x_1, t, H_1\left(x_1\right), H_1(t)\right) dt \ ,$$

$$I_3 = \int_0^1 F\left(x_1, t, H_2\left(x_2\right), H_1(t)\right) dt - \int_0^1 F\left(x_1, t, H_1(t), H_1\left(x_2\right)\right) dt$$

$$- \int_0^1 F\left(x_1, t, H_2\left(x_1\right), H_1(t)\right) dt - \int_0^1 F\left(x_1, t, H_1\left(x_1\right), H_1(t)\right) dt \ .$$

It is immediate that

$$I = I_1 + I_2 + I_3 \ .$$

We also get

$$|I_1| \le (l_3 + l_4) |x_2 - x_1| \ \|H_2 - H_1\|_X \ ,$$

$$|I_2| \le l_5 \int_0^1 |H_2(t) - H_1(t)| \ |H_2\left(x_2\right) - H_1\left(x_1\right)| \, dt$$

$$\leq l_5 \cdot R \, |x_2 - x_1|^P \, \|H_2 - H_1\|_X$$

(since $\|H\|_{X^P} \leq R$),

$$|I_3| \leq l_1 \, |H_2 \, (x_2) - H_1 \, (x_2) - H_2 \, (x_1) + H_1 \, (x_1)|$$

$$+ R l_6 \, |x_2 - x_1|^P \, \|H_2 - H_1\|_X \; .$$

From the above three inequalities we easily obtain

$$\|L \, (H_2) - L \, (H_1)\|_{X^P} \leq D \cdot \|H_2 - H_1\|_{X^P} \; .$$

That is by (i) T is a strict set-contraction operator.

Working exactly as in the second part of the proof of Theorem 2.31, we show that if $H \in U_r^{pR}$ then $T(H) \in U_r^p$, where r is defined by (f) and R is as defined in (2.129).

The result now follows as in Theorem 2.31.   ∎

THEOREM 2.33
Assume:

(j)   *conditions of Theorem 2.31 hold with* $K(x, t)\psi(t) \geq 0$ *for all* $x, t \in [0, 1]$;

(k)   *operator* $F(x, t, w, v)$ *is nondecreasing with respect to* w *and* v *and*

$$\int_0^1 [k(x, t)\psi(t) + F(x, t, 1, 1)]dt \geq 0 \quad for \; 0 \leq x \leq 1 \, . \qquad (2.130)$$

Define sequence $\{H_n\}$, $n = 0, 1, 2, \ldots$ *by*

$$H_0 = 1$$

$$H_{n+1} = T \, (H_n) = L \, (H_n) \, P \, (H_n) \, .$$

Then sequence $\{H_n\}$, $n = 0, 1, 2, \ldots$ *converges uniformly to a solution* $H \in U_r^R$ *of (2.113), where* r, R *are as defined in (f).*

PROOF   By (a) and (b), the sequences

$$Q = \{H_0, H_1, \ldots, H_n, \ldots\}$$

and

$$T(Q) = \{H_1, H_2, \ldots, H_{n+1}, \ldots\}$$

are both contained in U_r^R.

According to the proof of Theorem 2.31, there exists l, $0 < l < 1$, such that

$$m_{U_r^R}(T(Q)) \le l m_{U_r^R}(Q) .$$

Since $Q = H_0 \cup T(Q)$ the measurement of noncompactness of $T(Q)$ is just $m(Q)$, that is $m_{U_r^l}(Q) = 0$. Therefore, Q is relatively compact and there exists a subsequence which converges uniformly to a point H in U_r^R.

If we show that H_n is a nondecreasing sequence, then the entire sequence converges uniformly. We use induction because $H_1 \ge H_0$ by (2.130). Assume that $H_n \ge H_{n-1}$.

Then

$$H_{n+1} - H_n = T(H_n) - T(H_{n-1}) = \frac{y}{z} ,$$

where

$$z = \left(1 + \int_0^1 F(x, t, H_{n-1}(x), H_{n-1}(t)) \, dt\right)$$

$$\int_0^1 k(x, t)\psi(t)(H_n(t) - H_{n-1}(t)) \, dt$$

$$+ \left(1 - \int_0^1 k(x, t)\psi(t)H_{n-1}(t)dt\right) \int_0^1 (F(x, t, H_n(x), H_n(t))$$

$$- F(x, t, H_{n-1}(x), H_{n-1}(t))) \, dt ,$$

$$y = \left(1 - \int_0^1 k(x, t)H_n(t)\psi(t)dt\right)\left(1 - \int_0^1 k(x, t)\psi(t)H_{n-1}(t)dt\right) .$$

Now, $y > 0$ since H_n, $H_{n-1} \in U_r^R$ and $z \ge 0$ by the monotonicity of F and the induction hypothesis.

Therefore, $H_{n+1} \ge H_n$, that is the sequence H_n converges to H uniformly and $H \in U_r^R$ is a solution of (2.113). ∎

Example 2.15

Set

$$F = \varepsilon e^{-(w+v)} ,$$

then for all $w, v \in \mathbf{R}^+$ and $0 < \varepsilon < 1 - \frac{1}{4b}$, $0 < b < \frac{1}{4}$ the conditions of Theorem 2.32 are all satisfied for

$$0 < \varepsilon < \min\left(\left[(1 + R)\left(\frac{1}{2} + \frac{1}{r^2}\right) R\bar{c}\right]^{-1}, \frac{1}{4c} - 1\right),$$

where

$$\bar{c} = \sup_{x_1, x_2 \geq 0} \left| \int_0^1 (k(x_1, t) - k(x_2, t)) \psi(t) dt \, |x_1 - x_2|^{-p} \right|.$$

Also, note that

$$\sup_{H \in U_r^{pR}} \|P(H)\|_{X^p} \leq \frac{1}{r} + \frac{1}{r^2} R\bar{c}$$

where r and R are as in Theorem 2.31.

THEOREM 2.34
Assume:

(l)

$$2 \int_0^1 \psi(t) dt + 2 \int_0^1 \int_0^1 F(x, t, H(x), H(t)) dt dx < 1 \qquad (2.131)$$

where H is a solution of (2.113) obtained under the hypotheses of Theorem 2.31.

(m)   The following estimate is true:

$$\int_0^1 \frac{\psi(t)}{1 - t} H(t) dt > 1. \qquad (2.132)$$

Also,

(n)   there exist functions $\varphi_1, \varphi_2, \varphi_3 \in X$ such that

$$k(x, t)[\varphi_2(x)(1 - kt) - (1 + \varphi_1(t))] + \varphi_3(x) = 0, \qquad (2.133)$$

$$\varphi_1(x) + kx > 0, \quad \text{if } x \neq 0, \ \varphi_1(0) = 0 \qquad (2.134)$$

and

$$[1 + \varphi_1(x)]$$

$$\left[\varphi_2(x) \left(H(x) - 1 - \int_0^1 F(x, t, H(x), H(t)) dt \right) + \varphi_3(x) H(x) \right]$$

$$= (1 - kx) \left(H_1(x) - 1 - \int_0^1 F(x, t, H_1(x), H_1(t)) dt \right) \qquad (2.135)$$

for all $x, t \in [0, 1]$, where k is the unique number in $(0, 1)$ for which

$$\int_0^1 \frac{\varphi(t)}{1 - kt} H(t) dt = 1 . \qquad (2.136)$$

Here the function H_1 is given by

$$H_1(x) = \frac{1 + \varphi_1(x)}{1 - kx} H(x), \quad x \in [0, 1] . \qquad (2.137)$$

Then H_1 is a second solution of (2.113) such that

$$H_1(x) > H(x), \quad \text{for all } x \in [0, 1] .$$

PROOF   By the monotone convergence theorem

$$\lim_{k \to 1^-} \int_0^1 \frac{\psi(t)}{1 - kt} H(t) dt = \int_0^1 \frac{\psi(t)}{1 - t} H(t) dt$$

since $(1 - kt)^{-1}$ increases monotonically with $k, 0 < k < 1$.
  If (2.132) holds, we have

$$\int_0^1 \frac{\psi(t)}{1 - 0 \cdot t} H(t) dt = 1 - \left[1 - 2 \int_0^1 \psi(t) dt \right.$$

$$\left. - 2 \int_0^1 \int_0^1 F(x, t, H(x) H(t)) dt dx \right]^{1/2} < 1 ,$$

and since the function $f: (0, 1) \to \mathbf{R}$ defined by

$$f(k) = \int_0^1 \frac{\psi(t)}{1 - kt} H(t) dt$$

is strictly increasing, there exists a unique $k \in (0, 1)$, for which (2.136) holds. Let H_1 be defined as in (2.137). We now find that for each $x \in [0, 1]$

$$\int_0^1 k(x,t)\psi(t)H_1(t) = \varphi_2(x)\int_0^1 k(x,t)\psi(t)H(t)dt + \varphi_3(x)\int_0^1 \frac{H(t)}{1-kt}H(t)dt$$

$$= \varphi_2(x)\int_0^1 k(x,t)\psi(t)H(t)dt + \varphi_3(x)$$

$$= \varphi_2(x)\left[1 - \frac{1}{H(x)} - \frac{\int_0^1 F(x,t,H(x),H(t))dt}{H(x)}\right] + \varphi_3(x)$$

$$= 1 - \frac{1}{H_1(x)} - \frac{\int_0^1 F(x,t,H_1(x),H_1(t))dt}{H_1(x)} ,$$

[by (2.133), (2.135), and (2.136)] that is, H_1 satisfies (2.113). It now follows that

$$H_1(x) > H(x) \quad \text{for all } x \in [0, 1]$$

which completes the proof of the theorem.   ∎

Example 2.16
By choosing the kernel function $k(x, t)$ to be

$$k(x,t) = \frac{x}{x+t}, \quad x, t \in [0, 1]$$

we see that conditions (2.114) and (2.115) are satisfied and that Equation (2.113) reduces to Equation (1.22).

Moreover, if we choose $F \equiv 0$, for simplicity, then conditions (2.133), (2.134), and (2.135) are satisfied for

$$\varphi_1(x) = kx$$

$$\varphi_2(x) = \frac{1-kx}{1+kx}$$

and

$$\varphi_3(x) = \frac{2kx}{1+kx} .$$

Exercises

1. Investigate the applicability of the contraction mapping principle to the equivalent form of Equation (1.22) given by

$$x(s) = \left(1 - \lambda s \int_0^1 \frac{x(t)}{s+t} dt \right)^{-1}.$$

Construct an arithmetic model of this equation.

2. Derive an upper bound for $\|B\|$, where B is the bilinear integral operator in $C[0, 1]$ with the continuous kernel

$$B = B(s, t, u), \quad 0 \leq s, t, u \leq 1.$$

3. Give necessary and sufficient conditions for the bilinear operator

$$B = \begin{pmatrix} b_{111} & b_{112} & b_{121} & b_{122} \\ b_{211} & b_{212} & b_{221} & b_{222} \end{pmatrix}$$

to be singular in $\mathbf{R}^2$.

4. Show that any bilinear operator B from X into Y can be expressed as the sum of a symmetric bilinear operator, and a bilinear operator B_0 such that

$$B_0 xx = 0 \quad \text{for all } x \in X.$$

5. Prove Theorem 2.1.

6. Prove Theorem 2.9.

7. Verify (2.58).

8. Verify the details in Example 2.3.

9. Prove Lemma 2.1.

10. Prove Theorem 2.21.

11. Consider an operator $T: \mathbf{C}^3 \times \mathbf{C}^3 \to \mathbf{C}^4$, defined by $T(x, y) = (x_1 y_1, x_1 y_2, x_1 y_3 + x_3 y_1 + x_2 y_2, x_3 y_2 + x_2 y_1)$, where, e.g., $x = (x_1, x_2, x_3)$ and $y = (y_1, y_2, y_3)$.
Show:

(i) T is bilinear, continuous, surjective but not open at the origin.

(ii) The above examples go over unchanged if we replace $\mathbf{C}^3$ by $\mathbf{R}^3$ and $\mathbf{C}^4$ by $\mathbf{R}^4$.

(iii) Consider an operator

$$T_1 : \mathbf{C}^3 \oplus \mathbf{C} \oplus L^2 \times \mathbf{C}^3 \oplus L^2 \to \mathbf{C}^4 \oplus L^2$$

defined by $T_1(x_1, x_2, x_3, y_1, y_2) = (T(x_1, y_1), x_2 y_2)$, where $x_1, y_1 \in \mathbf{C}^3$, $x_3, y_2 \in L^2$, $x_2 \in \mathbf{C}$, and $x_2 y_2$ indicates scalar multiplication. Show (i) for the operator T_1.

(iv) Consider an operator $T_2 : \mathbf{C}^4 \times \mathbf{C}^4 \to \mathbf{C}^5$ given by $T_2(x, y) = (x_1 y_1, x_1 y_2 + x_2 y_1, x_1 y_3 + x_3 y_1, x_1 y_4 + x_4 y_1 + x_2 y_3 + x_3 y_2, x_2 y_2 + x_3 y_3)$. Show that T_2 maps $\mathbf{C}^4 \times \mathbf{C}^4$ onto $\mathbf{C}^5$, but on the subspaces $x_1 = 0$ and $y_1 = 0$, T_2 essentially reduces to our original T. Hence, show that the preimages of T_2 of points of the form $(0, a, a, a, 1)$ must "go to infinity" as $a \to 0$.

This exercise answers to the negative to whether "a continuous bilinear operator from the product of two Banach spaces onto a Banach space must be open at the origin".

12. Show that the space X with the maximum norm is a Banach algebra in Theorem 2.31.

13. Show that if for every uniformly bounded sequence $\{H_n\}$ $(n \geq 0)$ in a subset of a Banach algebra X, there is a p, $0 < p < 1$, such that $P(H_n) \in X^p$ (see Theorem 2.31 for the definitions) for every n and $\{K(H_n)\}$ $(n \geq 0)$ is bounded in the norm of X^p, then P is a compact operator.

14. Let X be the Banach space over the complex field F and let $L = L(X, X)$ be the noncommutative Banach algebra of all bounded linear operators from X to itself. It is assumed that the norm on L is such that $\|I\| = 1$, and we denote the spectrum of $A \in L$ by $\sigma(A)$. We consider functions $R : L \to L$, and we are interested in operators $X \in L$ for which $R(x) = 0$. In particular, we study polynomial function P defined by

$$P(x) = Ax^2 + Bx + C.$$

These types of problems are of interest in mechanics. If, for the moment, we assume that $A = I$ and that $(B^2 - 4C)^{1/2}$ exists and commutes with B, then show:

(a)

$$Z_{1,2} = -\frac{1}{2}B \pm \sqrt{\frac{1}{4}B^2 - C}$$

are two zeros of the equation $P(X) = 0$ and $Z_1 - Z_2 = (B^2 - 4C)^{1/2}$. We now say that if $Z_1 - Z_2$ has an inverse, then $Z_{1,2}$ is a complete pair of roots.

(b) Let P be as above and B be invertible. Suppose $h = 2\|B^{-1}A\| \cdot \|B^{-1}C\| \le \frac{1}{2}$, and define

$$t_1 = \frac{1}{2\|B^{-1}A\|}\left(1 - \sqrt{1 - 2h}\right),$$

$$t_2 = \frac{1}{2\|B^{-1}A\|}\left(1 + \sqrt{1 - 2h}\right).$$

Then the equation $P(X) = 0$ has a solution in $\bar{U}(t_1)$ which is unique in $\bar{U}(t_2)$.

(c) Let P be as above and A, B be invertible. Suppose $h_1 = 2\|B^{-1}A\| \cdot \|A^{-1}CB^{-1}A\| \le \frac{1}{2}$ and define t_1 and t_2 as above with $h = h_1$. Then $P(X)$ has a solution in $\bar{S}(t_1) = \bar{S} = \{X \mid \|X + A^{-1}B\| \le t_1\}$ which is unique in $S(t_2)$.

(d) Let P be as above with $A = I$ and B invertible. If $h_2 = 2\|B^{-1}\| \max(\|B^{-1}C\|, \|CB^{-1}\|) < \frac{1}{2}$, then the roots X_1, X_2, obtained in (b) and (c) form a complete pair.

(e) Let X be a Hilbert space. Suppose $\gamma = \inf \sigma(C^{1/2} + (C^{1/2})^*)/2 > 0$, $\|B\| < 2\gamma$, and $h_3 = 2\|BC^{1/2}\|/(2\gamma - \|B\|)^2 \le \frac{1}{2}$. Define

$$t_3 = \frac{2\gamma - \|B\|}{2}\left(1 - \sqrt{1 - 2h}\right),$$

$$t_4 = \frac{2\gamma - \|B\|}{2}\left(1 + \sqrt{1 - 2h}\right).$$

Then the equation $P(x) = 0$ has a solution in the ball $S_1 = \{X \mid \|X \mp iC^{1/2}\| < t_3\}$ which is unique in $S_2 = \{X \mid \|X \mp iC^{1/2}\| < t_4\}$.

15. Consider the (1.22) equation in the form

$$H(x) = 1 + xH(x)\int_0^1 \frac{g(y)H(y)}{x + y}dy \quad \text{for } x \in [0, 1] \qquad (2.138)$$

(a) Suppose that $H \in C[0, 1]$ is a solution of (2.138) above. Then, show that $\int_0^1 g(y)dy \le \frac{1}{2}$ and $H(x) \ge 1$ for all $x \in [0, 1]$.

(b) Suppose that $\int_0^1 g(x)dx \le \frac{1}{2}$ and that $H \in C[0, 1]$ is a positive solution of the equation

$$[H(x)]^{-1} = \left\{1 - 2\int_0^1 g(x)dx\right\}^{1/2} + \int_0^1 \frac{yg(y)H(y)}{x+y}dy$$

for $x \in [0, 1]$. Then, show that

$$\int_0^1 g(x)H(x)dx = 1 - \left\{1 - 2\int_0^1 g(x)dx\right\}^{1/2}$$

and H is a solution of Equation (2.138).

(c) Show that Equation (2.138) has an H solution if and only if

$$\int_0^1 g(x)dx \le \frac{1}{2}.$$

Show that all solutions H are such that $H(x) \ge 1$ for all $x \in [0, 1]$.

16. With the notation of the previous exercise, set

$$c = \left[1 - 2\int_0^1 g(t)dt\right]^{1/2}$$

and

$$A = \{h \in C[0, 1] \mid h(x) \ge c, x \in [0, 1]\}.$$

Define $S: A \to C[0, 1]$ by

$$Sh(x) = 1 + h(x)\int_0^1 \frac{x}{x+t}g(t)h(t)dt, \quad h \in A$$

and $T: A \to C[0, 1]$ by

$$Th(x) = c + \int_0^1 \frac{t}{x+t}g(t)(h(t))^{-1}dt, \quad h \in A.$$

(a) Show that A, B are continuous operators, S is isotone and T is anti-tone, and H is a solution of (2.138) if and only if $SH = H$.

(b) Show that Equation (2.138) has exactly one solution H satisfying

$$\int_0^1 g(t)H(t)dt = 1 - \left[1 - 2\int_0^1 g(t)dt \right]^{1/2} \qquad (2.139)$$

if and only if

$$\int_0^1 g(t)dt \leq \frac{1}{2} .$$

Furthermore show that the increasing sequence $\{S^n(1)\}\, n \geq 0$ converges to H and if the above inequality holds, the sequence $\{T^n(c)\}$ $n \geq 0$ converges to H^{-1} with

$$\left\| H^{-1}(x) - T^n(c)(x) \right\| \leq \left\| T^n(c)(x) - T^{n+1}(c)(x) \right\| \qquad n \geq 0 .$$

(c) Show that the sequence $\{S^n(1)\}\, n \geq 1$ is equicontinuous.

(d) Suppose that $\int_0^1 g(t) < \frac{1}{2}$ and let H be the unique solution of (2.138) and (2.139). Then show that Equation (2.138) has a solution H_1 satisfying

$$\int_0^1 g(t)H(t)dt = 1 + \left[1 - 2\int_0^1 g(t)dt \right]^{1/2}$$

if and only if

$$\int_0^1 \frac{g(t)}{1-t} H(t)dt > 1 . \qquad (2.140)$$

Furthermore, if (2.140) holds, then show that H_1 is the only solution of (2.138) and (2.140), and is given by

$$H_1(x) = \frac{1+kx}{1-kx}, \qquad x \in [0, 1]$$

where k is the unique number in $(0, 1)$ for which

$$\int_0^1 \frac{g(t)}{1-kt} H(t)dt = 1 .$$

(e)   Suppose that $\int_0^1 g(t)dt < \frac{1}{2}$. Then show that Equation (2.138) has exactly two solutions if and only if

$$\int_0^1 \frac{g(t)}{1-t^2}dt > \frac{1}{2} \ .$$

17.   Consider the (2.138) equation in the form

$$u(x) = g(x) + u(x) \int_0^1 \frac{x}{x+t} u(t)dt \qquad (2.141)$$

for the function $u(x) = g(x)H(x)$. Let (X, μ) be a σ-finite positive measure space with $\mu \neq 0$ and let $k(x, t)$ be a measurable function on the product measure space $(X \times X, \mu \times \mu)$ satisfying

(i) $0 < k(x, t) < 1, x, t \in X$,

(ii) $k(x, t) + k(t, x) = 1, x, t \in X$.

Let $L^1 = L^1(X, \mu)$, $L_\infty = L^\infty(X, \mu)$ and let K be the integral operator on L^1 with kernel $k(x, t)$

$$Ku(x) = \int k(x, t)u(t)d\mu(t), \quad u \in L^1 \ .$$

(a)   Show that K is a bounded linear operator from L^1 into L_∞ with $\|K\| \leq 1$.
      Then consider the nonlinear equation

$$u = g + uKu, \quad u \in L^1 \qquad (2.142)$$

where $g \in L^1$ is given. Define the operator $T = T(g)$ by

$$T = g + A(u, u), \quad A(u, v) = ukv \ .$$

(b)   Let

$$U_1 = \left\{ u \in Y : \|u\| \leq \|B + B^*\|^{-1} \right\} , \quad \|B + B^*\| > 0$$

$$U_2 = \left\{ y \in Y : \|y\| < \frac{1}{2} \|B + B^*\|^{-1} \right\}$$

where $B^*(x, y) = B(y, x)$ and B is a bounded bilinear operator on X. Then show that for each $g \in U_2$ the operator T_1 given by

$$T_1 = y + B(u, u)$$

maps U_1 into itself, has a unique fixed point $u \in U_1$ and for each $u \in U_1$ the sequence $\{T_1^n u\}$ converges to u. Moreover, show that the operator $y \to u$ of U_2 into U_1 is continuous.

(c) Let $U_3 = \{u \in L^1 \mid \|u\| < \frac{1}{2}\}$. Then show that for each $g \in U_3$ Equation (2.142) has a unique solution u in the closed unit ball U_1 of L^1, T maps U_1 into itself and for each $u \in U_1$, the sequence $\{T^n u\}$ converges to u in L^1. Moreover, show that the operator $g \to u$ is continuous on U_3.

(d) Let $U_4 = \{u \in L^1 \mid \|u\| \leq \frac{1}{2}, u \geq 0\}$. Then show that for each $g \in U_4$ Equation (2.142) has a unique solution u in the closed unit ball U_1 of L^1, $u \geq 0$, T maps U_1 into itself and for any $u \in U_1$ with $u \geq 0$ the sequence $\{T^n u\}$ converges to u in L^1. Moreover, show that the operator $g \to u$ is uniformly continuous on U_4.

(e) If $u_1 \in U_1$ and u_1 is a solution of (2.142), then show that u_1 is continuous whenever g is continuous. If g is continuous, $g \geq 0$ and $\|g\| \leq \frac{1}{2}$, then show that the unique solution u in U_1 of (2.142) is continuous and if u_1 is continuous, with $0 \leq u_1 \leq g$, then the sequence $\{T^n u_1\}$ consists of continuous functions converging monotonically and almost uniformly to u.

18. Consider the (1.22) equation in the form

$$f(x, \omega) = 1 + \omega f(x, \omega) \int_0^\infty \frac{t}{x + t} \varphi(t) f(t, \omega) dt . \qquad (2.143)$$

Here $\omega \in C$, φ is a measurable function on $(0, \infty)$. Assume:

(i) $\varphi \in L_p(0, \infty)$ for p, $1 \leq p < \infty$.

(ii) $\int_0^\infty \left| \frac{\varphi(t)}{t} \right| dt < \infty$.

(iii) Let $k(x) = \lim_{N \to \infty} \int_0^N c^{-|x|t} \varphi(t) dt$; then $k \geq 0$ and $k \in L_1(-\infty, \infty)$.

(iv) $\int_{-\infty}^\infty k(x) dx = 1$.

(v) $\lim_{N \to \infty} \int_0^N \varphi(t) dt = \frac{1}{2}$.

(vi) If f is a measurable, nonnegative decreasing function on $(0, \infty)$ and
$$0 < \sup_{0 < x \leq 1} |xf(x)| < \infty, \text{ then } \lim_{N \to \infty} \int_0^N \varphi(t)f(t)dt > 0.$$

For $\varepsilon \geq 0$; let $S(\varepsilon)$ be the space of complex-valued functions continuous on $[\varepsilon, \infty)$ and having finite limits at infinity. Show that $S(\varepsilon)$ is a Banach space under the sup norm. Moreover for $|\omega| \leq 1$, $x \geq 0$ define

$$H_0(x, \omega) = 1, \quad H_{n+1}(x) = 1 + \omega H_n(x, \omega) \int_0^\infty \frac{t}{x+t} H_n(t, \omega)\varphi(t)dt \quad (n \geq 0).$$

Then
(A) show for $0 < \varepsilon < 1$,

(a) H_n converges to H in $S(0)$ uniformly in ω for $|\omega| \leq 1 - \varepsilon$.

(b) H_n converges to H in $S(\varepsilon)$ uniformly in ω for $|\omega| \leq 1$.

(B) Let $0 < \varepsilon < 1$, $|\omega| \leq 1$, define for $x \geq 0$

$$K_0(x, \omega) = 1, \quad K_{n+1}(x, \omega) = \left[1 - \omega \int_0^\infty \frac{t}{x-t} \varphi(t) K_n(t, \varphi)dt\right]^{-1}.$$

Then show

(a) K_n converges to H in $S(0)$ uniformly in ω for $|\omega| \leq 1 - \varepsilon$;

(b) K_n converges to H in $S(\varepsilon)$ uniformly in ω for $|\omega| \leq 1$.

19. Let $C(\Omega)$ be the real Banach algebra of the space of real-valued continuous functions on the compact subset Ω in $\mathbf{R}^n$. We denote by $C_+(\Omega)$ the cone of nonnegative functions in $C(\Omega)$. Suppose $K: C_+(\Omega) \to C_+(\Omega)$ is compact. If $x_0 \in C_+(\Omega)$, then show that the equation

$$x_0 = x + xKx \tag{2.144}$$

has a solution x in $C_+(\Omega)$ with $x(t) \leq x_0(t)$, $t \in \Omega$. Furthermore, show that if x_0 is invertible and K linear, then x is the unique solution to Equation (2.144) in $C_+(\Omega)$. Apply the above result to the equation

$$1 = x(t) + x(t) \int_0^1 \frac{R(t, s)}{t^2 - s^2} x(s)ds$$

for a suitable choice of the function $R(t, s)$.

20. Let $\rho, v > 0$ and $n, m \in N$. If $k: [0, 1] \times [0, 1] \to \mathbf{R}$ is continuous and such that

(a) $k(t, s)(t - s) \geq 0$ for $t, s \in [0, 1]$ and

(b) $|k(t, s)| \leq \rho|t - s|^v(t + s)$ for $t, s \in [0, 1]$, then

$$x(t)^n - 1 + x(t)^m \int_0^1 \frac{k(t, s)}{t^2 - s^2} x(s) ds = 0$$

has a solution $x \in C([0, 1])$ with $0 < x(t) \leq 1$ for all $t \in [0, 1]$.

21.   Let $k: [0, 1] \times [0, 1] \rightarrow \mathbf{R}$ be measurable and such that $K: X = C([0, 1]) \rightarrow X$,

$$(Kx)(t) = \int_t^1 k(t, s)x(s) ds ,$$

is a well-defined compact linear operator $K \neq 0$. Then show that the following conclusions hold:

(a)   Fix $r > 0$. If $\lambda \in \mathbf{R}$ is such that $|\lambda| \leq 2r|K|^{-1}(r + |x_0|)^{-2}$, then equation

$$x(t) - x_0(t) - \frac{\lambda}{2} x(t) \int_t^1 k(t, s)x(s) ds = 0, \quad t \in [0, 1] \quad (2.145)$$

has a solution $x \in C([0, 1])$ with $|x(t) - x_0(t)| \leq r$ in $[0, 1]$.

(b)   If $x_0 \geq 0$ and if $k(t, s) \geq 0$ in $[0, 1]^2$, then (2.145) has a solution $x \in C([0, 1])$ with $x \geq 0$ whenever $\lambda \in (-\infty, 0]$.

(c)   For the special choice $x_0(t) = 1$ in $[0, 1]$ and $k(t, s) = \frac{t}{t+s}$ for $t, s \in [0, 1]$ with $t + s \neq 0$, Equation (2.145) has a solution $x \in C([0, 1])$ with $x \geq 0$ whenever $\lambda \in (-\infty, \frac{1}{2M})$, where

$$M = \max\{t \ln((1 + t/2t): t \in [0, 1]\} .$$

Chapter 3

Polynomial Equations in Banach Space

This chapter introduces existence and uniqueness results for a solution of a polynomial equation in a Banach space. Section 3.1 deals with contractive type results, whereas Section 3.2 deals with noncontractive ones [17, 18, 60]. Finally in Section 3.3 we deal with polynomial operator equations in ordered Banach spaces [175].

3.1   Polynomial Equations

CASE 3.1
We will need the definition:

DEFINITION 3.1   *An abstract polynomial operator P_k from X into Y of degree k defined by*

$$P(x) = P_k(x) = M_k x^k + M_{k-1} x^{k-1} + \cdots + M_2 x^2 + M_1 x + M_0 , \qquad (3.1)$$

is said to be bounded if its coefficients M_i, $i = 1, 2, \ldots, k$ are bounded i-linear operators from X into Y. From now on we assume P_k is bounded. The notation $P = P_k$ will be used conveniently.

We are seeking solutions x^* of the equation

$$P(x) = 0 . \qquad (3.2)$$

We assume that X and Y are Banach spaces.

DEFINITION 3.2   Let z be fixed in X and define the polynomial q_k of degree k on $\mathbf{R}^+$ by

$$q_k(r) = \|P_k(z) - z\| + \|M_1\| \left[\frac{(r + \|z\|) - \|z\|}{(r + \|z\|) - \|z\|} \right] r$$

$$+ \cdots + \|M_k\| \left[\frac{(r + \|z\|)^k - \|z\|^k}{(r + \|z\|) - \|z\|} \right] r . \qquad (3.3)$$

Note that by Descartes' rule of signs the equation $q(r) = q_k(r) - r = 0$ has two positive solutions $s_1 \leq s_2$ or none.

THEOREM 3.1
Assume that $q(r)$ has two positive solutions $s_1 < s_2$. Then P_k has a unique fixed point in the ball $\bar{U}(z, r)$, where $r \in (s_1', s_2') \subset (s_1, s_2)$.

PROOF

CLAIM 3.1
P_k maps $\bar{U}(z, r)$ into $\bar{U}(z, r)$.

$$\|P_k(x) - z\| = \|P_k(x) - P_k(z) + P_k(z) - z\| \leq \|P_k(x) - P_k(z)\|$$

$$+ \|P_k(z) - z\|$$

$$\leq \left\| M_1(x - z) + \cdots + M_k(x - z)x^{k-1} + M_k x^{k-2}(x - z)z \right.$$

$$\left. + \cdots + M_k z^{k-1}(x - z) \right\|$$

$$\leq \left[\|M_1\| + \|M_2\| (r + \|z\|) + \cdots + \|M_k\| ((r + \|z\|)^{k-1} \right.$$

$$\left. + (r + \|z\|)^{k-2} \|z\| + \cdots + \|z\|^{k-1} \right] r$$

$$+ \|P_k(z) - z\| \leq r$$

or

$$q(r) \leq 0 \quad \text{which is true by hypothesis.}$$

(Note that Claim 3.1 is true even if $s_1 = s_2$ and $r \in [s_1, s_2]$).

CLAIM 3.2
P_k *is a contraction operator on* $\bar{U}(z, r)$. *If* $x_1, x_2 \in \bar{U}(z, r)$, *then as in Claim 3.1,*

$$\|T(x_1) - T(x_2)\| \le q_k'(r) \|x_1 - x_2\|$$

but

$$q_k'(r) < 1 \quad \text{by hypothesis.}$$

The result now follows from the contraction mapping principle. ∎

DEFINITION 3.3   *Define the polynomial* $\tilde{q}_k(r)$ *of degree* k *on* $\mathbf{R}^+$ *by*

$$q_k'(r) = \|M_1 - N_1\| \, \|z\| + \|M_2 - N_2\| \, \|z\|^2 + \cdots + \|M_k - N_k\|$$

$$\cdot \|z\|^k + q_k(r) - \|P_k(z) - z\| . \tag{3.4}$$

Note that

$$\|P_k(z) - z\| = \left\| M_0 + M_1 z + M_2 z^2 + \cdots + M_k z^k - z \right\|$$

$$= \left\| M_0 - M_0 + M_1 z - N_1 z + \cdots + M_k z^k - N_k z^k + (F_k(z) - z) \right\|$$

$$\le \|M_1 - N_1\| \cdot \|z\| + \|M_2 - N_2\| \cdot \|z\|^2 + \cdots + \|M_k - N_k\| \cdot \|z\|^k$$

if z *is a fixed point of*

$$F(z) = F_k(z) = N_k z^k + N_{k-1} z^{k-1} + \cdots + N_2 z^2 + N_1 z + N_0 , \tag{3.5}$$

where the N_p*s are bounded p-linear operators on* X, $p = 1, 2, \ldots, k$.

The proof of the following theorem follows from Theorem 3.1 and the above observation.

THEOREM 3.2
Suppose that there exists a solution z *satisfying* (3.5) *and that* $\tilde{q}_k(r)$ *has two positive solutions* $s_1 < s_2$. *Then* P_k *has a unique fixed point in* $\bar{U}(z, r)$, *where* $r \in (s_1', s_2') \subset (s_1, s_2)$.

CASE 3.2

Let us now introduce the iterations for finding the solutions of the abstract polynomial equation of degree k in the form

$$P_k(x) = M_k(x)^k + M_{k-1}(x)^{k-1} + \cdots + M_2(x)^2 + M_1(x) + M_0 = 0 \quad (3.6)$$

given by

$$x_{n+1} = T_k(x_n) = - (M_2(x_n))^{-1} \left[M_0 + M_1(x_n) \right.$$

$$\left. + M_3(x_n)^3 + \cdots + M_k(x_n)^k \right] \quad (3.7)$$

or

$$x_{n+1} = x_n - (M_2(x_n))^{-1} P_k(x_n), \quad n = 0, 1, 2, \ldots$$

and

$$x_{n+1} = x_n - M_2(x_0)^{-1} P_k(x_n) \quad \text{for some } x_0 \in X. \quad (3.8)$$

Obviously in case of convergence the sequence $\{x_n\}, n = 0, 1, 2, \ldots$ converges to a solution $x^* \in X$ of (3.1). Note that the invertibility of $M_2(x_n)$ or $M_2(x_0)$ does not depend on the invertibility of $[P_k'(x_0)]$ or M_1.

We will need the lemma:

LEMMA 3.1

Let $z \neq 0$ be fixed in X. Assume that the linear operator $M_2(z)$ is invertible, then $M_2(x)$ is also invertible for all $x \in U(z, r)$, where $r \in (0, r_0)$ and $r_0 = [\|M_2\| \, \|M_2(z)^{-1}\|]^{-1}$.

PROOF   We have

$$\|M_2(x - z)\| \cdot \left\| M_2(z)^{-1} \right\| \leq \|M_2\| \cdot \|x - z\| \cdot \left\| M_2(z)^{-1} \right\|$$

$$\leq \|M_2\| \, \left\| M_2(z)^{-1} \right\| \cdot r$$

$$< 1$$

for $r \in (0, r_0)$. The result now follows from the Banach lemma on invertible operators. Also note that

$$\left\|(M_2(x))^{-1}\right\| \leq \frac{\left\|M_2(z)^{-1}\right\|}{1 - \left\|M_2(z)^{-1}\right\| \, \|M_2\| \cdot r} = c(r) = c . \quad \blacksquare \quad (3.9)$$

DEFINITION 3.4   *Let z be as in Lemma 3.1 and define the real functions q_k, p_k and $\tilde{p}_k$ on R^+ for $k > 2$ by*

$$q_k(r) = \left\|\bar{M}_k\right\| \left[(r + \|z\|)^k - \|z\|^k\right] + \cdots + \left\|\bar{M}_3\right\| \left[(r + \|z\|)^3 - \|z\|^3\right]$$

$$+ a \, \|M_2\| \cdot r^2 + \left(\left\|I - M_2(z)^{-1} M_1\right\| - 1\right) r + \left\|\bar{P}_k(z)\right\| ; \quad (3.10)$$

$$p_k(r) = c \left[\|M_1\| + 3 \, \|M_3\| \, (r + \|z\|)^2 + \cdots + k \, \|M_k\| \, (r + \|z\|)^{k-1}\right]$$

$$+ c^2 \, \|M_2\| \left(\|M_0\| + \|M_1\| \, (r + \|z\|) + \|M_3\| \, (r + \|z\|)^3\right.$$

$$+ \cdots + \|M_k\| \, (r + \|z\|)^k\Big) ;$$

$$\tilde{p}_k(r) = a \, (1 - a \, \|M_2\| \cdot r) \left[\|M_1\| + 3 \, \|M_3\| \, (r + \|z\|)^2 + \cdots\right.$$

$$\left. + k \, \|M_k\| \, (r + \|z\|)^{k-1}\right]$$

$$+ a^2 \, \|M_2\| \left[\|M_0\| + \|M_1\| \, (r + \|z\|) + \|M_3\| \, (r + \|z\|)^3\right.$$

$$+ \cdots + \|M_k\| \, (r + \|z\|)^k\Big] - (1 - a \, \|M_2\| \cdot r)^2 ,$$

where

$$a = \left\|M_2(z)^{-1}\right\|, \ \left\|\bar{M}_j\right\| = \left\|M_2(z)^{-1} M_j\right\|, \quad j = 3, 4, \ldots, k ;$$

and

$$\left\|\bar{P}_k(z)\right\| = \left\|M_2(z)^{-1} P_k(z)\right\| .$$

By computing on the right-hand side of (3.10) we can easily rewrite q_k as

$$q_k(r) = A_k r^k + A_{k-1} r^{k-1} + \cdots + A_2 r^2 + A_1 r + A_0$$

where

$$A_j > 0, \quad \text{for } j = 2, 3, \ldots, k .$$

In particular,

$$A_1 = k \left\| \bar{M}_k \right\| \, \|z\|^{k-1} + \cdots + 3 \left\| \bar{M}_3 \right\| \, \|z\|^2 + \left\| I + M_2(z)^{-1} M_1 \right\| - 1$$

and

$$A_0 = \left\| \bar{P}_k(z) \right\| - \left\| \bar{M}_k \right\| \, \|z\|^k - \cdots - \left\| \bar{M}_3 \right\| \, \|z\|^3 .$$

Therefore, it is not clear that $q_k(r) \geq 0$ for all $r > 0$ and $j = 0, 1, 2, \ldots, k$. For example, if for some $z \neq 0$

$$A_1 < 0 \quad \text{and} \quad A_0 > 0 \tag{3.11}$$

then by Descartes' rule of signs the equation

$$q_k(r) = 0 \tag{3.12}$$

has two or no positive solutions. If the two positive solutions $r_1 \leq r_2$ exist, then

$$q_k(r) \leq 0 \quad \text{for all } r \in [r_1, r_2] .$$

Moreover if z is such that

$$a \left(\|M_1\| + 3 \|M_3\| \cdot \|z\|^2 + \cdots + k \|M_k\| \cdot \|z\|^{k-1} \right) + a^2 \|M_2\| \left[\|M_0\| \right.$$

$$\left. + \|M_1\| \cdot \|z\| + \|M_3\| \, \|z\|^3 + \cdots + \|M_k\| \cdot \|z\|^k \right] < 1 , \tag{3.13}$$

then

$$\bar{p}_k(0) < 0.$$

By continuity there exists r_3 such that

$$\bar{p}_k(r) < 0 \quad \text{if } 0 < r < r_3.$$

Let $r_4 = \min(r_0, r_3)$ and set $I = (0, r_4) \cap [r_1, r_2]$.

THEOREM 3.3
Assume:

(a) *linear operator $M_2(z)$ is invertible for some $z \neq 0$, $z \in X$ satisfying (3.11) and (3.13);*

(b) *two positive solutions $r_1 \leq r_2$ of Equation (3.12) exist.*

(c) *There exist a nonempty closed interval $\bar{I} \subset I$.*

Then the operator

$$T_k(x) = -M_2(x)^{-1} \left[M_0 + M_1(x) + M_3(x)^3 + \cdots + M_k(x)^k \right] ,$$

$$k > 2 \tag{3.14}$$

has a unique fixed point x^* in the ball $\bar{U}(z, r)$ where $r \in \bar{I}$.

PROOF
   Claim 1. T_k maps $\bar{U}(z, r)$ into $\bar{U}(z, r)$.
   Let $x \in \bar{U}(z, r)$ with $r \in \bar{I}$, then using (3.14) we obtain

$$T_k(x) - z = -\left[I + M_2(z)^{-1} M_2(x - z) \right]^{-1} \left[\left(I + M_2(z)^{-1} M_1 \right) (x - z) \right.$$

$$+ M_2(z)^{-1} (M_3(x)^3 - M_3(z)^3)$$

$$\left. + \cdots + M_2(z)^{-1} \left(M_k(x)^k - M_k(z)^k \right) + M_2(z)^{-1} P_k(z) \right] .$$

Therefore, $\|T_k(x) - z\| \leq r$ for $x \in \bar{U}(z, r)$ if

$$\frac{1}{1 - a \|M_2\| \cdot r} \left\{ \left\| I + M_2(z)^{-1} M_1 \right\| r + \left\| \bar{M}_3 \right\| \left[(r + \|z\|)^2 + (r + \|z\|) \|z\| + \|z\|^2 \right] \right.$$

$$+ \cdots + \left\| \bar{M}_k \right\| \left[(r + \|z\|)^{k-1} + (r + \|z\|)^{k-2} \|z\| + \cdots + \|z\|^{k-1} \right]$$

$$\left. + \left\| M_2(z)^{-1} P_k(z) \right\| \right\} \leq r$$

of if $q_k(r) \leq 0$ which is true by the choice of r.
   Claim 2. T_k is a contraction operator on $\bar{U}(z, r)$. If $x_1, x_2 \in \bar{U}(z, r)$ with $r \in \bar{I}$ then for $i = 1, 2, \ldots, k$ we obtain

$$\left\|-M_2\,(x_1)^{-1}\,M_i\,(x_1)^i + M_2\,(x_2)^{-1}\,M_i\,(x_2)^i\right\|$$

$$= \left\|(M_2\,(x_1))^{-1}\left(M_i\,(x_1)^i - M_i\,(x_2)^i\right)\right.$$

$$\left. + \left((M_2\,(x_1))^{-1} - (M_2\,(x_2))^{-1}\right)M_i\,(x_2)^i\right\|$$

$$= \left\|(M_2\,(x_1))^{-1}\left(M_i\,(x_1 - x_2)\,(x_1)^{i-1} + M_i\,(x_1 - x_2)\,(x_1)^{i-2}\,(x_2)\right)\right.$$

$$+ \cdots + \left(M_i\,(x_1 - x_2)\,(x_2)^{i-1}\right) + (M_2\,(x_2))^{-1}$$

$$\left(M_2\,(x_2 - x_1)\,(M_2\,(x_1))^{-1}\right)M_i\,(x_2)^i\right\|$$

$$\le \left[ci\,(r + \|z\|)^{i-1}\,\|M_i\| + c^2\,\|M_2\|\,\|M_i\|\,(r + \|z\|)^i\right]\|x_1 - x_2\|$$

since,

$$\|x_1\| = \|x_1 - z + z\| \le \|x_1 - z\| + \|z\| \le r + \|z\|\,.$$

Similarly, we get

$$\|x_2\| \le r + \|z\|\,.$$

Also, we obtain

$$\left\|\left(M_2\,(x_2)^{-1} - M_2\,(x_1)^{-1}\right)M_0\right\| = \left\|M_2\,(x_1)^{-1}\left(M_2\,(x_1 - x_2)\,M_2\,(x_2)^{-1}\,M_0\right)\right\|$$

$$\le \|M_0\| \cdot \|M_2\|\,c^2\,\|x_1 - x_2\|\,.$$

Therefore, we can easily obtain

$$\|T_k\,(x_1) - T_k\,(x_2)\| \le p_k(r)\,\|x_1 - x_2\|\,.$$

Now T_k is a contraction on $\bar{U}(z, r)$ if $p_k(r) < 1$ or if $\tilde{p}_k(r) < 0$ since

$$\bar{p}_k(r) = (p_k(r) - 1)\,(1 - a\,\|M_2\| \cdot r)^2$$

which is true by choice of $r \in \bar{I}$. The result now follows from the contraction mapping principle. ∎

A similar theorem can be proved for the iteration (3.8). Let us now state the theorem for $k = 2$ and $M_1 = -I$, the identity operator on X. The proof follows exactly as the proof of Theorem 3.3.

THEOREM 3.4

Let $z \in X$, and assume that the operator $M_2(z)$ is invertible and that the following are true:

(a) $\quad \| M_2(z)^{-1}(I - M_2(z)) \| < 1$

(b) $\quad D = (\| M_2(z)^{-1}(I - M_2(z)) \| - 1)^2 - 4 \| M_2(z)^{-1} \| \, \| M_2 \| \, \| M_2(z)^{-1} P_2(z) \| > 0.$

Then the iteration (3.8) is well defined and it converges to a unique solution x^* of (3.6) for any $x_0 \in \bar{U}(z, r)$, where r is such that

$$c_1 \leq r < c_2$$

with

$$c_1 = \frac{1 - \| M_2(z)^{-1}(I - M_2(z)) \| - \sqrt{D}}{2 \| M_2 \| \cdot \| M_2(z)^{-1} \|}$$

$$c_2 = \frac{1 - \| M_2(z)^{-1}(I - M_2(z)) \|}{2 \| M_2 \| \cdot \| M_2(z)^{-1} \|}.$$

Note that condition (a) above corresponds to (3.11) and (b) guarantees the existence of two positive solutions of Equation (3.12). Finally (3.13) is satisfied for $r \in (0, c_2)$.

We now give an example of Theorem 3.4.

Example 3.1

Let $X = \mathbf{R} \times \mathbf{R}$ with max-norm and consider the equation on X,

$$\underline{x} = \underline{y} + \underline{x}^T \underline{E} \, \underline{x} \tag{3.15}$$

where $\underline{y} = \begin{bmatrix} m_1 \\ m_2 \end{bmatrix}$, $m_1 = 1.55$, $m_2 = -.85$, $\underline{E} = \begin{bmatrix} E_1 \\ E_2 \end{bmatrix}$ with

$$E_1 = \begin{bmatrix} -.45 & .01 \\ .9 & .02 \end{bmatrix}, \quad E_2 = \begin{bmatrix} .01 & -.7 \\ .02 & .5 \end{bmatrix},$$

the notation $\underline{x}^T \underline{E} \underline{x} = \begin{bmatrix} \underline{x}^T & E_1\underline{x} \\ \underline{x}^T & E_2\underline{x} \end{bmatrix}$ and $\underline{x} = \begin{bmatrix} x_1 \\ x_2 \end{bmatrix}$ is the unknown vector. Equation (3.15) can be written also as

$$x_1 = -.45x_1^2 + .91x_1x_2 + .02x_2^2 + 1.55$$

$$x_2 = .01x_1^2 - .68x_1x_2 + .5x_2^2 - .85 .$$

Here $\|M_2\| = 1.38$. Let $z = \begin{bmatrix} -2 \\ 1 \end{bmatrix}$, then $\|M_2(z) - I\| = .9$, $\|M_2(z)^{-1}\| = -.555555$, $\|P_2(z)\| = .05$. The requirements of Theorem 3.4 are satisfied for the above z in the ball $\bar{U}(z, r)$ for some r such that

$$.061321367 \le r < .326086956 .$$

We now use iteration (3.8) with $\underline{x}^0 = z$. If we allow an error $\varepsilon \le 5.10^{-3}$ then we need five iterations. More precisely

$$\underline{x}^{(1)} = \begin{bmatrix} -1.97222223 \\ .97368421 \end{bmatrix}, \quad \underline{x}^{(2)} = \begin{bmatrix} -1.996301957 \\ .976283584 \end{bmatrix}$$

$$\underline{x}^{(3)} = \begin{bmatrix} -1.99715663 \\ .96816564 \end{bmatrix}, \quad \underline{x}^{(4)} = \begin{bmatrix} -2.003524174 \\ .9654191 \end{bmatrix}$$

and

$$\underline{x}^{(5)} = \begin{bmatrix} -2.00038145 \\ .96224933 \end{bmatrix} .$$

CASE 3.3

Consider now the polynomial equation again but in the form

$$x = M_0 + M_1x + M_2x^2 + \cdots + M_kx^k .$$ (3.16)

We show how to approximate solutions of (3.16) using the solutions z_i of the equations

$$z_i = M_0 + M_1^i z_i + M_2^i z_i^2 + \cdots + M_k^i z_k^i$$ (3.17)

where $\{M_j^i\}$, $i = 1, 2, \ldots$ is a sequence of bounded j-linear operators converging (in norm) to M_j, $j = 1, 2, \ldots, k$.

DEFINITION 3.5 *Define the real polynomials*

$$P_i(r) = m_0^i + m_1^i r + m_2^i r^2 + \cdots + m_k^i r^k, \quad i = 0, 1, 2, \ldots,$$

where

$$m_0^i = \|M_0\|$$

$$m_j^0 = \|M_j\|, \quad i = 0, 1, 2, \ldots, \quad j = 1, 2, \ldots, k,$$

$$m_j^i = \left\|M_j^i\right\|.$$

Also, define the nonnegative numbers $\varepsilon_{n,m,j}$, $r_{n,m}$, $\varepsilon_{n,j}$, r_n by

$$\varepsilon_{n,m,j} = \left\|M_j^m - M_j^n\right\|$$

$$\varepsilon_{n,j} = \left\|M_j - M_j^n\right\|$$

$$r_{n,m} = r_{n,m}(z_m, z_n) = \|z_m - z_n\|$$

$$r_n = r_n(x, z_n) = \|x - z_n\|.$$

Note that by Descartes' rule of signs the equations

$$P_i(r) - r = 0 \quad \text{and} \quad P_j'(r) - 1 = 0$$

have, respectively, two positive solutions or none.
The following is an easy consequence of the contraction mapping principle.

THEOREM 3.5
Let $r^* > 0$ be such that

$$P_0(r^*) \leq r^* \quad \text{and} \quad P_0'(r^*) < 1.$$

Then (3.16) has a unique solution $z^* \in U(r^*)$.

DEFINITION 3.6   Let $\{M_j^i\}$, $i = 1, 2, \ldots$, $j = 1, 2, \ldots, k$ be a sequence of bounded j-linear operators on X. We say that the sequence $\{M_j^i\}$ converges in norm to the j-linear operator M_j on X if

$$\lim_{i \to \infty} \left\| M_j^i - M_j \right\| = 0, \quad j = 1, 2, \ldots, k \,.$$

We now prove the following, which is a useful modification of Vitali's theorem.

THEOREM 3.6

Let $\{M_j^i\}$, $j = 1, 2, \ldots, k$ be a sequence of j-linear operators converging (in norm) to M_j. For each i, assume z_i is the solution of Equation (3.17) given under the hypotheses of Theorem 3.2 and that

$$\sup \|z_i\| \le \|z^*\| \le r < r^*, \quad i = 1, 2, \ldots \,.$$

Then

(i)   the sequence $\{z_i\}$, $i = 1, 2, \ldots$ converges to z^*;

(ii) $r_n \le \dfrac{\varepsilon_{n,1} r + \varepsilon_{n,2} r^2 + \cdots + \varepsilon_{n,k} r^k}{1 - P_n'(r)}$.

PROOF   We have

$$z_m - z_n = M_1^m z_m - M_1^n z_m + M_2^m z_m^2 - M_2^n z_n^2 + M_3^m z_m^3 - M_3^n z_n^3$$

$$+ \cdots + M_k^m z_m^k - M_k^n z_n^k$$

$$= M_1^m (z_m - z_n) + \left(M_1^m - M_1^n\right) z_n + \cdots$$

$$+ M_k^m \left(z_m^k - z_n^k\right) + \left(M_k^m - M_k^n\right) z_n^k \,.$$

By taking norms in the above expression and using the triangle inequality we get

$$r_{n,m} \le \left\| M_1^m \right\| r_{n,m} + \varepsilon_{m,n,1} r + \cdots + k \left\| M_k^m \right\| r^{k-1} r_{n,m} + \varepsilon_{n,m,k} r^k$$

or

$$r_{n,m} \left(1 - P_m'(r)\right) \le \varepsilon_{n,m,1} r + \cdots + \varepsilon_{m,n,k} r^k \,.$$

Let $n, m \to \infty$, then the above expression becomes

$$\left(1 - P_0'(r)\right) \lim_{n,m \to \infty} r_{n,m} \leq 0$$

and since

$$1 - P_0'(r) > 0 \,,$$

$$\lim_{n,m \to \infty} r_{n,m} = 0.$$

Therefore, the sequence $\{z_n\}$, $n = 1, 2, \ldots$ is a Cauchy sequence in a Banach space X. Let $z \in X$ be such that

$$z = \lim_{n \to \infty} z_n = \lim_{n \to \infty} \left(M_0 + M_1^n z_n + M_2^n z_n^2 + \cdots + M_k^n z_n^k\right)$$

$$= M_0 + M_1 z + M_2 z^2 + \cdots + M_k z^k$$

so z is a solution of (3.16).

Part (ii) follows as part (i) if we take $M_j^m = M_j$ and $z_m = z^* = z$. The theorem is proved. ∎

Let us set λM_j instead of M_j in (3.16) for some given parameter λ. Then under the hypotheses of Theorem 3.5, the iteration

$$x_0 = M_0$$

$$x_1 = M_0 + \lambda \left(M_1 x_0 + M_2 x_0^2 + \cdots + M_k x_0^k\right)$$

$$\vdots$$

$$x_{n+1} = \sum_{k=0}^{n} \lambda^k v_k, \quad n = 0, 1, 2, \ldots$$

where

$$v_0 = M_0$$

and

$$v_{n+1} = M_1(v_n) + \sum_{i_2^1 + i_2^2 = n} M_2 v_{i_2^1} v_{i_2^2} + \sum_{i_3^1 + i_3^2 + i_3^3 = n} M_3 v_{i_3^1} v_{i_3^2} v_{i_3^3}$$

$$+ \cdots + \sum_{i_k^1 + \cdots + i_k^k = n} M_k v_{i_k^1} v_{i_k^2} \cdots v_{i_k^k}$$

converges to a solution x^* of (3.16). Therefore

$$x^* = \sum_{n=0}^{\infty} \lambda^n v_n.$$

However, the v_ns may be too difficult to evaluate in practice (e.g., see Chapters 1 and 2).

A similar power series representation for the solutions of (3.17) may be obtained in this case, i.e.,

$$z_i = \sum_{n=0}^{\infty} \lambda^n z_n^i.$$

The difficulties in evaluating the z_n^is may be similar to those of evaluating the v_ns. However, we showed in Section 2.4 that if the M_js are finite rank operators, then the z_is are easy to compute.

According to Section 2.4, the equation

$$x = F_0 + F_1 x + F_2 x^2 + \cdots + F_k x^k$$

can also be written as

$$x = F_0 + \sum_{i=1}^{n(1)} F_{1i}(x) b_{1i} + \sum_{i=1}^{n(2)} F_{2n(1)+i}(x) b_{2n(1)+i}$$

$$+ \cdots + \sum_{i=1}^{n(k)} F_{kn(k-1)+i}(x) b_{kn(k-1)+i} \qquad (3.18)$$

where the $F_{jn(j)}$s are j-linear functionals and the $b_{jn(j)}$s are bases for rang (F_j), $j = 1, 2, \ldots, k, i = 1, 2, \ldots, n(j)$.

Let c_{ji} denote scalars in S. Then, by the definition of the F_{ji}s, an expression of the form

$$F_{ji} \left[\sum_{i=1}^{n(j)} c_{ji} b_{ji} \right] = Q_{ji} \left(c_{j1}, c_{j2}, \ldots, c_{jn(j)} \right)$$

is a scalar polynomial Q_{ji} in $c_{j1}, c_{j2}, \ldots, c_{jn(j)}$ of degree j. The coefficients in Q_{ji} are specified once the b_{ji}s are given.

Let $N = n(1) + n(2) + \cdots + n(k)$.

We can now prove the following theorem.

THEOREM 3.7

The point $w \in X$ is a solution of the equation

$$x = F_0 + F_1 + F_2 x^2 + \cdots + F_k x^k \qquad (3.19)$$

if and only if

$$w = F_0 + \sum_{j=1}^{k} \sum_{i=1}^{n(j)} \xi_{jn(j-1)+i} b_{ji(j-1)+i} \qquad (3.20)$$

where the vector $\underline{\xi} = (\xi_{11}, \ldots, \xi_{1n(1)}, \ldots, \xi_{kn(k-1)+1}, \ldots, \xi_{kN}) \in S^N$ is a solution of the scalar equation

$$\xi_{jn(j-1)+i} = Q_{ji}(\underline{\xi}) \text{ in } S^N, \ j = 1, 2, \ldots, k, \ i = 1, 2, \ldots, n(j). \qquad (3.21)$$

PROOF   Assume that Equation (3.19), or equivalently (3.18), has a solution $w \in X$. Apply F_{ji}, $j = 1, 2, \ldots, k, i = 1, 2, \ldots, n(j)$ in turn to (3.18) to obtain

$$M_{jn(j-1)+i}(w) = Q_{ji}(\underline{w}) \qquad (3.22)$$

where $\underline{w} = (M_{11}(w), \ldots, M_{1n(1)}(w), \ldots, M_{kn(k-1)+1}(w), \ldots, M_{kN}) \in S^N$. Now set $\underline{w} = \underline{\xi}$ in (3.22) to obtain (3.21).

Conversely, assume that Equation (3.21) has a solution vector $\underline{\xi} \in S^N$.

Let $w \in X$ be defined as in Equation (3.20). Apply F_{ji} in turn to (3.20) to obtain

$$M_{jn(j-1)+j}(w) = Q_{ji}(\underline{\xi}) .$$

Now, since $\underline{\xi}$ satisfies (3.21), we have

$$\xi_{jn(j-1)+i} = Q_{ji}(\underline{\xi}) .$$

Comparing the last two equations we get

$$\xi = \underline{w}$$

so

$$w = F_0 + F_1 w + \cdots + F_k w^k .$$

Therefore, w is a solution of (3.19) and the theorem is proved.   ∎

It is clear that the map

$$\underline{\xi} \leftrightarrow w = F_0 + \sum_{j=1}^{k} \sum_{i=1}^{n(j)} \xi_{jn(j-1)+i} b_{jn(j-1)+i}$$

is a bijection between the solutions of (3.19) and those of (3.21).

Example 3.2
Let $X = C[0, 1]$ and consider the equation

$$x(s) = s + s \int_0^1 x^2(t) dt \tag{3.23}$$

where $s \in [0, 1]$. This equation is of the form (3.19), with rank $(F_2) = 1$,

$$F_0(x) = s$$

$$b_{21}(s) = s$$

$$F_2 x(s) y(s) = \int_0^1 x(t) y(t) dt .$$

Equation (3.23) can be written

$$x = F_0 + F_{21} x^2 b_{21} . \tag{3.24}$$

Set $\xi = F_{21} x^2$ in (3.24) and apply M_{21} to obtain

$$\xi = F_{21} \left(F_0 + F_{21} x^2 b_{21} \right)$$

$$= F_{21} \left(F_0 + \xi b_{21} \right)$$

$$= M_{21}M_0^2 + \xi M_{21}M_0 b_{21} + \xi M_{21}b_{21}M_0 + \xi^2 M_{21}b_{21}^2$$

or

$$\xi = 1 + \xi^2 \; ;$$

since $x = F_0 + \xi b_{21}$, we finally have that the solutions of (3.23) in this case are

$$x(s) = s + \left[\frac{1 \pm i\sqrt{3}}{2}\right] s$$

or

$$x(s) = \left[\frac{3 \pm i\sqrt{3}}{2}\right] s \; .$$

CASE 3.4

We now present the results of investigations on the continued fraction approach for the abstract polynomial operator equation of degree k given by (3.6).

We shall investigate the iterative procedure given by

$$x_{n+1} = \left[M_1 + M_2 x_n + \cdots + M_k x_n^{k-1}\right]^{-1}(y) \, ,$$

$$y = -M_0, \ n = 0, 1, 2, \ldots \tag{3.25}$$

where $x_0 \in X$. Note that if $x_n \in X$, then $\lfloor M_1 + M_2 x_n + \cdots + M_k x_n^{k-1}\rfloor \in L(X)$, so that if $x_n \in X$ and $[M_1 + M_2 x_n + \cdots + M_k x^{k-1}]^{-1}$ exists, it follows that $x_{n+1} \in X$. We shall say that x_{n+1} is defined if $[M_1 + M_2 x_p + \cdots + M_k x_p^{k-1}]^{-1}$ exists for $p = 0, 1, 2, \ldots, n$. Before proceeding, let us justify the iterative procedure given by (3.25).

THEOREM 3.8

If (3.25) is defined and if there exists an $x \in X$ such that $x_n \to x$ as $n \to \infty$, then x is a solution of (3.6).

PROOF   Since $x \in X$, $[M_1 + M_2 x + \cdots + M_k x^{k-1}] \in L(X)$. Consider $[(M_1 + M_2 x + \cdots + M_k x^{k-1})x_{n+1} - y]$. For $n \geq 1$,

$$\left\|\left(M_1 + M_2 x + \cdots + M_k x^{k-1}\right) x_{n+1} - y\right\|$$

$$= \left\|\left(M_1 + M_2 x + \cdots + M_k x^{k-1}\right)\left(M_1 + M_2 x_n + \cdots + M_k x_n^{k-1}\right)^{-1} y - y\right\|$$

$$= \left\| \left[M_1 \left(M_1 + M_2 x_n + \cdots + M_k x_n^{k-1} \right)^{-1} \right. \right.$$

$$+ \cdots + M_k x^k \left(M_1 + M_2 x_n + \cdots + M_k x_n^{k-1} \right)^{-1} - I \left. \right] y \right\|$$

$$= \left\| \left[\left(M_2 x - M_2 x_n \right) + \left(M_3 x^2 - M_3 x_n^2 \right) \right. \right.$$

$$+ \cdots + \left(M_k x_n^{k-1} - M_k x_n^{k-1} \right) \left. \right] x_{n+1} \right\|$$

$$\leq \left(\left\| M_2 x - M_2 x_n \right\| + \left\| M_3 x^2 - M_3 x_n^2 \right\| + \cdots \right.$$

$$+ \left\| M_k x^{k-1} - M_k x_n^{k-1} \right\| \left. \right) \left\| x_{n+1} \right\| .$$

Note,

$$\left\| M_p x^{p-1} - M_p x_n^{p-1} \right\| = \left\| M_p (x - x_n) x^{p-2} + M_p x_n (x - x_n) x^{p-2} \right.$$

$$+ \cdots + M_p x_n^{p-2} (x - x_n) \left. \right\|, \quad p = 2, \ldots, k$$

$$\leq \left\| M_p \right\| \left[\left(\left\| x - x_n \right\| + \left\| x_n \right\| \right)^{p-2} \right.$$

$$+ \cdots + \left\| x_n \right\|^{p-2} \left. \right] \left\| x - x_n \right\| \to 0 \quad \text{as } n \to \infty$$

since $\left\| x - x_n \right\| \to 0$ as $n \to \infty$ and $\{x_n\}$ is a bounded sequence, so

$$\left\| \left(M_1 + M_2 x + \cdots + M_k x^{k-1} \right) x_{n+1} - y \right\| \to 0 \quad \text{as } n \to \infty,$$

that is, $(M_1 + M_2 x + \cdots + M_k x^{k-1}) x_{n+1} \to y$ as $n \to \infty$. Since $(M_1 + M_2 x + \cdots + M_k x^{k-1})$ is continuous, we have

$$\left(M_1 + M_2 x + \cdots + M_k x^{k-1} \right) x = y,$$

$$M_1 x + M_2 x^2 + \cdots + M_k x^k = y$$

or, $P(x) = 0$, so x is a solution of (3.6). ∎

DEFINITION 3.7   *Define the function g_k on $\mathbf{R}^+$ such that*

$$g_k(r) = C_k e^{(k-1)r} + C_{k-1} e^{(k-2)r} + \cdots + C_2 e^r$$

where, for $p = 2, \ldots, k$,

$$C_p = \left\| M_1^{-1} \right\| \, \| M_p \| \cdot \| W \|^{p-1} , \quad w = M_1^{-1} y .$$

Assume that $y \neq 0$, $\| M_s \| \neq 0$ for at least one $s \in N$ such that $2 \leq s \leq k$. Then $g_k(r) > 0$ for all $r \subset \mathbf{R}^+$.

THEOREM 3.9
If

(a)
$$M_1^{-1} \quad \text{exists} , \tag{3.26}$$

and if

(b)   *there exists $r \in \mathbf{R}^+$ satisfying*

$$g_k(r) < \left(e^r - 1 \right) e^{-r} , \quad g_k'(r) < 1 , \tag{3.27}$$

and
$$\left(e^r - 1 \right) e^{-r} = d$$

with $d = \| M_1^{-1} (M_2 x_0 + \cdots + M_k x_0^{k-1}) \|$, then

(i)   *iteration (3.25) is defined;*

(ii)   *there exists $x \in X$ such that $x_n \to x$ as $n \to \infty$, and*

$$\| x_n - x \| \leq \frac{q^n}{1 - q} \, \| x_1 - x_0 \| , \tag{3.28}$$

where
$$0 \leq q = g_k'(r) < 1 .$$

[Note that (3.25) is defined even if (3.27) is not a strict inequality.]

PROOF   (i) $(M_1 + M_2x_n + \cdots + M_kx_n^{k-1})^{-1}$, $n = 0, 1, 2, \ldots$ exists if $(I + M_1^{-1}M_2x_n + \cdots + M_1^{-1}M_kx_n^{k-1})^{-1}$, $n = 0, 1, 2, \ldots$ exists since

$$\left(M_1 + M_2x_n + \cdots + M_kx_n^{k-1}\right)^{-1} = \left(I + M_1^{-1}M_2x_n + \cdots + M_1^{-1}M_kx_n^{k-1}\right)^{-1} M_1^{-1}.$$

The proof follows by induction on the inequality

$$\left\| M_1^{-1}M_2x_n + \cdots + M_1^{-1}M_kx_n^{k-1} \right\| \le d < 1.$$

For $n = 0$, equality holds. Assume

$$\left\| M_1^{-1}M_2x_n + \cdots + M_1^{-1}M_kx_m^{k-1} \right\| \le d \quad \text{for } m = 0, 1, 2, \ldots, n.$$

Then it is enough to show

$$\left\| M_1^{-1}M_2x_{n+1} + \cdots + M_1^{-1}M_kx_{n+1}^{k+1} \right\|$$

$$\le \left\| M_1^{-1} \right\| \left(\|M_2\| \, \|x_{n+1}\| + \cdots + \|M_k\| \cdot \|x_{n+1}\|^{k-1} \right)$$

$$\le \left\| M_1^{-1} \right\| \left[\|M_2\| \cdot \|w\| \cdot \frac{1}{1-d} + \cdots + \|M_k\| \cdot \|w\|^{k-1} \left(\frac{1}{1-d}\right)^{k-1} \right]$$

$$\le d < 1$$

(since

$$\|x_{n+1}\| = \left\| \left(I + M_1^{-1}M_2x_n + \cdots + M_1^{-1}M_kx_n^{k-1}\right) M_1^{-1}y \right\|$$

$$\le \frac{\|w\|}{1 - \left\| M_1^{-1}M_2x_n + \cdots + M_1^{-1}M_kx_n^{k-1} \right\|} \le \frac{\|w\|}{1-d})$$

which is true by (3.27).

(ii) By (i), x_n, $n = 0, 1, 2, \ldots$ is defined. For $n \ge 1$ consider

$$\|x_{n+1} - x_n\| = \left\| \left(M_1 + M_2 + \cdots + M_kx_n^{k-1}\right)^{-1} y - (M_1 + M_2x_{n-1} + \cdots \right.$$

$$+ M_k x_{n-1}^{k-1} \Big)^{-1} (y) \bigg\|$$

$$= \bigg\| \Big(I + M_1^{-1} M_2 x_n + \cdots + M_1^{-1} M_k x_n^{k-1} \Big) M_1^{-1}$$

$$\Big[\Big(M_k x_{n-1}^{k-1} - M_k x_{n-1}^{k-1} \Big) + \cdots + (M_2 x_n - M_2 x_n) \Big] x_{n+1} \bigg\|$$

$$\leq \frac{1}{1-d} \bigg[(k-1) C_{k-1} \Big(\frac{1}{1-d} \Big)^{k-1} + \cdots + C_1 \Big(\frac{1}{1-d} \Big) \bigg]$$

$$= q \, \|x_n - x_{n-1}\|$$

(since

$$\Big\| M_p x_{n-1}^{k-1} x_n - M_p x_n^{k-1} x_n \Big\| \cdot \Big\| M_1^{-1} \Big\| \leq \Big\| M_1^{-1} \Big\| \, \|M_p\| \Big(\|x_n\| \, \|x_{n-1}\|^{p-2} \Big)$$

$$+ \cdots + \|x_n\| \, \|x_n\|^{p-2} \Big) \cdot \|x_n - x_{n-1}\|$$

$$\leq (p-1) C_{p-1} \Big(\frac{1}{1-d} \Big)^{p-1} \|x_n - x_{n-1}\| \Big).$$

By induction, we get

$$\|x_{n+1} - x_n\| \leq q^n \, \|x_1 - x_0\| \, ,$$

so, if $n \geq 0$, $p = 1, 2, \ldots,$

$$\|x_{n+p} - x_n\| \leq \sum_{j=n}^{n+p-1} \|x_{j+1} - x_n\|$$

$$\leq \sum_{j=n}^{n+p-1} q^j \, \|x_1 - x_0\| \, ,$$

and since $0 \leq q < 1$,

$$\|x_{n+p} - x_n\| \leq \sum_{j=n}^{\infty} q^j \|x_1 - x_0\| = \frac{q^n}{1-q} \|x_1 - x_0\| .$$

Hence $\{x_n\}$ is a Cauchy sequence and as such it converges to some $x \in X$, i.e., $x_n \to x$ as $n \to \infty$. Letting $p \to \infty$, we obtain (3.28). ∎

The choice $x_0 = M_1^{-1} y$ is a usual one, but not always the best one. If it is possible to choose x_0 close to the actual solution, much quicker convergence is obtained as indicated by (3.28).

Suppose M_1^{-1} does not exist, then we might still hope to apply Theorem 3.4 if we set $x = x_1 + x_2$, where x_2 is a logical approximation to a solution of (3.6), then $\tilde{M}_1 = M_1 + 2M_2 x_2$ (for the $n = 2$ case) is the new linear operator which may be invertible. Any solution then of (3.6) will be given by $x = x_1 + x_2$.

Example 3.3

Let X be the space of real numbers with $\|x\| = |x|$, and consider the quadratic equation

$$ax^2 + bx + c = 0 \tag{3.29}$$

where $a, b, c \in \mathbf{R}$. Then $g_2(r) = \left|\frac{ac}{b^2}\right| e^r$, so (3.27) is satisfied if

$$\left|\frac{ac}{b^2}\right| e^r < (e^r - 1) e^{-r}$$

or

$$\left|\frac{ac}{b^2}\right| (e^r)^2 - e^r + 1 < 0 ,$$

which is true if $b^2 - 4|ac| > 0$ and

$$\left(1 - \sqrt{1 - 4\left|\frac{ac}{b^2}\right|}\right) \left(\left|\frac{ac}{b^2}\right|\right)^{-1} < e^r < \left[1 + \sqrt{1 - 4\left|\frac{ac}{b^2}\right|}\right] \left(\left|\frac{ac}{b^2}\right|\right)^{-1} .$$

According to Theorem 3.9 if $b \neq 0$, $a \neq 0$, $c \neq 0$ and $b^2 - 4|ac| > 0$, a solution x of (3.9) as given by (3.5) is the continued fraction

$$x = \frac{(-c)}{b} + \frac{a(-c)}{b} + \frac{a(-c)}{b} + \cdots .$$

The above makes the continued fraction approach to the solution of (3.6) apparent.

3.2 Noncontractive Results

CASE 3.5

In the first part of this section we classify the polynomial equations in Banach space in three distinct kinds by use of the Fréchet derivative. For the two more general kinds, necessary and sufficient conditions will be given for their solution by means of formulas involving the nth root of linear operators.

We consider the polynomial equation

$$P_n(x) = 0 \,, \tag{3.30}$$

where

$$P_n(x) = M_n x^n + M_{n-1} x^{n-1} + \cdots + M_2 x^2 + M_1 x + M_0 \tag{3.31}$$

or

$$P_n(x) = P_n(x_0) + P_n'(x_0)(x - x_0) + \frac{1}{2} P_n''(x_0)(x - x_0)^2$$

$$+ \cdots + \frac{1}{n!} P_n^{(n)}\left(x_0 (x - x_0)^n\right) \tag{3.32}$$

for any $x_0 \in X$, where the M_ks are k-linear operators on X, $k = 1, 2, \ldots, n$, M_0 is fixed in X and $P_n^{(n)}(x_0)$ denotes the nth Fréchet derivative of P_n at $x_0 \in X$.

Obviously (3.30) is a natural generalization of the scalar polynomial equation to the more abstract setting of a linear space. This class of abstract polynomial equations includes a number of interesting differential and integral equations, which contain nonlinearities consisting of powers or products of the unknown functions, mingled with linear or integral operators.

Let us assume that the operator P_n is k times Fréchet-differentiable (see Chapter 1). Then we need the following definition:

DEFINITION 3.8   *The operator*

$$P_n^{(k)}(x) = n(n-1)\ldots(n-k-1) M_n x^{n-k}$$

$$+ (n-1)(n-2)\ldots(n-k-2) M_{n-1} x^{n-k-1} + \cdots + k! M_k$$

is called the kth derivative of the abstract polynomial operator P_n, $k = 1, 2, \ldots, n$. Note that $P_n^{(k)}(x) \in L(X^k, Y)$, for $k = 1, 2, \ldots, n$, and that $P_n''(x)$, $P_n'''(x), \ldots, P_n^{(x)}(x)$ are symmetric multilinear operators. The computation of $P_n(x)$ and its derivatives at point $x = x_0$ may be accomplished by adapting Horner's algorithm for scalar polynomials to this purpose. An algebraic formulation of this algorithm may be obtained by setting

$$M_i^{(0)} = M_i, \quad i = 0, 1, \ldots, n,$$

$$M_n^{(j)} = M_n, \quad j = 1, 2, \ldots, n + 1,$$

and calculating

$$M_{n-k}^{(j+1)} = M_{n-k+1}^{(j+1)} x_0 + M_{n-k}^{(j)}$$

$$j = 0, 1, \ldots, n - 1; \quad k = 1, 2, \ldots, n - j.$$

The results of this calculation are

$$M_j^{(j+1)} = \frac{1}{j!} P_n^{(j)}(x_0),$$

$j = 0, 1, \ldots, n$, the notation $P_n^{(0)}(x_0)$ being used for $P(x_0)$.

Note that *Taylor's identity*

$$P_n(x) = P_n(x_0) + P_n'(x_0)(x - x_0) + \frac{1}{2} P_n''(x_0)(x - x_0)^2$$

$$+ \cdots + \frac{1}{n!} P_n^{(n)}(x_0)(x - x_0)^n$$

holds at any $x_0 \in X$.
From now on we assume that P_n is differentiable n-times on X.

PROPOSITION 3.1
If x^* is a solution of the equation $P_n(x) = 0$, then the equation has a second solution $x \neq x^*$ if and only if the equation

$$P_n'(x^*) h + \frac{1}{2} P_n''(x^*) h^2 + \cdots + \frac{1}{n!} P_n^{(n)}(x^*) h^n = 0$$

has a nonzero solution h.

PROOF   Let $x = x^* + h$, $h \neq 0$. Then x is a solution if and only if

$$0 = P_n(x) = P_n\left(x^* + h\right)$$

$$= P_n\left(x^*\right) + P_n'\left(x^*\right) h + \frac{1}{2} P_n''\left(x^*\right) h^2 + \cdots + \frac{1}{n!} P_n^{(n)}\left(x^*\right) h^n$$

$$= P_n'\left(x^*\right) h + \frac{1}{2} P_n''\left(x^*\right) h^2 + \cdots + \frac{1}{n!} P_n^{(n)}\left(x^*\right) h^n$$

since $P_n(x^*) = 0$.   ∎

PROPOSITION 3.2
Assume:

(a)   *there exists $x^* \in X$ such that $P_n(x^*) = 0$ and $P_n'(x^*)$ invertible;*

(b)   *there exists $h \neq 0$ such that $P_n'(x^* + h)$ is not invertible and*

$$\left\| P_n'\left(x^*\right)^{-1} \right\| \left[2 \|M_2\| \cdot \|h\| + \cdots + n \|M_n\| \cdot \|h\| \left(\|x^*\|^{n-2} \right. \right.$$

$$\left. \left. + \|x^*\|^{n-3} \|h\| + \cdots + \|h\|^{n-2} \right) \right] \geq 1 .$$

Then x^* is unique in the ball

$$U(r) = \left\{ x \in X \mid \|x^* - x\| < r, r = \|h\| \right\} .$$

PROOF   We have by (a) and (b)

$$\left\| P_n'\left(x^* + h\right) - P_n'\left(x^*\right) \right\| \geq \frac{1}{\left\| P_n'\left(x^*\right)^{-1} \right\|}$$

or

$$\left\| \left(2M_2\left(x^* + h\right) - 2M_2\left(x^*\right)\right) + \cdots + \left(nM_n\left(x^* + h\right)^{n-1} - nM_n x^{*n-1}\right) \right\|$$

$$\geq \frac{1}{\left\| P_n'\left(x^*\right)^{-1} \right\|}$$

or

$$\left\| P_n' \left(x^* \right)^{-1} \right\| \left[2 \left\| M_2 \right\| \, \|h\| + \cdots + n \left\| M_n \right\| \cdot \|h\| \left(\left\| x^* \right\|^{n-2} \right. \right.$$

$$\left. \left. + \left\| x^* \right\|^{n-3} + \|h\| + \cdots + \|h\|^{n-2} \right) \right] \geq 1 . \quad \blacksquare$$

DEFINITION 3.9   *The equation $P_n(x) = 0$ is said to be of*

(a)   *First kind, if there exists $x_0 \in X$ such that*

$$P_n^{(k)}(x_0) = 0_k, \quad k = 1, 2, \ldots, n-1$$

   where 0_k is the 0 k-linear operator on X.

(b)   *Second kind, if $P_n'(x_0) \neq 0_1$ for all $x_0 \in X$ and there exists x_0 such that $P_n'(x_0)$ is invertible.*

(c)   *Third kind, if the equation is not of first or second kind.*

Example 3.4
(a) The polynomial equations of ordinary algebra are of *first kind*, e.g., let $x_0 = \frac{1}{4}$
and
$$P(x) = 16x^4 - 16x^3 + 6x^2 - x - \frac{263}{16} .$$
It is easy to verify that
$$P'(x_0) = P''(x_0) = P'''(x_0) = 0 .$$

(b) *Second kind:* Let $n = 2$ for simplicity and consider the quadratic equation

$$P(x) = M_2 x^2 + M_1 x + M_0, \quad x \in \mathbf{R}^2$$

where M_2 is defined by

$$\left(\begin{matrix} \frac{1}{2} C_1 & 0 \\ 0 & 0 \end{matrix} \middle/ \begin{matrix} \frac{1}{2} C_2 & 0 \\ 0 & 0 \end{matrix} \right), \quad C_1, C_2 \in \mathbf{R}$$

and M_1 is defined by the matrix

$$\left(\begin{matrix} s_1 & s_2 \\ s_3 & s_4 \end{matrix} \right), \quad \text{with } s_1 s_4 \neq s_2 s_3 .$$

Then P_2 is a second degree polynomial on $\mathbf{R}^2$ and

$$P_2'(x_0) = 2M_2x_0 + M_1, \quad x_0 = \begin{pmatrix} x_1 \\ x_2 \end{pmatrix}$$

$$= \begin{pmatrix} c_1x_1 & 0 \\ c_2x_1 & 0 \end{pmatrix} + \begin{pmatrix} s_1 & s_2 \\ s_3 & s_4 \end{pmatrix} = \begin{pmatrix} c_1x_1 + s_1 & s_2 \\ c_2x_1 + s_3 & s_4 \end{pmatrix}$$

is nonzero for any $x_0 \in \mathbf{R}^2$ and is invertible for some $x_0 \in \mathbf{R}^2$; $x_0 = \begin{pmatrix} 0 \\ 1 \end{pmatrix}$, say.

(c) *Third kind:* Again, let

$$c_1 = c_2 \neq 0, \quad s_1 = s_2 = s_3 = s_4 = s \neq 0$$

then $P_2'(x_0) \neq 0$ for any $x_0 \in \mathbf{R}^2$, but $P_2'(x_0)$ is not invertible for any $x_0 \in \mathbf{R}^2$.

DEFINITION 3.10 *If the equation $P_n(x) = 0$ is of first kind then it obviously reduces to*

$$M_n h^n = z \tag{3.33}$$

or

$$\frac{1}{n!} P_n^{(n)}(x_0) h^n = z$$

where $z = -P_n(x_0)$ and $P_n^{(k)}(x_0) = 0_k$, $k = 1, 2, \ldots, n-1$. Equation (3.33) is then called the normal form of (3.32).

If the equation $P_n(x) = 0$ is of second kind, then by composing through both sides by $P_n'(x_0)^{-1}$ we obtain

$$\tilde{P}_n(x) = 0$$

where

$$\tilde{P}_n(x) = \tilde{M}_0 + I(x) + \tilde{M}_2 x^2 + \cdots + \tilde{M}_n X^n$$

with $\tilde{M}_0 = P_n'(x_0)^{-1} P_n(x_0)$ and

$$\tilde{M}_k = P_k'(x_0) M_k, \quad k = 1, 2, \ldots, n.$$

Finally denote by rad (M_k) the sets satisfying

$$M_k(x+h)^k = M_k h^k, \quad \text{for all } h \in X, x \in \text{rad}(M_k), \ k = 1, 2, \ldots, n.$$

If $k = 1$, rad $(M_1) = \text{Ker}(M_1)$. Denote by $R = \bigcap_{k=1}^{n} \text{rad}(M_k)$ and note that $R \neq \emptyset$ since $0 \in R$.

THEOREM 3.10

If the equation $P_n(x) = 0$ is of first kind and $x = x^*$ is a solution, then $x = x^* + w$ is also a solution for any $w \in R$.

PROOF   Let $x_1 = x^* - x_0$, then

$$0 = P_n(x_0 + x_1)$$

$$= P_n(x_0) + P_n'(x_0)x_1 + \cdots + \frac{1}{n!}P_n^{(n)}(x_0)x_1^n$$

$$= P_n(x_0) + \frac{1}{n!}P_n^{(n)}(x_0)x_1^n$$

and

$$P_n(x_0 + x_1 + w) = P_n(x_0) + P_n'(x_0)(x_1 + w) + \cdots + \frac{1}{n!}P_n^{(n)}(x_0)(x_1 + w)^n$$

$$= P_n(x_0) + \frac{1}{n!}P_n^{(n)}(x_0)x_1^n$$

$$= 0$$

since $w \in R$, so $x^* + w$ is a solution of $P_n(x) = 0$.   ∎

THEOREM 3.11

Assume (3.32) is of second kind for some $x_0 \in X$, then

    (a)   *If n is even, then $x = x_0 + h$ is a solution of (3.32) if and only if $x = x_0 - h$ is a solution.*

    (b)   *If n is odd and $P_n(x_0) = 0$, then $x = x_0 + h$ is a solution of (3.32) if and only if $x = x_0 - h$ is a solution.*

PROOF   As before if $x = x_0 + h$ is a solution of (3.32), then

$$0 = P_n(x)$$

$$= P_n (x_0 + h)$$

$$= P_n (x_0) + \frac{1}{n!} P_n^{(n)} (x_0) h^n \, , \tag{3.34}$$

and

$$P_n (x_0 - h) = P_n (x_0) + \frac{1}{n!} P_n^{(n)} (x_0) (-h)^n \, . \tag{3.35}$$

If n is even, then $P_n^{(n)}(x_0)(-h)^n = P_n^{(n)}(x_0)h^n$ and then by (3.34) and (3.35), $P_n(x_0 - h) = 0$, i.e., $x = x_0 - h$ is a solution of $P_n(x) = 0$.

If n is odd $P_n^{(n)}(x_0)(-h)^n = -P_n^{(n)}(x_0)h^n$, using again (3.34), (3.35) and the fact that $P_n(x_0) = 0$, we obtain $P_n(x_0 - h) = 0$, i.e., $x = x_0 - h$ is a solution of $P_n(x) = 0$. ∎

THEOREM 3.12

If $P_n(x) = 0$ is of second kind and for any $u, v \in X$, there exists $x = x(u, v)$ such that

$$P_n'(u) + P_n'(v) = P_n'(x) \tag{3.36}$$

then

$$P_n'(u) + P_n'(v) \neq 0_1 \quad \text{for any } u, v \in X \, .$$

PROOF Since the equation $P_n(x) = 0$ is of second kind $P_n'(x) \neq 0$ for all $x \in X$; therefore,

$$P_n'(u) + P_n'(v) \neq 0 \quad \text{for all } u, v \in X \, . \quad ∎$$

Note that (3.36) is a strong hypothesis; however, it is sometimes true. For example, take $n = 2$ and $x = \frac{u+v}{2}$ in (3.32).

THEOREM 3.13

Let $P_n(x) = 0$ be of second kind and $x = x^*$ be a solution. Then $x = x^* + w$ cannot be a solution for any nonzero $w \in R$.

PROOF Since $P_n(x) = 0$ is of second kind, there exists $x_0 \in X$ such that $P_n'(x_0)$ is invertible. Set $h = x^* - x_0$; then $P_n(x_0 + h) = P_n(x^*) = 0$, so

$$P_n' (x_0)^{-1} P_n (x_0) + h + \frac{1}{2} P_n' (x_0)^{-1} P_n'' h^2 + \cdots + \frac{1}{n!} P_n' (x_0)^{-1} P_n^{(n)} (x_0) h^n = 0 \, .$$

Suppose $x = x^* + w$ were a solution. Then $P_n(x_0 + h + w) = 0$, i.e.,

$$0 = P_n'(x_0)^{-1} P_n(x_0) + h + w + \frac{1}{2} P_n'(x_0)^{-1} P_n''(x_0) h^2$$

$$+ \cdots + \frac{1}{n!} P_n'(x_0)^{-1} P_n^{(n)}(x_0) h^n$$

$$= P_n'(x_0)^{-1} P_n(x_0) + h + w + \frac{1}{2} P_n'(x_0)^{-1} P_n''(x_0) h^2$$

$$+ \cdots + \frac{1}{n!} P_n'(x_0)^{-1} P_n^{(n)}(x_0) h^n .$$

So $w = 0$, contrary to hypothesis. The theorem now follows. ∎

DEFINITION 3.11   *Let E, L be linear operators on X; then E is called the nth root of L if $E^n x = Lx$ for all $x \in X$. We also write*

$$E^n = L \quad or \quad E = L^{1/n} .$$

DEFINITION 3.12   *The set*

$$F(M_n) = \left\{ x \in X \mid M_n x^{n-1} = \left[M_n \left(M_n x^n \right)^{n-1} \right]^{1/n} \right\}$$

is called the factor set of M_n.

Example 3.5

Let $X = \mathbf{R}^2$, then $x \in F(M_2)$ if $(M_2(x))^2 = M_2(M_2 x^2)$, $x = \begin{pmatrix} x_1 \\ x_2 \end{pmatrix}$, i.e.,

$$M_2 x M_2 x y = M_2 \left(M_2 x^2 \right) y \quad \text{for all } y = \begin{pmatrix} y_1 \\ y_2 \end{pmatrix} \in \mathbf{R}^2 .$$

Therefore,

$$M_2 (x, M_2 x y) = M_2 \left(M_2 x^2, y \right) . \tag{3.37}$$

Let us choose the array

$$\begin{pmatrix} 1 & 0 & 0 & 0 \\ 0 & 0 & 0 & 1 \end{pmatrix},$$

then $M_2 x y = M_2 y x = \begin{pmatrix} x_1 y_1 \\ x_2 y_2 \end{pmatrix}$. Therefore, M_2 is a symmetric bilinear operator on $\mathbf{R}^2$.

Now (3.37) becomes

$$M_2 \left(\begin{pmatrix} x_1 \\ y_1 \end{pmatrix}, \begin{pmatrix} x_1 y_1 \\ x_2 y_2 \end{pmatrix} \right) = M_2 \left(\begin{pmatrix} x_1^2 \\ x_2^2 \end{pmatrix}, \begin{pmatrix} y_1 \\ y_2 \end{pmatrix} \right)$$

i.e.,

$$x_1^2 y_1 = x_1^2 y_1$$

$$x_2^2 y_2 = x_2^2 y_2 .$$

Therefore, $F(M_2) = \mathbf{R}^2$.

Note that the equation $M_2 x^2 = z$, where $x = \begin{pmatrix} z_1 \\ z_2 \end{pmatrix}$, $z_1, z_2 \in \mathbf{R}^+$ has a solution $u = \begin{pmatrix} \sqrt{z_1} \\ \sqrt{z_2} \end{pmatrix}$, since $M_2 u^2 = \begin{pmatrix} (\sqrt{z_1})^2 \\ (\sqrt{z_2})^2 \end{pmatrix} = z$.

THEOREM 3.14

Assume that the equation $P_n(x) = 0$ is of first kind of some $x_0 \in X$, then $x = u \in F(M_n)$ is a solution if and only if $(M_n z^{n-1})^{1/n}$ exists and satisfies $(M_n z^{n-1})^{1/n} = M_n u^{n-1}$ and $(M_n z^{n-1})^{1/n} u = z$, where $z = -P_n(x_0)$.

PROOF   Let $u \in F(M_n)$ be a solution of $P_n(x) = 0$. According to Definition 3.10, we have $z = M_n u^n$. Now,

$$M_n z^{n-1} = M_n \left(M_n u^n \right)^{n-1}$$

$$= \left(M_n u^{n-1} \right)^n ,$$

since $u \in F(M_n)$ so $(M_n z^{n-1})^{1/n}$ exists, $(M_n z^{n-1})^{1/n} = M_n u^{n-1}$ and $(M_n z^{n-1})^{1/n} u = M_n u^{n-1} u = M_n u^n = z$.

Conversely, if $(M_n z^{n-1})^{1/n}$ exists, $(M_n z^{n-1})^{1/n} = M_n u^{n-1}$ and

$$z = \left(M_n z^{n-1} \right)^{1/n} u$$

then

$$z = \left(M_n z^{n-1} \right)^{1/n} u = M_n u^{n-1} u = M_n u^n$$

so u is a solution. Finally, we have

$$M_n u^{n-1} = \left(M_n z^{n-1}\right)^{1/n} = \left(M_n \left(M_n u^n\right)^{n-1}\right)^{1/n} ,$$

therefore $u \in F(M_n)$ and the theorem is proved.   ∎

A similar theorem has been proved in [167] for (3.32), when $n = 2$.

CASE 3.6
Consider again (3.30) and

$$M_p = 0_p, \quad p = 2, 3, \ldots, n-1$$

where 0_p denote the 0 p-linear operator on X.
We will give a sufficient condition for the uniqueness of a solution of a polynomial equation. A new local uniqueness condition is also presented for $n = 2$ together with an example of a polynomial differential equation with a unique solution.

DEFINITION 3.13   *Define the linear operators L_n, $\tilde{L}_n$, $n = 1, 2, \ldots$ on X by*

$$L_n = L_n (x_1, x_2)$$

$$= \left(M_n x_1^{n-1} + M_n x_1^{n-2} x_2 + \cdots + M_n x_2^{n-1}\right) + \cdots + M_1 , \quad (3.38)$$

for all $n = 1, 2, \ldots$ and all $x_1, x_2 \in X$,

$$\tilde{L}_n = \tilde{L}_n (z, z_1, \ldots, z_{n-1})$$

$$= \begin{cases} \tilde{P}_1' & \text{if } n = 1 \\ \tilde{P}_2''(z) & \text{if } n = 2 \\ \tilde{P}_n^{(n)} z_1 z_2 \ldots z_{n-1} + M_1 & \text{if } n = 3, 4, \ldots \end{cases} \quad (3.39)$$

for $z, z_1, \ldots, z_{n-1} \in X$.
L_n can also be defined by

$$L_n = P_n' (x_0) + \frac{1}{2} \left[P_n'' (x_0) (x_1 - x_0) + P_n'' (x_0) (x_2 - x_0) \right]$$

$$+ \cdots + \frac{1}{n!} \left[P_n^{(n)} (x_0) (x_1 - x_0)^{n-1} + P_n^{(n)} (x_0) (x_1 - x_0)^{n-2} (x_2 - x_0) \right.$$

$$+ \cdots + P_n^{(n)}(x_0)(x_2 - x_0)^{n-1} \Big] ,$$

for $x_0, x_1, x_2 \in X$, $n = 1, 2, \ldots$.

Note that

$$P_n(x_1) - P_n(x_2) = L_n(x_1 - x_2), \quad n = 1, 2, \ldots . \tag{3.40}$$

If L_n is nonsingular for all $x_1, x_2 \in X$ with $x_1 \neq x_2$ and $P_n(x) = 0$ for some $x \in X$, then it is obvious by (3.40) that (3.30) has a unique solution x in X.

The condition L_n being nonsingular imposes severe restrictions on X, but nevertheless it may sometimes be true. The following theorem which is trivially true for $n = 1$ is a positive result in this direction.

THEOREM 3.15
Assume:

(a) *the linear operators* $\tilde{L}_n$, $n = 2, 3, \ldots$ *are nonsingular for all* z, z_1, $\ldots, z_{n-1} \in X$.

(b) *there exist* $C_1^1, C_1^2, \ldots, C_1^{n-1}, C_2^1, C_2^2, \ldots, C_2^{n-1} \in F$ *such that*

$$M_n \left(C_1^1 x_1 + C_2^1 x_2 \right) \left(C_1^2 x_1 + C_2^2 x_2 \right) \ldots \left(C_1^{n-1} x_1 + C_2^{n-1} x_2 \right) (x_1 - x_2)$$

$$= M_n \left(x_1^n - x_2^n \right), \quad n = 2, 3, \ldots \tag{3.41}$$

for all $x_1, x_2 \in X$ *and*

(c) *there exists* $x \in X$ *such that* $\tilde{P}_n(x) = 0$. *Then*

(i) $\tilde{P}_n(x_1) - \tilde{P}_n(x_2) = \tilde{L}_n(x_1 - x_2)$, $n = 1, 2, \ldots$ *for all* $x_1, x_2 \in X$

(ii) *the solution* x *of (3.30) is unique in* X.

PROOF   We have

$$\tilde{P}_n(x_1) - \tilde{P}_n(x_2) = M_n x_1^n + M_1 x_1 - \left(M_n x_2^n + M_n x_2 \right)$$

$$= M_n \left(x_1^n - x_2^n \right) + M_1 (x_1 - x_2)$$

$$= \tilde{L}_n (x_1 - x_2) \quad \text{by (3.41)}$$

with, $z_k = C_1^k x_1 + C_2^k x_2$, $k = 1, 2, \ldots, n-1$ and $z = z_1$. If x' is another solution of (3.30), then by (3.41)

$$\tilde{L}_n (x - x') = 0$$

and by (a) $x - x' = 0$ or $x = x'$. ∎

REMARK 3.1 Equation (3.41) constitutes a system of $2(n-1)$ unknowns with $n+1$ equations with respect to $C_1^1, C_1^2, \ldots, C_1^{n-1}, C_2^1, C_2^2, \ldots, C_2^{n-1}$. The number of unknowns $U(n)$ is greater than the number of equations $E(n)$ if $n > 3$. This implies that (3.41) may not hold for $n > 3$. However, for $n = 2$ or $n = 3$, (3.41) becomes, respectively,

$$\left. \begin{array}{l} 2C_1^1 - 1 = 0 \\ 2C_2^1 - 1 = 0 \\ C_1^1 = C_2^1 \, . \end{array} \right\} \tag{3.42}$$

$$\left. \begin{array}{l} C_1^1 C_1^2 = 1 \\ C_2^1 C_2^2 = 1 \\ C_1^1 C_2^2 + C_2^1 C_2^2 - C_1^1 C_1^2 = 0 \\ C_2^1 C_2^2 = C_1^1 C_2^2 - C_2^1 C_1^2 = 0 \end{array} \right\} \tag{3.43}$$

System (3.42) has the solution

$$C_1^1 = C_2^1 = \frac{1}{2} \tag{3.44}$$

and system (3.43) has an infinity of solutions given by the equations

$$\left(C_1^1\right)^2 + \left(C_2^1\right)^2 = C_1^1 C_2^1$$

$$C_1^1 \neq 0, \quad C_2^1 \neq 0$$

$$C_1^2 = \frac{1}{C_1^1}, \quad C_2^2 = \frac{1}{C_2^1} \quad \text{in } F = \mathbf{C} \, .$$

The solution (3.44) suggests that (3.30) has a unique solution x in *star-shaped regions.* To prove that, we first need the following definitions. ∎

DEFINITION 3.14 *Let w be fixed in X. A set S_w is said to be star-shaped with* *respect to* $w \in X$ *if*

$$\{z : z = w + t(y - z), 0 \le t \le 1, y \in S_w\} \subseteq S_w .$$

The set $n S_w$ *defined by*

$$n S_w = \{z : z = w + t(y - z), y \in S_w, 0 \le t \le n\}, \quad n = 1, 2, \ldots$$

obviously contains S_w *and is likewise star-shaped with respect to* w.

Note that special cases of star-shaped regions are the convex sets.

THEOREM 3.16
If $\tilde{L}_2$ *is nonsingular for all* $z \in S_w$, *then*

(i) $\tilde{P}_2(x_1) \ne \tilde{P}_2(x_2)$ *for all* $x_2 \in 2S_w$.

(ii) *If* $\tilde{P}_2(z) = 0$, $z \in S_w$, *then* z *is the unique solution of the equation*

$$\tilde{P}_2(x) = 0 \quad in \; S_w .$$

PROOF   If $x_2 \in 2S_w$, then $x = \frac{1}{2}(x_1 + x_2) \in S_2$ and (i), (ii) now follow from Theorem 3.10.   ∎

Note that a similar theorem can be stated if S_w is replaced by a convex set $C \subset X$ in Theorem 3.16.

Also note the fact that $\tilde{P}_2(x_1) = \tilde{P}_2(x_2)$ implies that $\tilde{L}_2$ is singular at $x = \frac{1}{2}(x_1 + x_2)$ and is analogous to Rolle's theorem for real scalar functions.

An illustration of this situation will be given.

Example 3.6
Consider the differential equation

$$\frac{d^2 x}{dt^2} + t + 1 = 0, \quad t \in [0, \infty) \tag{3.45}$$

$$x(0) = 0$$

$$x'(0) = 0 .$$

As X take $C_0''[0, \infty)$, the space of all continuously differentiable (twice) real functions $x = x(t)$, $0 \le t < \infty$, such that $x(0) = 0$, $x'(0) = 0$, and as Y take the space $C[0, \infty)$ of all continuous real functions $M_0 = M(t)$ on $0 \le t < \infty$. Equation (3.45) is a quadratic equation of the form (3.30) with $n = 2$, with

$$M_2 x^2 = \frac{d^2 x}{dt^2}, \quad M_1 x = t, \; M_0 = 1 .$$

The derivative

$$\tilde{P}_2'(x) = 2M_2 x + M_1 \ne 0 \quad \text{for all } x \in X .$$

It is easy to verify that

$$x = x(t) = -\frac{1}{2} \left(\frac{t^3}{3} + t^2 \right)$$

is the unique solution of (3.45) in X.

CASE 3.7
We will now end this section by providing a global theorem for Equations (1.20) and (3.30), respectively.

The proof of the following proposition is similar to previous ones and therefore is omitted.

PROPOSITION 3.3
Let A denote the set defined by

$$A = \{L \in L(X)/B(L(y)) = B(y)L, \text{ with } B, y \text{ as in } (1.20)\} .$$

Then A is a vector subspace of $L(X)$.

Note that the problem of finding a solution x of (1.20) can be reduced to finding a pair (x, L_1), $x \in X$, $L_1 \in L(X)$ satisfying the system

$$B(x) + L = L_1 \tag{3.46}$$

$$L_1(x) + y = 0 . \tag{3.47}$$

We can now prove the result:

THEOREM 3.17
Assume:

(a) *the square root of the linear operator $L^2 - B(y)$ denoted by S exists;*

(b) *the operator $L_1 \in L(X)$ given by $L_1 = \frac{L+S}{2}$ is invertible on X;*

(c) *$L_1^{-1} \in A$; and*

(d) *the linear operators L and S commute, that is, $LS = SL$.*

Then (1.20) has a solution x given by

$$x = -2(L + S)^{-1}(y) . \tag{3.48}$$

PROOF   By the previous remark it is enough to show that the pair (x, L_1) satisfies (3.46) and (3.47). If x is given by (3.48), then (3.47) is satisfied. Also,

$$L_1 = \frac{L + S}{2} \Rightarrow$$

$$L_1^2 - LL_1 + B(y) = 0 \quad \text{(by (a), (d))} \Rightarrow$$

$$-B(y)L_1^{-1} + L = L_1 \quad \text{(by (b))} \Rightarrow$$

$$B(-L_1^{-1}(y)) + L = L_1 \quad \text{(by (c))} \Rightarrow$$

$$B(x) + L = L_1 .$$

That is, (x, L_1) satisfies (3.46) and (3.47). Therefore, x is a solution of (1.20).
∎

REMARK 3.2   (a) If the operator L_1 is not invertible, we can still make use of the theorem. Let $P'(x_0)$, $P''(x_0)$ denote the first and second Fréchet derivative of P at x_0. There is no loss of generality to assume that B is symmetric in (1.20). We then obtain using Taylor's theorem

$$P(x_0 + h) = P(x_0) + P'(x_0)(h) + \frac{1}{2}P''(x_0)(h, h) \tag{3.49}$$

where

$$P'(x_0) = 2B(x_0) + L \quad \text{and} \quad P''(x_0) = 2B . \quad ∎$$

Equation (3.49) can now be written as

$$P(x_0 + h) = P(x_0) + P'(x_0)(h) + B(h, h).$$

The linear operator $P'(x_0)$ may now be invertible. We can then set $y = P(x_0)$ and $L = P'(x_0)$. If the rest of the hypotheses of the theorem are now satisfied, we obtain a solution x in the form

$$x = x_0 + h$$

where h is given by (3.48).

REMARK 3.3   (b) If the linear operator S does not exist, the transformation (3.49) may produce an operator which is the square root of the new "discriminant" operator. However, if $B(x_0)$, $L \in A$ and live in a commutative algebra the transformation does not work, since

$$(2B(x_0) + L)^2 - 4B(P(x_0)) = 4B(x_0)^2$$

$$+L^2 + 2B(x_0)L + 2LB(x_0) - 4B(y)$$

$$-4B(L(x_0)) - 4B(B(x_0 x_0)) = L^2 - 4B(y). \quad \blacksquare$$

The real quadratic equation may serve as an example for Theorem 3.17. However, a more interesting example when $X = \mathbf{R}^2$ is now provided.

Example 3.7
Let $X = \mathbf{R}^2$ with the ordinary bases $\{(1, 0), (0, 1)\}$.
    Consider now the quadratic equation in X given by (1.20) where

$$B \approx \begin{bmatrix} 1 & -1 \\ 2 & 3 \end{bmatrix} \Big/ \begin{bmatrix} 2 & 2 \\ -1 & 1 \end{bmatrix},$$

$$L = \begin{bmatrix} 0 & -2 \\ \frac{3}{2} & \frac{1}{2} \end{bmatrix} \quad \text{and} \quad y = \begin{bmatrix} \frac{1}{4} \\ -\frac{1}{4} \end{bmatrix}.$$

Equation (1.20) can be written as

$$x_1^2 + x_1 x_2 + 3x_1^2 - 2x_2 + \frac{1}{4} = 0$$

$$2x_1^2 + x_2^2 + \frac{3}{2}x_1 + \frac{1}{2}x_2 - \frac{1}{4} = 0.$$

We now have that

$$\sqrt{L^2 - 4B(y)} = \begin{bmatrix} 1 & 2 \\ -\frac{3}{2} & \frac{1}{2} \end{bmatrix} = S$$

where

$$B(y) = \begin{bmatrix} -\frac{1}{4} & -1 \\ \frac{3}{4} & 0 \end{bmatrix}.$$

Also, $L + S = \begin{bmatrix} 1 & 0 \\ 0 & 1 \end{bmatrix}$. Therefore,

$$L_1^{-1} = 2 \begin{bmatrix} 1 & 0 \\ 0 & 1 \end{bmatrix} \in A$$

and finally we obtain the solution

$$x = \begin{bmatrix} x_1 \\ x_2 \end{bmatrix} = -2 \begin{bmatrix} 1 & 0 \\ 0 & 1 \end{bmatrix} \begin{bmatrix} \frac{1}{4} \\ -\frac{1}{4} \end{bmatrix} = \begin{bmatrix} -\frac{1}{2} \\ \frac{1}{2} \end{bmatrix}.$$

Note that this is a solution that we can find and not the only solution of (1.20).
The problem of finding a solution x^* of (3.30) can be reduced to finding a pair
(x, L), $x \in X$, $L \in (X)$ satisfying the system

$$M_n x^{n-1} + M_{n-1} x^{n-2} + \cdots + M_2 = L \tag{3.50}$$

$$Lx + M_0 = 0 \tag{3.51}$$

The proof of the following theorem, as similar to the proof of Theorem 3.17 is
left as an exercise.

THEOREM 3.18
Assume:

(a) *there exists an operator $L \in L(X)$ such that L^{-1} satisfies*

$$M_k \left(\left(L^{-1} \right)^{k-1} (M_0) \right)^{k-1} = M_k (M_0)^{k-1} \left(L^{-1} \right)^{k-1}, \quad k = 2, 3, \ldots, n$$

and

(b)   *the operator L satisfies the equation*

$$L^n - L^{n-1}M_1 + (-1)^{n+1}L^{n-2}M_2\,(M_0)$$

$$+ \cdots + (-1)^2 L M_{n-1}\,(M_0)^{n-2} - M_n\,(M_0)^{n-1} = 0\,.$$

Then, Equation (3.30) has a solution x* given by

$$x^* = -L^{-1}\,(M_0)\ .$$

3.3   Solving Polynomial Operator Equations in Ordered Banach Spaces

We are interested in solving the polynomial operator equation:

$$x = L_0 + L_1(x) + \cdots + L_m(x, \ldots, x) \tag{3.52}$$

for $x \in X$, where X is a partially ordered Banach space, $L_0 \in X$ is fixed, and $L_k \colon X \times \cdots \times X \to X$ is a bounded k-linear operator for $k = 1, \ldots, m$. We wish to exploit the order properties of X where possible and find positive solutions of (3.52) along with uniqueness results. Define the polynomial operator $Q \colon X \to X$ by

$$Q(x) = L_0 + L_1(x) + \cdots + L_m(x, \ldots, x) \tag{3.53}$$

and note that we are looking for fixed points of Q. Equations of this form appear often in applications, such as elasticity theory [9] and systems theory [158]. We are seeking fixed points of Q using the scheme

$$x_{n+1} = Q\,(x_n)\ . \tag{3.54}$$

In this study, we examine the case where the more general operator Q in (3.53) is a *decreasing* operator, and utilize method (3.54). This will enable us to obtain some new convergence and uniqueness results.

We write $a < b$ to mean $a \leq b$ and $a \neq b$. By the definition of an ordered Banach space, the positive cone $\mathcal{P}$ of X is assumed to be closed in the norm topology. In particular, closed *order intervals*

$$[a, b] = \{x \colon a \leq x \leq b\} \text{ and } [a, \infty) = \{x \in X \colon a \leq x\}$$

are closed in the norm topology. A cone $\mathcal{P}$ is said to be *regular* if every nondecreasing sequence $\{x_n\}$ which is bounded above ($x_n \leq a \ \forall n$) converges in norm. It is known that for a regular cone $\mathcal{P}$, there is a constant $\mu > 0$ for which $\|x\| \leq \mu \|y\|$ whenever $x \leq y$ and $x, y \in \mathcal{P}$.

We are interested in solving Equation (3.52) when the operator Q is "compatible" with the order properties of the space X. With this in mind, we give the following standard definition.

DEFINITION 3.15   *An operator $A: X \to Y$ is decreasing if $x \leq y$ implies that $A(x) \geq A(y)$. An operator $M: X \times \cdots \times X \to Y$ is negative on set S if $M(x_1, \ldots, x_m) \leq 0$ whenever $\{x_1, \ldots, x_m\} \subset S \subset X$.*

When L_k is a k-linear operator, we write $\hat{L}_k(x)$ instead of $L_k(\underbrace{x, \ldots, x}_{k \ \text{times}})$ for $k > 1$. We next list some useful properties of the operator Q.

PROPOSITION 3.4
Suppose that each L_k operator is negative on the positive cone $\mathcal{P}$ of X. Then the operator Q defined by (3.53) is decreasing on $\mathcal{P}$, and the Fréchet derivative Q' at b satisfies

$$Q'_b(a - b) \geq Q(a) - Q(b) \geq 0 . \tag{3.55}$$

whenever $0 \leq a \leq b$.

PROOF   Letting $x_i \geq y_i \geq 0$ for all i, we have

$$L_k(x_1 - y_1, x_2, \ldots, x_k) \leq 0$$

since L_k is negative on $\mathcal{P}$. The multilinearity of L_k implies that

$$L_k(x_1, x_2, \ldots, x_k) \leq L_k(y_1, x_2, \ldots, x_k) .$$

Continuing in this fashion for $i = 2, \ldots, k$ yields

$$L_k(x_1, x_2, \ldots, x_k) \leq L_k(y_1, y_2, \ldots, y_k) . \tag{3.56}$$

Hence, each $\hat{L}_k$ is decreasing, so $Q(x) = L_0 + \hat{L}_1(x) + \cdots + \hat{L}_m(x)$ must be decreasing on $\mathcal{P}$.

We need the following algebraic identity for k-linear mappings.

$$L_k(x_1, x_2, \ldots, x_k) - L_k(y_1, y_2, \ldots, y_k) = L_k(x_1 - y_1, x_2, \ldots, x_k)$$

$$+ L_k (y_1, x_2 - y_2, \ldots, x_k) + \cdots + L_k (y_1, y_2, \ldots, x_k - y_k) \ . \quad (3.57)$$

Hence, for $0 \leq a \leq b$ and $\hat{L}_k$ decreasing on $\mathcal{P}$ we have

$$0 \leq \hat{L}_k(a) - \hat{L}_k(b) = L_k(a - b, a, \ldots, a) + \cdots + L_k(b, b, \ldots, a - b)$$

$$\leq L_k(a - b, b, \ldots, b) + \cdots + L_k(b, b, \ldots, a - b)$$

$$= \hat{L}'_k(b)(a - b) \ .$$

Now $Q(x) = L_0 + \hat{L}_1(x) + \cdots + \hat{L}_m(x)$,
so this proves (3.55). ∎

We also need to recall a basic property of bounded polynomial operators on a Banach space X. It is known that

$$\mathcal{P}_n(X) = \left\{ \sum_{k=0}^{n} L_k : L_k \text{ is bounded and } k\text{-linear for } k \geq 1 \text{ and } L_0 \in X \right\}$$

is a Banach space with norm $\| \sum L_k \| = \sup\{ \sum \|L_k\| \, \|x\|^k : \|x\| \leq 1 \}$.

THEOREM 3.19
Suppose that X is a partially ordered Banach space with regular positive cone $\mathcal{P}$. Let $I = [0, L_0]$ and assume $L_k: I^k \to X$ is k-linear, bounded and $L_k \leq 0$ on $\mathcal{P}$ for all $k \geq 1$. Suppose there exists an initial guess x_0 for procedure (3.54) for which

1. *$0 \leq x_{j+1} \leq x_{j+2} \leq x_j \leq L_0$ for some j, and*
2. *$\mu \| Q'_{x_{j+2}} \| < 1$.*

Then (3.52) has a solution $x^* \in [x_{j+1}, x_{j+2}]$ and the iterative method (3.54) converges to x^*. Moreover, this solution x^* is unique in the order intervals $[0, x_j]$ and $[x_{j+1}, \infty)$.

PROOF   To prove that the iterative method (3.54) converges, assume WLOG that $j = 0$ in part 1 of the theorem. We show that

$$x_{2k-2} \geq x_{2k} \geq x_{2k+1} \geq x_{2k-1} \quad (3.58)$$

by induction on k. Since $x_2 \geq x_1$ by hypothesis and Q is decreasing, we have

$$x_3 - x_2 = Q(x_2) - Q(x_1) \leq 0.$$

This proves the case $k = 1$.

Now assume that (3.58) holds for general k. We must show

$$x_{2k} \geq x_{2k+2} \geq x_{2k+3} \geq x_{2k+1}. \tag{3.59}$$

The first inequality in (3.59) follows from $x_{2k+1} \geq x_{2k-1}$ and

$$x_{2k} - x_{2k+2} = Q(x_{2k-1}) - Q(x_{2k+1}) \geq 0 \tag{3.60}$$

by (3.58). The last inequality in (3.59) is proved by observing that

$$x_{2k+3} - x_{2k+1} = Q(x_{2k+2}) - Q(x_{2k}) \geq 0$$

from (3.60). We note that

$$x_{2k+2} \geq x_{2k+1} \tag{3.61}$$

since

$$x_{2k+2} - x_{2k+1} = Q(x_{2k+1}) - Q(x_{2k}) \geq 0$$

by (3.58). Thus

$$x_{2k+2} - x_{2k+3} = Q(x_{2k+1}) - Q(x_{2k+2}) \geq 0$$

which yields

$$x_{2k+2} \geq x_{2k+3}.$$

This completes the induction proof of (3.58).

It follows from (3.58) that the sequence (x_{2k}) is bounded below by x_{2k-1} for all k and is monotonic, so $\mathcal{P}$ regular implies that (x_{2k}) converges to some x_R with $x_R \geq x_{2k-1}$ for all k. Similarly, the sequence (x_{2k-1}) is bounded above by x_{2k} for all k and is monotonic, so (x_{2k-1}) converges to some $x_L \leq x_{2k}$ $\forall k$. We note that $x_L \leq x_R$.

We now show that $x_L = x_R$. Using Proposition 3.4 we have

$$x_R - x_L = \lim_{k \to \infty} x_{2k} - x_{2k+1} = \lim_{k \to \infty} Q(x_{2k-1}) - Q(x_{2k})$$

$$\leq \lim_{k \to \infty} Q'_{x_2}(x_{2k-1} - x_{2k}) = Q'_{x_2}(x_L - x_R).$$

Thus
$$\|x_R - x_L\| \le \mu \left\| Q'_{x_2} \right\| \|x_R - x_L\| .$$

Since $\mu \|Q'_{x_2}\| < 1$, $\|x_R - x_L\|$ must be zero. We next put $x^* = x_L = x_R$ and verify that x^* is a solution to (3.52). By the continuity of Q

$$Q(x^*) - x^* = \lim_{k \to \infty} Q(x_{2k}) - x_{2k} = \lim_{k \to \infty} x_{2k+1} - x_{2k} = x_L - x_R = 0 .$$

Our last task is to demonstrate the uniqueness claims for x^*. Suppose that α is a solution in $[0, x^*]$ or in $[x^*, \infty)$. Then $\alpha - x^* = Q(\alpha) - Q(x^*)$. If $\alpha \in [x^*, \infty)$, then Q decreasing on $\mathcal{P}$ yields

$$0 \le \alpha - x^* = Q(\alpha) - Q(x^*) \le 0 ,$$

whence $\alpha = x^*$.

Similarly, if $\alpha \in [0, x^*]$, then

$$0 \ge \alpha - x^* = Q(\alpha) - Q(x^*) \ge 0 .$$

Thus $\alpha = x^*$ is the unique solution of (3.52) in the order intervals $[0, x^*]$ and $[x^*, \infty)$.

Now suppose there is a positive solution $\bar{x}$ for which $\bar{x} \le x_j$. Now $\bar{x} = Q^{(2n)}(\bar{x}) \le Q^{(2n)}(x_j) = x_{2n+j}$ for all n, so $\bar{x} \le x^*$. Since x^* is unique in $[0, x^*]$, $\bar{x} = x^*$. This shows that x^* is unique in the order interval $[0, x_j] \supset [0, x^*]$. A similar argument shows that x^* is unique in the order interval $[x_{j+1}, \infty) \supset [x^*, \infty)$.

∎

COROLLARY 3.1

Suppose that $Q(L_0) \ge 0$, $\mu \|Q'_{L_0}\| < 1$, and assume the hypotheses of Theorem 3.19 for Q. Then (3.52) has a solution x^*, and x^* is the unique positive solution.

PROOF   An inspection of the proof of Theorem 3.19 shows that conditions 1 and 2 can be replaced by $0 \le x_j \le x_{j+2} \le x_{j+1} \le L_0$, and $\mu \|Q'_{x_{j+1}}\| < 1$, respectively. Let $j = 0$ and choose $x_0 = 0$. Then $L_0 = x_1 = Q(0) > Q(L_0) = x_2 \ge 0 = x_0$. Hence, a positive solution x^* must exist by the theorem. Now suppose that z is also a positive solution. Then $z = Q^{(2n)}(z) \ge Q^{(2n)}(0) = x_{2n+1}$ for all n, so $z \ge x^*$. But x^* is unique in $[x^*, \infty)$ by the theorem, so we must have $z = x^*$.   ∎

REMARK 3.4   For many situations, the uniqueness results given above compare favorably with those obtained using series or contraction methods. These approaches yield uniqueness in balls of various norm radii depending on Q. For example, it has been shown in Chapter 1 that if $Q(x) = L_0 + \hat{L}_2(x)$, $\mu = 1$, and

$$4 \left\| \hat{L}_2 \right\| \, \left\| L_0 \right\| < 1 , \tag{3.62}$$

then the solution is unique in the ball about L_0 of radius

$$r = \left(1 - 2 \left\| \hat{L}_2 \right\| \, \left\| L_0 \right\| \right) \left(2 \left\| \hat{L}_2 \right\| \right)^{-1} . \quad \blacksquare$$

However, if (3.62) holds, then $\| Q'_{L_0} \| < 1$. If $Q(L_0) \geq 0$ as well, then we can apply Corollary 3.1 and get uniqueness on all of $\mathcal{P}$.

We now apply Theorem 3.19 to solve

$$x(t) = 1 + \lambda \int_t^1 x(s - t) x(s) ds , \tag{3.63}$$

which arises in statistical mechanics, and to solve

$$x(t) = 1 + \lambda x(t) \int_0^1 \frac{t}{s + t} x(s) ds , \tag{3.64}$$

which occurs in the study of radiative transfer [77]. We shall see that these equations have some interesting common characteristics. With $\lambda \leq 0$, we search for positive solutions $x \in X = L^1[0, 1]$, where $L^1[0, 1]$ is equipped with the usual norm and the standard ordering $f \geq g$ if $f(x) > g(x)$ a.e. on $[0, 1]$. We note that this makes the positive cone of X regular and $\mu = 1$. We also seek the range of negative λ values for which these equations have positive solutions.

For Equation (3.63), define $L_0 = 1 \in X$ and the bilinear operator $B: X \times X \to X$ by

$$B(f, g)(t) = \lambda \int_t^1 f(s - t) g(s) ds . \tag{3.65}$$

It can easily be seen that B is not compact in $L^1[0, 1]$ or in $C[0, 1]$, so standard arguments using the compactness of operators will not work here. It can also be shown that

$$\| B \| = |\lambda|/2 \quad \text{and} \quad \| B'_x \| = |\lambda| \, \| x \| \quad \text{for } x \geq 0 . \tag{3.66}$$

For $\lambda < 0$, B satisfies the assumptions of Theorem 3.19, so we need only find an appropriate initial guess x_0.

For $\lambda \in [-1, 0]$ and $x_0 = L_0 = 1$, some simple calculations show that $0 < x_1 < x_2 < x_0$ and $\|B'_{x_2}\| < 1$. This shows that (3.63) has a positive solution x_λ for any $\lambda \in [-1, 0]$. Corollary 3.1 and (3.66) show that x_λ is the unique positive solution for $\lambda \in (-1, 0]$. We note in passing that the condition (3.62) will only guarantee a unique solution for $-1/2 < \lambda < 1/2$.

We also note that if $f(t)$ is an increasing function, then $1 + \lambda B(f, f)(t)$ is also increasing for $\lambda \leq 0$. Hence, x_n will be increasing for all n if x_0 is increasing. This and the continuity of B force the solution x_λ to be an increasing function for each $\lambda \in [-1, 0]$.

More careful choices of x_0 lead to positive, increasing solutions for $\lambda < -1$. For example, some basic calculus (and involved calculations expedited by *Maple*) show that the Theorem 3.19 assumptions for x_0 ($j = 0$) are satisfied when

$$x_0(t) = -0.274t^3 + 0.199t^2 + .831t + .241 \text{ and } \lambda = -1.45,$$

leading to a positive, increasing solution x_λ for $\lambda = -1.45$.

For λ values in $[-1, 0]$ the situation is similar for Equation (3.64). Define a bilinear operator $B: X \times X \to X$ by

$$B(f, g)(t) = \lambda f(t) \int_0^1 \frac{t}{s + t} g(s) ds . \tag{3.67}$$

It can easily be seen that (3.66) holds for this operator. For $\lambda < 0$, B satisfies the assumptions of Theorem 3.19, so we need only find an appropriate x_0.

As with Equation (3.63), for $\lambda \in [-1, 0]$ and $x_0 = L_0 = 1$, some calculations show that $0 < x_1 < x_2 < x_0$ and $\|B'_{x_2}\| < 1$. This shows that (3.63) has a positive solution x_λ for any $\lambda \in [-1, 0]$. Corollary 3.1 and (3.66) show that x_λ is the unique positive solution for $\lambda \in (-1, 0]$.

It may be that we seek solutions to (3.52) in a Banach space Y, but useful information about the equation may be gathered by viewing it in a "larger" space X. This is the purpose of the following theorem, which will be applied to both Equations (3.63) and (3.64) in order to extend the range of negative λ values for which they have positive solutions.

THEOREM 3.20
Let X and Y be Banach spaces for which the injection map $Y \hookrightarrow X$ is continuous. Suppose that a polynomial operator Q satisfies $Q \in \mathcal{P}_n(X)$ and $Q \in \mathcal{P}_n(Y)$ for some $n \in \mathbb{N}$.

1.   Consider $f \in Y$ such that $I - Q'_f$ is $1 - 1$ viewed as a map of X into itself, and such that $I - Q'_f: Y \to Y$ can be written as $I - Q'_f = A + K$ where

> *A is a linear homeomorphism on Y and K is compact on Y. Then $I - Q'_f$
> is a linear homeomorphism on Y.*
>
> 2. *Under the hypotheses of 1 above, suppose further that f is a fixed point
> of Q in Y. Then there is a neighborhood V of Q in $\mathcal{P}_n(Y)$ for which each
> $P \in V$ has a unique fixed point p in a neighborhood N_P of f. Moreover,
> the map $T: V \to Y$ defined by $T(P) = p$ is continuous.*

PROOF   1. The injection map $Y \hookrightarrow X$ is continuous so $I - Q'_f$ is $1-1$ viewed
as a map of Y into itself. Writing $I - Q'_f = A(I + A^{-1}K)$, we see that $I + A^{-1}K$
is $1 - 1$. Now $A^{-1}K$ is compact, so by the Fredholm Alternative $I + A^{-1}K$ is
onto and therefore a linear homeomorphism on Y.

2. Since Q is a polynomic operator, P'_f will be close to Q'_f for P sufficiently
close to Q. The set of all linear homeomorphisms is open in $L(Y, Y)$, so $I - P'_f$
will be a linear homeomorphism for P sufficiently close to Q. Define the map
$Z: Y \to Y$ by $Z(y) = y - P(y)$ for each such P and note that Z'_f is a linear
homeomorphism. Now $Z(f) = (Q - P)(f)$ since f is a fixed point of Q. By
choosing P sufficiently close to Q in $\mathcal{P}_n(Y)$, we can make $\|Z(f)\|$ arbitrarily close
to zero. Then by the Inverse Function Theorem for Banach space [114], for P in
a sufficiently small neighborhood V of Q, the equation $Z(y) = 0$ has a unique
solution p in a neighborhood N_P of f. The Inverse Function Theorem ensures
that the map T is continuous. ∎

We now apply this result to our polynomial equations and extend the range of
negative λ values found in the last section.

THEOREM 3.21
*Equations (3.63) and (3.64) have unique continuous, positive solutions x_λ for each
$\lambda \in (-1.5, 0]$.*

PROOF   As noted earlier, Corollary 3.1 can be used to prove that unique positive
solutions for both equations exist in $L^1[0, 1]$ for $\lambda \in (-1, 0]$. For both of these
equations, it can be shown that $L^1[0, 1]$ solutions must be in $C[0, 1]$.

To extend the range of λ values we wish to apply Theorem 3.20. Let $Q(x) =
1 + \lambda B(x, x)$, $Y = C[0, 1]$ and $X = L^1[0, 1]$. Characteristics of both spaces will
come in handy. A useful common trait of positive solutions x_λ of Equations (3.63)
and (3.64) is that they all satisfy

$$\lambda G_\lambda^2 - 2G_\lambda + 2 = 0 \tag{3.68}$$

where $G_\lambda = \int_0^1 x_\lambda = \|x_\lambda\|_1$. Thus, $\|Q'_{x_\lambda}\|_1 = -\lambda G_\lambda$ from (3.66). Manipulating (3.67) then yields

$$\|Q'_{x_\lambda}\|_1 < 1 \tag{3.69}$$

for all $\lambda > -1.5$. Note that we are forced to settle for a smaller range of λ values if we work in $C[0, 1]$ only. We shall prove in Lemma 3.2 below that $I - Q'_{x_\lambda} = A + K$ as in part 1 of Theorem 3.20, so $I - Q'_{x_\lambda}$ is a homeomorphism for $\lambda > -1.5$. Then, by part 2 of Theorem 3.20, for α values sufficiently near λ, both equations have solutions x_α in $Y = C[0, 1]$. However, it is not clear from this information that said solutions are positive. To see this, first note that positive solutions to Equations (3.63) and (3.64) clearly satisfy $x_\alpha \leq 1$. Then consider the following calculations for $-1.5 < \alpha < 0$ and solutions $0 \leq x_\alpha \leq 1$. For Equation (3.63)

$$x_\alpha(t) = 1 + \alpha \int_t^1 x_\alpha(s - t)x_\alpha(s)ds \geq 1 + \alpha \int_0^1 x_\alpha(s)ds$$

$$= 1 + \alpha G_\alpha > 0 \tag{3.70}$$

for all $t \in [0, 1]$. For Equation (3.64) we have

$$x_\alpha(t) \geq 1 + \alpha \int_0^1 \frac{t}{s + t} x_\alpha(s)ds \geq 1 + \alpha \int_0^1 x_\alpha(s)ds$$

$$= 1 + \alpha G_\alpha > 0 \tag{3.71}$$

for all $t \in [0, 1]$.

These calculations (3.70) and (3.71) show that solutions x_α are in fact *strictly* positive. The map T in Theorem 3.20 is continuous, so solutions corresponding to αs sufficiently near λ must also be strictly positive. We can therefore conclude that $I = \{\lambda: x_\lambda \text{ is a strictly positive solution and } \lambda > -1.5\}$ is open in $\mathbf{R}$ for both (3.63) and (3.64).

Next we show that $I = (-1.5, 0)$. Let $L = \inf\{x: (x, 0) \subset I\}$. We have already shown that $L \leq -1$. Suppose for the sake of contradiction that $L > -1.5$. Let (λ_n) be a sequence converging to L from above, and let (x_n) be the sequence of corresponding solutions: $x_n = 1 + \lambda_n B(x_n, x_n)$. From the bilinearity of B and $x_n \leq 1$ we have

$$\|x_n - x_m\| \leq \left\|\lambda_n B_{\frac{x_n + x_m}{2}}(x_n - x_m)\right\| + |\lambda_n - \lambda_m| \|B(x_m, x_m)\|$$

$$\leq 0.5 \left(\|\lambda_n B'_{x_n}\| + \|\lambda_m B'_{x_m}\|\right) \|x_n - x_m\|$$

$$+ |\lambda_n - \lambda_m| \|B(1, 1)\| . \tag{3.72}$$

From (3.69) and $L > -1.5$ there is some $\delta < 1$ for which $\|\lambda_n B'_{x_n}\| \leq \delta$ for all n, so rearranging (3.72) yields

$$\|x_n - x_m\| \leq \frac{|\lambda_n - \lambda_m|}{1 - \delta} \|B(1, 1)\| \ .$$

Since (λ_n) converges to L, (x_n) converges to some x_L. The operator B is continuous, so $x_L = 1 + \lambda B(x_L, x_L)$. Hence, there is a solution x_L corresponding to L. All solutions for $\lambda \geq L$ are strictly positive, so by the continuity of the map T in Theorem 3.20 this solution x_L is nonnegative. But calculations (3.70) and (3.71) show x_L must be strictly positive, so $L \in I$. Since I is open, this contradicts the definition of L. Hence, $L = -1.5$.

To prove the uniqueness assertion, for the sake of contradiction let x and z be distinct positive solutions for either (3.63) or (3.64), and some $\lambda \in (-1.5, 0]$. The bilinearity of B yields

$$x - z = \lambda B'_{\frac{x+z}{2}} (x - z) = (1/2) \left(\lambda B'_x + \lambda B'_z\right) (x - z) \ .$$

This, (3.69), and (3.66) force $\|x - z\| \leq (1/2)\|\lambda B'_x + \lambda B'_z\| \, \|x - z\| < \|x - z\|$, a contradiction. ∎

The following lemma is needed to complete the proof above.

LEMMA 3.2
For Equations (3.63) and (3.64), and positive solutions x_λ, $I - Q'_{x_\lambda}$ can be written in the form $A + K$ where A is a linear homeomorphism on $C[0, 1]$ and K is compact on $C[0, 1]$.

PROOF   For (3.63), $Q'_{x_\lambda}(f)(t) = \lambda \int_t^1 x_\lambda(s-t) f(s)ds + \lambda \int_t^1 f(s-t)x_\lambda(s)ds$, which is the sum of two Volterra operators and is thus compact. Letting A be the identity map, we have $I - Q'_{x_\lambda}$ in the desired form.

We must work a bit harder for Equation (3.64). In this case, we have

$$Q'_{x_\lambda}(f)(t) = \lambda f(t) \int_0^1 \frac{t}{s+t} x_\lambda(s)ds + \lambda x_\lambda(t) \int_0^1 \frac{t}{s+t} f(s)ds \ ,$$

where the second term is a Fredholm operator we shall use for compact operator K. We need only show that

$$Af(t) := f(t) - \lambda f(t) \int_0^1 \frac{t}{s+t} x_\lambda(s)ds$$

is a homeomorphism on $C[0, 1]$. For this, observe that $Af = 0$ iff $f = 0$ whenever $\lambda < 0$ since $x_\lambda \geq 0$. Hence, A is $1 - 1$. It is easy to see that A is onto, for $Af = g$ iff

$$f(t) = \frac{g(t)}{1 - \lambda \int_0^1 \frac{t}{s+t} x_\lambda(s) ds} .$$

Now A is clearly continuous, so A^{-1} is continuous by the Open Mapping Theorem.
∎

Exercises

1. Give a power series solution for

$$x = y + \lambda T x x x ,$$

where T is a trilinear operator in a Banach space X.

2. Use Newton's method to calculate a zero of

$$p(z) = z^4 - 3z^3 + 20z^2 + 44z + 54 ,$$

located near $z_0 = 2.5 + 4.5i$.

3. For the polynomial

$$p(x) = a_0 + a_1 x + \cdots + a_n x^n, \quad a_n \neq 0$$

define $c = (|a_0| + |a_1| + \cdots + |a_{n-1}|)/|a_n|$. Show that every root x of $p(x) = 0$ satisfies
$$|x| \leq \max \left\{ c, \sqrt[n]{c} \right\} .$$

4. Show that the abstract polynomial operator

$$P(x) = A_n x^n + \cdots + A_2 x^2 + A_1 x + A_0 ,$$

where $A_i, i = 1, 2, \ldots, n$ are i-linear and symmetric operators on a Banach space X has the derivative

$$P'(x_0) = n A n x_0^{n-1} + \cdots + 2 A_2 x_0 + A_1 .$$

5. Prove Theorem 3.2.

6. Verify the claims made in Definition 3.4.

7. Prove Theorem 3.4.

8. Verify the computations in Example 3.1.

9. Prove Theorem 3.5.

10. Verify the claim made in Definition 3.4.

11. Verify the computations in Example 3.3.

12. Prove Proposition 3.3.

13. Prove Theorem 3.18.

14. Let E denote a real or complex Banach space. Also let A_1, A_2 be continuous homogeneous polynomial operators on E of degrees a_1, a_2, respectively, with $1 \leq a_1 < a_2$. Assume, any continuous homogeneous polynomial operator A of degree $a > 1$ on E is assigned two norms, namely its uniform norm $\|A\| = \sup\{\|A(x)\|/\|x\|^a, x \neq 0\}$ and its polar norm $|A| = \|A^v\|$, i.e., the uniform norm

$$\|A^v\| = \sup\left\{\|A^v(x_1, \ldots, x_a)\| / \|x_1\| \cdots \|x_a\| :\right.$$

$$\left. x_1 \neq 0, \ldots, x_a \neq 0 \right\} .$$

of the polar A^v of A (the unique continuous symmetric multilinear operator $A^v \colon E^a \to E$ such that $A^v(x, \ldots, x) = A(x)$ for all $x \in E$). We have seen that

$$\|A\| \leq |A| \leq b_a \|A\| ,$$

where we can take $b_a = \frac{a^a}{a!}$ in every case, and $= 1$ if E is a real Hilbert space.

(i) Show that the polynomial operator in E

$$x = y + \lambda_1 A_1(x) + \lambda_2 A_2(x) \tag{3.73}$$

has a unique small (i.e., tending to 0 with y) solution $x = B(y)$ whenever

$$\|y\| \leq Y(\lambda_1 |A_1|, \lambda_2 |A_2|) = \inf\{|f(z)|/|z| = \gamma\} \tag{3.74}$$

where $f(z) = z - (\lambda_1|A_1|^{a_1} + \lambda_2|A_2|^{a_2})$ $(z \in \mathbb{C})$ and γ is the least number $|z_0|$ such that $z_0 \neq 0$ and $f'(z_0) = 0$. In the case in which $\lambda_2 A_2 = 0$, and

$a = a_1 \geq 2$, $A = A_1$, $\lambda = \lambda_1$ show that the condition (3.74) becomes

$$|\lambda| \cdot |A| \cdot \|y\|^{a-1} < (a-1)^{a-1}/a^a .$$

(ii) Show that

$$B(y) = \sum_{(k_1,k_2) \in N^2} \lambda_1^{k_1} \lambda_2^{k_2} B^{(k_1,k_2)}(y)$$

where the operators $B^{(k_1,k_2)}$ are given by the (a_1, a_2)th-order algorithm:

$$B^{(0,0)} = I, \quad B^{(1,0)} = A_1, \quad B^{(0,1)} = A_2$$

and for all other indices $(k_1, k_2) \in N^2$,

$$B^{(k_1,k_2)} = \sum_{[k_1-1,k_2]} A_1 \left(B^{(k_{11},k_{21})} \cdots B^{(k_{1a_1},k_{2a_1})} \right)$$

$$+ \sum_{[k_1,k_2-1]} A_2 \left(B^{(k_{11},k_{21})} \cdots B^{(k_{1a_1},k_{2a_2})} \right) . \quad (3.75)$$

The summation convention here is this: the first indicated summation is vacuous if $k_1 = 0$ but otherwise is over all a_1-lists (k_{1i}, k_{2i}), $1 \leq i \leq a_1$ in $(N^2)^{a_1}$ satisfying

$$\sum_{i=1}^{a_1} (k_{1i}, \ell_{2i}) = (k_1 - 1, k_2)$$

(coordinatewise addition); the second is defined analogously. If A, B_1, ..., B_a are continuous homogeneous polynomial operators on E of degrees $a, b_1, \ldots, b_a$, then the notation $A\langle B_1 \cdots B_a \rangle$ is used to denote the "polar-composition" of the operators in question, that is to say, the continuous homogeneous polynomial operator T such that

$$T(x) = A^v (B_1(x), \ldots, B_a(x)) \quad (x \in E) ;$$

show that each $B^{(k_1,k_2)}$ is a continuous homogeneous polynomial operator on E of degree $k_1(a_1 - 1) + k_2(a_2 - 1) + 1$, and such that

$$\left\| B^{(k_1,k_2)}(y) \right\| \leq C_{k_1,k_2} |A_1|^{k_1} |A_2|^{k_2} \|y\|^{k_1(a_1-1)+k_2(a_2-1)+1} \quad (y \in E) ,$$

where the "Lagrangean" ("generalized hypergeometric") coefficients C_{k_1,k_2} are defined by this complex analogue of the above algorithms: $C_{0,0} = 1$, $C_{1,0} = a_1$, $C_{0,1} = a_2$,

$$C_{k_1,k_2} = \sum_{[k_1-1,k_2]} \prod_{i=1}^{a_1} C_{k_{1i},k_{2i}} + \sum_{[k_1,k_2-1]} \prod_{j=1}^{a_2} C_{k_{1j},k_{2j}} .$$

Then show that series (3.75) is absolutely convergent whenever the majorizing real series

$$\sum_{(k_1,k_2)\in N^2} C_{k_1,k_2} (|\lambda_1| \, |A_1|)^{k_1} (|\lambda_2| \, |A_2|)^{k_2} \|y\|^{k_1(a_1-1)+k_2(a_2-1)+1}$$

is convergent. As an example, consider the complex equation

$$y = f(z) = z - \left(\lambda_1 |A_1| z^{a_1} + \lambda_2 |A_2| z^{a_2}\right) .$$

Then it is well known that the above has a unique small solution given by

$$z = 1 + \sum_{n=1}^{\infty} c_n y^n$$

where each $c_n = 0$ in $n \neq k_1(a_1 - 1) + k_2(a_2 - 1)$ for some $(k_1, k_2) \in N^2$, and otherwise,

$$c_n = c_{k_1,k_2} (\lambda_1 |A_1|)^{k_1} (\lambda_2 |A_2|)^{k_2}$$

with

$$c_{k_1,k_2} = \frac{(k_1 a_1 + k_2 a_2)!}{[k_1(a_1 - 1) + k_2(a_2 - 1)]! k_1! k_2!} .$$

Then show that the Lagrange series solution is then absolutely convergent whenever

$$|y| < Y(\lambda_1 |A_1|, \lambda_2 |A_2|) .$$

15. Consider the polynomial equation

$$P(x) = 0 \quad \text{with} \quad P(x) = M_n x^n + \cdots + M_2 x^2 + h + M_0$$

in a Banach space X, where each M_k is a bounded k-linear operator on X, $k = 0, 1, 2, \ldots, n$.
(i) Set

$$x = F(x) \tag{3.76}$$

where
$$F(x) = -\left(M_n x^n + \cdots + M_2 x^2 + M_0\right)$$

and define the iteration
$$x_0 = 0, \quad x_{n+1} = F(x_n) \quad (n \geq 0)$$

Assume F is contractive, i.e., there exists $c \in [0, 1)$ such that
$$\|F(x) - F(y)\| \leq c\|x - y\| \quad \text{for all } x, y \in \bar{U}(r) ,$$

and set $n_0 = \|x_1 - x_0\| = \|M_0\|$. Show: If F is contractive on $\bar{U}(r)$ and
$$r \geq \frac{n_0}{1 - c} , \tag{3.77}$$

then the sequence $\{x_n\}$, $n \geq 0$ converges to a fixed point x^* of F which is unique in $\bar{U}(r)$ with
$$\|x^* - x_n\| \leq \frac{c^n}{1 - c} n_0 \quad (n \geq 0) .$$

(ii) Set $c_i = \|M_i\|$, $i = 1, 2, \ldots, n$ and define the scalar majorant polynomial
$$f(r) = c_n r^n + \cdots + c_2 r^2 + n_0 .$$

Show that
$$\|F'(x)\| \leq f'(r)$$

and
$$\|F(x) - F(y)\| \leq f'(r)\|x - y\| \quad \text{for all } x, y \in \bar{U}(r) .$$

Let $r = R > 0$ be such that
$$f'(R) = 1 .$$

Then show that for $0 \leq r < r$, the function
$$g(r) = \frac{n_0}{1 - f'(r)}$$

is positive (assuming $n_0 > 0$), strictly convex, monotonically increasing, and tends to $+\infty$ as $r \to R$. Also show that the equation
$$r = \frac{n_0}{1 - f'(r)}$$

has two or no positive solutions. If r_1, r_2 denote the positive solutions with $r_1 \leq r_2$ show that the condition (3.77) is satisfied for

$$r_1 \leq r \leq r_2 .$$

(iii) Show that if positive solutions $r_1 \leq r_2$ exist, then a solution x^* of Equation (3.76) exists and is unique in $\bar{U}(r_2)$. Furthermore show that the sequence $\{x_n\}$ ($n \geq 0$) converges to x^* with

$$\|x^* - x_n\| \leq \frac{(f'(r_1))^n \eta_0}{1 - f'(r_1)} .$$

(iv) Let us consider Newton's method with $x_0 = 0$ for solving Equation (3.76). Then show that if

$$K \eta_0 \leq \frac{1}{2}, \quad \|F''(x)\| \leq K \quad \text{for all } x \in \bar{U}(r)$$

and

$$r \geq r_3 = \frac{1 - \sqrt{1 - 2K \eta_0}}{K} ,$$

then a solution x^* of (3.76) exists in $\bar{U}(r_3)$, and the Newton sequence $\{x_n\}$ ($n \geq 0$) exists and converges to x^*, with

$$\|x^* - x_n\| \leq \frac{(2K \eta_0)^{2^n - 1}}{2^{n-1}} \eta_0 .$$

If

$$r \geq r_4 = \frac{1 + \sqrt{1 - 2K \eta_0}}{K}$$

then show x^* is unique in $U(r_4)$ if $K \eta_0 < \frac{1}{2}$, or is in $\bar{U}(\frac{1}{K}) = \bar{U}(2\eta_0)$ if $K \eta_0 = \frac{1}{2}$.

(v) Show that if positive solutions $r_3 \leq r_4$ of equation

$$f''(r)r^2 - 2r + 2\eta_0 = 0$$

exist, then Equation (3.76) has a solution $x^* \in \bar{U}(r_3)$, to which the Newton sequence $\{x_n\}$ ($n \geq 0$) converges, with

$$\|x_n - x^*\| \leq \frac{[2\eta_0 f''(r_1)]^{2^n - 1}}{2^{n-1}} \eta_0 \quad (n \geq 0) .$$

If $r_4 > 2\eta_0$, then show x^* is unique in $U(r_4)$. If $r_4 = 2\eta_0$, then show x^* is unique in $\bar{U}(2\eta_0)$.

(vi) Derive results similar to (d) and (e) above for the modified Newton method.

16. Consider the famous Riccati equation

$$\frac{dy}{dx} + Q(x)y + R(x)y^2 = S(x)$$

$$y(0) = c \ .$$

If the coefficient functions Q, R, S are assumed to be continuous, then

$$P(x) = \frac{dy}{dx} + Q(x)y + R(x)y^2 - S(x)$$

may be regarded as an operator from the space $C^1[0, R]$ of continuously differentiable functions on $0 \le x \le R$ into the space $C[0, R]$ of continuous real functions on the same interval. In order for us to apply the results of the previous exercise we must bring the equation

$$P(x) = 0$$

in the form (3.76) of Exercise 15.

(i) Choose $y_0 = y_0(x)$ in $C^1[0, R]$ such that $y_0 = c$, and set $y(x) = y_0(x) + z(x)$. Then show that we can get

$$R(x)z^2 + \frac{dz}{dt} + [Q(x) + 2R(x)y_0(x)]z$$

$$+ \left(\frac{dy_0}{dx} + Q(x)y_0 + R(x)y_0^2 - S(x) \right) = 0 \ ,$$

$$z(0) = 0 \ . \tag{3.78}$$

(ii) Show that the linear operator

$$P'(y_0) = \frac{d}{dx} + [a(x) + 2R(x)y_0(x)]\,I$$

has as its inverse the linear integral transform with kernel

$$K(x, t) = \exp\{\mu(t) - \mu(x)\} \ ,$$

where

$$\mu(s) = \int_0^s [Q(u) - 2R(u)y_0(u)]\,du \ .$$

Hence, show that Equation (3.78) is equivalent to the nonlinear Volterra integral equation

$$0 = \int_0^x K(x,t)z^2(t)dt + z(x) + g(x) \ , \quad 0 \le x \le R \qquad (3.79)$$

where

$$g(x) = \int_0^x K(x,t)\left[\frac{dy_0(t)}{dt} + Q(t)y_0(t) + R(t)y_0^2(t) - S(t)\right]dt \ .$$

Equation (3.79) may be considered to be in the space $C_0^1[0, R]$ of continuously differentiable functions on $0 \le x \le R$ which vanish at $x = 0$. Choose as a norm

$$\|f\| = \max_{[0,R]} \left\{|f(x)| + |f'(x)|\right\} \ .$$

Show that the scalar equation defined in (v) of Exercise 15 will now have two positive solutions $r_3 \le r_4$ provided that R is chosen sufficiently small.

17. Consider the multilinear equation

$$x = y + M_k x^k \qquad (3.80)$$

where M_k is a bounded k-linear operator on X and $y \in X$ is fixed. Show:

(i) If $k(k-1)2^{k-1}\|y\|^{k-1}\|M_k\| \le 1$, then Equation (3.80) has a unique solution x^* in a certain ball centered at 0.

(ii) Assume:

$$k\left(\frac{k}{k-1}\right)^{k-1}\|y\|^{k-1}\|M_k\| < 1,$$

and

there exists $r > 0$ such that

$$\frac{k\|y\|}{k-1} \le r < \sqrt[k-1]{\frac{1}{k\,\|M_k\|}} \quad k > 1$$

Then show that Equation (3.80) has a unique solution x^* in $\bar{U}(r)$.

(iii) For $k \geq 3$, show that

$$\left(\frac{k}{k-1}\right)^{k-1} < (k-1)2^{k-1} .$$

18. Consider the iteration

$$x_{n+1} = \left[M_k \, (x_n)^{k-1}\right]^{-1} (x_n - y) \quad (n \geq 0) \, (k \geq 2)$$

to approximate a solution x^* of Equation (3.80) of Exercise 17 such that $\|x^*\| \geq p$, where

$$P = \frac{1}{2} \sqrt[k-1]{\frac{\|M_k\|}{k(k-1)}} .$$

Use the contraction mapping principle. Compare this result with the one obtained if you use the iteration

$$y_{n+1} = y + M_k y_n^k \quad (n \geq 0) .$$

In particular show that the solution y^* of Equation (3.80) satisfies

$$\|y^*\| \leq p$$

provided that

$$\|y\| \leq \frac{1}{2} p .$$

(Also see Section 2.2.)

19. Show that the operator B defined in (3.65) is not compact in $L^1[0, 1]$ or in $C[0, 1]$.

20. Show estimates (3.66).

21. Show that estimates (3.66) hold for the operator B given by (3.67).

Chapter 4

Integral and Differential Equations

In this chapter we have included several results involving equations of Hammerstein-type, equations appearing in neutron transport, and differential equations of polynomial type. The polynomial nature is what all applied equations have in common in this chapter. We apply the general techniques developed in Chapters 1, 2, and 3, or techniques not mentioned earlier to solve these equations [22, 23, 33, 60, 76, 77, 115, 116, 127, 143, 184, 190].

4.1 Equations of Hammerstein Type

Consider the nonlinear equation

$$x = y + B(A(x), A(x)) \tag{4.1}$$

in a Banach space X, where $y \in X$ is fixed, B is a bounded bilinear operator on X, and A is a nonlinear operator on X such that there exist two real constants c_1 and c_2 with $c_2 \geq c_1$ such that

$$\|A(w) - A(v) - \lambda(w - v)\| \leq d(\lambda)\|w - v\| \tag{4.2}$$

where

$$d(\lambda) = \max[(\lambda - c_1), (c_2 - \lambda)] \tag{4.3}$$

for any real constant λ and all $w, v \in X$.

Equation (4.1) is sometimes referred to as an equation of Hammerstein type [190]. It appears in the study of nonlinear physical systems [190]. A simpler form of (4.1) has already been studied in [151, 190]. However, the technique applied in [167] does not apply here.

There is a large class of nonlinear operators that satisfy (4.2). Consider as $X = L^p(n)$ the space of real measurable n-vector valued functions of t defined on **R**. Let $x = (x_1, x_2, \ldots, x_n)$ and define a norm on X by

$$\|x\| = \left[\int_{-\infty}^{\infty} \sum_{j=1}^{n} |x_j(t)|^p dt \right]^{1/p} \quad \text{if } 1 \le p < \infty$$

$$\|x\| = \sum_{j=1}^{\infty} \sup_{-\infty < t < \infty} |x_j(t)| \quad \text{if } p = \infty.$$

Let $a_1(s, t), a_2(s, t), \ldots, a_n(s, t)$ denote real valued functions for $(s, t) \in \mathbf{R} \times \mathbf{R}$ such that:

(a)   there exist real numbers c_1 and c_2 with

$$c_1 \le \frac{a_j(s_1, t) - a_j(s_2, t)}{s_1 - s_2} \le c_2$$

   for all $t \in \mathbf{R}$ and all $s_1, s_2 \in \mathbf{R}$ such that $s_1 \neq s_2$.

(b) $a_j(0, t) \in L^p(1), j = 1, 2, \ldots, n.$

(c) $a_j(s(t), t)$ is a measurable function of t whenever $s(t)$ is measurable $j = 1, 2, \ldots, n.$

Then the mapping A defined on X by

$$(ax)_j(t) = a_j(x_j(t), t), \quad t \in \mathbf{R}, \ j = 1, 2, \ldots, n$$

in which $(ax)_j$ is the jth component of (ax) clearly satisfies (4.2) for every $\lambda \in \mathbf{R}$ and all $w, v \in X$.

The nonlinear operator A arises in both the study of feedback systems containing an arbitrary finite number of time-varying amplifiers and the study of electrome-chanical networks containing an arbitrary finite number of time-varying nonlinear dissipative elements.

DEFINITION 4.1   *Define the operators P, T on X by*

$$P(x) = y + B(A(x), A(x)) - x$$

and

$$T(x) = y + B(A(x), A(x)) .$$

Define the real function f by

$$f(r) = ar^2 + br + c$$

where

$$a = 2\|B\|(|\lambda| + d(\lambda))^2$$

$$b = 2\|B\|\|A(z)\|(|\lambda| + d(\lambda)) - 1, \quad and$$

$$c = \|P(z)\| .$$

Let r_1, r_2 with $r_1 \leq r_2$ denote the real solutions of $f(r) = 0$ if $b^2 - 4ac \geq 0$.

We can now prove the theorem.

THEOREM 4.1
Let (z, λ) be fixed in $(X, \mathbf{R})$. Assume:

(a) $b < 0$

(b) $b^2 - 4ac > 0$.

Then

(i) Equation (4.1) has a unique solution x^* in $U\left(z, \frac{r_1 + r_2}{2}\right)$;

(ii) moreover x^* is in $\bar{U}(z, r_1)$.

PROOF   Let r be such that

$$r_1 \leq r < \frac{r_1 + r_2}{2} . \tag{4.4}$$

CLAIM 4.1
T is a contraction on $\bar{U}(z, r)$. Let $w, v \in \bar{U}(z, r)$, then

$$T(w) - T(v) = (y + B(A(w), A(w))) - (y + B(A(v), A(v)))$$

$$= B(A(w), A(w)) - B(A(v), A(v))$$

$$= B(A(w), A(w) - A(v)) + B(A(w) - A(v), A(v)). \quad (4.5)$$

Now,

$$B(A(w), A(w) - A(v)) = B(A(w), A(w) - A(v) - \lambda(w - v))$$

$$+ B(A(w), \lambda(w - v))$$

$$= B(A(w) - A(z) - \lambda(w - z), A(w) - A(v) - \lambda(w - v))$$

$$+ \lambda B(w - z, A(w) - A(v) - \lambda(w - v))$$

$$+ B(A(z), A(w) - A(v) - \lambda(w - v))$$

$$+ B(A(w) - A(z) - \lambda(w - z), \lambda(w - v))$$

$$+ \lambda B(w - z, \lambda(w - v)) + B(A(z), \lambda(w - v)) . \quad (4.6)$$

A similar expression can be derived for $B(A(w) - A(v), A(v))$.
  We now take norms in (4.5), using (4.6) to get

$$\|T(w) - T(v)\| \le 2\|B\|(|\lambda| + d(\lambda))[(|\lambda| + d(\lambda))r + \|A(z)\|]\|w - v\| .$$

The operator T is a contraction on $\bar{U}(z, r)$ if

$$0 < q(\lambda, r) < 1$$

where
$$q(\lambda, r) = 2\|B\|(|\lambda| + d(\lambda))[(|\lambda| + d(\lambda))r + \|A(z)\|]$$

which is true by (a) and the claim is proved.

CLAIM 4.2
  T maps $\bar{U}(z, r)$ into $\bar{U}(z, r)$. Let $x \in \bar{U}(z, r)$, then it is enough to show

$$\|T(x) - z\| = \|y + B(A(x), A(x)) - z\|$$

$$= \|y + B(A(x), A(x)) - z - B(A(z), A(z)) + B(A(z), A(z))\|$$

$$= \|B(A(x), A(x)) - B(A(z), A(z)) + P(z)\| \le r .$$

Using the triangle inequality, as in Claim 4.1 we get

$$\|T(x) - z\| \le 2\|B\|(|\lambda| + d(\lambda))^2 r^2$$

$$+ [2\|B\|\|P(z)\|(|\lambda| + d(\lambda)) - 1]r + \|P(z)\| \le r$$

which is true by (b) and the claim is proved.

The result now follows from the contraction mapping principle. ∎

In applications it is desirable to determine a value for λ for which the contraction function $q(\lambda)$ is minimized (assuming r, z are fixed for simplicity).

Let $E = \{\lambda \in \mathbf{R} \mid q(\lambda) < 1\}$. We want to find $\bar\lambda$ such that

$$\bar\lambda = \inf_{\lambda \in E} q(\lambda).$$

We now assume $c_1, c_2, \lambda > 0$ for simplicity and state the relevant proposition.

PROPOSITION 4.1

Let $\bar\lambda$ be such that $q(\bar\lambda) < 1$. Then $\frac{1}{2}(c_1 + c_2) \in E$ and

(i) $q\left(\frac{1}{2}(c_2 + c_2)\right) < q(\bar\lambda)$ if $\bar\lambda > \frac{c_1 + c_2}{2}$;

(ii) $q\left(\frac{1}{2}(c_1 + c_2)\right) = q(\bar\lambda)$ if $\bar\lambda \le \frac{c_1 + c_2}{2}$.

PROOF   By the definition of $q(\lambda)$ it is enough to show

$$d\left(\frac{1}{2}(c_1 + c_2)\right) + \frac{1}{2}|c_1 + c_2| \le d(\bar\lambda) + |\bar\lambda|. \tag{4.7}$$

We consider two cases.

CASE 4.1

Let $\bar\lambda - c_2 > c_2 - \bar\lambda$ or $\bar\lambda > \frac{c_1 + c_2}{2}$. Then (4.7) becomes

$$\bar\lambda - c_1 + \bar\lambda > c_2 \quad or \quad \bar\lambda > \frac{c_1 + c_2}{2}$$

which is true in this case and (i) is proved.

CASE 4.2
$\bar{\lambda} - c_1 \le c_2 - \bar{\lambda}$ or $\bar{\lambda} \le \frac{c_1 + c_2}{2}$; then (4.7) becomes

$$c_2 - \bar{\lambda} + \bar{\lambda} = c_2 \quad or \quad c_2 = c_2$$

and (ii) is proved.   ∎

Example 4.1
Let X be the space of real valued almost periodic functions of t with

$$\|x\| = \sup_{-\infty < t < \infty} |x(t)|, \quad x \in X .$$

Consider the equation

$$x(s) = y(s) + \int_{-\infty}^{\infty} \text{sech}(t + s)(k(x(t)))^2 dt . \qquad (4.8)$$

Equation (4.8) is of the form (4.1), where

$$B(k(w(s)), k(v(s))) = \int_{-\infty}^{\infty} \text{sech}(t + s)k(w(t))k(v(t))dt$$

and k is a real valued function of the real variables s, t such that

$$c_1 \le \frac{k(s_1, t) - k(s_2, t)}{s_1 - s_2} \le c_2$$

for some real c_1, c_2 with $c_1 \le c_2$ and for all $s_1, s_2 \in \mathbf{R}$ such that $s_1 \ne s_2$.
    Now,

$$\|B\| = \int_{-\infty}^{\infty} \text{sech}\, r dr = \lim_{t \to -\infty} \int_t^0 \text{sech}\, r dr + \lim_{s \to \infty} \int_0^s \text{sech}\, r dr$$

$$= \lim_{t \to -\infty} \arctan|\sinh r|_t^0 + \lim_{s \to +\infty} \arctan|\sinh r|_0^s$$

$$= \lim_{t \to \infty} \arctan \left(\frac{1}{2}\right) (e^{-t} - e^t) + \lim_{s \to \infty} \arctan \left(\frac{1}{2} (e^s - e^{-s})\right)$$

$$= \frac{\pi}{2} + \frac{\pi}{2} = \pi .$$

Choose $\lambda \le \frac{c_1 + c_2}{2}$ and $z = y$, then condition (a) in Theorem 4.1 becomes

$$2\pi \|k(y)\| c_2 - 1 < 0 ,$$

or

$$c_2 < \frac{1}{2\pi \|k(y)\|} \cdot \qquad (4.9)$$

Condition (b) in Theorem 4.1 becomes

$$(2\pi \|k(y)\| c_2 - 1)^2 - 8\pi c_2^2 \|B(k(y), k(y))\| > 0$$

since $\|B(k(y), k(y))\| \le \|B\| \cdot \|k(y)\|^2 = \pi \|k(y)\|^2$ it is enough to have

$$(2\pi \|k(y)\| c_2 - 1)^2 - 8\pi^2 \|k(y)\|^2 c_2^2 > 0$$

or

$$4\pi^2 \|k(y)\|^2 c_2^2 + 4\pi \|k(y)\| c_2 - 1 < 0$$

which is true if

$$c_2 < \frac{\sqrt{2} - 1}{2\pi \|k(y)\|} \cdot \qquad (4.10)$$

By (4.9) and (4.10) it is enough to choose c_2 to satisfy (4.10).

Therefore, (4.8) has a unique solution in $\bar{U}(z, r_1)$ if we choose $z = y, \lambda \le \frac{c_1 + c_2}{2}$, where c_1, c_2 are such that

$$c_1 \le c_2 < \frac{\sqrt{2} - 1}{2\pi \|k(y)\|} \cdot$$

The solution $x(s)$ can be obtained approximately using the iteration scheme

$$x_0(s) = y(s)$$

$$x_{n+1}(s) = y(s) + B(k(x_n(s)), k(x_n(s))), \quad n = 0, 1, 2, \ldots .$$

The sequence $\{x_n(s)\}, n = 0, 1, 2, \ldots$ converges to $x(s)$ in such a way that

$$\|x_n(s) - x(s)\| \le \frac{q^n}{1 - q} \|y - T(y)\|$$

where, according to the theorem

$$q = 2\pi c_2 [c_2 r_1 + \|k(y)\|] .$$

If the desired tolerance of the error is ε then the number k of iterations to obtain the inequality $\|x_k(s) - x(s)\| \leq \varepsilon$ is given in Theorem 1.3.

4.2 Radiative Transfer Equations

CASE 4.3

Several different techniques are developed here to solve quadratic integral equations appearing in neutron transport. The results in this section constitute our work [15, 17, 22, 23, 33]. We first apply the method of continuation to study the structure of the solutions of quadratic integral equations of the form

$$x(s) = y(s) + \frac{1}{2}\lambda x(s) \int_s^1 k(s,t)x(t)dt \qquad (4.11)$$

in the space $C[s, 1]$ of all functions continuous on the interval $0 \leq s \leq 1$, with norm

$$\|x\| = \max_{0 \leq s \leq 1} |x(s)|\ .$$

Here we assume that λ is a real number and the kernel $k(s,t)$ is a continuous function of two variables s, t with $0 \leq s, t \leq 1$ and satisfying

(i) $0 < k(s,t) < 1,$ $0 \leq s, t \leq 1$

(ii) $k(s,t) + k(t,s) = 1,$ $0 \leq s, t \leq 1, k(0,0) = 1$.

                                                                (4.12)

The function $y(s)$ is a continuous given function on $[s, 1]$ and finally $x(s)$ is the unknown function sought in $C[s, 1]$.

There exists an extensive literature on equations such as (4.11) under various assumptions on the kernel $k(s, t)$ if the lower limit of integration in (4.11) is zero and λ is a real or complex number.

Here, we use the method of continuation to find solutions of (4.11) in order, on the one hand, to suggest a new method for solving equations such as (4.11) and, on the other hand, to improve our previous results.

For $\lambda = 0$, $x(s) \equiv y(s)$ is the unique solution of (4.11). Using this simple observation, we show how we can extend this solution for $\lambda \in (\lambda_1, 1)$ where $\lambda_1 < 0$.

The result in Corollary 1.1 for $k(s,t) = \frac{s}{s+t}$, $0 \leq s, t \leq 1$ requires

$$0 < \lambda < .72134 \cdots \qquad (4.13)$$

whereas, our result in Chapter 1 requires

$$0 < \lambda < .848108 \cdots \qquad (4.14)$$

for the existence of a solution of (4.11).

Here we prove a number of new results for (2.68). We first introduce the following definition.

DEFINITION 4.2   *Let B be a symmetric bilinear operator on X. Let x_0 be a solution of (2.68) corresponding to $\lambda = \lambda_0$, where λ_0^{-1} is not a characteristic value of the linear operator $2B(x_0)$ on X. Fix $\lambda_1 \in \mathbf{R}$. Define the numbers k_1, k_2, $\bar{r}$ and the real function $f_1(r)$, $f_2(r)$, $f_3(k)$ or $\mathbf{R}$ by*

$$k_1 = \left\| (I - 2\lambda_0 B(x_0))^{-1} \right\| ,$$

$$k_2 = \frac{1}{2 \left(|\lambda_1| \, \|B(x_0)\| + \sqrt{|\lambda_1| \, \|B\| \cdot \|B(x_0, x_0)\| \, |\lambda_0 + \lambda_1|} \right)} ,$$

$$\bar{r} = \frac{1 - 2|\lambda_1| \cdot k_1 \, \|B(x_0)\|}{2 \, |\lambda_0 + \lambda_1| \, \|B\|} ,$$

$$f_1(r) = a_2 r^2 + b_1 r + c_1 ,$$

$$f_2(r) = a_2 r + b_2 , \quad and$$

$$f_3(r) = a_3 k^2 + b_3 k + c_3$$

where

$$a_1 = \|\lambda_0 + \lambda_1\| \, \|B\| \, k_1 ,$$

$$b_1 = 2 |\lambda_1| \cdot \|B(x_0)\| \, k_1 - 1 ,$$

$$c_1 = |\lambda_1| \, \|B(x_0, x_0)\| \, k_1 ,$$

$$a_2 = 2 |\lambda_0 + \lambda_1| \, \|B\| k_1 ,$$

$$b_2 = -1 + 2 |\lambda_1| \, \|B(x_0)\| \, k_1 ,$$

$$a_3 = 4 |\lambda_1| \left(|\lambda_1| \, \|B(x_0)\|^2 - \|B(x_0, x_0)\| \cdot \|B\| \, |\lambda_0 + \lambda_1| \right) ,$$

$$b_3 = -4 |\lambda_1| \, \|B(x_0)\|, \quad and$$

$$c_3 = 1 .$$

It is easy to check that $f_1(r) \le 0$ and $f_2(r) < 0$ if $r \in [r_1, \bar{r})$ and $k_1 \in (0, k_2)$, where r_1 is the small solution of the equation $f_1(r) = 0$.

Obviously, for $\lambda = 0$, $x(s) \equiv y(s)$ is the unique solution of Equation (2.68). The above solution motivates us to introduce the following theorem on the extension of solutions.

THEOREM 4.2
Let x_0 be a solution of Equation (2.68) corresponding to $\lambda = \lambda_0$, where λ_0^{-1} is not a characteristic value of $2B(x_0)$. Fix $\lambda_1 \in \mathbf{R}$. Assume that b_1, k_1 are such that

$$b_1 < 0 \quad and \quad k_1 \in (0, k_2)$$

and choose $r \in [r_1, \bar{r})$, where $r_1, \bar{r}$ and k_2 are as in Definition 4.2.
Then

(i)   there exists a unique $w \in \bar{U}(r)$ such that

$$x = x_0 + w$$

is a solution of (2.68) corresponding to

$$\lambda = \lambda_0 + \lambda_1 .$$

(ii) The element $w \in X$ is continuous in norm as a function of λ_1 and $\|w\| \to 0$ as $\lambda_1 \to 0$.

PROOF   The element $x_0 + w \in X$ is a solution of (2.68) if

$$x_0 + w = y + (\lambda_1 + \lambda_0) B(x_0 + w, x_0 + w) ,$$

or if

$$w = (I - 2\lambda_0 B(x_0))^{-1} [\lambda_1 B(x_0, x_0) + 2\lambda_1 B(x_0, w) + (\lambda_0 + \lambda_1) B(w, w)]$$

$$= T(w) \,,$$

since x_0 is a solution of (2.68) and λ_0^{-1} is not a characteristic value of $2B(x_0)$. Let $w_1, w_2 \in \bar{U}(r)$.

CLAIM 4.3
T is a contraction operator on $\bar{U}(r)$.

We have

$$\|T(w_1) - T(w_2)\|$$

$$= \left\| (I - 2\lambda_0 B(x_0))^{-1} [2\lambda_1 B(x_0, w_1 - w_2) + (\lambda_0 + \lambda_1) B(w_1 - w_2, w_1 + w_2)] \right\|$$

$$\leq k_1 \left(2\,|\lambda_1|\,\|B(x_0)\| + |\lambda_0 + \lambda_1|\,\|B\| \cdot 2r \right) \|w_1 - w_2\|$$

so, T is a contraction on $\bar{U}(r)$ if

$$f_2(r) < 0$$

which is true, since $r \in [r_1, \bar{r})$ and $k_1 \in (0, k_2)$.

CLAIM 4.4
T maps $\bar{U}(r)$ into $\bar{U}(r)$.

Let $w \in \bar{U}(r)$, then $\|T(w)\| \leq r$ if

$$\|T(w)\| \leq k_1 \left(|\lambda_1|\,\|B(x_0, x_0)\| + 2\,|\lambda_1|\,\|B(x_0)\|\,r + |\lambda_0 + \lambda_1| \right) \|B\| \cdot r^2 \leq r \,,$$

or, $f_1(r) \leq 0$, which is true since, $r \in [r_1, \bar{r})$ and $k \in (0, k_2)$.

The result (i) now follows from the contraction mapping principle (see Theorem 1.3).

By the assumption on b_1 and k_1 there exists $\varepsilon > 0$ such that

$$|\lambda_1| \leq \varepsilon \,.$$

Let $\Delta\lambda$ be such that

$$|\lambda_1 + \Delta\lambda| < \varepsilon \,,$$

then

$$\left\| w_{\lambda_1+\Delta\lambda} - w_{\lambda_1} \right\| = \left\| T_{\lambda_1+\Delta\lambda} \left(w_{\lambda_1+\Delta\lambda} \right) - T_{\lambda_1} \left(w_{\lambda_1} \right) \right\|$$

$$\leq \left\| T_{\lambda_1+\Delta\lambda} \left(w_{\lambda_1+\Delta\lambda} \right) - T_{\lambda_1+\Delta\lambda} \left(w_{\lambda_1} \right) \right\|$$

$$+ \left\| T_{\lambda_1+\Delta\lambda} \left(w_{\lambda_1} \right) - T_{\lambda_1} \left(w_{\lambda_1} \right) \right\|$$

$$\leq (f_2(r) - 1) \left\| w_{\lambda_1+\Delta\lambda} - w_{\lambda_1} \right\|$$

$$+ \left\| T_{\lambda_1+\Delta\lambda} \left(w_{\lambda_1} \right) - T_{\lambda_1} \left(w_{\lambda_1} \right) \right\| ,$$

so that

$$\left\| w_{\lambda_1+\Delta\lambda} - w_{\lambda_1} \right\| \leq (2 - f_2(f))^{-1} \left\| T_{\lambda_1+\Delta\lambda} \left(w_{\lambda_1} \right) - T_{\lambda_1} \left(w_{\lambda_1} \right) \right\| \to 0$$

as $\Delta\lambda \to 0$. Moreover, $\|w\| \to 0$. This proves (ii) and the proof of the theorem is completed. ∎

REMARK 4.1   We observe from Theorem 4.2 that w, λ_2 are such that

$$|\lambda_1| \leq \varepsilon \quad \text{and} \quad \|w\| \leq r_1 . \quad ∎$$

Assuming as in Theorem 4.2 that $[I - 2\lambda_0 B(x_0)]^{-1}$ exists, we can choose $\varepsilon > 0$ and $r_1 > 0$ small enough so that $[I - 2(\lambda_0 + \lambda_1)B(x_0 + w)]^{-1}$ exists, provided $|\lambda_1| \leq \varepsilon$, $\|w\| \leq r_1$. The solution x of Equation (4.11) given by Theorem 4.2 is such that $(I - 2\lambda B(x))^{-1}$ exists for any

$$\lambda \in [\lambda_0 - \delta, \lambda_0 + \delta] .$$

We can now apply Theorem 4.2 for $\lambda = \lambda_0 \pm \delta$. This will extend the solution to two adjacent intervals overlapping the original one until we can find a pair (λ_c, x_c) such that $x_c \in X$ and λ_c is a characteristic value of $2B(x_c)$.

We now return to Equation (4.11). We work in the Banach space $C[s, 1]$. We define a bounded linear operator L and a bounded bilinear operator $\bar{B}$ on $C[s, 1]$ by

$$(Lv)(s) = \frac{1}{2} \int_s^1 k(s, t)v(t)dt ,$$

$$\bar{B}(v, w)(s) = v(s)(Lw)(s) + w(s)(Lv)(s)$$

where $v, w \in C[s, 1], 0 \le s \le 1$. It is standard to show that $\bar{B}(v, w) \in C[s, 1]$.
Let λ be a fixed real number, define an operator F on $C[s, 1]$ by

$$F(v) = v - y - \frac{1}{2}\lambda\bar{B}(v, v) \, .$$

Then, Equation (4.11) can be written as

$$F(x) = 0 \, .$$

Finally, define the Fréchet derivative of F with respect to v by

$$F'(v)w = w - \lambda\bar{B}(v, w) \, . \tag{4.15}$$

Note that $F'(v)$ is a bounded linear operator on $C[s, 1]$.
We now prove the following useful lemma.

LEMMA 4.1
Let x be a solution of (4.11) for some fixed λ. Set

$$A_s = \int_s^1 x(s)ds, \quad 0 \le s < 1$$

and

$$D_s = \int_s^1 y(s)ds, \quad 0 \le s < 1 \, .$$

Then,

$$\frac{1}{2}\lambda A_s^2 - 2A_s + 2D_s = 0 \, . \tag{4.16}$$

PROOF Integrating (4.11) and using the identity

$$k(s, t) = \frac{1}{2}\{1 + k(s, t) - k(t, s)\}$$

we have

$$A_s = D_s + \frac{1}{4}\lambda A_s^2 + \frac{1}{2}\int_s^1\int_s^1 \{k(s, t) - k(t, s)\}x(s)x(t)dsdt \, ,$$

and the last integral is obviously zero. ∎

From now on we take $y(s) = 1$, i.e., $D_s = 1 - s$ for simplicity.

Since $\lambda A_s^2 - 4A_s + 4D_s = 0$ has no real solutions when $\lambda > \frac{1}{1-s}$, a fundamental interruption in the continuation of the solution $x_\lambda(s)$ must occur at some boundary point $\lambda_c \leq \frac{1}{1-s} \leq 1$. Equation (4.11) can have no real solutions for $\lambda > \frac{1}{1-s}$.

Moreover, we can show the following:

THEOREM 4.3

Let $x_{\lambda_0}(s)$ be any solution of (4.11) with $\lambda = \lambda_0$. Then the linear operator $F'(x_{\lambda_0})$ given by (4.15) is an invertible operator on $C[s, 1]$ with bounded inverse if $\lambda_0 \neq 1$.

PROOF   Let w be such that

$$F'(v)(w) = 0, \quad v, w \in C[s, 1].$$

If the lower limit of integration in (4.11) is different than zero, we set

$$x_{\lambda_0}(t) = 0$$

$$v(t) = 0 \qquad \text{for } 0 \leq t < s.$$

$$w(t) = 0$$

Then by integrating (4.15), we obtain

$$\int_0^1 w(s)ds = \frac{1}{2}\lambda_0 \int_0^1 B(x_{\lambda_0}, w)(s)ds$$

$$= \frac{1}{2}\lambda_0 \left[\int_0^1 x_{\lambda_0}(s)ds \right] \left[\int_0^1 w(s)ds \right].$$

If $\lambda_0 \neq 1$, using Lemma 4.1 in the above equation, we get

$$\int_0^1 w(s)ds = 0.\tag{4.17}$$

Let $p(s) = \frac{w(s)}{x_{\lambda_0}(s)}$, to obtain

$$p = \frac{1}{2}\lambda_0 x_{\lambda_0} L(x_{\lambda_0} p)$$

and

$$\int_0^1 p(s) x_{\lambda_0}(s) ds = 0 ,$$

$$p = -\frac{1}{2} \lambda x_{\lambda_0} Q\left(x_{\lambda_0} p\right) \qquad (4.18)$$

where

$$(Qw)(s) = \int_0^1 \frac{t}{s+t} w(t) dt .$$

Define a subspace S of $C[0, 1]$ by

$$S = \left\{ p \in C[0, 1] \mid \int_0^1 p(s) x_{\lambda_0}(s) ds = 0 \right\} .$$

We will show that (4.18) has only the trivial solution in S. Let $q \in C[0, 1]$. Define $E(q)$ by

$$E(q) = -\frac{1}{2} \lambda \left(1 - \frac{1}{2} \lambda r\right)^{-1} \int_0^1 x_{\lambda_0}(s) s q(s) ds ,$$

where r is a root of (4.16) for the lower limit of integration in (4.11) being zero. The functional E is well defined since $1 - \frac{1}{2} \lambda r \neq 0$.

Define the operator R on $C[0, 1]$ by

$$R(q(s)) = s(sq(s) - E(q)) .$$

Since

$$x_{\lambda_0} L\left(x_{\lambda_0} R(p)\right) = x_{\lambda_0} L\left[s\left(sx_{\lambda_0} p(s) - x_{\lambda_0} E(p)\right)\right]$$

$$= x_{\lambda_0} Q\left[sx_{\lambda_0}(sp(s) - E(p))\right] ;$$

(4.17) gives

$$sp(s) = -\frac{1}{2} \lambda s \left(x_{\lambda_0} Q\left(x_{\lambda_0} p(s)\right)\right) = -\frac{1}{2} \lambda s x_{\lambda_0}(Q\left(x_{\lambda_0} p(s)\right))$$

$$= -\frac{1}{2} \lambda x_{\lambda_0} L\left(sx_{\lambda_0} p(s)\right)$$

$$= \left(1 - \frac{1}{2}\lambda r\right) E(p(s))x_{\lambda_0} + \frac{1}{2}\lambda x_{\lambda_0} Q\left(sx_{\lambda_0}\right) . \tag{4.19}$$

Hence, as x_{λ_0} is a solution of (4.11),

$$\frac{1}{2}\lambda x_{\lambda_0} Q(x_{\lambda_0}) = \frac{1}{2}\lambda r x_{\lambda_0} - \frac{1}{2}\lambda x_{\lambda_0} L x_{\lambda_0} = \frac{1}{2}\lambda r x_{\lambda_0} + 1 - x_{\lambda_0}$$

$$= 1 - \left(1 - \frac{1}{2}\lambda r\right) x_{\lambda_0}$$

we have, by (4.18)

$$\frac{1}{2}\lambda x_{\lambda_0} Q\left[x_{\lambda_0}(sp(s) - E(p))\right] = sp(s) - \left(1 - \frac{1}{2}\lambda r\right) E(p)x_{\lambda_0}$$

$$- E(p)\frac{1}{2}\lambda x_{\lambda_0} Q(x_{\lambda_0})$$

$$= sp(s) - E(p) .$$

Therefore,

$$\frac{1}{2}\lambda x_{\lambda_0} L\left(x_{\lambda_0} R(p)\right) = \left[\frac{1}{2}\lambda x_{\lambda_0} Q\left(s\left(sx_{\lambda_0}p - E(p)\right)\right)\right]$$

$$= s(sp(s) - E(p)) = R(p) .$$

Therefore, if $\lambda_0 \neq 1$ and p is a solution of (4.17), then $p \in S$, $R(g) \in S$ and $R(g)$ is a solution of (4.17). Now, as L is compact, (4.17) has a finite dimensional solution space. Hence, there exist an integer $n \geq 1$ and numbers $\{c_i\}_{i=1}^{n}$ with $c_n \neq 0$, so that

$$\sum_{i=1}^{n} c_i R^i(g) = 0 .$$

The above equation implies that there exist polynomials D_1 and D_2 so that

$$D_2\left(s^2\right) p(s) = sD_1\left(s^2\right) .$$

The linearity of R implies that

$$\sum_{i=1}^{n} c_i R^{i+1}(p) = D \, .$$

Hence $R(p) = p$, so

$$p(s) = \frac{s E(p)}{s^2 - 1} \, ,$$

that is, $p = 0$ as $p \in C[0, 1]$. The proof of the theorem is now completed. ∎

PROPOSITION 4.2

There exists a sheet of solutions x_λ for λ such that $0 \le \lambda \le \lambda_c$ and $x_0 \equiv 1$. The sheet x_λ is continuous as a function of λ uniformly over $0 \le s \le 1$.

PROOF Obviously, for $\lambda_0 = 0$, $x_0 \equiv 1$ is the unique solution of (4.11). Since $I - 2\lambda_0 B(x_0) = I$ is nonsingular, we can apply Theorem 4.2 to generate x_λ. The element w is a continuous function of λ in each extension. Therefore, x_λ is continuous as a function of λ uniformly over $0 \le s \le 1$. ∎

A similar proposition can be stated for some negative λ values.

PROPOSITION 4.3

The following are true:

(i) *The solutions x_λ, $\lambda > 0$ are such that*

$$x_\lambda(s) > 0, \quad 0 \le s \le 1 \, ;$$

(ii) *for fixed s, $0 \le s \le 1$, $x_\lambda(s)$ is a monotone increasing function of λ.*

PROOF (i) For $\lambda = 0$ and for small positive λ since $x_0(s) \equiv 1$, the result is true by continuity. Let $p_\lambda = \inf x_\lambda(s)$. Then

$$|p_{\lambda+\Delta\lambda} - p_\lambda| \le |\sup (p_{\lambda+\Delta\lambda} - p_\lambda)| \le \sup |p_{\lambda+\Delta\lambda} - p_\lambda| \to 0 \text{ as } \Delta\lambda \to 0$$

that is, p_λ is a continuous function of λ. If x_λ becomes nonpositive, then there exists $\lambda_0 > 0$ such that

$$p_{\lambda_0} \equiv 0 \, .$$

Therefore,

$$x_{\lambda_0}(s) \geq 0 \,,$$

with zero actually attained. Moreover, at $\lambda = \lambda_0$

$$x_{\lambda_0}(s) = 1 + \frac{1}{2}\lambda_0 x_{\lambda_0}(s) \int_s^1 x_{\lambda_0}(t)k(s,t)dt \geq 1 \,,$$

which provides a contradiction.

(ii) For $0 \leq \lambda \leq \lambda_c < 1$, $[I - 2\lambda B(x_\lambda)]^{-1}$ does exists and

$$[I - 2\lambda B\,(x_\lambda)]^{-1} = I + 2\lambda B\,(x_\lambda) + 4\lambda^2\,(B\,(x_\lambda))^2 + \cdots \,.$$

By (i) the above series, i.e., the above inverse, is a positive operator. Moreover, w is the limit of the iterations

$$w_{n+1} = T\,(w_n), \quad n = 0, 1, 2, \dots \,,$$

where w_0 might be chosen positive. Hence, $w > 0$ at each extension step of Proposition 4.2. That proves (ii). ∎

PROPOSITION 4.4
The solution x_λ of Equation (4.11) exists and is positive at least for $\lambda \in (-1, 0]$.

PROOF   By Proposition 4.3, $x_\lambda > 0$ for small $|\lambda|$ by continuity. Suppose either that continuation ceases or that p_λ vanishes at $\lambda_1 < 0$. Therefore, x_λ exists and is positive on $(\lambda_1, 0)$. We have

$$x_\lambda(s) = 1 - \frac{1}{2}|\lambda|x_\lambda(s) \int_s^1 k(s,t)x(t)dt, \quad \lambda_1 < \lambda \leq 0 \qquad (4.20)$$

so, $\|x_\lambda(s)\|$ is finite if λ_1 is a leftward limit of continuation and $\lambda \to \lambda_1$.

CLAIM 4.5
$\lambda \leq -1$.

If $\lambda_1 > -1$, or $|\lambda_1| < 1$. Since $p_{\lambda_1} = 0$, $x_{\lambda_1}(s)$ has a zero s_1. At $s = s_1$, we have

$$0 = x_\lambda(s_1) = 1 - \frac{1}{2}|\lambda_1|x_\lambda(s_1) \int_{s_1}^1 k\,(s_1, t)\,x_\lambda(t)dt$$

$$> 1 - \frac{1}{2} x_\lambda (s_1) \int_{s_1}^{1} k (s_1, t) \, x_\lambda(t) dt \geq 0 \,,$$

since

$$\frac{1}{2} x_\lambda(s) \int_{s}^{1} k(s, t) x_\lambda(t) dt \leq 1 \,,$$

by (4.20). This contradiction justifies the claim and the proposition is proved.
∎

REMARK 4.2 In Chapter 1 we saw that the range for λ is

$$0 \leq \lambda < .72134 \cdots$$

$$0 \leq \lambda < .848108 \cdots$$

respectively, for

$$k(s, t) = \frac{s}{s + t}, \quad 0 \leq s, t \leq 1$$

the lower limit of integration in (4.11) being zero and $y(s) = 1$.

Here, it follows from Theorems 4.2 and 4.3, and Proposition 4.2 that the range for λ is at least such that

$$\lambda_1 \leq \lambda < 1 \,,$$

where λ_1 by Proposition 4.4 is such that

$$\lambda_1 \leq -1 \,.$$

That is, there exists solutions x_λ by continuation such that

$$.848108 \cdots \leq \lambda < 1, \quad \text{for example} \,.$$

This justifies the claim made at the introduction.   ∎

CASE 4.4

We now investigate quadratic integral equations using a different technique. In the theories of radiative transfer and neutron transport an important role is played by nonlinear integral equations of the form

$$H(x) = 1 + x H(x) \int_{0}^{1} \frac{\psi(t) H(t)}{x + t} dt \,. \tag{4.21}$$

The known function ψ is assumed to be nonnegative, bounded, and measurable on [0, 1], and a positive, continuous solution H of (4.21) is sought.

The first proof however of the existence of a solution of (4.21) was given by Crum, who considered the equation in the complex plane [85]. Crum also showed that if $\int_0^1 \psi(t)dt \leq \frac{1}{2}$, then (4.21) has at most two solutions which are bounded in [0, 1] and in case $\int_0^1 \psi(t)dt = \frac{1}{2}$, there is only one such solution. Fox [99] solved simpler equations in order to prove existence of solutions of (4.21). But the solution of Fox's equation are not necessarily solutions of (4.21). Stuart [184] gave nonconstructive existence proof for (4.21) using the Leray–Schauder degree theory but did not discuss the number or location of solutions. Cahlon and Eskin [74] used a theorem of Darbo [87] for a set contraction map to prove a nonconstructive existence theorem for (4.21).

Finally, Kelley [116] has solved some interesting generalizations of (4.21) using the solutions of finite rank approximations of solutions of (4.21).

Here we consider the generalized equation:

$$H(x) = 1 + H(x) \int_0^1 k(x, t)\psi(t)H(t)dt . \qquad (4.22)$$

The known kernel function $k(x, t)$ is a measurable function on $[0, 1] \times [0, 1]$ satisfying

(a) $0 < k(x, t) < 1 \; \forall x, t \in [0, 1]$, $k(0, 0) = 1$,

and

(b) $k(x, t) + k(t, x) = 1 \; \forall x, t \in [0, 1]$.

We show that whenever $\int_0^1 \psi(t)dt \leq \frac{1}{2}$, a minimal solution H can be found using a specific iteration.

Finally, under the same assumption we provide a way of constructing new non-minimal solutions H of (4.21) in terms of the minimal solution.

We denote by $C[0, 1]$ the Banach space of all real continuous functions on [0, 1] with the maximum norm

$$\|u\| = \max_{0 \leq t \leq 1} |u(t)| .$$

We now list the following well-known theorem whose proof can be found in [77, pp. 106–107].

THEOREM 4.4
If H is a solution of (4.22), then either

$$\int_0^1 \psi(t)H(t)dt = 1 - \left[1 - 2\int_0^1 \psi(t)dt\right]^{1/2} \qquad (4.23)$$

or

$$\int_0^1 \psi(t) H(t) dt = 1 + \left[1 - 2 \int_0^1 \psi(t) dt \right]^{1/2} . \tag{4.24}$$

A necessary condition that (4.22) has a solution is that

$$\int_0^1 \psi(t) dt \le \frac{1}{2} . \tag{4.25}$$

A function $H \in C[0, 1]$ satisfies the equation

$$H(x)^{-1} = \left[1 - 2 \int_0^1 \psi(t) dt \right]^{1/2} + \int_0^1 k(t, x) \psi(t) H(t) dt \tag{4.26}$$

if and only if H satisfies (4.22) and (4.23).

Chandrasekhar in [77], after proving that a solution H satisfies either (4.23) or (4.24), claims that, in fact, H must satisfy (4.23). This claim is not true because, as we show, there always exists a solution H satisfying (4.23), but in many cases there exists a second solution $\tilde{H}$ satisfying (4.24) and not (4.23).

Let $\le$ be the natural partial ordering on $C[0, 1]$, that is, if $p_1, p_2 \in C[0, 1]$, then $p_1 \le p_2$ if $p_1(x) \le p_2(x)$ for all $x \in [0, 1]$ and define the following:

$$d = \left[1 - 2 \int_0^1 \psi(t) dt \right]^{1/2} ,$$

$$D = \{ p \in C[0, 1] / p(x) \ge d, x \in [0, 1] \} ,$$

the operator $R : D \to C[0, 1]$ by

$$R(p(x)) = 1 + p(x) \int_0^1 k(x, t) \psi(t) p(t) dt , \quad p \in D$$

and for $d > 0$, define the operator $F : D \to C[0, 1]$ by

$$F(p(x)) = d + \int_0^1 k(t, x) \psi(t) (p(t))^{-1} dt , \quad p \in D .$$

It is routine to verify that R is isotone, that is, if $p_1 \le p_2$ then $R(p_1) \le R(p_2)$ and F is antitone, that is, if $p_1 \le p_2$ then $F(p_2) \le F(p_1)$.

Finally, denote by 1 (respectively, d) the function with constant value 1 (respectively, d).

We can now prove the proposition:

PROPOSITION 4.5

Assume that the kernel function $k(x, t)$ is as in the introduction and satisfies the condition

$$|k(x, t) - k(y, t)| \le b(t)|x - y| \quad \text{for all } x, y, t \in (0, 1]$$

and some positive function $b(t)$ such that

$$tb(t) \le b \text{ for some } b > 0 \text{ and all } t \in (0, 1] .$$

Then the sequence $R^n(1)$, $n = 1, 2, \ldots$ is equicontinuous.

PROOF   Let H be a solution satisfying (4.22) and (4.23) and set $A = \{p \in C[0, 1]/1 \le p \le H\}$. Define $Q : A \to C[0, 1]$ by

$$Q(p(x)) = \int_0^1 k(x, t)\psi(t)p(t)dt, \quad x \in [0, 1], \; p \in A .$$

Let $\varepsilon > 0$, then there exists a, $0 < a < 1$, such that $\int_0^a \psi(t)H(t)dt < \frac{\varepsilon}{4}$ and $\int_a^1 \psi(t)H(t)dt > 0$. Then for $x, y \in [0, 1]$,

$$|Q(p(x)) - Q(p(y))| = \left| \int_0^1 (k(x, t) - k(y, t))\psi(t)p(t)dt \right|$$

$$\le \int_0^a (|k(x, t)| + |k(y, t)|)\psi(t)H(t)dt$$

$$+ \int_a^1 |k(x, t) - k(y, t)|\psi(t)H(t)dt$$

$$\le 2 \int_0^a \psi(t)H(t)dt + \frac{b}{a} \int_a^1 |x - y|\psi(t)H(t)dt$$

$$< \frac{\varepsilon}{2} + \frac{\varepsilon}{2} = \varepsilon ,$$

by choosing $|x - y| < \delta = \frac{a\varepsilon}{2b} \left(\int_a^1 \psi(t) H(t) dt \right)^{-1}$. That is the set $Q(A)$ is equicontinuous.

Let $p \in A$ and $x \in [0, 1]$, then

$$Q(p(x)) = \int_0^1 k(x, t) \psi(t) p(t) dt \leq \int_0^1 k(x, t) \psi(t) H(t) dt$$

$$< \int_0^1 \psi(t) H(t) dt = 1 - d \leq 1 .$$

Therefore, there exists c, $0 < c < 1$, such that $Q(p(x)) < c$ for all $p \in A$ and $x \in [0, 1]$.

For any $\varepsilon_0 > 0$, there exists $\delta_0 > 0$ such that for every $g \in Q(A)$, $|g(x) - g(y)| \leq \|H\|^{-1}(1-c)\varepsilon_0$ if $|x - y| < \delta_0$ (since $Q(A)$ is equicontinuous). The function $R(1)$ is continuous and hence uniformly continuous; therefore, there exists $0 < \delta_1 \leq \delta_0$ such that

$$|R(1(x)) - R(1(y))| < \varepsilon_0 \quad \text{if } |x - y| < \delta_1 .$$

We shall show that the same δ_1 works for ε_0 and $R^{n+1}(1)$ if

$$|R^k(1(x)) - R^k(1(y))| < \varepsilon_0 \text{ if } |x - y| < \delta_1 \text{ for } k = 1, 2, \ldots, n .$$

Set $p = R^n(1)$. Then if $|x - y| < \delta_1$,

$$|R(p(x)) - R(p(y))| = |p(x) Q(p(x)) - p(y) Q(p(y))|$$

$$\leq |p(x) Q(p(x)) - p(x) Q(p(y))| + |p(x) Q(p(y)) - p(y) Q(p(y))|$$

$$\leq p(x) |Q(p(x)) - Q(p(y))| + Q(p(y)) |p(x) - p(y)|$$

$$\leq \|H\| \|H\|^{-1} (1 - c) \varepsilon_0 + c \varepsilon_0 = \varepsilon_0 ,$$

that is,

$$|R^{n+1}(1(x)) - R^{n+1}(1(y))| < \varepsilon_0 \quad \text{if } |x - y| < \delta_1$$

which completes the induction and the proof of the proposition. ∎

THEOREM 4.5

Assume that the kernel function $k(x, t)$ is as in Proposition 4.5. Then the following are true:

(i)   Equation (4.22) has exactly one solution H satisfying (4.23) if and only
      if (4.25) holds. Moreover, the increasing sequence $R^n(1)$, $n = 0, 1, 2, \ldots$
      converges to H; and

(ii) if inequality holds in (4.25), the sequence $F^n(d)$, $n = 0, 1, 2, \ldots$ converges
      to H^{-1} and

$$|H^{-1}(x) - F^n(d(x))| \leq |F^n(d(x)) - F^{n+1}(d(x))| ,$$

$$x \in [0, 1] . \tag{4.27}$$

PROOF   (i). If (4.22) has a solution H, then by Theorem 4.4, $\int_0^1 \psi(t)dt \leq \frac{1}{2}$.

CASE 4.5
Assume $\int_0^1 \psi(t)dt \leq \frac{1}{2}$. It can easily be verified that since F is antitone:

$$d \leq F^2(d) \leq F^4(d) \leq F^6(d) \leq \cdots \leq F^7(d) \leq F^5(d) \leq F^3(d) \leq F(d) .$$

Working as in the proposition we can easily show that the bounded set

$$N = \{F(p)/d \leq p \leq F(d)\}$$

is equicontinuous. Then the sequences $F^{2n}(d)$, $n = 1, 2, \ldots$ and $F^{2n+1}(d)$, $n = 0, 1, 2, \ldots$ have convergent subsequences converging to the functions v and w respectively. From the monotonicity of the above sequences and the continuity of F we obtain

$$F^{2n}(d) \to v,$$

$$F^{2n+1}(d) \to w,$$

$$d \leq v \leq w,$$

$$F(v) = w$$

and

$$F(w) = v .$$

The function v has a minimum value greater than zero, so that there exists a largest number q, $0 < q \leq 1$, with $qw \leq v$. If $q = 1$, then $w \leq v \leq w$, that is,

$v = w$. If $q < 1$, define on the domain of F the operator F_1 by

$$F_1(p) = F(p) - d .$$

Then we get

$$v = d + F_1(w) \geq d + F_1\left(q^{-1}v\right) = d + q F_1(v)$$

$$= (1 - q)d + q(d + F_1(v)) = (1 - q)d + qw$$

$$\geq ew \mid qw = (e + q)w ,$$

for some $e > 0$. But this contradicts the maximality of q. Therefore,

$$F(v) = v = w,$$

$$H = v^{-1}$$

is a solution of (4.22), satisfying (4.23) and the sequence $F^n(d)$, $n = 0, 1, 2, \ldots$ converge to H^{-1}. Inequality (4.27) follows from the fact that $F^{2k}(d) \leq H^{-1} \leq F^{2k+1}(d)$, for $k = 1, 2, 3, \ldots$.

CASE 4.6
Assume that $\int_0^1 \psi(t)dt = \frac{1}{2}$. Let $\{c_n\}$, $n = 1, 2, \ldots$ be a strictly increasing sequence of positive numbers converging to 1, and consider the functions $c_n\psi$, $n = 1, 2, 3, \ldots$.
  Since $\int_0^1 c_n\psi(t)dt = \frac{1}{2}c_n < \frac{1}{2}$, it follows from Case 4.5 that the equation

$$H(x) = 1 + H(x) \int_0^1 k(x, t)c_n\psi(t)H(t)dt$$

has a solution H_n for $n = 1, 2, 3, \ldots$. Then for each $x \in [0, 1]$ $h_n(x) \geq 1$ and

$$(H_n(x))^{-1} = \left[1 - 2\int_0^1 c_n\psi(t)dt\right]^{1/2} + \int_0^1 k(t, x)c_n\psi(t)H_n(t)dt$$

$$\geq c_n \int_0^1 k(t, x)\psi(t)dt \geq c_1 \int_0^1 k(t, x)\psi(t)dt .$$

Therefore, there exists $r > 0$ such that $(H_n(x))^{-1} \geq r$ for each $x \in [0, 1]$ and each $n = 1, 2, 3, \ldots$.

Set $M = \{p \in C[0, 1]/r \leq p(x) \leq 1, x \in [0, 1]\}$. Then $H_n^{-1} \in M$, $n = 1, 2, \ldots$. Define $F : M \to C[0, 1]$ by

$$F(p(x)) = \int_0^1 k(t, x)\psi(t)(p(t))^{-1}dt, \quad p \in M.$$

It is easy to verify that the set $F(M)$ is bounded and equicontinuous. Also, for each n,

$$H_n^{-1}(x) = \left[1 - 2\int_0^1 c_n\psi(t)dt\right]^{1/2} + \int_0^1 k(t, x)c_n\psi(t)H_n(t)dt$$

$$= \left[1 - 2\int_0^1 c_n\psi(t)dt\right]^{1/2} + c_n F\left(H_n^{-1}\right)(x).$$

Since $F(H_n^{-1}) \in F(M)$ for each n, some subsequence $F(H_{n_j}^{-1})$, $j = 1, 2, \ldots$, of $F(H_n^{-1})$, $n = 1, 2, \ldots$ converges in $C[0, 1]$ to some point H_0^{-1}, so that $H_{n_j}^{-1} \to H_0^{-1}$. Then the sequence $F(H_{n_j}^{-1})$, $j = 1, 2, \ldots$ converges to $F(H_0^{-1})$ and H_0^{-1}; that is,

$$F\left(H_0^{-1}\right) = H_0^{-1}.$$

Then H_0 satisfies (4.22), (4.23), and (4.24).

Therefore, there exists a positive function H satisfying (4.22) and (4.23) whenever ψ satisfies (4.25).

(ii) Assume (4.25) holds, and suppose H satisfies (4.22) and (4.23). Since $1 \leq H$ and $1 \leq R(1)$, it follows from the fact that R is isotone that

$$1 \leq R(1) \leq R^2(1) \leq R^3(1) \leq \cdots \leq H.$$

Since the sequence $R^n(1)$, $n = 1, 2, \ldots$ is uniformly bounded and equicontinuous there is a convergent subsequence, say $R^{n_k} \to h \leq H$, and, since the sequence $R^n(1)$, $n = 0, 1, 2, \ldots$ is nondecreasing, the entire sequence converges to h. It follows from the continuity of R that $R(h) = h$. Now h must satisfy either (4.23) or (4.24), and since $0 \leq h \leq H$, h must satisfy (4.23). Therefore, for $x \in [0, 1]$,

$$h^{-1}(x) = \left[1 - 2\int_0^1 \psi(t)dt\right]^{1/2} + \int_0^1 k(t,x)\psi(t)h(t)dt$$

$$\leq \left[1 - 2\int_0^1 \psi(t)dt\right]^{1/2} + \int_0^1 k(t,x)\psi(t)H(t)dt = H^{-1}(x),$$

that is, $h^{-1} \leq H^{-1}$. Together with the inequality $h \leq H$, this implies $h = H$.

We have proved that H is the only function satisfying both (4.22) and (4.23) and that the increasing sequence $R^n(1)$, $n = 0, 1, 2, \ldots$ converges to H which completes the proof of the theorem. ∎

COROLLARY 4.1
Suppose that ψ_1 and ψ_2 are nonnegative, bounded, measurable functions on $[0, 1]$ such that $\psi_1(t) \leq \psi_2(t)$ almost everywhere in $[0, 1]$ and such that $\int_0^1 \psi(t)dt \leq \frac{1}{2}$, $i = 1, 2$. Let H_i be the unique solution of Equations (4.22) and (4.23) corresponding to $\psi = \psi_i$, $i = 1, 2$.
Then,
$$H_1 \leq H_2 .$$

PROOF Define $R_i : C[0, 1] \rightarrow C[0, 1]$, $i = 1, 2$, by

$$R_i(p(x)) = 1 + p(x)\int_0^1 k(x,t)\psi_i(t)p(t)dt, \quad p \in C[0, 1] .$$

If p_1 and p_2 are nonnegative functions in $C[0, 1]$ with $p_1 \leq p_2$, then $R_1(p_1) \leq R_2(p_2)$. Hence, $R_1(1) \leq R_2(1)$, $R_1^2(1) \leq R_2^2(1)$, and in general, $R_1^n(1) \leq R_2^n(1)$. Since the increasing sequence $R_i^n(1)$, converges to H_i, $i = 1, 2$, it follows that $H_1 \leq H_2$. ∎

Note that if $\int_0^1 \psi(t)dt = \frac{1}{2}$, it follows from the previous results that the function H satisfying (4.22) and (4.23) is the unique solution of (4.22), since, in this case (4.23) and (4.22) reduce to the same equation. However, if $\int_0^1 \psi(t)dt < \frac{1}{2}$, Equation (4.22) may have two distinct solutions.

THEOREM 4.6
Assume:

(a) $\int_0^1 \psi(t)dt < \frac{1}{2}$ *and H is the unique solution of (4.22) and (4.23);*

(b)   *the following estimate is true*

$$\int_0^1 \frac{\psi(t)}{1-t} H(t)dt > 1 \tag{4.28}$$

and

(c)   *there exist functions* $\varphi_1, \varphi_2, \varphi_3 \in C[0, 1]$ *such that*

$$k(x, t) [\varphi_2(x)(1 - kt) - (1 + \varphi_1(t))] + \varphi_3(x) = 0,$$

$$\text{for all } x, t \in [0, 1] \tag{4.29}$$

$$\varphi_1(x) + kx > 0 \quad \text{for all } x \in (0, 1], \quad \varphi_1(0) = 0 \tag{4.30}$$

and

$$(1 + \varphi_1(x)) [\varphi_2(x)(H(x) - 1) + \varphi_3(x)H(x)]$$

$$= (H_1(x) - 1)(1 - kx) \tag{4.31}$$

for all $x \in [0, 1]$ *where k is the unique number in* $(0, 1)$ *for which*

$$\int_0^1 \frac{\psi(t)}{1 - kt} H(t)dt = 1 \tag{4.32}$$

and the function H_1 *is given by*

$$H_1(x) = \frac{1 + \varphi_1(x)}{1 - kx} H(x), \quad x \in [0, 1] . \tag{4.33}$$

Then H_1 *is a solution of (4.22) and (4.23) and*

$$H_1(x) > H(x), \quad x \in (0, 1], \quad H_1(0) = H(0) .$$

PROOF   By the monotone convergence theorem

$$\lim_{k \to 1^-} \int_0^1 \frac{\psi(t)}{1 - kt} H(t)dt = \int_0^1 \frac{\psi(t)}{1 - t} H(t)dt$$

since $(1 - kt)^{-1}$ increases monotonically with k, $0 < k < 1$.

If (4.28) holds, since

$$\int_0^1 \frac{\psi(t)}{1 - 0 \cdot t} H(t)dt = 1 - \left[1 - 2\int_0^1 \psi(t)dt\right]^{1/2} < 1 ,$$

and since the function $f : (0, 1) \to \mathbf{R}$ defined by

$$f(k) = \int_0^1 \frac{\psi(t)}{1 - kt} H(t)dt$$

is strictly increasing, there exists a unique $k \in (0, 1)$, for which (4.32) holds. Let H_1 be defined as in (4.33). We can now find that for each $x \in [0, 1]$

$$\int_0^1 k(x, t)\psi(t)H_1(t)dt = \varphi_2(x)\int_0^1 k(x, t)\psi(t)H(t)dt$$

$$+ \varphi_3(x)\int_0^1 \frac{H(t)}{1 - kt}\psi(t)dt$$

$$= \varphi_2(x)\int_0^1 k(x, t)\psi(t)H(t)dt + \varphi_3(x)$$

$$= \varphi_2(x)\left[1 - \frac{1}{H(x)}\right] + \varphi_3(x) = 1 - \frac{1}{H_1(x)} ,$$

[by (4.29) and (4.31)] that is, H_1 satisfies (4.22). Since H_1 must satisfy either (4.23) or (4.24) and since $H_1(x) > H(x)$, $x \in [0, 1]$ [by (4.30)], H_1 satisfies (4.24) and the proof of the theorem is completed. ∎

REMARK 4.3   By choosing the kernel function $k(x, t)$ to be $k(x, t) = \frac{x}{x+t}$, $x, t \in [0, 1]$ we observe that the conditions (a) and (b) in the introduction are satisfied and that Equation (4.22) reduces to Equation (4.21).

Moreover, the function $b(t)$ introduced in the proposition can be chosen to be

$$b(t) = \frac{1}{t}, \quad t \in (0, 1]$$

and therefore

$$b = 1 .$$

Finally, the conditions (4.29), (4.30), and (4.31) can then be satisfied if we choose

$$\varphi_1(x) = kx ,$$

$$\varphi_2(x) = \frac{1-kx}{1+kx} , \quad \text{and}$$

$$\varphi_3(x) = \frac{2kx}{1+kx} . \quad \blacksquare$$

CASE 4.7
We consider the equation

$$H(x) = 1 + xH(x) \int_0^1 k(x,t)\psi(t)H(t)dt$$

$$+ \int_0^1 F(x,t,H(x),H(t))dt , \qquad (4.34)$$

which is a generalization of the classical Chandrasekhar equation (1.22) for $k(x,t) = \frac{x}{x+t}$ and $F \equiv 0$ for all $x, t \in [0, 1]$.

The kernel function $k(x,t)$ is known on $[0, 1] \times [0, 1]$ and satisfies the conditions
(a)
$$0 < k(x,t) < 1, k(0,0) = 1, x+t \neq 0 , \qquad (4.35)$$

and
(b)
$$k(x,t) + k(t,x) = 1, x+t \neq 0 , \qquad (4.36)$$

for all $x, t \in [0, 1]$.

The operator F is a perturbation of the Chandrasekhar H-equation (1.22).

We prove the existence of two positive distinct solutions of (4.34) in the spaces $X = C[0, 1]$ and $X^p = C^p[0, 1]$, $(0 < p < 1)$, respectively, under certain assumptions on F and ψ.

The main idea is a fixed point theorem of Darbo [87] (see Theorem 2.1) for a set contraction map used to prove an existence theorem for an equation of the form (2.1).

For $r > 0$ and $R > 0$, let

$$U_r = \left\{ H(x) \in X_+ \mid 1 - \int_0^1 k(x,t)\psi(t)H(t)dt \geq r \right\} ,$$

$$U_r^R = \{H(x) \in U_r \mid \|H\|_X \le R\} \,,$$

$$U_r^P = \left\{ H(x) \in X_+^P \mid 1 - \int_0^1 k(x,t)\psi(t)H(t)dt \ge r \right\} \,,$$

and

$$U_r^{PR} = \left\{ H(x) \in U_r^P \mid \|H\|_{X^P} \le R \right\} \,,$$

where X_+^P denotes the cone of nonnegative functions in X^P (see Section 2.1). We can now prove the theorem:

THEOREM 4.7

Suppose that the functions ψ, F and $k(x,t)$ satisfy the following conditions:

(a) $|F(x,t,w_1,v_1) - F(x,t,w_2,v_2)| \le l_1|w_1 - w_2| + l_2|v_1 - v_2|$ *for all* $x,t \in [0,1]$ *and all* $w_i, v_i \in \mathbf{R}^+$, $i = 1,2$;

(b) $|F(x,t,w,v)| \le \varepsilon$;

(c) $c = \sup_{0 \le x \le 1} \left| \int_0^1 k(x,t)\psi(t)dt \right| \le \frac{1}{4(1+\varepsilon)}$;

(d) *there exist* $p, q_3 > 0$ *with* $0 < p < 1$ *such that*

$$\frac{1}{|x_1 - x_2|} \left| \int_0^1 (k(x_1,t) - k(x_2,t))\,\psi(t)dt \right| \le q_3$$

for all $x_1, x_2 \in U_r$, *where r is as defined in (e) and (f).*

Then for all $r > 0$ such that

(e) $\frac{1}{r}(l_1 + l_2) < 1$; *and*

(f) $r_1 \le r \le r_2$, *where*

$$r_1 = \frac{1 - \sqrt{1 - 4c(1+\varepsilon)}}{2}, \quad r_2 = \frac{1 + \sqrt{1 - 4c(1+\varepsilon)}}{2}$$

and for all $R > 0$ such that $R_1 \le R \le R_2$, where

$$R_1 = \frac{1 - \sqrt{1 - 4c(1+\varepsilon)}}{2c}, \quad R_2 = \frac{1 + \sqrt{1 - 4c(1+\varepsilon)}}{2c}$$

there exists a solution H to Equation (4.34) such that $H \in U_r^R$.

PROOF   Equation (4.34) can equivalently be written in the form

$$H(x) = \frac{1 + \int_0^1 F(x, t, H(x), H(t))dt}{1 - \int_0^1 k(x, t)\psi(t)H(t)dt} \tag{4.37}$$

or

$$T(H) = L(H)P(H) \tag{4.38}$$

where

$$L(H)(x) = 1 + \int_0^1 F(x, t, H(x), H(t))dt \tag{4.39}$$

and

$$P(H)(x) = \left[1 - \int_0^1 k(x, t)\psi(t)H(t)dt \right]^{-1}. \tag{4.40}$$

The space X with the maximum norm is a Banach algebra.

We will first show that P is a compact operator on U_r. We know [100] that if for every uniformly bounded sequence $\{H_n\}$ in a subset of X, there is a $p, 0 < p < 1$, such that $P(H_n) \in X^p$ for every n and $\{K(H_n)\}$ is bounded in the norm of X^p, then P is a compact operator.

Let $H_n \in U_r$, where r satisfies (f) and such that $\|H_n\|_X \leq q$, for some $q > 0$. Consider the sequence h_n defined by

$$h_n(x) = \int_0^1 k(x, t)\psi(t)H_n(t)dt , \tag{4.41}$$

then

$$\frac{|h_n(x_1) - h_n(x_2)|}{|x_1 - x_2|^p} = \frac{1}{|x_1 - x_2|^p} \left| \int_0^1 (k(x_1, t) - k(x_2, t))\psi(t)H_n(t)dt \right| .$$

It can easily be seen using (d) that there exists $q_1 > 0$ such that

$$\frac{|h_n(x_1) - h_n(x_2)|}{|x_1 - x_2|^p} \leq q_1 \quad \text{for } x_1, x_2 \in [0, 1] . \tag{4.42}$$

Therefore, by (4.42), $h_n \in X^p$ and

$$\|P(H_n)\|_{X^p} \leq \frac{1}{r} + \frac{q_1}{r^2} = q_2 . \tag{4.43}$$

From (4.43) we get that P is a compact operator on U_r and

$$\sup_{H \in U_r} \|P(H)\|_X \leq \frac{1}{r} \,. \tag{4.44}$$

Using condition 1, we obtain the estimate

$$\|L(H_1) - L(H_2)\|_X = \max_{0 \leq x \leq 1} \left| \int_0^1 (F(x, t, H_1(x), H_1(t)) - F(x, t, H_2(x), H_2(t))) \, dt \right|$$

$$< (l_1 + l_2) \|H_1 - H_2\|_X \,. \tag{4.45}$$

Let $H \in U_r^R$. Then $T(H) \in U_r^R$ if

$$1 - \int_0^1 k(x, t)\psi(t) \left(1 + \int_0^1 P(x, t, H(x), H(t)) dt \right) \cdot$$

$$\cdot \left[1 - \int_0^1 k(x, t)\psi(t) H(t) dt \right]^{-1} \geq r \tag{4.46}$$

and

$$\|T(H)\|_X = \|L(H)P(H)\|_X$$

$$\leq (1 + \varepsilon) \left[1 - \int_0^1 k(x, t)\psi(t) H(t) dt \right]^{-1} \leq R \,. \tag{4.47}$$

The inequalities (4.46) and (4.47) are now satisfied if

$$1 - \frac{1}{r}(1 + \varepsilon)c \geq r \tag{4.48}$$

and

$$\frac{1}{1 - cR}(1 + \varepsilon) \leq R \tag{4.49}$$

hold, respectively. However, inequalities (4.48) and (4.49) are satisfied by the choice of r and R in condition (f).

Hence, U_r^R is an invariant bounded convex set under T. Moreover, from (4.44), (4.45), condition (e) and Theorem 4.6, T is a strict set-contraction operator.

Therefore, T has a fixed point $H \in U_r^R$. That is, H satisfies (4.37) and (4.34).

That completes the proof of the theorem.   ∎

THEOREM 4.8
Assume:

(a)   *the conditions (a)–(d) and (f) in Theorem 4.7 are satisfied;*

(b)   *the partial derivatives* $\frac{\partial^2 F}{\partial x \partial w}$, $\frac{\partial^2 F}{\partial x \partial v}$, $\frac{\partial^2 F}{\partial w \partial v}$, $\frac{\partial^2 F}{\partial v^2}$ *exist and*

$$l_3 = \sup \left| \frac{\partial^2 F}{\partial x \partial w} \right| ,$$

$$l_4 = \sup \left| \frac{\partial^2 F}{\partial x \partial v} \right| ,$$

$$l_5 = \sup \left| \frac{\partial^2 F}{\partial w \partial v} \right| ,$$

$$l_6 = \sup \left| \frac{\partial^2 F}{\partial v^2} \right| ,$$

are finite, and that

(c) $\sup \| P(H) \|_{X^p} \cdot D < 1,$ *where*

$$D = 2l_1 + l_2 + l_3 + l_4 + R (l_5 + l_6) .$$

Then there exists a solution $H \in U_r^{pR}$ *of (4.34). Here r is as defined in (f) and*

$$R_1 \leq R \leq R_2 \tag{4.50}$$

where

$$R_1 = \frac{1 - (ac + db) - \sqrt{\Delta}}{2ad} ,$$

$$R_2 = \frac{1 - ac - bd + \sqrt{\Delta}}{2ad} ,$$

$$\Delta = 1 + (ac + bd)^2 - 2ac - 2bd ,$$

$$a = \frac{m}{r^2}$$

$$m = \sup_{x_1, x_2 \in (0,1)} \left| \int_0^1 (k(x_1, t) - k(x_2, t)) \, \psi(t) dt \right| ,$$

(assuming that it is finite)

$$b = \frac{1}{r} ,$$

$$c = 1 + \varepsilon + l_0 ,$$

$$d = l_1 ,$$

$$l_0 = \sup \left| \frac{\partial F}{\partial x} \right| .$$

For m and l_1 sufficiently small $0 < R_1 < R_2$.

PROOF   As in Theorem 4.7 we show that if $H_n \in X^{p_1}$ and $\|H_n\|_{X^{p_1}} \le q$, then $\|P(H)\|_{X^p} \le q_1$ for all $0 < p_1 < p < 1$. Then, P is a compact operator on U_r^p.

Let $H_1, H_2 \in U_r^{pR}$ and set

$$I = \int_0^1 F(x_2, t, H_2(x_2), H_2(t)) \, dt - \int_0^1 F(x_2, t, H_1(x_2), H_1(t)) \, dt$$

$$- \int_0^1 F(x_1, t, H_2(x_1), H_2(t)) \, dt + \int_0^1 F(x_1, t, H_1(x_1), H_1(t)) \, dt ,$$

$$I_1 = \int_0^1 F(x_2, t, H_2(x_2), H_2(t)) \, dt - \int_0^1 F(x_1, t, H_2(x_2), H_2(t)) \, dt$$

$$- \int_0^1 F(x_2, t, H_1(x_2), H_1(t)) \, dt + \int_0^1 F(x_1, t, H_1(x_2), H_1(t)) \, dt ,$$

$$I_2 = \int_0^1 F(x_1, t, H_2(x_2), H_2(t)) \, dt - \int_0^1 F(x_1, t, H_2(x_2), H_1(t)) \, dt$$

$$- \int_0^1 F\left(x_1, t, H_2\left(x_1\right), H_2(t)\right) dt + \int_0^1 F\left(x_1, t, H_1\left(x_1\right), H_1(t)\right) dt \, ,$$

$$I_3 = \int_0^1 F\left(x_1, t, H_2\left(x_2\right) H_1(t)\right) dt - \int_0^1 F\left(x_1, t, H_1(t), H_1\left(x_2\right)\right) dt$$

$$- \int_0^1 F\left(x_1, t, H_2\left(x_1\right), H_1(t)\right) dt + \int_0^1 F\left(x_1, t, H_1\left(x_1\right), H_1(t)\right) dt \, .$$

It is immediate that

$$I = I_1 + I_2 + I_3 \, .$$

As in [74], we get

$$|I_1| \leq (l_3 + l_4) |x_2 - x_1| \, \|H_2 - H_1\|_X \, ,$$

$$|I_2| \leq l_5 \int_0^1 |H_2(t) - H_1(t)| \, |H_2\left(x_2\right) - H_1\left(x_1\right)| \, dt$$

$$\leq l_5 \cdot R \, |x_2 - x_1|^P \, \|H_2 - H_1\|_X$$

(since $\|H\|_{X^P} \leq R$),

$$|I_3| \leq l_1 \, |H_2\left(x_2\right) - H_1\left(x_2\right) - H_2\left(x_1\right) + H_1\left(x_1\right)| + R l_6 \, |x_2 - x_1|^P \, \|H_2 - H_1\|_X \, .$$

From the above three inequalities we easily obtain

$$\|L\left(H_2\right) - L\left(H_1\right)\|_{X^P} \leq D \cdot \|H_2 - H_1\|_{X^P} \, .$$

That is by (c) T is a strict set-contraction operator.

Working exactly as in the second part of the proof of Theorem 4.7, we show that if $H \in U_r^{pR}$ then $T(H) \in U_r^P$, where r is defined by (f) and R is as defined by (4.34).

The result now follows as in Theorem 4.7. ∎

THEOREM 4.9
Assume:

(a) *the conditions of Theorem 4.7 hold with $k(x, t)\psi(t) \geq 0$ for all $x, t \in$
[0, 1];*

(b) the operator $F(x, t, w, v)$ is nondecreasing with respect to w and v and

$$\int_0^1 [k(x, t)\psi(t) + F(x, t, 1, 1)]dt \geq 0 \ \text{for} \ 0 \leq x \leq 1 . \qquad (4.51)$$

Define the sequence $\{H_n\}$, $n = 0, 1, 2, \ldots$ *by*

$$H_0 = 1$$

$$H_{n+1} = T(H_n) = L(H_n) P(H_n) .$$

Then the sequence $\{H_n\}$, $n = 0, 1, 2, \ldots$ *converges uniformly to a solution* $H \in U_r^R$ *of (4.34), where* r, R *are as defined in (f).*

PROOF   By (a) and (b), the sequences

$$Q = \{H_0, H_1, \ldots, H_n, \ldots\}$$

and

$$T(Q) = \{H_1, H_2, \ldots, H_{n+1}, \ldots\}$$

are both contained in U_r^R.

According to the proof of Theorem 4.7 there exists l, $0 < l < 1$, such that

$$m_{U_r^R}(T(Q)) \leq l m_{U_r^R}(Q) .$$

Since $Q = H_0 \cup T(Q)$ the measurement of noncompactness of $T(Q)$ is just $m(Q)$, that is $m_{U_r^R}(Q) = 0$. Therefore, Q is relatively compact and there exists a subsequence which converges uniformly to a point H in U_r^R.

If we show that H_n is a nondecreasing sequence then the entire sequence converges uniformly. We use induction because $H_1 \geq H_0$ by (4.51), assume that $H_n \geq H_{n-1}$.

Then

$$H_{n+1} - H_n = T(H_n) - T(H_{n-1}) = \frac{y}{z} ,$$

where

$$z = \left(1 + \int_0^1 F(x, t, H_{n-1}(x), H_{n-1}(t)) \, dt\right) \int_0^1 k(x, t)\psi(t) (H_n(t) - H_{n-1}(t)) \, dt$$

$$+ \left(1 - \int_0^1 k(x,t)\psi(t)H_{n-1}(t)dt \right) \int_0^1 (F(x,t,H_n(x),H_n(t))$$

$$- F(x,t,H_{n-1}(x),H_{n-1}(t))\,dt\;,$$

$$y = \left(1 - \int_0^1 k(x,t)H_n(t)\psi(t)dt \right) \left(1 - \int_0^1 k(x,t)\psi(t)H_{n-1}(t)dt \right).$$

Now, $y > 0$ since H_n, $H_{n-1} \in U_r^R$ and $z \geq 0$ by the monotonicity of F and the induction hypothesis.

Therefore, $H_{n+1} \geq H_n$, that is the sequence H_n converges to H uniformly and $H \in U_r^R$ is a solution of (4.34).   ∎

Example 4.2

Set

$$F = \varepsilon e^{-(w+v)}\;,$$

then for all $w, v \in \mathbf{R}^+$ and $0 < \varepsilon < 1 - \frac{1}{4b}$, $0 < b < \frac{1}{4}$ the conditions of Theorem 4.9 are all satisfied for

$$0 < \varepsilon < \min\left(\left[(1+R)\left(\frac{1}{2}+\frac{1}{r^2}\right)R\bar{c} \right]^{-1}, \frac{1}{4c} - 1 \right),$$

where

$$\bar{c} = \sup_{x_1,x_2 \geq 0} \left| \int_0^1 (k(x_1,t) - k(x_2,t))\,\psi(t)dt\,|x_1 - x_2| \right|^{-p}.$$

Also, note that

$$\sup_{H \in U_r^{pR}} \|P(H)\|_{X^p} \leq \frac{1}{r} + \frac{1}{r^2}R\bar{c}$$

where r and R are as in Theorem 4.9.

THEOREM 4.10

Assume:

(a)

$$2\int_0^1 \psi(t)dt + 2\int_0^1\int_0^1 F(x,t,H(x),H(t))dtdx < 1 \qquad (4.52)$$

where H is a solution of (4.34) obtained under the hypotheses of Theorem 4.7;
  (b) the following estimate is true:

$$\int_0^1 \frac{\psi(t)}{1-t} H(t)dt > 1 .$$

(4.53)

Also,
  (c) there exist functions $\varphi_1, \varphi_2, \varphi_3 \in X$ such that

$$k(x, t) [\varphi_2(x)(1 - kt) - (1 + \varphi_1(t))] + \varphi_3(x) = 0 ,$$

(4.54)

$$\varphi_1(x) + kx > 0, \quad if \ x \neq 0, \ \varphi_1(0) = 0$$

(4.55)

and

$$[1 + \varphi_1(x)] \left[\varphi_2(x) \left(H(x) - 1 - \int_0^1 F(x, t, H(x), H(t))dt \right) + \varphi_3(x)H(x) \right]$$

$$= (1 - kx) \left(H_1(x) - 1 - \int_0^1 F(x, t, H_1(x), H_1(t)) \, dt \right)$$

(4.56)

for all $x, t \in [0, 1]$, where k is the unique number in $(0, 1)$ for which

$$\int_0^1 \frac{\varphi(t)}{1 - kt} H(t)dt = 1 .$$

(4.57)

Here the function H_1 is given by

$$H_1(x) = \frac{1 + \varphi_1(x)}{1 - kx} H(x), \quad x \in [0, 1] .$$

(4.58)

Then H_1 is a second solution of (4.34) such that

$$H_1(x) > H(x), \quad for \ all \ x \in [0, 1] .$$

PROOF   By the monotone convergence theorem

$$\lim_{k \to 1^-} \int_0^1 \frac{\psi(t)}{1 - kt} H(t)dt = \int_0^1 \frac{\psi(t)}{1 - t} H(t)dt$$

since $(1 - kt)^{-1}$ increases monotonically with k, $0 < k < 1$.

If (4.53) holds, we have

$$\int_0^1 \frac{\psi(t)}{1 - 0 \cdot t} H(t)dt = 1 - \left[1 - 2 \int_0^1 \psi(t)dt \right.$$

$$\left. - 2 \int_0^1 \int_0^1 F(x, t, H(x)H(t))dtdx \right]^{1/2} < 1 ,$$

and since the function $f : (0, 1) \to \mathbf{R}$ defined by

$$f(k) = \int_0^1 \frac{\psi(t)}{1 - kt} H(t)dt$$

is strictly increasing, there exists a unique $k \in (0, 1)$, for which (4.57) holds. Let H_1 be defined as in (4.58). We find that for each $x \in [0, 1]$

$$\int_0^1 k(x, t)\psi(t)H_1(t) = \varphi_2(x) \int_0^1 k(x, t)\psi(t)H(t)dt + \varphi_3(x) \int_0^1 \frac{H(t)}{1 - kt} H(t)dt$$

$$= \varphi_2(x) \int_0^1 k(x, t)\psi(t)H(t)dt + \varphi_3(x)$$

$$= \varphi_2(x) \left[1 - \frac{1}{H(x)} - \frac{\int_0^1 F(x, t, H(x), H(t))dt}{H(x)} \right] + \varphi_3(x)$$

$$= 1 - \frac{1}{H_1(x)} - \frac{\int_0^1 F(x, t, H_1(x), H_1(t))dt}{H_1(x)} ,$$

[by (4.54), (4.56), and (4.57)] that is, H_1 satisfies (4.34). By (4.35) it follows that

$$H_1(x) > H(x) \quad \text{for all } x \in [0, 1]$$

which completes the proof of the theorem. ∎

Example 4.3

By choosing the kernel function $k(x, t)$ to be

$$k(x, t) = \frac{x}{x + t}, \quad x, t \in [0, 1]$$

we see that conditions (4.35) and (4.36) are satisfied and that Equation (4.34) reduces to Equation (1.22).

Moreover if we choose $F \equiv 0$, for simplicity, then conditions (4.54), (4.55), and (4.56) are satisfied for

$$\varphi_1(x) = kx$$

$$\varphi_2(x) = \frac{1 - kx}{1 + kx}$$

and

$$\varphi_3(x) = \frac{2kx}{1 + kx} .$$

4.3 Differential Equations

In this section we study the existence and cardinality of solutions of multilinear differential equations giving upper bounds on the number of solutions. These equations appear in pursuit and bending of beams problems [16].

Let $n(i)$, $i = 1, 2, \ldots, m$ be positive integers such that $n(1) \geq n(2) \geq \cdots \geq n(m)$ and let $L_i = \sum_{j=0}^{n(i)} C_{ij} D^j$, $i = 1, 2, \ldots, m$ be regular linear differential operators defined on $C^{n(1)}(I)$, where $I = [a, b]$ usually (but not necessarily). The coefficient functions C_{ij}, $i = 1, 2, \ldots, m$, $j = 0, 1, 2, \ldots, n(i)$ are never vanishing real and continuous on I.

We study the branching of solutions $u \in C^{n(1)}(I)$ to the multilinear equation

$$Mu = (L_1 u)(L_2 u) \cdots (L_m u) = 0 . \tag{4.59}$$

Equation (4.59) is related with the null set $N(M)$

$$N(M) = \left\{ u \in C^{n(1)}(I) : Mu = 0 \right\}$$

which can be infinite dimensional.

We give necessary and sufficient conditions for a $(m-1)$-tuple $(\alpha_1, \alpha_2, \ldots, \alpha_{m-1})$ to be a multiple ordinary branching of a solution to (4.62) where $\alpha_e \in I$, $e = 1, 2, \ldots, m-1$.

We also study the existence and cardinality of solutions to the initial value problem

$$D^{n(1)}u(z) = z_i, \quad i = 1, 2, \ldots, n(1) - 1 \tag{4.60}$$

where $z, z_i \in I$, giving upper bounds on the number of solutions with n multiple branchings.

A few special cases of applications (e.g., pursuit problems and bending of beams [16]) may be formulated in the form (4.59).

Finally, we study the problem

$$\frac{dM}{dx} - \lambda M = 0 \tag{4.61}$$

when it assumes the form (4.59) for some function λ.

DEFINITION 4.3   *Let* $B_1, B_2, \ldots$ *and* B_m *denote bases for* $N(L_1)$, $N(L_2)$, $\ldots$ *and* $N(L_m)$, *respectively, where*

$$B_i = \{U_{1i}, U_{2i}, \ldots, U_{n(i)i}\} \quad with \; \dim(B_i) = n(i), \; i = 1, 2, \ldots, m$$

and let

$$E_j = \left(B_j \cap C^{n(1)}(I)\right) - B_{j-1} \quad with \; \dim\left(E_j\right) = \bar{n}(j) < n(j), \; j = 2, 3, \ldots, m \;.$$

Obviously $N(L_1) \cup N(L_2) \cup \cdots \cup N(L_m) \subset N(M)$. *We will seek solutions* $u \in N(M)$ *of the form*

$$u_{\alpha_1 \alpha_2 \cdots \alpha_{m-1}}(x) = u(x)$$

$$= \begin{cases} \displaystyle\sum_{j=1}^{n(1)} c_{1j}u_{1j}(x) = u_{\alpha_1}(x) & a \le x \le \alpha_1 \\[2em] \vdots \\[1em] \displaystyle\sum_{j=1}^{n(e)} c_{ej}u_{ej} & \alpha_{e-1} \le x \le \alpha_e \\[2em] \displaystyle\sum_{j=1}^{n(e+1)} c_{e+1j}u_{e+1j} & \alpha_e \le x \le \alpha_{e+1} \\[2em] \vdots \\[1em] \displaystyle\sum_{j=1}^{n(m)} c_{mj}u_{mj} = u_{\alpha_m}(x) & \alpha_{m-1} \le x \le b \end{cases} \tag{4.62}$$

for $\alpha_e \in I$, $e = 1, 2, \ldots, m-1$ and $\alpha_e \notin N(L_e) \cup N(L_{e+1})$. A function of the form (4.62) in $N(M)$ will be said to have a single ordinary branching at $x = \alpha_e$, on $[\alpha_{e-1}, \alpha_{e+1}]$. A function of the form (4.62) will be said to have a multiple ordinary branching at $(\alpha_1, \alpha_2, \ldots, \alpha_{m-1})$ on $I = [a, b]$ with $\alpha_e \le \alpha_{e+1}$, $e = 1, \ldots, m-2$.

Denote the Wronskian $W_e(u_{1i}, u_{2i}, \ldots, u_{n(i)i}, u_{1(i+1)})(x_0)$ by $W_e(x_0)$, $e = 1, 2, \ldots, m-1$.

The following theorem shows when $N(M)$ will contain functions having a multiple ordinary branching.

THEOREM 4.11

Assume that

$$n(e) - \bar{n}(e) + n(e+1) \ge n(1) + 1, \quad e = 2, \ldots, m-1 \tag{4.63}$$

and if

(a) E_j has just one function $u_{1j}(x)$, $j = 2, \ldots, m$, then there exists $u \in N(M)$ having a multiple ordinary branching at $(\alpha_1, \ldots, \alpha_{m-1})$ if and only if

$$W_e(\alpha_e) = 0, \ e = 1, \ldots, m-1$$

$$\Longleftrightarrow \left(L_i u_{1(i+1)}\right)(\alpha_i) = 0, \ i = 1, \ldots, m-1. \tag{4.64}$$

(b) $\dim(E_j) \ne 1$, $j = 2, \ldots, m$, then for every $(\alpha_1, \ldots, \alpha_{m-1})$ with $\alpha_e \in \text{int} I$, and $\alpha_e \geq \alpha_{e+1}$, $e = 1, \ldots, m - 2$ there exists a $u \in N(M)$ having a multiple ordinary branching at $(\alpha_1, \alpha_2, \ldots, \alpha_{m-1})$.

PROOF   It is enough to find numbers,

$$C_{11}, \ldots, C_{1n(1)}, C_{21}, \ldots, C_{2n(2)}, \ldots, C_{m1}, \ldots, C_{mn(m)},$$

so that $U \in C^{n(1)}(I)$. Therefore, we must have

$$\left\{
\begin{array}{l}
\displaystyle\sum_{j=1}^{n(1)} c_{1j} u_{1j}^{(k)}(\alpha_1) = \sum_{j=1}^{\bar{n}(2)} c_{2j} u_{2j}^{(k)}(\alpha_1) \\[2em]
\displaystyle\sum_{j=1}^{n(2)} c_{2j} u_{2j}^{(k)}(\alpha_2) = \sum_{j=1}^{\bar{n}(3)} c_{3j} u_{3j}^{(k)}(\alpha_2), k = 0, 1, \ldots, n(1) \\[2em]
\qquad\qquad\qquad\vdots \\[1em]
\displaystyle\sum_{j=1}^{n(m-1)} c_{m-1j} u_{m-1j}^{(k)}(x) = \sum_{j=1}^{\bar{n}(m)} c_{mj} u_{mj}^{(k)}(\alpha_{m-1})
\end{array}
\right\} \qquad (4.65)$$

CASE 4.8
In this case, (4.65) becomes

$$\sum_{j=1}^{n(e)} c_{ej} u_{ej}^{(k)}(\alpha_e) - c_1^{n(e+1)} u_{1n(e+1)}(\alpha_e) = 0$$

$$e = 1, 2, \ldots, m - 1, \ k = 0, 1, \ldots, n(1) \qquad (4.66)$$

where $c_{1n(e+1)} \neq 0$ *(we take* $c_{1n(e+1)} = 1$*). The homogeneous equation (4.66) has a nontrivial solution if and only if (4.64) holds.*

Note that it is easy to verify that

$$W_e(\alpha_e) = W_e\left(u_{1e}, u_{2e}, \ldots, u_{n(e)e}, u_{1(e+1)}\right)(\alpha_e)$$

$$= a_{en(e)}^{-1}(\alpha_e) W_e\left(u_{1e}, u_{2e}, \ldots, u_{n(e)e}(\alpha_e)\left(L_e u_{1(e+1)}\right)(\alpha_e)\right)$$

$e = 1, 2, \ldots, m - 1$.

CASE 4.9

If $(L_e u_{s_e(e+1)})(\alpha_e) = 0$, $e = 1, 2, \ldots, m-1$ we let $c_{s_e(e+1)} = 1$ and the rest of the coefficients are zero. We then work as in Case 4.8. Otherwise we write (4.65) as

$$\sum_{j=1}^{n(e)} c_{ej} u_{ej}^{(k)} (\alpha_e) - c_{1n(e+1)} u_{in(e+1)} (\alpha_e) = \sum_{j=2}^{\bar{n}(e+1)} c_{jn(e+1)} u_{jn(e+1)} (\alpha_e)$$

$e = 1, 2, \ldots, m-1$, $k = 0, 1, \ldots, n(1)$.

Note now that the rank of the coefficients matrix on the left-hand side is $(n(1)+1)$ and thus we have a unique solution for the coefficients on the left-hand side for any choice of the coefficients on the right-hand side and for any $a_e \in I$, $e = 1, 2, \ldots, m-1$. ∎

The next theorem characterizes the conditions which the coefficients in (4.62) must satisfy in order that multiple branching can occur at $(\alpha_1, \alpha_2, \ldots, \alpha_{m-1})$ with $\alpha_e \le \alpha_{e+1}$, $e = 1, \ldots, m-2$ and $\alpha_e \in I$.

THEOREM 4.12

The following are equivalent:

$$u \in N(M) \text{ on } [c, d] \subset I \quad \text{and } u \text{ is as in (4.65)} \tag{4.67}$$

$$\left(L_e \left(\sum_{j=1}^{n(e+1)} c_{e+1j} u_{e+1j} \right) \right) (\alpha_e) = 0, \quad e = 1, \ldots, m-2 \tag{4.68}$$

$$D^{k_e} \left(L_{e+1} \left[\sum_{j=1}^{n(e)} c_{ej} u_{ej} \right] \right) (\alpha_e) = 0,$$

$$k_e = 0, 1, \ldots, n(e+1) - n(e) \tag{4.69}$$

In particular, (4.68) with $c_{e+1j} \ne 0$ for at least one $u_{e+1j} \in E_j$ and (4.69) with $c_{ej} \ne 0$ for at least one $u_{ej} \in B_e - E_{e+1}$ are both necessary and sufficient conditions for $U \in N(M)$ to have a multiple branching at $(\alpha_1, \alpha_2, \ldots, \alpha_{m-1})$ on $[c, d]$.

PROOF   If $B_e \cap E_{e+1} \ne 0$, $e = 1, 2, \ldots, m-2$ the result is trivially true.

Otherwise, as in Theorem 4.11, we have that $u \in N(M)$ if and only if

$$\sum_{j=1}^{n(e)} c_{ej} u_{ej}^{(k)} (\alpha_e) = \sum_{j=1}^{n(e+1)} c_{e+1j} u_{e+1j}^{(k)} (\alpha_e) ,$$

$$k = 0, 1, \ldots, n(1), \quad e = 1, 2, \ldots, m-1 .$$

The above can be written in the form

$$\sum_{j=1}^{n(e)} c'_{ej} u_{ej}^{(k)} (\alpha_e) = \sum_{j=1}^{\tilde{n}(e+1)} c_{e+1j} u_{e+1j}^{(k)} (\alpha_e) , \qquad (4.70)$$

$k = 0, 1, \ldots, n(1), e = 1, 2, \ldots, m-1$ where $\{u_{e+1j}\}_{j=1}^{\tilde{n}(e+1)} = E_{e+1}$ and at least one $c_{e+1j} \neq 0$. Here $c'_{ej} = c_{ej} - c_{e+1j}$ if $u_{e+1j} \in B_e \cap E_{e+1}$, $c'_{ej} = c_{ej}$ otherwise. Now set $c'_{e(n(e)+1)} = -1$ and

$$u_{e(n(e)+1)}(x) = \sum_{j=1}^{\tilde{n}(e+1)} c_{e+1j} u_{e+1j}^{(k)}(x)$$

and (4.70) can be written

$$\sum_{j=1}^{n(e)+1} c'_{ej} u_{ej}^{(k)} (\alpha_e) = 0, \quad k = 0, 1, \ldots, n(1), \quad e = 1, 2, \ldots, m-1 . \qquad (4.71)$$

Now, (4.71) has a nontrivial solution for c'_{ej} if and only if

$$W_e \left(u_{1i}, u_{2i}, \ldots, u_{n(e)i}, u_{n(e)+1i} \right) (\alpha_e) = 0 ,$$

but

$$W_e \left(u_{1i}, u_{2i}, \ldots, u_{n(e)i}, u_{n(e)+1i} \right) (\alpha_e)$$

$$= a_{en(e)}^{-1} (\alpha_e) W_e \left(u_{1i}, u_{2i}, \ldots, u_{n(e)i} \right) (\alpha_e) L_e u_{e(n(e)+1)} (\alpha_e) ,$$

i.e., if and only if (4.68) holds and at least one $c_{p+1j} \neq 0$.

On the other hand, u has a nontrivial branching at $(\alpha_1, \ldots, \alpha_{m-1})$ if and only if (4.70) has a nontrivial solution for the coefficients on the right-hand side. As before we set $c_{e(\bar{n}(e+1)+1)} = -1$ and

$$u_{e(\bar{n}(e+1)+1)}(x) = \sum_{j=1}^{n(e)} c'_{ej} u_{ej}^{(k)}(x)$$

and (4.70) can now be written as

$$\sum_{j=1}^{\bar{n}(e+1)+1} c_{e+1j} u_{e+1j}^{(k)}(\alpha_e) = 0, \quad k = 0, 1, 2, \ldots, n(1), \quad e = 1, \ldots, m-1 \quad (4.72)$$

or

$$A_e \bar{d}_e = \bar{0} \quad (4.73)$$

in matrix form, where A_e is the coefficient matrix in (4.72) and $\bar{d}_e$ the unknown vector. There will exist a nontrivial solution $\bar{d}_e \neq \bar{0}$, $e = 1, 2, \ldots, m-1$ if and only if the rank of A_e, $e = 1, 2, \ldots, m-1 \leq \bar{n}(e+1)$. But the $\bar{n}(e+1) \times \bar{n}(e+1)$ principle submatrix of A_e is the Wronskian matrix evaluated at α_e. Hence, the rank of $A_e \geq \bar{n}(e+1)$. Therefore, (4.73) will have a nontrivial solution if and only if the rank of A_e is $\bar{n}(e+1)$. Now elementary row operations on A_e show that this is equivalent to (4.69). ∎

We now show that $N(M)$ may contain infinitely many linearly independent functions.

THEOREM 4.13
Assume that either case 4.8 holds in theorem for infinitely many $(\alpha_{1i}, \alpha_{2i}, \ldots, \alpha_{m-1i})$, $i = 1, 2, \ldots$ or case 4.9 holds. In either case, there is a sequence $\{u_{\alpha_{1i}\alpha_{2i}\ldots\alpha_{m-1i}}\}_{i=1}^{\infty} \subset N(M)$ such that $u_{\alpha_{1i}\alpha_{2i}\ldots\alpha_{m-1i}}$ has a multiple branching at $(\alpha_{1i}, \alpha_{2i}, \ldots, \alpha_{m-1i})$ with $\alpha_{ei} < \alpha_{e+1i}$, $e = 1, \ldots, m-2$, $i = 1, 2, \ldots$ and the set $\{u_{\alpha_{1i}\alpha_{2i}\ldots\alpha_{m-1i}}\}_{i=1}^{n}$ is linearly independent on I for every n.

PROOF We proceed by induction. We may assume without loss of generality that $\alpha_{ei} < \alpha_{ei+1}$, $i = 1, 2, \ldots$, $e = 1, 2, \ldots, m-2$. Choose $u_{e+1j}s \subset E_{e+1}$ then

$$L_p \left(\sum_{j=1}^{\bar{n}(e+1)} c_{e+1j} u_{e+1j} \right)(x) \neq 0, \quad x \in \left(\alpha_{e1}, \alpha_{(e+1)} \right).$$

Hence, $u_{\alpha_{e1}}(x) \neq 0$ on $[\alpha_{e1}, \alpha_{(e+1)1}]$, so

$$u_{\alpha_{11}\alpha_{21}...\alpha_{m-11}}(x) \neq 0 \text{ on } I = [a, b] .$$

Now suppose that $u_{\alpha_{1i}\alpha_{2i}...\alpha_{m-1i}}, i = 1, 2, \ldots, n$ are linearly independent. Suppose that there exist constants $d_k, i = 1, 2, \ldots, n+1$:

$$\sum_{i=1}^{n+1} d_i u_{\alpha_{1i}\alpha_{2i}...\alpha_{m-1}}(x) = 0 .$$

If $d_{n+1} = 0$, then $d_i = 0, i = 1, 2, \ldots, n$ and $\{u_{\alpha_{1i}\alpha_{2i}...\alpha_{m-1}}\}_{i=1}^{n+1}$ is linearly independent. If $d_{n+1} \neq 0$

$$u_{\alpha_{1n+1}\alpha_{2n+1}...\alpha_{m-1n+1}}(x) = d_{n+1}^{-1} \sum_{i=1}^{n} d_i u_{\alpha_{1i}\alpha_{2i}...\alpha_{m-1i}}$$

for all $x \in I$ in particular for each $x \in (\alpha_{e-1i}, \alpha_{e+1i})$, but

$$L_e u_{\alpha_{en+1}}(x) = 0, \quad x \in (\alpha_{en-1}, \alpha_{en+1})$$

whereas

$$\left(d_{n+1}^{-1} \sum_{i=1}^{n} d_i u_{ei} \right) \in \text{span} E_{e+1} \text{ when } x \in (\alpha_{en}, \alpha_{en+1}) ,$$

so $L_e u_{\alpha_{en+1}}(x) \neq 0$ for some $x \in (\alpha_{en}, \alpha_{en+1})$ a contradiction. ∎

DEFINITION 4.4   *Define the set S_i by setting*

$$S_i = \{x \in I / (L_i u)(x) = 0\} .$$

Then since $L_i u, i = 1, 2, \ldots, m$ are continuous functions on I the $S_i s$, $i = 1, 2, 3, \ldots, m$ are closed sets and $S_1 \cup S_2 \cup \cdots \cup S_m = I$. In particular, any point $\alpha_e \in [\alpha_{e-1}, \alpha_{e+1}]$ at which an ordinary branching occurs on $[\alpha_{e-1}, \alpha_{e+1}]$, $e = 1, 2, \ldots, m-1$ must belong to $S_{e-1} \cap S_{e+1}$ together with any limit point of the set of points at which ordinary branching occurs since $S_{e-1} \cap S_{e+1}$ is closed.

We now show that $S_{e-1} \cap S_{e+1}$ is nowhere dense in $[\alpha_{e-1}, \alpha_{e+1}]$, $e = 1, 2, \ldots, m-1$.

THEOREM 4.14

Assume that $u \in N(M)$ as in (4.62) and $B_e \cap E_{e+1} = \emptyset$. Then $S_{e-1} \cap S_{e+1}$ is nowhere dense in $[\alpha_{e-1}, \alpha_{e+1}]$, $e = 1, 2, \ldots, m - 1$.

PROOF   Suppose that $S_{e-1} \cap S_{e+1}$, $e = 1, 2, \ldots, m - 1$ contains a maximal closed interval $[\alpha'_{e-1}, \alpha'_{e+1}]$ with $|\alpha'_{e+1} - \alpha'_{e-1}| \neq 0$, $e = 1, 2, \ldots, m - 1$. Then

$$u(x) = \sum_{j=1}^{n(e+1)} c_{ej} u_{ej} \quad \text{for } x \in [\alpha'_{e-1}, \alpha'_{e+1}] \, .$$

Now let $(\alpha''_{e-1}, \alpha''_{e+1}) \subset [\alpha'_{e-1}, \alpha'_{e+1}]$. Then by Case 4.8 in Theorem 4.11 there exist constants $c_{ej}^{(1)}$, $c_{ej}^{(2)}$, such that

$$u\left(\alpha''_{e-1}, \alpha''_{e+1j}(x)\right) = \begin{cases} \displaystyle\sum_{j=1}^{n(e)} c_{ej}^{(1)} u_{ej}(x) & \alpha_{e-1} \leq x \leq \alpha''_{e-1} \\[2ex] \displaystyle\sum_{j=1}^{n(e+1)} c_{ej} u_{ej}(x) & \alpha''_{e-1} \leq x \leq \alpha''_{e+1} \\[2ex] \displaystyle\sum_{j=1}^{n(e)} c_{ej}^{(2)} u_{ej}(x) & \alpha''_{e+1} \leq x \leq \alpha_{e+1} \end{cases}$$

belongs to $N(M)$ since

$$L_e\left(\sum_{j=1}^{n(e+1)} c_{ej} u_{ej}\right)(z) = 0 \quad \text{at } z = \alpha''_{e-1}, \alpha''_{e+1} \, .$$

But $L_e\left(\sum_{j=1}^{n(e+1)} c_{ej} u_{ej}\right)(x) = 0$ on $[\alpha''_{e-1}, \alpha''_{e+1}]$. Hence $U(\alpha''_{e-1}, \alpha''_{e+1j})(x) \in N(L_e)$. Since

$$\sum_{j=1}^{n(e+1)} c_{ej} u_{ej}(x) \notin N(L_e) \, ,$$

$e = 1, 2, \ldots, m - 1$ the proof of Theorem 4.13 shows that the set $B_e \cup \{u(\alpha''_{e-1}, \alpha''_{e+1j}(x)\}$ is linearly independent. But this contradicts $d(L_e) = n(e)$, $e = 1, 2, \ldots, m$.

We now assume that $n(1) = n(2) = \cdots = n(m)$ for simplicity (the other cases can be dealt analogously) and consider the following problem: given

$(z_0, z_1, \ldots, z_{n(1)-1}) \in \mathbf{R}^{n(1)}$ and $z \in I$ find u such that

$$M_u = (L_1 u)(L_2 u) \ldots (L_m u) = 0$$

$$D^{n(1)} u(z) = z_i, \quad i = 0, 1, \ldots, n(1) - 1 \tag{4.74}$$

if $N(L_e) \neq N(L_{e+1})$, $e = 1, 2, \ldots, m - 1$, then we have at least m solutions, the unique solutions belonging to $N(L_e)$, $e = 1, 2, \ldots, m - 1$. In addition, according to Theorems 4.11 and 4.12 we may have solutions with one or many multiple ordinary branchings.

In the event that L_e, $e = 1, 2, \ldots, m$ have constant coefficients, we proceed as follows: let s_{ej}, $j = 1, 2, \ldots, n(1)$, $e = 1, 2, \ldots, m$ denote the solutions of the characteristic equation L_e and assume $u \in N(L_e)$ on some subinterval $I(z)$ of I containing z, then the restriction $\bar{u}$ of u on $I(z)$ can be written

$$\bar{u}(x) = \sum_{j=1}^{n(1)} c_{je} e^{s_{je} x}$$

where $\{e^{s_{je} x}\}_{j=1}^{n(1)}$ spans $N(L_e)$ and c_{je} are uniquely determined by (4.74). By (4.69) we must have

$$L_{e+1}(\bar{u}(\alpha_e)) = 0, \quad e = 1, 2, \ldots, m - 1.$$

It follows that

$$\sum_{j=1}^{n(1)} d_{je} e^{t_{je} \alpha_e} = 0 \tag{4.75}$$

where

$$d_{je} = c_{je} \sum_{i=0}^{n(1)} c_{ie+1} t_{ie}^i, \quad i = 1, 2, \ldots, n(1) \tag{4.76}$$

$$t_{ie} = s_{ie} - s_{n(1)e}, \quad e = 1, 2, \ldots, m. \tag{4.77}$$

Note that each one of the equations in (4.75) can have at most $n(1) - 1$ real solutions if the d_{je}s and t_{je}s are all real [126].

Denote by $\alpha_{p1}, \alpha_{p2}, \ldots, \alpha_{pn(1)-1}$ the solutions obtained in the pth equation in (4.75), $p = 1, 2, \ldots, m - 1$ and assume that (the other cases can be dealt analogously)

$$\alpha_{11} \leq \alpha_{12} \leq \alpha_{13} \leq \cdots \leq \alpha_{1n(1)-1} \leq$$
$$\alpha_{21} \leq \alpha_{22} \leq \alpha_{23} \leq \cdots \leq \alpha_{2n(1)-1} \leq$$
$$\vdots$$
$$\alpha_{m-11} \leq \alpha_{m-12} \leq \alpha_{m-13} \leq \cdots \leq \alpha_{m-1n(1)-1}. \tag{4.78}$$

Inequality (4.78) shows that we can have at most $(n(1) - 1)^{m-1}$ ordinary multiple branchings, e.g., $(\alpha_{11}, \alpha_{21}, \ldots, \alpha_{m-11})$ is one of them. We have thus proved the following theorem. ∎

THEOREM 4.15

If L_i, $i = 1, 2, \ldots, m$ have constant coefficients, then there exists a solution $u \in N(M)$ [u as in (4.62)] to the initial value problem (4.74) having a multiple ordinary branching $(\alpha_1, \alpha_2, \ldots, \alpha_{m-1})$ with $\alpha_e \in I$, $e = 1, 2, \ldots, m - 1$ if and only if α_e is a root of the exponential polynomial (4.75), where the $d_{je}s$ and $t_{je}s$ are all real and they are given by (4.76) and (4.77).

Moreover if (4.78) holds there are at most $(n(1) - 1)^{m-1}$ solutions $u \in N(M)$ [u as in (4.62)].

THEOREM 4.16

Assume that the hypotheses of Theorem 4.15 are satisfied. Then there are at most

$$(m - 1)(n(1) - 1)(n(1) - 2)^{n-1} \tag{4.79}$$

solutions u [u as in (4.62)] to the initial value problem having exactly n multiple branchings $(\alpha_1, \alpha_2, \ldots, \alpha_{m-1})$ in I where any $m - 2$ of the $\alpha_e s$ are fixed $e = 1, 2, \ldots, m - 1$.

Moreover in this case if there are no solutions with $n + 1$ multiple branchings, then the total number of solutions to the problem $Mu = 0$ is bounded by

$$(m - 1)(n(1) - 1) \sum_{j=0}^{n-1} (n(1) - 2)^j . \tag{4.80}$$

PROOF Without loss of generality we can assume that α_{11} denotes the first point at which a branching occurs and $u \in N(L_1)$ on some subinterval $I(z) = [z, \alpha_{11}]$. Then $u \in N(L_2)$ on $[\alpha_{11}, \alpha_{11} + \varepsilon)$, for some $\varepsilon > 0$. Then there are at most $m - 1$ possible values for α_{11}. Suppose $w > \alpha_{11}$ is the next point at which a multiple branching of u occurs. Then $u \in N(L_2)$ on $[\alpha_{11}, w]$. Hence, there exist uniquely determined $c_{j2}(\alpha_{11})$, $j = 1, 2, \ldots, n(1)$ such that

$$u(x) = \sum_{j=1}^{n(1)} d_{j2}(\alpha_{11}) u_{j2}(x) \quad \text{on } [\alpha_{11}, w]$$

where $\{u_{j2}\}_{j=1}^{n(1)}$ span $N(L_2)$. By Theorem 4.12,

$$\left[L_1 \left(\sum_{j=1}^{n(1)} d_j(\alpha_{11}) u_{j2} \right) \right](v) = 0 \quad \text{at } v = \alpha_{11} \text{ and } v = w .$$

Hence, there are $m - 2$ possible ws with $w > \alpha_{11}$. This argument applies again for the next branching. Since this argument can be applied in any of the $m - 1$ rows in (4.78), this proves (4.79).

Finally (4.79) can easily be proved if we use (4.79) for $j = 0, 1, 2, \ldots, n$ and add the results. ∎

REMARK 4.4   (a) We can assume in Theorem 4.16 that any h points $h \in \{1, 2, \ldots, m - 1\}$ are fixed from $(\alpha_1, \alpha_2, \ldots, \alpha_{m-1})$. Then proceeding as in Theorem 4.15 we can prove that the corresponding relations for (4.79) and (4.80) are, respectively,

$$(m - 1 - (h - 1))(m(1) - 1)^h (n(1) - 2)^{h-1} \tag{4.81}$$

and

$$(m - 1 - (h - 1))(n(1) - 1)^h \sum_{j=0}^{n-1} (n(1) - 2)^h . \tag{4.82}$$

(b) Up till now we obtained the cardinality results in Theorems 4.15, 4.16, and in (a) above by assuming that (4.78) is true and u as in (4.62). But (4.62) can be written in $(m - 1)!$ different ways by interchanging the role of the L_is, $i = 1, 2, \ldots, m$. Therefore, in general, all the cardinality results obtained up till now can be multiplied by $(m - 1)!$.

(c) If the L_i, $i = 1, 2, \ldots, m$ are nonconstant but continuous (as in the Introduction) we can restate Theorem 4.15 and (4.62). However, the conclusions and the proofs are going to be exactly analogous. ∎

We now provide examples for Theorems 4.14, 4.16, and Equation (4.60).

Example 4.4
Let $m = 2$ and consider the function f defined by

$$f(x) = \begin{cases} x^8 \ln \frac{1}{x^2}, & x \neq 0 \\ 0, & x = 0 \end{cases}$$

$u_1(x) = e^{f(x)}$, $u_2(x) = e^{2f(x)}$, $L_1 u = u' - f'(x)u$, $L_2 u = u' - 2f'(x)u$. Then $u_1 \in N(L_1)$, $u_2 \in N(L_2)$ and $u \in N(M)$ can be written as

$$u(x) = \begin{cases} c e^{f(x)}, & -\varepsilon \leq x \leq 0 \\ d e^{2f(x)}, & 0 \leq x \leq \varepsilon \end{cases}, \varepsilon > 0$$

or

$$u(x) = \begin{cases} c e^{2f(x)}, & -\varepsilon \leq x \leq 0 \\ d e^{f(x)}, & 0 \leq x \leq \varepsilon \end{cases}, \varepsilon > 0$$

That is, 0 is a limit point of branching points of u.

In the event that the characteristic equations of $L_i, i = 1, 2, \ldots, m$ have complex roots (4.75) may have infinite solutions to the initial value problem on $(-\infty, \infty)$ even if we have one ordinary multiple branching in $(-\infty, \infty)$.

Example 4.5
Let $m = 2$, $L_1 = D^2 + 1$, $L_2 = D^2 + 4$, $u(0) = 0$, $u'(0) = 1$. Let $u \in N(L_1)$ on $[-\varepsilon, \varepsilon]$ for some $\varepsilon > 0$. Then

$$u_1(x) = -\frac{i}{2} e^{ix} + \frac{i}{2} e^{-ix} = \sin x$$

and (4.75) due to (4.76) and (4.77) becomes

$$e^{2i\alpha} = 1 .$$

Therefore, $\alpha_n = n\pi$, $n = 0, 1, 2, \ldots$

$$u_2(x) = \frac{1}{4i} e^{2ix} - \frac{1}{4i} e^{-2ix} = \frac{1}{2} \sin 2x .$$

Example 4.6
Consider the equation
$$\frac{dM}{dx} - \lambda M = 0 .$$

Let $L_1 = (D-1)(D-2)(D-3)$, $L_2 = (D-4)(D-5)(D-6)$, $L_3 = (D-7)(D-8)(D-9)$ and $\lambda = 0$. Then

$$u_1(x) = 2e^x - 3e^{2x} + e^{3x}$$

$$u_2(x) = 5e^{4x} - 9e^{5x} + 4e^{6x}$$

$$u_3(x) = 8e^{7x} - 15e^{8x} + 7e^{9x}$$

and

$$L_2 u_1(x) = 0 \rightarrow -120e^x + 72e^{2x} - 6e^{3x} = 0 \rightarrow \alpha = \ln 2, \ln 10$$

$$L_1 u_2(x) = 0 \rightarrow 30e^{4x} - 216e^{5x} + 240e^{6x} = 0 \rightarrow \alpha = \ln\left(\frac{108 + \sqrt{4464}}{240}\right)$$

$$L_3 u_2(x) = 0 \rightarrow -300e^{4x} + 216e^{5x} - 24e^{6x} = 0 \rightarrow \alpha = \ln\left(\frac{54 \pm \sqrt{1116}}{12}\right)$$

$$L_2 u_3(x) = 0 \rightarrow 48e^{7x} - 360e^{8x} + 420e^{9x} = 0 \rightarrow \alpha = \ln\left(\frac{90 \pm \sqrt{3060}}{210}\right).$$

So we can have multiple branchings at

$$\left(\ln 2, \ln\left(\tfrac{54+\sqrt{1116}}{12}\right)\right),$$

$$\left(\ln\left(\tfrac{108+\sqrt{4464}}{240}\right), \ln\left(\tfrac{54+\sqrt{1116}}{12}\right)\right),$$

$$\left(\ln\left(\tfrac{108+\sqrt{4465}}{240}\right), \ln\left(\tfrac{54-\sqrt{1116}}{12}\right)\right),$$

$$\left(\ln\left(\tfrac{108-\sqrt{4464}}{240}\right), \ln\left(\tfrac{54+\sqrt{1116}}{12}\right)\right),$$

$$\left(\ln\left(\tfrac{108-\sqrt{4464}}{240}\right), \ln\left(\tfrac{54-\sqrt{1116}}{12}\right)\right),$$

and

$$\left(\ln\left(\frac{108 - \sqrt{4464}}{240} \right), \quad \ln\left(\frac{90 + \sqrt{3060}}{210} \right) \right).$$

For example we can have the solution $u \in N(M)$ given by

$$u(x) = \begin{cases} 2e^x - 3e^{2x} + e^{3x} & -\infty < x \le \ln 2 \\ 5e^{4x} - 9e^{5x} + 4e^{6x} & \ln 2 \le x \le \ln\left(\frac{54+\sqrt{1116}}{12} \right) \\ 8e^{7x} - 15e^{8x} + 7e^{9x} & \ln\left(\frac{54+\sqrt{1116}}{12} \right) \le x < +\infty \end{cases}$$

etc.

The above are solutions corresponding to the order (L_1, L_2, L_3). But we can obtain additional solutions corresponding to (L_1, L_3, L_2), (L_2, L_1, L_3), (L_2, L_3, L_1), (L_3, L_1, L_2), and (L_3, L_2, L_1).

For a further study we refer the reader to [2, 6, 83, 89, 104, 126].

4.4 Integrals on a Separable Hilbert Space

In this section we give a survey of the properties of the ideals of the space of bounded linear operators on a separable Hilbert space. This section constitutes published and unpublished work of ours. Let H be a complex Hilbert space. Throughout, we assume H is separable, that is, $H \cong C^m$ for some $m \in N$, or H contains a Hilbert basis $\{u_n, n \in N\}$ such that $H = \overline{\text{span}\{u_n, n \in N\}}$, thus $H \cong \ell^2(N, C)$.

Denote by $B(H)$ the space of bounded linear operators on H, which is thought to be equipped with the usual norm, i.e., if $x \in H$ then $\|x\|^2 = (x, x)$.

We identify the ideals of $B(H)$ and discuss the properties and the relationship between them. Some examples are also provided. Finally in this exposition, we tried to simplify the already existing proofs of most of the theorems.

DEFINITION 4.5 *Let $L \in B(H)$, then rank L is obviously a vector subspace. Define by rank(L) the dimension of the rang L and set*

$$\mathcal{F}(H) = \{L \in B(H) \mid \text{rank}(L) \text{ is finite}\}.$$

For an example, fix $b \in H$, $w \in H$. Then $L : x \to (x, w)b$ is obviously linear and bounded since

$$\|Lx\| = |(x, w)| \cdot \|b\| \le \|x\| \cdot \|w\| \cdot \|b\|, \text{ i.e., } \|L\| \le \|w\| \cdot \|b\|.$$

Also $rang(L) = \{0\}$ if $b = 0$, otherwise rang L is spanned by $\{b\}$. So $rank(L) = 0$ or 1 and $L \in \mathcal{F}(H)$. Denote this L as $w \otimes b$.

PROPOSITION 4.6
$\|w \otimes b\| = \|w\| \cdot \|b\|$.

PROOF   Trivial if $w = 0$, so let $w \neq 0$;

$$\|w \otimes b\| = \sup_{\|x\|=1} \|(x, w)b\| \geq \left\|\left(\frac{w}{\|w\|}, w\right)b\right\| = \|w\| \cdot \|b\| \geq \|w \otimes b\| . \quad \blacksquare$$

THEOREM 4.17
$\mathcal{F}(H)$ *is an ideal in $B(H)$, proper if $dim(H) = \infty$.*

PROOF   Let $L_1, L_2 \in \mathcal{F}(H)$, $\lambda \in C$, then $L_1 + L_2 \in \mathcal{F}(H)$; hence vector subspace. Let $L \in B(H)$;

   (i) $rang(L_1 L) \subseteq rang(L_1) < \infty \Rightarrow L_1 L \in \mathcal{F}(H)$;

   (ii) $dim(rang(LL_1)) \leq dim(rang(L_1)) \Rightarrow LL_1 \in \mathcal{F}(H)$.

Let $dim(H) \neq$ finite $\Rightarrow I \in B(H)$ with $dim(ran(I)) = dim H \neq$ finite $\Rightarrow L \notin \mathcal{F}(H)$, hence $\mathcal{F}(H)$ proper; otherwise $\mathcal{F}(H) = B(H)$.

Let $H \otimes H$ denote the vector subspace generated in $B(H)$ by all operators of form $w \otimes b$ $(w, b \in H)$. Thus $L \in H \otimes H \Leftrightarrow L = \sum_{i=1}^n w_i \otimes b_i$ say, $n \in N$, $w_i, b_i \in H$. $\blacksquare$

THEOREM 4.18
$H \otimes H = \mathcal{F}(H)$.

PROOF   We first show $H \otimes H \subseteq \mathcal{F}(H)$. If $L \in H \otimes H \Rightarrow rang(L) \subseteq rang(w_1 \otimes b_1) + \cdots + rang(w_n \otimes b_n) \Rightarrow rank(L) \leq n \Rightarrow L \in \mathcal{F}(H)$. We now show that $\mathcal{F}(H) \subseteq H \otimes H$. Let $L \in \mathcal{F}(H) \Rightarrow rang(L) = $ [subspace $\cong C^n$] $\Rightarrow rang(L)$ has a complete orthonormal basis $\{u_1, \ldots, u_n\}$, say $\Rightarrow (\forall x \in H)$

$$Lx = \sum_{k=1}^n \xi_k u_k = \sum_{k=1}^n (Lx, u_k) u_k = \sum_{k=1}^n (x, L^* u_k) u_k$$

$$= \sum_{k=1}^n (x, w_k) u_k \text{ (say)} = \sum_{k=1}^n (w_k \otimes u_k) x \Rightarrow L = \sum_{k=1}^n w_k \otimes b_k$$

$\in H \otimes H$. ∎

THEOREM 4.19

If $J(H)$ is a nonzero ideal in $B(H)$ then $\mathcal{F}(H) \subseteq J(H) \subseteq B(H)$.

PROOF   Trivial if $\dim(H) =$ finite ($\mathcal{F} = B$ then). Let $\dim(H) = \infty$. Take $L \in \mathcal{F}(H) \setminus \{0\}$; then $\exists x \neq 0$ such that $Lx = y \neq 0$. Take any $u, v \in H$ and find some $B \in B(H)$ with $By = v$, so $B(Lx) = v$. Then $(\forall z \in H)$ $(u \otimes v)z = (z, u)v = (z, u)(BLx) = (BL)((z, u)x) = (BL) \circ (u \otimes x)(z) \Rightarrow u \otimes v = B \circ L \circ (u \otimes x) \Rightarrow u \otimes v \in \mathcal{F}(H)$ (since $L \in \mathcal{F}(H)$). Therefore, every generator $u \otimes v$ of $\mathcal{F}(H)$ lies in $J(H)$, so $\mathcal{F}(H) \subseteq \mathcal{F}(H)$. ∎

COROLLARY 4.2

$\mathcal{F}(H)$ is the smallest nonzero ideal in $B(H)$.

THEOREM 4.20

If $\dim(H) = \infty$, the $\mathcal{F}(H), \| \cdot \|_{op}$ is not a closed ideal.

PROOF   Let $\{u_n, n \in N\}$ be a Hilbert basis for H. Thus $\forall x \in H$, $x = \sum_{n=1}^{\infty}(x, u_n)u_n$ ($\Leftrightarrow H = \overline{\text{span}}\{u_n, n \in N\} \Leftrightarrow \|x\|^2 = \sum_{n=1}^{\infty}|(x, u_n)|^2$; Parseval's formula).

Define operator $L_p = \sum_{n=1}^{p} \lambda_n u_n \otimes u_n$ $(\lambda_k \in C)$, $\in \mathcal{F}(H)$ so,

$$L_p(x) = \sum_{n=1}^{p} \lambda_n (x, u_n) u_n,$$

$$\|L_p - L_{p+q}\|_{op} = \left\| \sum_{n=p+1}^{p+q} \lambda_n u_n \otimes u_n \right\| \leq \sum_{n=p+1}^{p+q} |\lambda|.$$

So choose $(\lambda_n)_{n \in N}$ such that $\sum_n |\lambda_n|$ is convergent; then $(L_p)_{p \in N}$ is Cauchy in complete $B(H), \| \cdot \|_{op}$; hence convergent to some $L \in B(H)$. We now set

$$L = \lim_{p \to \infty} \sum_{n=1}^{p} \lambda_n u_n \otimes u_n = \sum_{n=1}^{\infty} \lambda_n u_n \otimes u_n.$$

Now, $L_p \xrightarrow{u} L \Rightarrow L_p \xrightarrow{p \cdot w} L$, i.e., $(\forall x \in H)\ L_p x = Lx$. But

$$L_p x = \left(\sum_{n=1}^{p} \lambda_n u_n \otimes u_n \right) x = \sum_{n=1}^{\infty} (\lambda_n u_n \otimes u_n) x = \sum_{n=1}^{p} \lambda_n (x, u_n) u_n \ .$$

Also,

$$L_p x \to Lx = \left[\sum_{n=1}^{\infty} \lambda_n u_n \otimes u_n \right] x \quad \text{as } p \to \infty \ .$$

This shows that

$$\lim_{p \to \infty} \left[\sum_{n=1}^{p} [\lambda_n (x, u_n) u_n] \right]$$

also exists in H and we can define it to be

$$\sum_{n=1}^{\infty} [\lambda_n (x, u_n) u_n] = Lx \ .$$

Plainly, if infinitely many $\lambda_n \neq 0$, then $\text{rang}(L)$ has infinite dimension so $L \notin \mathcal{F}(H)$. Thus $\mathcal{F}(H) \neq \overline{\mathcal{F}(H)}$ (closure with respect to $\| \cdot \|_{op}$). ∎

What is $\overline{\mathcal{F}(H)}$ in $B(H)$? The ideal $K(H)$ of compact operators in $B(H)$, $\| \cdot \|_{op}$. Moreover $K(H)$ is the largest, indeed the only, proper uniformly closed ideal in $B(H)$, $\| \cdot \|_{op}$ (provided H is separable), so $K(H)$ is the only maximal ideal. We are now going to describe $K(H)$ and eventually prove that $K(H) = \overline{\mathcal{F}(H)}$.

DEFINITION 4.6   *Let $L \in B(H)$, then L is compact (or completely continuous) if and only if for every bounded sequence $\{x_n\}_{n \in N}$ in H, $\{Lx_n\}$ contains a convergent subsequence in H.*

PROPOSITION 4.7
$\mathcal{F}(H) \subseteq K(H)$.

PROOF   Let $L \in \mathcal{F}(H)$ and $\{x_n\}_{n \in N}$ be a bounded sequence in H. Then $\{Lx_n\}$ forms a bounded set in the subspace ran L which lies in a subspace $\cong C^m$ (some $m \in N$) but every bounded set in C^m contains a convergent subsequence. ∎

Obviously $K(H)$ is a vector subspace $\subseteq B(H)$.

THEOREM 4.21

$K(H)$ *is a uniformly closed ideal in* $B(H)$. *It is proper if* $dim(H) = \infty$.

PROOF (i) Let $L \in K(H)$, $B \in B(H)$ and $\{x_n\}_{n\in N}$ be a bounded sequence. Then,

$$LB (x_n) = L (Bx_n) = L (y_n)$$

and $\{y_n\}_{n\in N}$ is a bounded sequence. So $\{Ly_n\}$ contains a convergent subsequence, so LB is compact. Also, $BL(x_n) = B(Lx_n)$ contains a convergent subsequence $\{By_m\}_{m\in N}$ (since Lx_n contains a convergent subsequence $\{y_m\}$ say), so BL compact. Hence LB, $BL \in K(H)$.

(ii) If $dim(H) = \infty \Rightarrow I$ is not a compact operator since if $\{u_n, n \in N\}$ is a Hilbert basis for H, then $\|u_n - u_m\|^2 = (u_n - u_m, u_n - u_m) = (u_n, u_n) + (u_m, u_m) = 2$, so $\{u_n\}_{n\in N}$ can contain no convergent subsequence.

(iii) $K(H)$ is $\| \cdot \|_{op}$-closed.

"Diagonal proof". Let $L_n \in K(H)$, $(n \in N)$, and $L \in B(H)$ such that $\|L_n - L\|_{op} \to 0$ $(n \to \infty)$. We show $L \in K(H)$: Let $\{x_n\}_{n\in N}$ be a bounded sequence, $\|x_n\| \le b$ $(\forall n \in N)$, say.

Step 1. Take L_1 (compact): Let $\{x_{n1}\}$ be a subsequence of $\{x_n\}$ such that $\{L_1 x_{n1}\}$ is convergent with limit v_1, say.

Step 2. Take L_2: Let $\{x_{n2}\}$ be a subsequence of $\{x_{n1}\}$ such that $\{L_2 x_{n2}\}$ is convergent to v_2, say.

$\vdots$

Step m. Take L_m: Let $\{x_{nm}\}$ be a subsequence of $\{x_{nm-1}\}$ such that $\{L_m x_{nm}\}$ is convergent, to v_m, say.

$\vdots$

"Step ∞". Pick out "diagonal sequence" $x_{11}, x_{22}, \ldots, x_{mm}, \ldots$ one from each step above. For each m, the sequence $(x_{mm}, x_{(m+1)m+1}, \ldots, x_{kk}, \ldots)$ is itself a subsequence of a convergent sequence in Step m; so $\lim L_m x = v_m, m = 1, 2, \ldots$.

We show that $\{Lx_{kk}\}_{k\in N}$ is Cauchy: $\forall \varepsilon > 0$ $\exists m_0$ such that $\|L - L_m\| \le \varepsilon$ $\forall m \ge m_0$. Hence, $\forall k, j \ge m_0$:

$$\left\| Lx_{kk} - Lx_{jj} \right\| \le \left\| (L - L_m) x_{kk} \right\| + \left\| L_m x_{kk} - L_m x_{jj} \right\| + \left\| (L_m - L) x_{jj} \right\|$$

$$\le \varepsilon b + \left\| L_m x_{kk} - L_m x_{jj} \right\| + \varepsilon b .$$

But, $\| L_m x_{kk} - L_m x_{jj} \| \to 0$ as k's $\to \infty$, i.e., $\exists n_0$ such that

$$\left\| L_m x_{kk} - L_m - L_m x_{jj} \right\| < \varepsilon b .$$

Hence $\| Lx_{kk} - Lx_{jj} \| < 3\varepsilon b$ whenever $k, j \ge \max\{m_0, n_0\}$.

Thus the diagonal sequence images $\{Lx_{k^k}\}_{k \in N}$ form a Cauchy sequence in complete H; hence, convergent sequence. Thus $\{Lx_n\}_{n \in N}$ contains a convergent subsequence, and L is compact. ∎

COROLLARY 4.3
No compact operator on infinite-dimensional Hilbert space can be invertible.

COROLLARY 4.4
$K(H)$ is a closed subalgebra.

THEOREM 4.22
Let $L \in B(H)$. The following are equivalent:

(a)   *L is a compact operator;*

(b)   *Given any orthonormal sequence $\{u_n\}_{n \in N}$ in H, $(Lu_n, u_n) \to 0$ $(n \to \infty)$;*

(c)   *There exists a sequence $\{F_n\}_{n \in N}$ of finite rank operators $F_n \in \mathcal{F}(H)$ uniformly convergent to L.*

PROOF   (a) $\Rightarrow$ (b): Given (a), suppose (b) false: then $\exists \{u_n\}_{n \in N}$ and $\delta > 0$ such that

$$(Lu_m, u_m) \geq 2\delta \quad \text{for infinitely many } ms . \tag{4.83}$$

Change notation, let $m = 1, 2, 3, \ldots$ for convenience, so $\exists \{u_m\}$ with property (4.1). But L is compact: Hence $\exists$ convergent subsequence $\{L_{m'}\}$ with limit $v \in H$, so deleting at most finitely many terms, and relabelling the rest we have $\|Lu_m - v\|$, say for all m. Now

$$|(Lu_m, u_m) - (x, u_m)| = |(Lu_m - x, u_m)|$$

$$\leq \|Lu_m - x\| \cdot 1 < \delta \quad (\forall m) .$$

Also,
$$|(Lu_m, u_m) - (x, u_m)| \geq |(Lu_m, u_m)| - |(x, u_m)| .$$

Rearranging, and using (4.83), we get

$$|(x, u_m)| \geq \delta > 0 \quad (\forall m) . \tag{4.84}$$

But Bessel's inequality implies $\sum_{n=1}^{\infty} |(x, u_m)|^2 \leq \|x\|^2$, which contradicts (4.84).

(b) $\Rightarrow$ (c): Given $k \in N$, let S be the collection of all sequences $\bar{U}$ in H (including $u = \emptyset$ and finite $u = \{u_1, u_2, \ldots, u_m\}$) such that

$$(Lu, u) \geq \frac{1}{4k} \quad (\forall u \in \bar{U}) .$$

Condition (b) implies that each $u \in S$ must be a finite set. Since the union of a strictly increasing (with respect to $\underline{c}$) sequence of sets in S is again a member of S (and hence a finite set), each such sequence terminates (if not, its union would contain infinitely many elements). Hence, S has a maximal element, w say. Let M be the finite dimensional vector subspace generated by w; then $|(Lx, x)| < \frac{1}{4k}$ whenever $x \in M^\perp$ and $\|x\| = 1$ (otherwise S would contain $w \cup \{x\}$, contradicting the maximality of w). Hence $|(Lx, x)| < \frac{1}{k}$ whenever $x \in M^\perp$ and $\|x\| \leq 2$. Hence, $\forall z, w \in M^\perp$ with $\|z\|, \|w\| \leq 1$,

$$|(Lz, w)| \leq \frac{1}{4}|(L(z + w), z + w) - \text{etc.} \cdots|$$

$$\leq \frac{1}{4} \cdot 4 \cdot \frac{1}{k} = \frac{1}{k}$$

(use Gen. Polarization roll, and $\|z + w\| \leq \|z\| + \|w\| \leq 2$, etc.).
 Thus

$$|(Lz, w)| \leq \frac{1}{k} \quad \left(\forall z, w \in M^\perp, \|z\|, \|w\| \leq 1\right) . \qquad (4.85)$$

Now let $P : H \to M$ be the orthogonal projection onto M, with $I - P$ the orthogonal projection onto $M^\perp$; and take $z = (I - P)x$, $w = (I - P)y$ $\forall x, y \in H$ with $\|x\|, \|y\| \leq 1$. Then by (4.85),

$$|((I - P)L(I - P)x, y)| = |(L(I - P)x, (I - P)y)|$$

$$= |(Ax, w)| \leq \frac{1}{k} \quad (I - P \text{ is Hermitian}) .$$

Hence

$$\|(I - P)L(I - P)\|_{op} \leq \frac{1}{k} .$$

So, put $F_K = PL + LP - PLP$, then $F_K \in \mathcal{F}(H)$, since P has rank equal to the $\dim(M) < \infty$, so $P \in \mathcal{F}(H)$ and $\mathcal{F}(H)$ is an ideal; and $L - F_k = (I - P)L(I - P)$.
 (c) $\Rightarrow$ (a). We know that $\mathcal{F}(H) \subseteq K(H)$. Now,

$$\bar{\mathcal{F}}(H) \subseteq \overline{K(H)} = K(H) \quad \text{(by Proposition 4.7)} .$$

Hence $\|L - F_K\| \to 0$, $L \in \overline{\mathcal{F}(H)}$ so $L \in K(H)$. ∎

COROLLARY 4.5
$\overline{\mathcal{F}(H)} = K(H)$. *(This is often taken to be the definition of $K(H)$.)*

PROOF   Given $L \in K(H)$, by (c), $\bar{L} \in \overline{\mathcal{F}(H)}$ so $K(H) \subseteq \overline{\mathcal{F}(H)} \subseteq K(H)$ (by Proposition 4.7). Hence equality. ∎

PROPOSITION 4.8
$(w \otimes b)^* = b \otimes w$.

PROOF $((w \otimes b)x, y) = ((x, w)b, y) = (x, w)(b, y) = (x, (b, y)w) =$ $(x, (y, v)w) = (x, (b \otimes w)y)$ $(\forall x, y \in H)$, so $(w \otimes b)^* = b \otimes w$. ∎

COROLLARY 4.6
If $F \in \mathcal{F}(H)$ such that $F = \sum_{i=1}^{n} b_i \otimes w_i$, the $F^* \in \mathcal{F}(H)$; i.e., $\mathcal{F}(H)$ is a Hermitian (s self-adjoint) ideal in C^*-algebra $B(H)$.

THEOREM 4.23
$K(H)$ is a Hermitian ideal in C^*-algebra $B(H)$.

PROOF   Let $L \in K(H) \Leftrightarrow L = \lim_{n \to \infty} F_n (F_n \in \mathcal{F}(H))$, so $\|L - F_n\| \to 0$ (as $n \to \infty$). Hence $\|L^* - F_n^*\| \to 0$ $(n \to \infty) \Rightarrow L^* \in K(H)$ (since each $F_n^* \in \mathcal{F}(H)$), by Proposition 4.8. ∎

THEOREM 4.24
Let $L \in B(H)$; let $\{u_n\}$, $\{v_n\}$ be any Hilbert bases for H.

(a) *If any one of the four real series below is convergent, then all are, and all have the same sum:*

$$\sum_m \|Lu_n\|^2, \sum_m \|L^* v_m\|^2, \sum_n \sum_m |(Lu_n, v_m)|^2, \sum_m \sum_n |(Lu_n, v_m)|^2.$$

(b) *Then this sum is independent of the choice of bases, i.e., it is an invariant of the operator L, so write it as $\||L|\|^2$.*

(c) $\||L|\| = \||L^*|\|$.

PROOF (a). Use Parseval: $(\forall n \in N)\ \|Lu_n\|^2 = \sum_m |(Lu_n, u_n)|^2 <$ ∞ so $\sum_n \|Lu_n\|^2$ and $\sum_n \sum_m |(Lu_n, u_m)|^2$ are both convergent (or not) together, and then their sums are equal. Similar remarks apply to $\sum_m \|A^* v_m\|^2$ and $\sum_m \sum_n |(L^* v_m, u_n)|^2$. Now using the fact that $(\forall n, m)\ |(Lu_n, v_m)|^2 = |(L^* v_m, u_n)|^2$ and the fact that for positive real series "$\sum_{n=1}^{\infty} \sum_{m=1}^{\infty} \equiv \sum_{m=1}^{\infty} \sum_{n=1}^{\infty}$". (a) follows.

(b). Choosing a third Hilbert basis $\{w_n\}$ for H we get

$$\sum_{n=1}^{\infty} \|Lu_n\|^2 = s < \infty \Leftrightarrow \sum_{m=1}^{\infty} \|L^* v_m\|^2 = s < \infty$$

$$\Leftrightarrow \sum_{n=1}^{\infty} \|Lw_n\|^2 = s < \infty,$$

(c). Since the choice of Hilbert bases is immaterial, use $\{u_n\}$ in place of $\{v_m\}$ in the proof of (4.84); this gives

$$\sum_{n=1}^{\infty} \|Lu_n\|^2 = \sum_{n=1}^{\infty} \|L^* u_n\|^2 \quad \text{(when } < \infty\text{)},$$

i.e., $\||A\||^2 = \||A^*\||^2$. ∎

DEFINITION 4.7 *Define by $S(H)$ the set $S(H) = \{L \in B(H) \mid \||L\|| < \infty\}$, the so-called Hilbert–Schmidt operators.*

Note (1): All integral operators L on $H = C[0, 1]$, $\mathcal{L}^2[0, 1]$, etc., with $\mathcal{L}^2$-kernel $k(s, t)$ such that $\||A\||^2 = \int_0^1 \int_0^1 |k(s, t)|^2 ds dt < \infty$ are Hilbert–Schmidt operators. Hence their importance.

Note (2): $S(H)$ is Hermitian ($L \in S(H) \Leftrightarrow L^* \in S(H)$) by (c) above.

Example 4.7

(1) Let $H = \ell^2$; $L : \ell^2 \to \ell^2$ given by infinite matrix $a = [a_{rk}]$ such that

$$\|a\|^2 = \sum_r \sum_k |a_{rk}|^2 \equiv \sum_r \|a_r\|^2 < \infty$$

(a_r row r in $a \in \ell^2$) via:

$$Lx = y, \quad x = (x_k) \in \ell^2, \quad y = (y_r) \in \ell^2$$

(Have to show this!)

$$y_r = \sum_{k=1}^{\infty} a_{rk} x_k \equiv (a_r, x)$$

(this is well defined since):

$$|y_r| = |(a_r, x)| \leq \|a_r\| \cdot \|x\| \leq \|a\| \cdot \|x\| \Rightarrow y_r \in C$$

and

$$\|Lx\|^2 = \|y\|^2 = \sum_{r=1}^{\infty} \|y_r\|^2 \leq \|x\|^2 \sum_{r=1}^{\infty} \|a_r\|^2 = \|x\|^2 \cdot \|a\|^2 .$$

So L is linear and bounded, $\|L\| \leq \|a\|$. This L is Hilbert–Schmidt because if $\{u_n = (0, 0, \ldots, 1_n, 0, \ldots)\}$ is a standard basis for ℓ^2, then

$$|||A|||^2 \overset{\text{def}}{=} \sum_r \sum_k |(Lu_r, u_k)|^2 .$$

But $(Lu_r, u_k) = $ element a_{rk} in matrix a representation for L, so

$$|||L|||^2 = \sum_r \sum_k |a_{rk}|^2 \overset{\text{def}}{=} \|a\|^2 < \infty .$$

Thus $\|L\|_{op} \leq \|a\| = |||L|||_{H-S}.$

(2) $w \otimes z$: Rank 0 or 1 operator on H with Hilbert basis $\{u_n\}$.

$$\|w \otimes z\|_{op} = \|w\| \cdot \|z\| .$$

Now $\|w\|^2 = \sum_{n=1}^{\infty} |(w, u_n)|^2$ (Parseval)

$$\Rightarrow \|w\|^2 \|z\|^2 = \sum_{n=1}^{\infty} |(u_n, w)|^2 \|z\|^2$$

$$= \sum_{n=1}^{\infty} \|(u_n, w) z\|^2 = \sum_{n=1}^{\infty} \|(w \otimes z) u_n\|^2 = |||w \otimes z|||^2 .$$

Hence $\|w \otimes z\|_{op} = |||w \otimes z|||.$

We now provide a sketch of the proof of the theorem.

THEOREM 4.25
$S(H)$, $||| \cdot |||$, is a Banach space, and $\|L\| \leq |||L||| \; \forall L \in S(H)$.

PROOF (sketch) (a) $S(H)$ is a vector space. Use Minkowski's inequality and get Δ-inequality for $||| \cdot |||$.
(b)

$$\|Lx\|^2 = \sum_n |(Lx, u_n)|^2 = \sum_n \left|(x, L^* u_n)\right|^2 \leq \|x\|^2 \sum_n \|L(u_n^*)\|^2 = \|x\|^2 |||L^*|||^2$$

$$\Rightarrow \cdots \Rightarrow \|L\| \leq |||L||| . \tag{4.86}$$

(c) $||| \cdot |||$ is a norm. Use (4.86) to get $|||L||| = 0 \Leftrightarrow L = 0$,
(d) If $||| \cdot |||$-Cauchy $\overset{(4.86)}{\Longleftrightarrow} \| \cdot \|$-Cauchy in $B(H) \Rightarrow \| \cdot \| - \lim L_n = L \in B(H)$.
Now show

(i) $L \in S(H)$;

(ii) $|||L - L_n|||^2 \to 0$. ∎

THEOREM 4.26
$S(H)$ is a Hermitian ideal in $B(H)$.

PROOF (i) $L \in S(H)$, $B \in B(H)$, $\{u_n; n \in N\}$ is a Hilbert basis for H then $(\forall p \in N) \sum_{n=1}^p \|BLu_n\|^2 \leq \|B\|^2 \sum_{n=1}^p \|Lu_n\|^2$, let $p \to \infty$ then $|||BL|||^2 \leq \|B\|^2 |||L|||^2 < \infty$, so $BL \in S(H)$ (and $|||BL||| \leq \|B\| \cdot |||L||| \leq |||B||| \cdot |||L|||$).
(ii) $B^* \in B(H)$, $L^* \in S(H)$ [by Definition 4.5, note (4.84)]. Hence, by (i), $B^* L^* \in S(H) \Rightarrow (LB)^* \in S(H) \Rightarrow LB \in S(H)$ and

$$|||LB||| \equiv \|B^* L^*\| \leq \|B^*\| \cdot |||L^*||| = \|B\| \cdot |||L|||$$

$$\leq |||L||| \cdot |||B||| . \quad ∎$$

The proof of the following theorem is similar to Theorem 4.19.

THEOREM 4.27
$\mathcal{F}(H) \subseteq S(H)$ proper if $dim(H) = \infty$.

THEOREM 4.28
The following are true:

(i) $S(H) \subseteq K(H)$.

(ii) $S(H)$ is proper if $\dim(H) = \infty$.

PROOF   (i) $x = \sum_{n=1}^{\infty} (x, u_n) u_n$,

$$L \in S(H) \Rightarrow Lx = \sum_{n=1}^{\infty} (x, u_n)(Lu_n) = \sum_{n=1}^{\infty} (u_n \otimes Lu_n) x \,,$$

so define "pth cutoff operators" $L_p = \sum_{n=1}^{p} u_n \otimes Lu_n \in \mathcal{F}(H)$; observe ($\forall m \in N$)

$$L_p u_m = L u_m \quad (1 \le m \le p)$$

$$= 0 \quad (p > m)$$

$$\Rightarrow \|L - L_p\|^2 \le \||L - L_p|\|^2 = \sum_{n=1}^{\infty} \|(L - L_p) u_n\|^2 = \sum_{n=1}^{\infty} \|Lu_n - L_p u_n\|^2$$

$$= \sum_{n=p+1}^{\infty} \|Lu_n\|^2 \to 0 \quad (p \to \infty)$$

since $\||L|\|^2 = \sum_{n=1}^{\infty} \|Lu_n\|^2$ is convergent $\Rightarrow L = \lim_{p \to \infty} L_p (\| \cdot \|_{op}) \in \bar{\mathcal{F}}(H) = K(H)$.

(ii) Let $H = \ell^2$. Let $L : \ell^2 \to \ell^2$ be given by an infinite diagonal matrix

$$a \equiv \begin{bmatrix} 1 & & & \\ & 1/\sqrt{2} & 0 & \\ & & 1/\sqrt{3} & a_{rr} = \frac{1}{r^r} \\ 0 & & & \ddots \end{bmatrix},$$

$a_{rk} = 0 \ (r \ne k)$,

$$L \in B\left(\ell^2\right) \left\{ Lx = \left(x_1, \frac{1}{\sqrt{2}}x_2, \frac{1}{\sqrt{2}}x_3, \ldots\right) \right. ;$$

$$\left. \|Ax\|^2 = \sum \frac{1}{n} |x_n|^2 \le \sum |x_n|^2 = \|x\|^2; \|L\| \le 1 \right\} .$$

Define "cutoff" L_p to be given by a with all rows equal, 0 beyond row p, so $L_p \in \mathcal{F}(\ell^2)$ and hence $L_p \in S(\ell^2) \subseteq K(\ell^2)$. But, we show $L \notin \mathcal{F}(\ell^2)$ whilst $L \in K(\ell^2)$;

$$\|Lx - L_p x\|^2 = \left\| \left(0, \ldots, 0, \frac{1}{\sqrt{p+1}} x_{p+1}, \frac{1}{\sqrt{p+2}} x_{p+2}, \ldots \right) \right\|^2$$

$$= \sum_{n=p+1}^{\infty} \frac{1}{n} |x_n|^2 \leq \frac{1}{p} \sum_{n=1}^{\infty} |x_n|^2 = \frac{1}{p} \|x\|^2 .$$

So,

$$\|(L - L_p) x\|^2 \leq \frac{1}{\sqrt{p}} \|x\| \quad \left(\forall p \geq 1, \forall x \in \ell^2 \right)$$

$$\Rightarrow \|L - L_p\|_{op} \leq \frac{1}{\sqrt{p}} \to 0 \quad (p \to \infty)$$

$$\Rightarrow L = \lim_{p \to \infty} L_p \left(\| \cdot \|_{op} \right) \Rightarrow L \in K \left(\ell^2 \right) .$$

But

$$\||L|\|^2 = \sum_{n=1}^{\infty} \|L u_n\|^2 \quad (L u_n \equiv n\text{th row of matrix } a)$$

$$= \sum_{n=1}^{\infty} \left| \frac{1}{\sqrt{n}} \right|^2 = \sum_{n=1}^{\infty} \frac{1}{n} = \infty \Rightarrow L \notin S \left(\ell^2 \right) .$$

Hence $S(\ell^2) \subsetneq K(\ell^2)$. ∎

REMARK 4.5   (a) If H is a separable Hilbert space of dimension ∞, then the following ideals are all distinct:

$$\{0\} \subset S(H) \subset K(H) \subset B(H) .$$

(b) It can be shown that the above ideals of operators, correspond as normed

Banach spaces to the sequence spaces

$$\{0\} \subset \ell_0 \subset \ell^2 \subset C_0 \subset C\ell^\infty .$$

(c) It can also be shown that just as there are other sequence spaces ℓ^p, $1 \le p < \infty$, $\ell_0 \subset \ell^1 \subset \ell^p \subset \ell^{p'} \subset C_0 \subset \ell^\infty$ ($1 \le p < p' < \infty$) so there are other ideals—the von Neumann–Schatten ideals C^p such that

$$\mathcal{F}(H) \subset C^1(H) \subset C^p(H) \subset C^{p'}(H) \subset K(H) \subset B(H)$$

$$\left(1 < p < p' < \infty\right) .$$

They are constructed via norms

$$|||A|||_p = \left[\sum_r \sum_k |(Au_r, u_k)|^p \right]^{1/p}$$

just as Hilbert–Schwartz operators were. The class $C^1(H)$ is thought to be even more important than $C^2(H) = S(H)$; it consists of the so-called *Nuclear* or *Trace-class* operators.

(d) It will be very important to know how by following these ideas can we extend to bilinear (or polynomial) operators on a separable Hilbert space! ∎

4.5 Approximation of Solutions of Some Quadratic Integral Equations in Transport Theory [116]

We consider the integral equation

$$H(x, \mu) = 1 + c \int_0^1 \psi(v) \int_x^\tau H(t, \mu) H(t, v) \omega(t)$$

$$\times \exp\left[-(t - x) \left(\frac{1}{\mu} + \frac{1}{v} \right) \right] dt \frac{dv}{v} . \tag{4.87}$$

In Equation (4.87), ψ and ω are given nonnegative functions, c is a complex parameter and $0 < \tau \leq \infty$, and H is the function to be found. We assume

$$\omega \in L^1[0, \tau) \cap L^\infty[0, \tau) \tag{4.88}$$

and

$$\psi \in L^1[0, 1]; \qquad \int_0^1 \psi(v)dv = 1/2 . \tag{4.89}$$

Equation (4.87) is a generalization of the Chandrasekhar equation (1.22) and arises in the study of radiative transfer and neutron transport through inhomogeneous media.

Numerical work on Equation (4.87) has been done by many authors under assumptions on ω more restrictive than (4.88).

Of the many possible solutions to Equation (4.87) only one is of physical importance; it is characterized by analyticity in the parameter c for $|c|$ sufficiently small. It was shown [120, 143] that H is analytic in $|c|$ for $|c| < c_0 = 1/\lambda_0$, where λ_0 is the spectral radius of the integral operator K on $L^1[0, \tau]$ given, for $f \in L^1[0, \tau]$, by

$$Kf(x) = \int_0^\tau \left[\int_0^1 e^{-|x-y|/v} \psi(v) \frac{dv}{v} \right] \omega(y) f(y) dy . \tag{4.90}$$

If $\psi = 1/2$, the connection between H and radiative transfer problems may be simply described. We consider the equation for monoenergetic radiative transfer in an inhomogeneous slab with isotropic scattering

$$\mu \frac{\partial}{\partial x} I(x, \mu) + I(x, \mu) = \frac{c\omega(x)}{2} \int_{-1}^1 I\left(x, \mu'\right) d\mu' . \tag{4.91}$$

We impose boundary conditions

$$I(0, \mu) = I_1(\mu), \qquad \mu \geq 0 ,$$

$$\lim_{x \to \tau} I(x, -\mu) = 0, \qquad \mu \geq 0 . \tag{4.92}$$

The exit distribution problem is to find $I(0, -\mu)$ and, if $\tau < \infty$, $I(\tau, \mu)$. These quantities may be expressed in terms of H by

$$I(0, -\mu) = \frac{c}{2\mu} \int_0^1 I_1(v) \int_0^\tau \exp\left[-t \left(\frac{1}{\mu} + \frac{1}{v} \right) \right]$$

$$\times H(t, \mu) H(t, v) \omega(t) dt dv , \qquad (\mu > 0) \tag{4.93}$$

and, if $\tau < \infty$,

$$I(\tau, \mu) = e^{-\tau/\mu} I_1(\mu) + \frac{c}{2\mu} \int_0^1 I_1(v) \int_0^\tau \exp\left[-(\tau - t)\left(\frac{1}{\mu} + \frac{1}{v}\right)\right]$$

$$\times H(t, -\mu) H(t, v) \omega(t) dt dv \,. \tag{4.94}$$

If we write Equation (4.87) as

$$H = 1 + cB(H, H) \,, \tag{4.95}$$

where B is a bilinear form, it was shown [120] that the iteration scheme

$$H_0 = 1,$$

$$H_{n+1} = 1 + cB(H_n, H_n)$$

converges to H in an appropriate Banach space sense for $|c| < c_0$. However, this iteration scheme is difficult to carry out in practice because many double integrals must be done at each stage in the iteration.

The purpose of this section is to give results on alternative methods of approximate solution to Equation (4.87). We will prove some general results on approximation of solutions to quadratic equations. Later we show how these results apply to Equation (4.87) and give some numerical results.

We first give results that apply both to Equation (4.87) and to the more general problems.

Let X be a Banach space of functions from $\Omega \subseteq \mathbf{R}^m$ to $\mathbf{C}^n$. We say $f \in X$ is real if for all $x \in \Omega$, the vector $f(x)$ has real entries. We say that $f \geq g$ if both f and g are real and for each $\omega \in \Omega$, $f(x) \geq g(x)$ componentwise.

We let $[f]_i$ denote the complex-valued function on Ω whose value is the ith entry of the vector f. We define $|f|$ by

$$[|f|]_i = |[f]_i| \,. \tag{4.96}$$

DEFINITION 4.8   *A continuous bilinear form B on X is positive, if for all $f, g \in X$ with $f \geq 0$ and $g \geq 0$,*

$$0 \leq B(f, g) \,. \tag{4.97}$$

If B is positive we write $B \geq 0$.

The norm of B is given by

$$\|B\| = \sup_{\|f\|=\|g\|=1} \|B(f, g)\| . \tag{4.98}$$

If A and B are continuous positive bilinear forms on X we say $A \geq B$ if for all $f, g \in X$ with $f \geq 0$ and $g \geq 0$ we have

$$B(f, g) \leq A(f, g) . \tag{4.99}$$

Note that if A and B are positive continuous bilinear forms and $A \geq B$ we have for all $f, g \in X$

$$|B(f, g)| \leq B(|f|, |g|) \leq A(|f|, |g|) \tag{4.100}$$

and hence

$$\|B\| \leq \|A\| . \tag{4.101}$$

Of interest here are quadratic equations of the form

$$f = g + cB(f, f) . \tag{4.102}$$

In (4.102), $g \in X$, $g \geq 0$, c is a complex parameter, B is a positive continuous bilinear form on X, and a solution, $f \in X$ is sought.

For $|c|$ sufficiently small it is known [120] that there is a solution $f(c)$ to Equation (4.102) that is analytic in c near $c = 0$. We call this solution the *A-solution.* The A-solution is the unique solution to Equation (4.102) in the class of X-valued functions of c that are analytic near $c = 0$.

If we write the A-solution to Equation (4.102) in a Taylor series about $c = 0$,

$$f(c) = \sum_{m=0}^{\infty} \gamma_m c^m , \tag{4.103}$$

we have [see (1.38)]

$$\gamma_0 = g ,$$

$$\gamma_m = \sum_{k+l=m-1} B(\gamma_k, \gamma_l) , \quad m \geq 1 . \tag{4.104}$$

The radius of convergence of this series will depend on g and B. We denote this radius by $\rho(g, B)$. As $\gamma_m \geq 0$ for all m we have, for $|c| < c_1 < \rho(g, B)$,

$$|f(x)| = \left| \sum_{m=0}^{\infty} \gamma_m c^m \right| \leq \sum_{m=0}^{\infty} \gamma_m |c|^m = f(|c|) \leq f(c_1) . \tag{4.105}$$

Also, if $h(c) = \sum_{m=0}^{\infty} \delta_m c^m$ is an analytic X-valued function of c with $0 \leq \delta_m \leq \gamma_m$ we have, for $|c| < \rho(g, B)$,

$$|f(c) - h(c)| \leq \sum_{m=0}^{\infty} (\gamma_m - \delta_m) |c|^m = f(|c|) - h(|c|) \tag{4.106}$$

and hence

$$\| f(c) - h(c) \| \leq \| f(|c|) - h(|c|) \| . \tag{4.107}$$

Inequalities of the type (4.105) through (4.107) based on the positivity of power series coefficients will be used frequently in the remainder of this section.

Useful throughout this paper will be a special case of a theorem of Vitali. We state the version required here as a lemma:

LEMMA 4.2
Let $f_i(c) = \sum_{m=0}^{\infty} \gamma_m^i c^m$ be a sequence of X-valued functions of c, analytic in c for $|c| < c_1$. Assume that

$$\lim_{i \to \infty} \gamma_m^i = \gamma_m \tag{4.108}$$

exists in X for all $m \geq 0$. Moreover assume that there is $M > 0$ such that

$$\| f_i(c) \| \leq M \quad \text{for } |c| < c_1 . \tag{4.109}$$

Then the function

$$f(c) = \sum_{m=0}^{\infty} \gamma_m c^m \tag{4.110}$$

is analytic for $|c| < c_1$ and

$$\lim_{i \to \infty} \| f(c) - f_i(c) \| = 0 \tag{4.111}$$

uniformly for c in compact subsets of $\{c \mid |c| < c_1\}$.

THEOREM 4.29

Let $g \geq 0$ and let B be a positive continuous bilinear form on X. Let $\rho(g, B) = \bar{\rho}$. Then the iteration scheme

$$f_0 = g,$$

$$f_{n+1} = g + cB(f_n, f_n), \quad n \geq 0, \tag{4.112}$$

converges to the A-solution, $f(c)$, of Equation (4.102) in the sense that

$$\lim_{n \to \infty} \|f(c) - f_n(c)\| = 0 \tag{4.113}$$

uniformly in compact subsets of $\{c \mid |c| < \bar{\rho}\}$.
  Conversely, if $\{f_n\}$ is given by (4.112) and for some $\rho^* > 0$ and $M > 0$

$$\sup_{n \geq 0, |c| \leq \rho^*} \|f_n(c)\| \leq M, \tag{4.114}$$

then $\rho(g, B) \geq \rho^*$ and Equation (4.113) holds uniformly in c for c in compact subsets of $\{c \mid |c| < \rho^*\}$.

PROOF Note that $\rho(g, B)$ is exactly the radius of convergence of the series

$$\sum_{m=0}^{\infty} \gamma_m c^m, \tag{4.115}$$

where γ_m is given by Equation (4.104).
  For any $n \geq 0$ we have

$$f_n(c) = \sum_{m=0}^{2^n - 1} \gamma_m^n c^m, \tag{4.116}$$

where

$$\gamma_0 = g,$$

$$\gamma_m^n = \sum_{l+k=m-1} B\left(\gamma_k^{n-1}, \gamma_l^{n-1}\right). \tag{4.117}$$

Inducting on n, we have

$$0 \leq \gamma_m^n \leq \gamma_m, \quad m \geq 0, n \geq 0, \tag{4.118}$$

and

$$\gamma_m^n = \gamma_m, \quad m \leq n, \ n \geq 0. \tag{4.119}$$

Hence, for $0 \leq |c| \leq c_1 < \bar{\rho}$,

$$\| f_n(c) \| \leq \| f(|c_1|) \|. \tag{4.120}$$

This completes the proof of the first part of the theorem by Lemma 4.2 as $c_1 < \bar{\rho}$ was arbitrary.

The converse is an immediate consequence of Lemma 4.2. ∎

DEFINITION 4.9   Let $\{B_i\}_{i=1}^{\infty}$ *be a sequence of continuous bilinear forms on* X. *We say that the sequence* $\{B_i\}$ *converges strongly to the bilinear form* B *on* X *if for all* $f, g \in X$

$$\lim_{i \to \infty} B_i(f, g) = B(f, g). \tag{4.121}$$

We say the sequence $\{B_i\}$ *is uniformly bounded, if there is* $M > 0$ *such that*

$$\| B_i \| \leq M \quad \text{for all } i \geq 1. \tag{4.122}$$

We require:

LEMMA 4.3
Let B *be a positive continuous bilinear form on* X, *and let* $\{B_i\}$ *be a sequence of positive continuous bilinear forms converging strongly to* B. *Assume* $\{B_i\}$ *is uniformly bounded. Then if* $\{\gamma_m\}_{m=0}^{\infty}$ *by Equation (4.104) and* $\{\gamma_m^i\}_{m=0}$ *is given for* $i \geq 1$ *by*

$$\gamma_0^i = g \geq 0,$$

$$\gamma_m^i = \sum_{k+l=m-1} B_i\left(\gamma_k^i, \gamma_l^i\right), \tag{4.123}$$

then, for all m,

$$\lim_{i \to \infty} \gamma_m^i = \gamma_m. \tag{4.124}$$

PROOF   As $\{B_i\}$ is uniformly bounded there is $M > 0$ such that, for all $i \geq 1$

$$\|B_i\| \leq M . \tag{4.125}$$

Strong convergence of $\{B_i\}$ to B implies that, for $f, g \in X$

$$\|B(f, g)\| = \lim_{i \to \infty} \|B_i(f, g)\| \leq M\|f\|\|g\| \tag{4.126}$$

and hence

$$\|B\| \leq M . \tag{4.127}$$

We proceed inductively. Equation (4.124) is trivially true for $m = 0$. Assume now that Equation (4.124) holds for $0 \leq m \leq n$. Let K be given by

$$K = 4(n + 1)M \left[\max_{0 \leq r \leq n} \|\gamma_r\| + 1 \right] \tag{4.128}$$

and let $\varepsilon \in (0, K/2)$ be arbitrary. We show that there is I so that if $i > I$ we have

$$\left\| \gamma_{n+1}^i - \gamma_{n+1} \right\| < \varepsilon . \tag{4.129}$$

By the induction hypothesis there is i_0 so that if $i > i_0$

$$\max_{0 \leq r \leq n} \left\| \gamma_r^i - \gamma_r \right\| < \varepsilon/K . \tag{4.130}$$

As $\{B_i\}$ converges strongly to B, there is i_1 so that $i > i_1$,

$$\max_{0 \leq r,s \leq n} \|B_i(\gamma_r, \gamma_s) - B(\gamma_r, \gamma_s)\| < \varepsilon M/K . \tag{4.131}$$

Therefore, for $i > I = \max(i_0, i_1)$ and $0 \leq k, l \leq n$, we have, as $\varepsilon < K/2$,

$$\left\| B_i\left(\gamma_k^i, \gamma_l^i\right) - B(\gamma_k, \gamma_l) \right\|$$

$$= \left\| B_i\left(\gamma_k^i - \gamma_k, \gamma_l^i - \gamma_l\right) + B_i\left(\gamma_k^k - \gamma_k, \gamma_l\right) \right.$$

$$\left. + B_i\left(\gamma_k, \gamma_l^i - \gamma_l\right) + B_i(\gamma_k, \gamma_l) - B(\gamma_k, \gamma_l) \right\|$$

$$\leq \|B_i\| \left(\left\| \gamma_k^i - \gamma_k \right\| \left\| \gamma_l^i - \gamma_l \right\| + \left\| \gamma_k^i - \gamma_k \right\| \|\gamma_l\| \right.$$

$$\left. + \left\| \gamma_l^i - \gamma_l \right\| \|\gamma_k\| \right) + \|B_i(\gamma_k, \gamma_l) - B(\gamma_k, \gamma_l)\|$$

$$\leq M \left(\frac{\varepsilon^2}{K^2} + \frac{\varepsilon}{K} \|\gamma_l\| + \frac{\varepsilon}{K} \|\gamma_k\| \right) + \frac{\varepsilon M}{K} < \frac{\varepsilon}{(n+1)} . \qquad (4.132)$$

Therefore, by Equations (4.104) and (4.123), we have, for $i > I$

$$\left\| \gamma_{n+1}^i - \gamma_{n+1} \right\| \leq \sum_{k+l=n-1} \left\| B_i \left(\gamma_k^i, \gamma_l^i \right) - B(\gamma_k, \gamma_l) \right\| < \varepsilon . \qquad (4.133)$$

The proof of the lemma is now complete.   ■

The main result of this section is the following theorem, which is an immediate consequence of Lemmas 4.2 and 4.3.

THEOREM 4.30
Let $g \geq 0$ and let B and $\{B_i\}_{i=1}^{\infty}$ be as in the statement of Lemma 4.3. Assume there exists $\rho^* \leq \rho(g, B) = \bar{\rho}$ such that

$$\rho(g, B_i) \geq \rho^* . \qquad (4.134)$$

If $f^i(c)$ is the A-solution of

$$f^i = g + cB_i \left(f^i, f^i \right) \qquad (4.135)$$

and there is a constant K such that for i sufficiently large

$$\left\| f^i(c) \right\| \leq K \quad \text{for all } c , \ |c| < \rho^* , \qquad (4.136)$$

then, if $f(c)$ is the A-solution to Equation (4.102), we have

$$\lim_{i \to \infty} \left\| f(c) - f^i(c) \right\| = 0 \qquad (4.137)$$

uniformly in compact subsets of $\{c \mid |c| < \rho^*\}$.

We now make the additional assumption that ω is piecewise continuous with at most finitely many discontinuities on $[0, \tau)$.

In this section B will denote the bilinear map given by Equation (4.87). H will denote the A-solution to Equation (4.87). If $\tau = \infty$, the space X is

$$X = \mathcal{C}([0, \infty] \times [0, 1]) . \tag{4.138}$$

Recall that a function F is continuous on $[0, \infty] \times [0, 1]$ if it is continuous on $[0, \infty) \times [0, 1]$ and $\lim_{x \to \infty} F(x, \mu) = F(\infty, \mu)$ exists uniformly in μ for $\mu \in [0, 1]$.

If $\tau < \infty$, values of $H(x, \mu)$ for $-1 \le \mu < 0$ are important. In order to specify X in this case we first choose $0 < \eta < 1$ and let

$$U = [-1, -\eta] \cup [0, 1] . \tag{4.139}$$

The space X, for $\tau < \infty$, is

$$X = \mathcal{C}([0, \tau] \times U) . \tag{4.140}$$

For convenience we set $U = [0, 1]$ if $\tau = \infty$.

It was shown [120] that in either of these situations

$$\rho(1, B) = c_0 , \tag{4.141}$$

where $c_0 = 1/\lambda_0$ and λ_0 is the spectral radius of the integral operator K given by Equation (4.90).

For $f, g \in X$, we have

$$|B(f, g)| \le \|\omega\|_\infty \|f\|_\infty \|g\|_\infty R(x, \mu) \tag{4.142}$$

where

$$R(x, \mu) = \int_0^1 \frac{\mu}{\mu + \upsilon} \left(1 - \exp\left[-(\tau - x) \left(\frac{1}{\mu} + \frac{1}{\upsilon} \right) \right] \right) \psi(\upsilon) d\upsilon . \tag{4.143}$$

It is easy to see that if $\tau = \infty$ and $\mu \ge 0$,

$$R(x, \mu) \le \int_0^1 \psi(\upsilon) d\upsilon = \frac{1}{2} . \tag{4.144}$$

For $\tau < \infty$ and $-1 \le \mu \le -\eta$ we have

$$\frac{1}{v} \int_x^\tau \exp\left[-(t-x)\left(\frac{1}{\mu} + \frac{1}{v}\right)\right] dt \le e^{\tau/\eta} \left(1 - e^{-(\tau-x)/v}\right) ;$$

hence, for $\tau < \infty$ and $-1 \le \mu \le -\eta$,

$$R(x, \mu) \le e^{\tau/\eta} \int_0^1 \left(1 - e^{-(\tau-x)/v}\right) \psi(v) dv \le \frac{1}{2} e^{\tau/\eta} . \qquad (4.145)$$

Therefore,

$$\|B\| \le \frac{1}{2} \|\omega\|_\infty, \quad \tau = \infty , \qquad (4.146)$$

$$\|B\| \le \frac{1}{2} \left(1 + e^{\tau/\eta}\right) \|\omega\|_\infty, \quad \tau < \infty . \qquad (4.147)$$

We now describe our approximation to B. Let $s(x) = \sqrt{\omega(x)}$. Let Δ be a partition of the interval $[0, \tau)$ into intervals $[t_{n-1}, t_n)$ for $1 \le n \le N + 1$ such that

$$0 = t_0 < t_1 < \cdots < t_N < t_{N+1} = \tau . \qquad (4.148)$$

Let χ_n be the characteristic function of the interval $[t_{n-1}, t_n)$ and let a_n, b_n, c_n, $d_n \in [t_{n-1}, t_n]$.

Define B_Δ on X by

$$B_\Delta(f, g)(x, \mu) = \int_0^1 \psi(v) \int_x^\tau \sum_{\rho=1}^{N+1} \chi_\rho(t) \left[s\left(a_\rho\right) f\left(b_\rho, \mu\right) s\left(c_\rho\right) g\left(d_\rho, v\right)\right.$$

$$\left. \times \exp\left[-(t-x)\left(\frac{1}{\mu} + \frac{1}{v}\right)\right]\right] dt \frac{dv}{v} . \qquad (4.149)$$

We first show that as the partition Δ becomes "finer" the bilinear forms B_Δ converge strongly to B. We make this precise.

Let $\{\Delta_i\}$ be a sequence of partitions of the type described above with intervals $[t_{n-1}^i, t_n^i)$, $1 \le n \le N_i + 1$. Assume that the following hypothesis holds:
(H) For every $L > 0$

$$\lim_{i \to \infty} \left(\max_{t_n^i \le L} \left|t_{n-1}^i - t_n^i\right|\right) = 0 .$$

If $\tau < \infty$, we may of course take $L = \tau$.

LEMMA 4.4

Let $\{\Delta_i\}$ be a sequence of partitions of $[0, \tau)$ satisfying (H). Let $B_i = B_{\Delta_i}$. Then B_i converges strongly to B.

PROOF For $f, g \in X$ the functions $s(t)f(t, \mu)$ and $s(t)g(t, \mu)$ are continuous functions of t and μ except at the values of t at which ω is discontinuous. Hypothesis (H) and the fact that ω has at most finitely many discontinuities imply that the sequence $\{h_i\}$ given by

$$h_i(t, \mu, v) = \sum_{p=1}^{N_i+1} \chi_n^i(t) s\left(a_p^i\right) f\left(b_p^i, \mu\right) s\left(c_p^i\right) g\left(d_p^i, v\right) \tag{4.150}$$

converges to $\omega(t)f(t, \mu)g(t, v)$ in the sense that for all finite L, $L \leq \tau$,

$$\lim_{i \to \infty} \left[\max_{\mu, v \in U} \int_0^L |h_i(t, \mu, v) - \omega(t)f(t, \mu)g(t, v)|^2 dt \right] = 0 . \tag{4.151}$$

If $\tau < \infty$, Equation (4.151) implies that $\lim_{i \to \infty} B_i(f, g) = B(f, g)$. To see this set $F_i(t, \mu, v) = h_i(t, \mu, v) - \omega(t)f(t, \mu)g(t, v)$ and note that

$$|B_i(f, g) - B(f, g)|$$

$$\leq \|\psi\|_\infty \int_0^1 \int_x^\tau |F_i(t, \mu, v) \exp\left[-(t-x)\left(\frac{1}{\mu} + \frac{1}{v}\right)\right] dt \frac{dv}{v} . \tag{4.152}$$

We apply the Cauchy–Schwarz inequality to the inner integral in Equation (4.152) to obtain

$$|B_i(f, g) - B(f, g)|$$

$$\leq \|\psi\|_\infty \int_0^1 \left[\int_x^\tau |F_i(t, \mu, v)|^2 dt \right]^{1/2}$$

$$\times \left[\int_x^\tau \exp\left[-2(t-x)\left(\frac{1}{\mu} + \frac{1}{v}\right)\right] dt \right]^{1/2} \frac{dv}{v}$$

$$\leq \|\psi\|_\infty \int_0^1 \left[\int_0^\tau |F_i(t, \mu, v)|^2 \, dt \right]^{1/2}$$

$$\times \left[\frac{\mu v}{2(\mu + v)} \left(1 - \exp\left[-2(\tau - x) \left(\frac{1}{\mu} + \frac{1}{v} \right) \right] \right) \right]^{1/2} \frac{dv}{v}$$

$$\leq \|\psi\|_\infty \left[\max_{\mu, v \in U} \int_0^\tau |F_i(t, \mu, v)|^2 \, dt \right]^{1/2} \frac{1}{\sqrt{2}} \int_0^1 \left| \frac{\mu}{\mu + v} \right|^{1/2}$$

$$\times \left| 1 - \exp\left[-2(\tau - x) \left(\frac{1}{\mu} + \frac{1}{v} \right) \right] \right|^{1/2} \frac{dv}{\sqrt{v}} . \qquad (4.153)$$

If $\mu \geq 0$, the integral in Equation (4.153) may be estimated simply by

$$\int_0^1 \left(\frac{\mu}{\mu + v} \right)^{1/2} \left(1 - \exp\left[-2(\tau - x) \left(\frac{1}{\mu} + \frac{1}{v} \right) \right] \right)^{1/2} \frac{dv}{\sqrt{v}}$$

$$\leq \int_0^1 v^{-1/2} dv = 2 . \qquad (4.154)$$

If $-1 \leq \mu \leq -\eta$, we estimate this integral by

$$\int_0^1 \left| \frac{\mu}{\mu + v} \right|^{1/2} \left| 1 - \exp\left[-2(\tau - x) \left(\frac{1}{\mu} + \frac{1}{v} \right) \right] \right|^{1/2} \frac{dv}{\sqrt{v}}$$

$$\leq \int_0^{-\mu} \left| \frac{\mu}{\mu + v} \right|^{1/2} \frac{dv}{\sqrt{v}} + \int_{-\mu}^1 \left| \frac{\mu}{\mu + v} \right|^{1/2}$$

$$\exp\left[-(\tau - x) \left(\frac{1}{v} + \frac{1}{\mu} \right) \right] \frac{dv}{\sqrt{v}} . \qquad (4.155)$$

For $0 \leq x \leq \tau$ and $-\mu < v \leq 1$ we have

$$\exp\left[-(\tau - x) \left(\frac{1}{\mu} + \frac{1}{v} \right) \right] = \exp\left[(\tau - x) \left(\frac{1}{|\mu|} - \frac{1}{v} \right) \right]$$

$$\leq e^{2\tau/|\mu|} \leq e^{2\tau/\eta} ,$$

and

$$\left|\frac{\mu}{\mu + v}\right|^{1/2} v^{-1/2} \le \left(\frac{1}{v - |\mu|}\right)^{1/2} .$$

For $0 \le v < -\mu$ we have

$$\mu^{1/2} v^{-1/2} (|\mu| - v)^{-1/2} \le v^{-1/2} + (|\mu| - v)^{-1/2} .$$

Therefore, for $-1 \le \mu \le -\eta$ we have

$$\int_0^1 \left|\frac{\mu}{\mu + v}\right|^{1/2} \left|1 - \exp\left[-2(\tau - x)\left(\frac{1}{\mu} + \frac{1}{v}\right)\right]\right|^{1/2} \frac{dv}{\sqrt{v}}$$

$$\le \int_0^{-\mu} \frac{1}{\sqrt{v}} + \frac{1}{\sqrt{|\mu| - v}} dv + e^{2\tau/\eta} \int_{-\mu}^1 \frac{dv}{\sqrt{v - |\mu|}}$$

$$= 4|\mu|^{1/2} + 2e^{2\tau/\eta} \left(\sqrt{1 - |\mu|}\right) \le e\left(1 + e^{2\tau/\eta}\right) . \qquad (4.156)$$

Inequalities (4.154) and (4.156) imply that if $\tau < \infty$,

$$\|B_i(f, g) - B(f, g)\|$$

$$\le \|\psi\|_\infty 4\left(1 + e^{2\tau/\eta}\right) \left[\max_{\mu, v \in U} \int_0^\tau |F_i(t, \mu, v)|^2 dt\right]^{1/2} . \qquad (4.157)$$

Hence, by Equation (4.151), B_i converges strongly to B if $\tau < \infty$.

If $t = \infty$, let $\varepsilon > 0$ and $f, g \in X$ be given. As ω, f, and g have limits at infinity there is L such that if $t, x \ge L$, the following inequality holds for all $v, \mu \in U = [0, 1]$:

$$|s(t)g(t, v)s(x)f(x, \mu) - \omega(\infty)g(\infty, v)f(\infty, v)| < \varepsilon/4 . \qquad (4.158)$$

These inequalities imply that $x \ge L$, $|F_i(x, \mu, v)| < \varepsilon/2$ for all i and hence,

$$|B_i(f, g) - B(f, g)| \le \frac{\varepsilon}{2} \int_0^1 \psi(v) \int_x^\tau \exp\left[-(t - x)\left(\frac{1}{\mu} + \frac{1}{v}\right)\right] dt \frac{dv}{v}$$

$$= (\varepsilon/2)R(x, \mu) \le (\varepsilon/4) . \qquad (4.159)$$

For $x < L$ we have

$$|B_i(f, g) - B(f, g)|$$

$$= \int_0^1 \psi(v) \int_x^L |F_i(t, \mu, v)| \exp\left[-(t - x)\left(\frac{1}{\mu} + \frac{1}{v}\right)\right] dt \frac{dv}{v}$$

$$+ \int_0^1 \psi(v) \int_L^\infty |F_i(t, \mu, v)| \exp\left[-(t - x)\left(\frac{1}{\mu} + \frac{1}{v}\right)\right] dt \frac{dv}{v}$$

$$\leq \int_0^1 \psi(v) \int_x^L |F_i(t, \mu, v)|$$

$$\exp\left[-(t - x)\left(\frac{1}{\mu} + \frac{1}{v}\right)\right] dt \frac{dv}{v} + \frac{\varepsilon}{4} . \tag{4.160}$$

An analysis identical to that used to obtain (4.153) and (4.154) implies that there is $I > 0$ such that if $i > I$,

$$\int_0^1 \psi(v) \int_x^L |F_i(t, \mu, v)| \exp\left[-(t - x)\left(\frac{1}{\mu} + \frac{1}{v}\right)\right] dt \frac{dv}{v} < \frac{3\varepsilon}{4} \tag{4.161}$$

and therefore

$$\|B_i(f, g) - B(f, g)\| < \varepsilon . \tag{4.162}$$

This completes the proof. ∎

An estimate identical to that used in the derivation of Equation (4.146) gives, for all i,

$$\|B_i\| \leq \frac{1}{2}\|\omega\|_\infty , \quad \tau = \infty , \tag{4.163}$$

$$\|B_i\| \leq \frac{1}{2}\left(1 + e^{\tau/\eta}\right) \|\omega\|_\infty, \quad \tau < \infty . \tag{4.164}$$

THEOREM 4.31

Let $\rho^* < \rho$. For i sufficiently large, $\rho(1, B_i) \geq \rho^*$. Moreover, the A-solutions H_i of

$$H^i = 1 + cB_i\left(H^i, H^i\right) \tag{4.165}$$

converge to the A-solution H of Equation (4.87), in the sense that

$$\lim_{i \to \infty} \left\| H^i(c) - H(c) \right\| = 0 \qquad (4.166)$$

uniformly for c in compact subsets of $\{c \mid |c| < \rho\}$.

PROOF For $0 < c_1 < \rho^* < \rho$ we show that there is $I > 0$ so that if $i > I$, $(1, B_i) \geq c_1$ and for $|c| < c_1$

$$\left\| H^i(c) \right\| \leq (\rho^*/c_1)^{1/2} \left\| H(\rho^*) \right\| . \qquad (4.167)$$

This will complete the proof as Theorem 4.31 will imply that (4.166) holds in compact subsets of $\{c \mid |c| < c_1]$ and $c_1 < \rho^* < \rho$ were arbitrary.

We consider the iterates

$$H_0^i = 1,$$

$$H_n^i = 1 + B_i \left(H_{n-1}^i, H_{n-1}^i \right) . \qquad (4.168)$$

As B_i converges strongly to B there is I such that if $i > I$, we have

$$\left\| B \left(H(\rho^*), H(\rho^*) \right) - B_i \left(H(\rho^*), H(\rho^*) \right) \right\| < \hat{\varepsilon} , \qquad (4.169)$$

where

$$\hat{\varepsilon} = \left((\rho^*/c_1)^{1/2} - 1 \right) / (1 + \rho^*) . \qquad (4.170)$$

We show that if $i > I$ and $|c| < c_1$,

$$\left\| H_n^i(c) \right\| \leq (\rho^*/c_1)^{1/2} \left\| H(\rho^*) \right\| . \qquad (4.171)$$

By Theorem 4.31, this implies that $\rho(1, B_i) \geq c_1$ and that the inequality (4.167) holds. As the power series coefficients of all functions involved are nonnegative it suffices to show that, for $i > I$ and $n \geq 0$,

$$H_n^i(c_1) \leq (\rho^*/c_1)^{1/2} H(\rho^*) . \qquad (4.172)$$

We prove (4.172) by induction. The inequality is trivially satisfied for $n = 0$. If the inequality holds for $n = N \geq 0$, then

$$H_{N+1}^i (c_1) = 1 + c_1 B_i \left(H_N^i (c_1), H_N^i (c_1) \right)$$

$$\leq 1 + c_1 \left(\rho^*/c_1 \right) B_i \left(H \left(\rho^* \right), H \left(\rho^* \right) \right) . \qquad (4.173)$$

Equation (4.169) implies

$$H_{N+1}^i (c_1) \leq 1 + \rho^* B \left(H \left(\rho^* \right), H \left(\rho^* \right) \right) + \rho^* \hat{\varepsilon} . \qquad (4.174)$$

The H-equation itself and the definition of $\hat{\varepsilon}$ then imply

$$H_{N+1}^i (c_1) \leq \left(1 + \rho^* \hat{\varepsilon} \right) + \left(H \left(\rho^* \right) - 1 \right) \rho^* = \rho^* \hat{\varepsilon} + H \left(\rho^* \right)$$

$$\leq \left(1 + \rho^* \hat{\varepsilon} \right) H \left(\rho^* \right) \leq \left(\rho^*/c_1 \right) H \left(\rho^* \right) . \qquad (4.175)$$

Hence, (4.172) holds for $n = N + 1$. This completes the proof.   ∎

We now consider the specific selection of B_i such that

$$a_n^i = b_n^i = t_{n-1}^i,$$

$$c_n^i = d_n^i = t_n^i . \qquad (4.176)$$

For this choice of the a, b, c, and d and $|c| < \rho^*$ and i sufficiently large, the equation $H^i = 1 + cB_i(H^i, H^i)$ has a unique solution that is specified once the functions $\{H^i(t_n^i, \mu)\}_{n=0}^{N_i+1}$ are given. These functions may be found by recursion. We have

$$H^i \left(t_{N+1}^i, \mu \right) = H^i(\tau, \mu) = 1 \qquad (4.177)$$

and, for $1 \leq n \leq N_i + 1$,

$$H \left(t_{n-1}^i, \mu \right) = 1 + c \int_0^1 \psi(v) \sum_{p=n-1}^{N_i+1} H^i \left(t_p^i, \mu \right) s \left(t_p^i \right) H^i \left(t_{p+1}^i, v \right)$$

$$\times s \left(t_{p+1}^i \right) K_i(n, p, \mu, v) dv , \qquad (4.178)$$

where, in Equations (4.178)

$$K_i(p, n, \mu, v) = \frac{1}{v} \int_{t_p^i}^{t_{p+1}^i} \exp\left[-\left(t - t_{n-1}^i\right)\left(\frac{1}{\mu} + \frac{1}{v}\right)\right] dt$$

$$= \frac{\mu}{\mu + v}\left[\exp\left[-\left(t_p^i - t_{n-1}^i\right)\left(\frac{1}{\mu} + \frac{1}{v}\right)\right]\right.$$

$$\left. - \exp\left[-\left(t_{p+1}^i - t_{n-1}^i\right)\left(\frac{1}{\mu} + \frac{1}{v}\right)\right]\right]. \qquad (4.179)$$

If the functions $\{H^i(t_p^i, \mu)\}_{p=n}^{N_i+1}$ are known, then we may solve Equation (4.178) for $H^i(t_{n-1}^i, \mu)$ by

$$H^i\left(t_{n-1}^i, \mu\right) = M^i(n, \mu)/Q^i(n, \mu), \qquad (4.180)$$

where

$$M^i(n, \mu) = 1 + c \int_0^1 \psi(v) \sum_{p=n}^{N_i} H\left(t_p^i, \mu\right) H\left(t_{p+1}^i, v\right) s\left(t_p^i\right) s\left(t_{p+1}^i\right)$$

$$\times K(n, p, \mu, v)dv \qquad (4.181)$$

and

$$Q^i(n, \mu) = 1 - cs\left(t_{n-1}^i\right) s\left(t_n^i\right) \int_0^1 \psi(v) H^i\left(t_n^i, v\right)$$

$$K(n, n-1, \mu, v)dv. \qquad (4.182)$$

Example 4.8

We conclude by giving the results of some numerical experiments. We consider the transport equation, Equation (4.91) with $I_1(\mu) \equiv 1$ and

$$\omega(x) = e^{-xa}, \qquad (4.183)$$

where a is a parameter. In this case we will have

$$\psi \equiv 1/2. \qquad (4.184)$$

For this choice of ω and any $\tau \leq \infty$ we will have $\rho(1, B) > 1$. Hence, we may approximate H using Equation (4.178) and then the exit distributions by Equations (4.93) and (4.94). We may then calculate the albedo

$$A^* = 2 \int_0^1 \mu I(0, -\mu) d\mu \tag{4.185}$$

and, for $\tau < \infty$, the transmission

$$B^* = 2 \int_0^1 \mu I(\tau, \mu) d\mu . \tag{4.186}$$

Our choice of the points $\{t_n^i\}$ can be described as follows. For each i we subdivide the interval $[0, 1]$ into i subintervals and let $\{x_n^i\}_{n=20(i-1)+1}^{20i}$ be the abscissas for 20 point Gaussian quadratures on the ith interval. With $N_i = 20i$ and $t_{N_i+1}^i = \tau$ and $t_0^i = 0$ we have, for $1 \leq i \leq N_i$,

$$t_n^i = x_n^i / \left(\left(\frac{1+\tau}{\tau} \right) - x_n^i \right) \quad (\tau < \infty) , \tag{4.187}$$

$$t_n^i = \frac{x_n^i}{1 - x_n^i} \quad (\tau = \infty) . \tag{4.188}$$

This choice of $\{t_n^i\}$ clearly satisfies the hypothesis (H).

We give our results in the following tables. Tables 4.1 and 4.2 give our approximate values of A^* and B^* for the problem,

$$\tau = 1, \quad c = .7 \tag{4.189}$$

for various values of a. All integrals in the μ and v variables were calculated with 20 point Gaussian quadratures. These results apply to all piecewise continuous ω having finitely many discontinuities on $[0, \tau]$ that satisfy (4.88).

Note that convergence of the approximate values for B^* is substantially slower than those for A^*. Our numerical experiments indicate that this is a property of our method.

Table 4.3 gives approximate values of A^* for the problem

$$\tau = \infty, \quad c = .7 \tag{4.190}$$

for various values of a. Integrals in the μ and v variables were calculated as in the previous example.

Table 4.1 A^* for $\tau = 1, c = .7, \omega(x) = e^{-xa}, I_1(\mu) \equiv 1$
Numerical solution of Chandrasekhar's equation III

a	1	2	3	6	[77]
1	0.1519	0.1520	0.1521	0.1521	0.1521
10^{-1}	0.2116	0.2116	0.2116	0.2116	0.2116
10^{-2}	0.2210	0.2210	0.2210	0.2210	0.2210
10^{-3}	0.2220	0.2220	0.2220	0.2220	0.0220

Table 4.2 B^* for $\tau = 1, c = .7, \omega(x) = e^{-xa}, I_1(\mu) = 1$
Numerical solution of Chandrasekhar's equation IV

a	1	2	3	6	[77]
1	0.2926	0.2931	0.2933	0.2935	0.2936
10^{-1}	0.3553	0.3565	0.3570	0.3576	0.3582
10^{-2}	0.3665	0.3678	0.3684	0.3691	0.3698
10^{-3}	0.3677	0.3691	0.3696	0.3703	0.3711

Table 4.3 A^* for $\tau = \infty, c = .7, \omega(x) = e^{-xa}, I_1(\mu) \equiv 1$
Numerical solution of Chandrasekhar's equation V

a	1	2	3	6	[77]
1	0.1545	0.1550	0.1552	0.1552	0.1552
10^{-1}	0.2347	0.2353	0.2354	0.2354	0.2354
10^{-2}	0.2539	0.2540	0.2540	0.2540	0.2540
10^{-3}	0.2563	0.2563	0.2563	0.2563	0.02563

4.6   Multipower Equations

Here we examine the solvability of multipower equations. We then apply our results to nonlinear integral equations [176].

We seek the solution $x \in X$, where X is a Banach space to the multipower equation

$$y = x + \lambda L(\underbrace{x, \ldots, x}_{k\,\text{times}}) \tag{4.191}$$

Here, $y \in X$ and $L: X^k \mapsto X$ is a k-linear operator. We shall consider for $k \geq 2$

the linear map

$$L_x(\cdot) := L(\underbrace{x, \ldots, x}_{k-1\,\text{times}}, \cdot)$$

and observe that

$$\|L_x\| \le \|L\|\|x\|^{k-1} . \tag{4.192}$$

The following lemma gives a bound that is useful when measuring the effectiveness of our iterative scheme. The proof of the lemma is omitted.

LEMMA 4.5
Let $L: X^k \mapsto X$ be a k-linear operator with $k \ge 2$. For $w, z \in X$ with $\|w\|, \|z\| \le M$ we have

$$\|L_w - L_z\| \le \left(\|L\| M^{k-2}(k-1) \right) \|w - z\| . \tag{4.193}$$

In order to approximate solutions to Equation (4.191), we consider the fixed point problem

$$Q(x) = x , \tag{4.194}$$

where, for appropriate $\lambda \in \mathbf{R}$ and $x \in X$, the map $Q: X \mapsto X$ is given as follows:

$$Q(x) = (\lambda L_x + I)^{-1} (y) . \tag{4.195}$$

The difficulty lies with inverting an infinite dimensional linear operator, so the standard approach is to use successive subspaces V_n and approximate the solution to the problem in finite-dimensional settings. Uniformly contractive systems will be developed to show that these finite-dimensional approximations do indeed converge to the true solution of Equation (4.194).

To formulate the finite-dimensional approximating scheme, we first assume that X has a Schauder basis $\{e_k\}_{k=1}^{\infty} \subset X$. Then each $x \in X$ has a unique representation

$$x = \sum_{k=1}^{\infty} \langle e_k', x \rangle e_k ,$$

where $e_k' \in X'$ satisfy $\langle e_k', e_j \rangle = \delta_{kj}$.

Next, let $\{k_n\} \subset \mathbf{N}$ be an increasing sequence. We then define the projection operators S_n as follows:

$$S_n(x) = \sum_{j=1}^{k_n} \langle e_j', x \rangle e_j . \tag{4.196}$$

Since X is complete, $\sup_{n \in \mathbf{N}} \|S_n\| < \infty$. For convenience, we assume that $\|S_n\| = 1$. We then take as our finite-dimensional subspaces $V_n = S_n(X)$, and define the linear map $L_x^n: X \mapsto V_n$ as

$$L_x^n(\cdot) = S_n(L_x(\cdot)) \, . \tag{4.197}$$

We recall that a k-linear operator $L: X^k \mapsto X$ is *compact* if for any bounded set $B \subset X$, the set $L(B^k)$ is relatively compact.

The next result illustrates that compactness is sufficient to ensure that $L_x^n \mapsto L_x$.

LEMMA 4.6
Suppose that X has a Schauder basis.

   (i)   If L is compact in its kth variable, then for each $x \in X$,

$$\lim_{n \to \infty} \left\| L_x^n - L_x \right\| = 0 \, . \tag{4.198}$$

   (ii) If L is compact, then

$$\lim_{n \to \infty} \left\| L^n - L \right\| = 0 \, . \tag{4.199}$$

Here, $L^n = S_n L$.

PROOF   It is clear that S_n converges uniformly to the identity map I on relatively compact sets. Since L_x is a compact map, for any bounded set $B \subset X$, the set $L_x(B)$ is relatively compact. Now L_x is linear, so

$$\left\| L_x^n - L_x \right\| = \| (S_n - I) L_x \| = \sup_{\|w\| \le 1} \| (S_n - I) L_x(w) \| \to 0$$

which proves assertion (i).

If L is compact, then $L(B^k)$ is relatively compact for any bounded set $B \subset X$, so

$$\lim_{n \to \infty} \left\| L^n - L \right\| = \sum_{n \to \infty} \| (S_n - I) L \| = 0$$

which is assertion (ii).   ■

The preceding result will be used later when we derive conditions that guarantee convergence of the sequence of solutions obtained in finite-dimensional subspaces to the true fixed point solution of Equation (4.194). Sufficient conditions on the operator L for this convergence compare favorably with those given in Chapter 1.

In order to obtain the results we have found it convenient to develop the notion of a uniformly contractive system. Such a system is defined and is a useful framework with which to show that $z_n \to z$, where z solves Equation (4.194) and the z_n are the fixed points of the map $Q_n: X \to V_n$.

We conclude the case with a section of examples illustrating the application of our iterative scheme to approximating solutions of certain multipower equations. We consider approximating solutions to the

(i) Hammerstein equation

$$y(s) = x(s) + \lambda \int_a^b K(s, t)(x(t))^2 dt \tag{4.200}$$

where $X = L^2[a, b]$ and the

(ii) Chandrasekhar (1.22) equation where $X = L^2[0, 1]$.

In both cases, we use a sequence of closed nested subspaces $V_0 \subset \ldots \subset V_n \subset X$. Such sequences of subspaces have been found to be particularly useful in many applications when the V_n form a so-called multiresolution analysis. We may then employ a sequence of wavelet bases $\{\psi_{nk}\} \subset V_n$ for providing approximate solutions to the multipower equations. These bases are orthonormal and compactly supported. Such properties are desirable in view of the number of integrals that must be computed when we devise a scheme for obtaining approximate solutions to Equations (i) and (ii) above. A discussion of the algorithm used for obtaining approximate solutions is also included.

We need a lemma that will be of use later.

LEMMA 4.7
For $k \geq 2$ and $0 < \alpha < \frac{(k-1)^{k-1}}{k^k}$, there exists some $D > 0$ such that

$$\frac{\alpha(1 + D)^{k-1}}{1 - \alpha(1 + D)^{k-1}} < D < \frac{1}{k - 1} . \tag{4.201}$$

PROOF   Observe that the left inequality in (4.201) is equivalent to

$$\alpha(1 + D)^k - (1 + D) + 1 < 0 .$$

Now let $f(v) = \alpha v^k - v + 1$. Clearly $f(1) > 0$ and since $\alpha < \frac{(k-1)^{k-1}}{k^k}$, we find that $f(\frac{k}{k-1}) < 0$. Thus, there exists some $v = 1 + D$ satisfying $1 < v < \frac{k}{k-1}$ so that the desired inequality (4.191) holds.   ∎

The following theorem gives conditions on λ, L, and y to ensure that the iteration scheme

$$x_{n+1} = Q(x_n) \tag{4.202}$$

converges to the true fixed point of (4.194). Such conditions also lead to the definition of a sphere S wherein any initial guess x_0 will lead to the unique fixed point in some sphere $U \subset S$.

THEOREM 4.32

If

$$0 < |\lambda| \|L\| \|y\|^{k-1} = \alpha < \frac{(k-1)^{k-1}}{k^k}, \qquad (4.203)$$

then there exists a solution x_s to Equation (4.191), unique in the open sphere

$$S = \left\{ z \in X \left| \|z - y\| < \frac{\|y\|}{k-1} \right. \right\}. \qquad (4.204)$$

If the initial guess $x_0 \in S$, then

$$\lim_{n \to \infty} x_n = x_s \qquad (4.205)$$

where $x_{n+1} = Q(x_n)$. The solution x_s is contained in the closed sphere

$$U = \left\{ z \in X \left| \|z - y\| \le D\|y\| < \frac{\|y\|}{k-1} \right. \right\}$$

where D is given in Lemma 4.7.

PROOF Let $x_0 \in S$ and set $\alpha = |\lambda| \|L\| \|y\|^{k-1}$. By Lemma 4.7, we can choose D so that

$$\frac{\|x_0 - y\|}{\|y\|} < D < \frac{1}{k-1} \quad \text{and} \quad \frac{\alpha(1+D)^{k-1}}{1 - \alpha(1+D)^{k-1}} < D.$$

Note that x_0 is in the closed sphere U. We claim that $Q(U) \subset U$. To verify the claim, let $x \in U$. Then

$$Q(x) - y = (\lambda L_x + I)^{-1} (I - (\lambda L_x + I)) (y)$$

so that

$$\|Q(x) - y\| \le \frac{\|\lambda L_x\|}{1 - \|\lambda L_x\|} \|y\|$$

$$\leq \frac{\|\lambda L\|((1+D)\|y\|)^{k-1}}{1-\|\lambda L\|((1+D)\|y\|)^{k-1}}\|y\|$$

$$= \frac{\alpha(1+D)^{k-1}}{1-\alpha(1+D)^{k-1}}\|y\|$$

$$\leq D\|y\|\,.$$

Thus $Q(x) \in U$, which proves the claim.

Next we show that $Q: U \mapsto U$ is a contraction mapping, with contraction factor

$$r = \frac{\alpha(1+D)^{k-2}(k-1)}{(1-\alpha(1+D)^{k-1})^2} < 1$$

(to see that $r < 1$, note that replacing D with $\frac{1}{k-1}$ and α with $\frac{(k-1)^{k-1}}{k^k}$ yields $r = 1$). In order to prove that Q is indeed a contraction, let $x, w \in U$. Then

$$Q(x) - Q(w) = (\lambda L_x + I)^{-1}(\lambda L_w - \lambda L_x)(\lambda L_w + I)^{-1}y$$

which along with Lemma 4.5 and repeated use of (4.192) yields

$$\|Q(x) - Q(w)\| \leq \frac{\|\lambda L\|\left(\|y\|^{k-2}(1+D)^{k-2}(k-1)\right)\|w-x\|}{(1-\|\lambda L_x\|)(1-\|\lambda L_2\|)}\|y\|$$

$$\leq \frac{\alpha(1+D)^{k-2}(k-1)\|w-x\|}{\left(1-\|\lambda L\|((1+D)\|y\|)^{k-1}\right)^2} = r\|w-x\|\,. \quad (4.206)$$

Now since Q is a contraction mapping with $Q(U) \subset U$, we can apply the contraction mapping principle to the iterative scheme $x_{n+1} = Q(x_n)$ and conclude that the iterates must converge to the unique fixed point $x_s \in U$ of Q. Since D can be chosen arbitrarily close to $\frac{1}{k-1}$, the solution must be unique in the open sphere S.

∎

COROLLARY 4.7
The bound $\frac{(k-1)^{k-1}}{k^k}$ on $|\lambda|\|L\|\|y\|$ in (4.203) is optimal for $X = \mathbf{R}$.

PROOF   Consider the equation $1 = x - \frac{(k-1)^{k-1}}{k^k}x^k$.   ∎

Some remarks are in order before we conclude this section. We first note that the iterative method described in Theorem 4.32 can be generalized slightly to solve

equations of the form
$$y = Ax + \lambda L(x, \ldots, x), \tag{4.207}$$

where A and A^{-1} are linear and bounded. Putting $y^a = A^{-1}y$ and $L^a = A^{-1}L$, we have
$$y^a = x + \lambda L^a(x, \ldots, x),$$

which is of the form (4.191) and can thus be solved using Theorem 4.32.

From Lemma 4.7 we observe that for λ near 0, we can pick D near 0 and conclude that the solution to Equation (4.191) is close to y, thus improving our choice of the initial guess in the iterative scheme (4.202).

In the special case $k = 2$, it is useful to compare our results with those obtained elsewhere. McFarland [129] considered Equation (4.207) with A invertible and linear (see also Exercise 37 in Chapter 1). McFarland showed in his Theorem 3 that the iterative scheme
$$x_{n+1} = (A + \lambda Lx_n)^{-1} y = (I + \lambda L^a x_n)^{-1} y^a$$

converges to a solution of Equation (4.207) if
$$0 < \left\| A^{-1} \right\| |\lambda| \|L\| \|y^a\| \le \delta < \frac{1}{4} \tag{4.208}$$

and if
$$\frac{1 - (1 - 4\delta)^{1/2}}{2} \le \|\lambda L^a x_0\| < \frac{1}{2}. \tag{4.209}$$

If our condition (4.203) with $k = 2$ is satisfied, then so is McFarland's condition (4.208). Note that McFarland's requirement (4.209) on the initial condition x_0 may be more difficult to verify than the condition (4.204) $\|x_0 - y\| < \|y\|$. McFarland does not use the contraction mapping principle in his proof, and obtains no uniqueness results.

In [167], Rall solves Equation (4.191) subject to (4.203), both with $k = 2$ (see Chapter 1). He uses a series approach and shows in Theorem 19 that the solution x is unique in the sphere
$$\left\{ z \in X \,\middle|\, \|z - x\| < \frac{\sqrt{1 - 4|\lambda| \|L\| \|y\|}}{2|\lambda| \|L\|} \right\}. \tag{4.210}$$

When $4|\lambda| \|L\| \|y\|$ is near 1, Rall's uniqueness condition (4.210) does not give as much information as condition (4.204). On the other hand, for λ near 0, Rall's sphere is much larger than our sphere S in (4.204).

In Chapter 1, Argyros uses a different iterative method and an auxiliary quadratic equation to obtain several existence and uniqueness results for Equation (4.191) when $k = 2$.

In Corollary 1.1, Argyros' uniqueness ball (i) is bigger than (4.204) but his existence ball (ii) is the same as ours. We summarize this in the following corollary.

COROLLARY 4.8
In the case $k = 2$, the solution x_s of Equation (4.191) is in the closed sphere

$$U = \left\{ z \in X \mid \|z - y\| \leq \tilde{D}\|y\| \right\}$$

$$= \left\{ z \in X \,\middle|\, \|z - y\| \leq \frac{1 - 2|\lambda|\|L\|\|y\| - (1 - 4|\lambda|\|L\|\|y\|)^{1/2}}{2|\lambda|\|L\|} \right\} .$$

PROOF   It is easy to see that there exists a unique root $v_0 \in (1, \frac{k}{k-1})$ for the function $f(v)$ given in the proof of Lemma 4.7. Then for each $v \in (v_0, \frac{k}{k-1})$, $D = v - 1$ satisfies inequality (4.201) and so by Theorem 4.32

$$\|y - x_s\| \leq D\|y\| = (v - 1)\|y\| .$$

Hence $\|y - x_s\| \leq (v_0 - 1)\|y\|$. In the case $k = 2$, the quadratic formula yields

$$(v_0 - 1)\|y\| \leq \frac{1 - 2|\lambda|\|L\|\|y\| - (1 - 4|\lambda|\|L\|\|y\|)^{1/2}}{2|\lambda|\|L\|}$$

and the assertion is proved.   ∎

The linearity of L in the first $k - 1$ variables is not critical for the results of this section, and we can generalize them somewhat.

PROPOSITION 4.9
Let $L: X \times X \to X$ satisfy the following conditions:

(i) $L_x := L(x, \cdot)$ *is a bounded linear operator for all $x \in X$.*

(ii)   *There exist $\mu \geq 2$ and $C > 0$ such that*

$$\|L(x, v)\| \leq C\|x\|^{\mu-1}\|v\|$$

and

$$\|L_x - L_v\| \leq C(\mu - 1)(\max(\|x\|, \|v\|))^{\mu-2}\|x - v\|$$

for all $x, v \in X$.

Then Theorem 4.32 and Corollary 4.8 hold for this operator L with $\|L\|$ and k replaced by C and μ, respectively, throughout the theorem and corollary.

PROOF   Note that condition (ii) is just a minor generalization of inequalities (4.11) and (4.12). The proofs of Theorem 4.32 and Corollary 4.8 are valid with these adjustments. ∎

It is clear from Lemma 4.5 that a bounded k-linear operator will satisfy the hypotheses of this proposition. The next example is an important example of an operator of this type that is not k-linear.

Example 4.9
Define $L: C[0, 1] \times C[0, 1] \to C[0, 1]$ by

$$L(x, v)(s) = \int_0^1 h(s, t)x(t)^{\mu-1}v(t)dt ,$$

where $h \in C([0, 1]^2)$ and $\mu \geq 2$. Then L is linear in v and

$$\|L(x, v)\|_\infty \leq \|x\|_\infty^{\mu-1}\|v\|_\infty \left\|\int_0^1 h(s, t)dt\right\|_\infty .$$

Moreover,

$$\|(L_x - L_w)(v)\|_\infty = \left\|\int_0^1 h(s, t)\left((x(t))^{\mu-1} - (w(t))^{\mu-1}\right)v(t)dt\right\|_\infty$$

$$\leq (\mu - 1)M^{\mu-2}\left\|\int_0^1 h(s, t)dt\right\|_\infty \|x - w\|_\infty\|v\|_\infty ,$$

where $M = \max(\|x\|_\infty, \|w\|_\infty)$, by the lemma below.

LEMMA 4.8
For $x, w \geq 0$ and $\mu \geq 2$,

$$\left|x^{\mu-1} - w^{\mu-1}\right| \leq (\mu - 1)(\max(x, \mu))^{\mu-2}|x - w| . \qquad (4.211)$$

PROOF   Without loss of generality, assume $w < x$. Fix w and let

$$f(x) = (\mu - 1)x^{\mu-2}(x - w) \quad \text{and} \quad g(x) = x^{\mu-1} - w^{\mu-1} .$$

Then

$$f'(x) = (\mu - 1)x^{\mu-2} + (x - w)(\mu - 1)(\mu - 2)x^{\mu-3}$$

$$\geq (\mu - 1)x^{\mu-2} = g'(x) . \tag{4.212}$$

Since $f(w) = 0 = g(w)$, inequality (4.212) yields (4.211). ∎

REMARK 4.6

The class of boundary value problems

$$x''(t) + \lambda a(t)x^{\mu}(t) = f(t)$$

with $t \in [0, 1]$, $\mu \geq 2$ and appropriate boundary conditions can be transformed into

$$x(t) = \lambda \int_0^1 K(s, t)a(s)x^{\mu}(s)ds + F(t) . \quad ∎$$

Then Proposition 4.9 will apply with suitable restrictions on λ and $a(t)$.
For the case $y = 0$ in Equation (4.11), our iterative scheme

$$x_n = \left(\lambda L_{x_n} + I\right)^{-1}(y)$$

will yield only the trivial solution $x = 0$. We show below that this is the only "small" solution. To obtain "large" solutions, schemes such as the Newton–Kantorovich method (see Chapters 1 through 3) can be used. For many problems, the Newton–Kantorovich method will be faster than our iterative scheme. However, if the Fréchet derivative $\lambda L'(x, \ldots, x)$ is not defined or is significantly more expensive to numerically compute than λL_x, then our scheme is preferable.

PROPOSITION 4.10

Equation (4.191) has at most one solution $x \in X$ for which

$$\|x\| < \left(\frac{1}{k\|L\|}\right)^{\frac{1}{k-1}} . \tag{4.213}$$

PROOF For the sake of reaching a contradiction, suppose that u and v are distinct solutions to Equation (4.191) satisfying (4.213). Then

$$u - v = L(u, \ldots, u) - L(v, \ldots, v)$$

$$= L(u - v, u, \ldots, u) + L(v, u - v, u, \ldots, u) + \cdots + L(v, \ldots, v, u - v)$$

so by (4.213) we have

$$\|u - v\| \leq k\|u - v\|\|L\|(\max(\|u\|, \|v\|))^{k-1} < \|u - v\|,$$

which is a contradiction. ∎

We now introduce the notion of a uniformly contractive system. The role of such a system is to provide a general framework for obtaining iterative solutions of operator equations that involve contraction mappings. In particular, we will use the concept of the uniformly contractive system in conjunction with the method discussed above to construct approximate solutions to certain multipower equations.

DEFINITION 4.10 *Let X be a Banach space, $\{V_n\}$ a sequence of closed subspaces of X such that*

$$\lim_{n \to \infty} \text{dist}(V_n, x) = 0$$

for each $x \in X$. Let U be a closed set in X, and, for each $n \in \mathbf{N}$, let Q_n be an operator with $Q_n : X \mapsto V_n$, and define the set $U_n = V_n \cap U$. We say that $\{U_n, Q_n\}$ is a uniformly contractive system if conditions (C1) and (C2) below hold.

(C1) There exists a $c \in \mathbf{R}$, $0 < c < 1$, and an $N \in \mathbf{N}$ such that if $n \geq N$ and $x, y \in U$, then

$$Q_n(U) \subset U_n \quad \text{and} \quad \|Q_n(x) - Q_n(y)\| \leq c\|x - y\|. \quad (4.214)$$

(C2) For any $x, y \in U$ and $\varepsilon > 0$, there exists an $N \in \mathbf{N}$ such that if $k \geq j \geq N$, then

$$\|Q_k(x) - Q_j(y)\| \leq c\|x - y\| + \varepsilon. \quad (4.215)$$

Note that the subspaces V_n need not be nested, so that the finite element method can be used within the context of a uniformly contractive system.

THEOREM 4.33
Let $\{U_n, Q_n\}$ *satisfy condition (C1) above. Then condition (C2) is equivalent to the existence of a contraction map* $Q: U \mapsto U$, *defined by* $Q(x) = \lim_{n \to \infty} Q_n(x)$, *such that*

$$\|Q(x) - Q(y)\| \le c\|x - y\|$$

for $x, y \in U$.

PROOF   Assume condition (C2) holds. We first show that the map Q is well defined. Fix $x \in U$ and $\varepsilon > 0$. Choose N as prescribed in condition (C2) and set $y = x$ in (4.215). Then for $k \ge j \ge N$ we have

$$\|Q_k(x) - Q_j(x)\| \le \varepsilon .$$

Thus $\{Q_j(x)\}_j$ is a Cauchy sequence. Since X is complete, $\lim_{n \to \infty} Q_n(x)$ exists. Noting that U is closed and $Q_n(x) \in U_n \subset U$ for all n yields

$$\lim_{n \to \infty} Q_n(x) = Q(x) \in U .$$

Now let $x, y \in U$ and $\varepsilon > 0$. Choose $N \in \mathbf{N}$ so that

$$\|Q(x) - Q_N(x)\| < \varepsilon \quad \text{and} \quad \|Q(y) - Q_N(y)\| < \varepsilon .$$

Then

$$\|Q(x) - Q(y)\| \le \|Q(x) - Q_N(x)\| + \|Q_N(x) - Q_N(y)\| + \|Q_N(y) - Q(y)\|$$

$$\le \varepsilon + c\|x - y\| + \varepsilon . \tag{4.216}$$

Since ε is arbitrary, we have

$$\|Q(x) - Q(y)\| \le c\|x - y\| .$$

Next we assume the existence of the map Q, and fix $x, y \in U$ and $\varepsilon > 0$. By the definition of Q, there exists an $N \in \mathbf{N}$ such that if $k \ge j \ge N$, then

$$\|Q_k(x) - Q_j(y)\| \le \|Q_k(x) - Q(x)\| + \|Q(x) - Q(y)\| + \|Q(y) - Q_j(y)\|$$

$$\le \varepsilon + c\|x - y\| + \varepsilon . \tag{4.217}$$

Thus, condition (C2) is satisfied.   ∎

We observe that the equations $Q_n(x) = x$ all have unique fixed points $z_n \in U$ by the contraction mapping principle (see Chapter 1). Our next result shows that these fixed points converge to z_s, the unique fixed point of the map $Q \in U$.

THEOREM 4.34
Let $\{U_n, Q_n\}$ be a uniformly contractive system. Then

$$\lim_{n \to \infty} z_n = z_s, \quad \text{where } Q(z_s) = z_s .$$

PROOF Let $\varepsilon > 0$ and $0 < c < 1$ be the contractive constant for $\{U_n, Q_n\}$. By Theorem 4.33, we can choose $N \in \mathbf{N}$ so that $n \geq N$ implies that

$$\| Q(z_s) - Q_n(z_s) \| \leq (1 - c)\varepsilon .$$

Then

$$\| z_s - z_n \| = \| Q(z_s) - Q_n(z_n) \|$$

$$\leq \| Q(z_s) - Q_n(z_s) \| + \| Q_n(z_s) - Q_n(z_n) \|$$

$$< (1 - c)\varepsilon + c \| z_s - z_n \| .$$

So

$$\| z_s - z_n \| (1 - c) < (1 - c)\varepsilon \quad \text{whence } \| z_s - z_n \| < \varepsilon .$$

Thus, $\lim_{n \to \infty} z_n = z_s$. ∎

THEOREM 4.35
Let $\{U_n, Q_n\}$ be a uniformly contractive system such that U is bounded and $\{Q_n\}$ converges to Q uniformly on U. Let $N \in \mathbf{N}$ be given as per condition (C1). Beginning with any $k \geq N$ and initial guess $x_k \in U_k$, the iterative scheme

$$x_{n+k+1} = Q_n(x_{n+k}) \tag{4.218}$$

will converge to the fixed point of Q in U: $\lim_{n \to \infty} x_{n+k} = z_s = Q(z_s)$.

REMARK 4.7 We note at this time that to numerically implement (4.218), a hierarchical basis, such as one provided by a multiresolution analysis, is required.

∎

PROOF   (of Theorem 4.35) Fix $\varepsilon > 0$. By the uniform convergence of $\{Q_n\}$, there is some $M_1 \geq k \geq N$ such that for $n \geq m \geq M_1$, we have

$$\|x_n - x_m\| = \|Q_{n-1}(x_{n-1}) - Q_{m-1}(x_{m-1})\|$$

$$\leq \|Q_{n-1}(x_{n-1}) - Q_{m-1}(x_{n-1})\|$$

$$+ \|Q_{m-1}(x_{n-1}) - Q_{m-1}(x_{m-1})\| \qquad (4.219)$$

$$\leq \varepsilon + c\|x_{n-1} - x_{m-1}\| \ .$$

Now choose $M_2 > M_1$ such that

$$c^{M_2 - M_1} < \frac{\varepsilon}{2\mathrm{diam}\,U} \ .$$

Then for any $n \geq m \geq M_2$, we can repeat the iteration (4.219) $m - M_1$ times to obtain

$$\|x_n - x_m\| \leq \varepsilon \sum_{j=0}^{m-M_1-1} c^j + c^{m-M_1} \|x_{n-(m-M_1)} - x_{M_1}\|$$

$$\leq \varepsilon \sum_{j=0}^{m-M_1-1} c^j + 2c^{m-M_1}\mathrm{diam}\,U \qquad (4.220)$$

$$\leq \varepsilon \left(\sum_{j=0}^{m-M_1-1} c^j + 1 \right) \ .$$

Thus $\{x_n\}$ is a Cauchy sequence, with $\lim_{n\to\infty} x_n = z \in U$.

Now $\{U_n, Q_n\}$ is a uniformly contractive system, so for $n \geq N$,

$$\|Q(z) - z\| \leq \|Q(z) - Q_n(z)\| + \|Q_n(z) - Q_n(x_n)\| + \|Q_n(x_n) - z\|$$

$$\leq \|Q(z) - Q_n(z)\| + c\|z - x_n\| + \|x_{n+1} - z\| \ .$$

For n sufficiently large, we have

$$\|Q(z) - z\| \leq \|Q(z) - Q_n(z)\| + c\|z - x_n\| + \|x_{n+1} - z\| \leq \varepsilon$$

since $\{Q_n\}$ converges to Q pointwise and $\lim_{n\to\infty} x_n = z$. As ε is arbitrary, we have $Q(z) = z$. Since Q has a unique fixed point in U, we conclude that $z_s = z$.

∎

REMARK 4.8   The convergence rate for the scheme (4.218) to the solution z of the fixed point problem (4.194) will be governed by the size of the contraction constant c of the uniformly contractive system, as well as the diameter of U and the uniform convergence of the operators Q_n on U. To be precise, for any given $\varepsilon > 0$, there are $M_1, M_2 \in \mathbf{N}$ such that

$$\|Q_n(x) - Q_m(x)\| < \varepsilon$$

for all $m, n \geq M_2 \geq M_1$ and $x \in U$. Then

$$\|z - x_m\| \leq \varepsilon \left(\sum_{j=0}^{M_2 - M_1} c^j + 1 \right)$$

for $m \geq M_2$.   ∎

Let X be a Banach space with Schauder basis $\{e_k\} \subset X$. Consider the operator

$$R = \left(\lambda L_x^n + I \right) S_n$$

as a map from V_n into V_n. Note that $R^{-1} : V_n \mapsto V_n$ exists when $\|\lambda L_x^n S_n\| \leq \|\lambda L_x\| < 1$. In matrix terms, with respect to the basis $\{e_k\}$, R^{-1} is formed by inverting the principal submatrix (corresponding to V_n) of the matrix representation of the linear operator $(\lambda L_x^n + I)S_n$.

Let $J_n : V_n \mapsto X$ denote the natural injection operator. Define $(\lambda L_x^n + S_n)^{-1} : X \mapsto X$ by

$$\left(\lambda L_x^n + S_n \right)^{-1} = J_n R^{-1} S_n$$

and define $Q_n : X \to V_n$ to be

$$Q_n(x) = \left(\lambda L_x^n + S_n \right)^{-1} y . \tag{4.221}$$

We now give convergence conditions for the finite-dimensional operators $\{Q_n\}$.

THEOREM 4.36
Suppose that X has a Schauder basis $\{e_k\}$.

(a) *If* $L: X^k \mapsto X$ *is compact in the kth variable, then* Q_n *converges to* Q *pointwise on*

$$\{x \in X \mid \|\lambda L x\| < 1\}.$$

(b) *If* $L: X^k \mapsto X$ *is compact and* $\delta < 1$, *then* Q_n *converges to* Q *uniformly on*

$$U_\delta = \left\{x \in X \mid \|\lambda L\| \|x\|^{k-1} \leq \delta\right\}.$$

PROOF   Let $x \in \{x \in X: \|\lambda L_x\| < 1\}$ and observe that

$$Q(x) - Q_n(x) = (\lambda L_x + I)^{-1} y - \left(\lambda L_x^n + S_n\right)^{-1} S_n(y)$$

$$= (\lambda L_x + I)^{-1} (y - S_n(y))$$

$$+ (\lambda L_x + I)^{-1} \left(\left(\lambda L_x^n + S_n\right) - (\lambda L_x + I)\right)$$

$$\left(\lambda L_x^n + S_n\right)^{-1} S_n(y).$$

Since $(\lambda L_x^n + S_n)^{-1} S_n(y) \in V_n$, we have

$$(S_n - I) \left(\lambda L_x^n + S_n\right)^{-1} S_n(y) = 0.$$

Thus

$$Q(x) - Q_n(x) = (\lambda L_x + I)^{-1} (y - S_n(y))$$

$$+ (\lambda L_x + I)^{-1} \left(\lambda L_x^n - \lambda L_x\right) \left(\lambda L_x^n + S_n\right)^{-1} S_n(y)$$

which yields

$$\|Q(x) - Q_n(x)\|$$

$$\leq \frac{1}{1 - \|\lambda L_x\|} \left(\|y - S_n(y)\| + \|\lambda L_x^n - \lambda L_x\| \left\|\left(\lambda L_x^n + S_n\right)^{-1} S_n(y)\right\|\right)$$

$$\leq \frac{\|y - S_n(y)\|}{1 - \|\lambda L_x\|} + \frac{\|\lambda L_x^n - \lambda L_x\| \|y\|}{(1 - \|\lambda L_x\|)^2}.$$

To prove assertion (A), we note that for each $x \in X$, L_x is a compact linear map. Now $\|\lambda L_x^n - \lambda L_x\| \to 0$ by Lemma 4.6 (i) so that $\|Q(x) - Q_n(x)\| \to 0$ thus proving (A).

Next assume that L is compact and that $x \in U_\delta$. Then

$$\|Q(x) - Q_n(x)\| \le \frac{\|y - S_n(y)\|}{1 - \delta} + \frac{\|\lambda L^n - \lambda L\| \|y\|}{(1 - \delta)^2} \frac{\delta}{\|\lambda L\|},$$

and since $\|\lambda L^n - \lambda L\| \to 0$ uniformly on U_δ by Lemma 4.6 (ii), Q_n converges to Q uniformly on U_δ. This proves assertion (B). ∎

REMARK 4.9 For the sake of notation, we have chosen L to be compact in the kth variable. The result holds as long as L is compact in at least one variable. ∎

We now state and prove our main result.

THEOREM 4.37
Suppose that X has a Schauder basis $\{e_k\}$ and that

$$0 < |\lambda| \|L\| \|y\|^{k-1} = \alpha < \frac{(k-1)^{k-1}}{k^k}.$$

Then:

(a) *If L is compact in at least one variable, then $\{U_j, Q_j\}$ is a uniformly contractive system.*

(b) *If L is compact, then the iterative scheme (4.218) given in Theorem 4.35 converges.*

PROOF The first part of the proof is needed for both assertions (a) and (b). Choose $K \in \mathbb{N}$ so that if $j \ge K$, then

$$\|S_j(y) - y\| + \frac{\alpha(1 + D)^{k-1}}{1 - \alpha(1 + D)^{k-1}} \le D$$

where the existence of D is guaranteed by Lemma 4.7. Consider the closed ball

$$U = \{z \in X \mid \|z - y\| \le D\|y\|\}.$$

Then, arguing similarly to the proof of Theorem 4.32, for $x \in U$ and $j \geq K$, we have

$$\left\| Q_j(x) - y \right\| = \left\| \left(\left(\lambda L_x^j + S_j \right)^{-1} - S_j \right) y + \left(S_j - I \right) y \right\|$$

$$= \left\| \left(\lambda L_x^j + S_j \right)^{-1} \left(S_j - \left(\lambda L_x^j + S_j \right) \right) y + \left(S_j - I \right) y \right\|$$

$$\leq \frac{\| \lambda L_x \|}{1 - \| \lambda L_x \|} \| y \| + \left\| \left(S_j - I \right) y \right\|$$

$$\leq \frac{\alpha (1 + D)^{k-1}}{1 - \alpha (1 + D)^{k-1}} \| y \| + \left\| \left(S_j - I \right) y \right\|$$

$$\leq D \| y \| \, .$$

Hence $Q_j(U) \subset U_j$. The proof that each Q_j is a contractive map with the same contraction factor r as Q is very similar to that given for Q in the proof of Theorem 4.32 and is omitted. Thus, condition (C1) is satisfied.

To complete the proof of part (a), note that $U \subset \{ x \in X \mid \| \lambda L_x \| < 1 \}$ so Theorem 4.36 (a) applies. Then by Theorem 4.33, condition (C2) is satisfied. If L is compact, then Q_n converges to Q uniformly on U by Theorem 4.36 (b). Therefore, Theorem 4.35 applies. ∎

We consider the case $k = 2$ so that Theorem 4.37 applies to the quadratic equation (1.37).

Observe that Argyros in Theorem 2.18 requires uniform convergence of the operators F_n to B, while our Theorem 4.37 (a) assumes only that B is compact in one variable in order to guarantee that the fixed point solutions z_n converge to the solution z_s of Equation (1.37).

Also note that Argyros' theorem requires that the bilinear operators F_n be symmetric, while this is not needed for Theorem 4.37 (a). This fact is quite important for an operator $B(x, w)$ that is compact in only one variable, for if B is "symmetrized" using the formula

$$\bar{B}(x, w) = \frac{B(x, w) + B(w, x)}{2} \, ,$$

then the compactness in one variable is destroyed. These points should be kept in mind for Example 4.10 below.

We employ an iterative scheme and the previous results to obtain approximate solutions to the two classes of integral equations given in the introduction. In both examples, we work in $X = L^2[a, b]$, $-\infty \leq a < b \leq \infty$. While we have considered different finite-dimensional subspaces of $L^2(\mathbf{R})$ in our examples, we have found that it is quite beneficial to utilize the closed subspaces $V_n \subset L^2[a, b]$ with $V_n \subset V_{n+1}$ ($n \in \mathbf{N}$), that form a multiresolution analysis (MRA) of $L^2[a, b]$. The multiresolution analysis gives rise to a so-called *wavelet basis* $\{\psi_{nk}\}$, where for fixed n, $\{\psi_{nk}\}_k$ forms an orthonormal basis for the space W_n, $V_{n+1} := V_n \oplus W_n$. For $n, k \in \mathbf{Z}$, $\{\psi_{nk}\}$ forms an orthonormal basis for $L^2[a, b]$.

It is desirable to use an orthonormal basis in the subsequent computations since each iterative step involves solving a linear finite-dimensional system. In addition, the compact support properties of the wavelet basis greatly reduces the number of numerical integrations that must be performed when we project our operator into finite-dimensional subspaces.

Example 4.10

Consider the *Chandrasekhar* (1.22) integral equation in the form

$$1 = H(s) + \frac{1}{2}\lambda s H(s) \int_0^1 \frac{H(t)}{s+t} dt . \qquad (4.222)$$

We consider solving (4.222) in $L^2[0, 1]$ for the function $H(s)$. While this equation is generally solved in $C[0, 1]$, it can be shown that an L^2-solution to Equation (4.222) is in fact in $C[0, 1]$. If we put $y(s) = 1$ and define $B: L^2[0, 1] \times L^2[0, 1] \mapsto L^2[0, 1]$ by

$$B(G, H)(s) = \frac{1}{2} s G(s) \int_0^1 \frac{H(t)}{s+t} dt , \qquad (4.223)$$

then the Chandrasekhar equation (4.222) can be expressed in the form (1.37). To obtain a bound on $\|B\|$, note that

$$\|B(G, H)\|^2 = \frac{1}{4} \int_0^1 \left(s G(s) \int_0^1 \frac{H(t)}{s+t} dt \right)^2 ds$$

$$\leq \frac{1}{4} \int_0^1 |G(s)|^2 \left(\int_0^1 \left(\frac{s}{s+t} \right)^2 dt \int_0^1 (H(t))^2 dt \right) ds \qquad (4.224)$$

by the Cauchy-Schwarz inequality. Then

$$\|B(G, H)\|^2 \le \frac{1}{4}c^2\|G\|^2\|H\|^2 \quad \text{where} \quad c^2 = \sup_{0 \le s \le 1} \int_0^1 \left(\frac{s}{s+t}\right)^2 dt = \frac{1}{2}$$

so $\|B\| \le \frac{1}{2\sqrt{2}}$. Note that the linear operator

$$B(G, \cdot)(s) = sG(s)\int_0^1 \frac{\cdot}{s+t}dt$$

is compact for each $G(s) \in L^2[0, 1]$. Hence, Theorem 4.37 (a) applies for appropriate λ and U, and any orthonormal basis of $L^2[0, 1]$. It is worthwhile to note that Theorem 4.36 (b) does not apply, since the operator B is only compact in one variable. It is not clear that Theorem 1.12 of Argyros can be applied to this example since B is only compact in one variable.

This example is an illustration of and not a rationale for Theorem 4.37. Using more information about Equation (4.222) than is used here, it can be proved that (see [77]) Equation (4.222) has a solution for $0 \le \lambda \le 1$, while our results apply for $|\lambda| \le \frac{1}{\sqrt{2}}$. Our numerical experiments indicate that this iterative scheme converges for $0 \le \lambda \le 1$, but we choose a λ value below that is justified by Theorem 4.37 (a).

In order to compute approximate solutions to Equation (4.222), we use the iterative scheme similar to (4.194):

$$Q_n\left(x^k\right) = x^{k+1} \quad (k \in \mathbf{N}) \tag{4.225}$$

where Q_n was defined by (4.193), and approximate the fixed point z_n of (4.225). We then choose larger and larger V_n spaces and repeat the iterative process. Finally, we appeal to Theorem 4.34 to conclude that the z_n approach the true solution.

Suppose we wish to compute the fixed point z_n of (4.225). Let $\{e_1, \ldots, e_N\}$ be an orthonormal basis for V_n, set $y(t) = 1$, and consider projecting (1.37) into V_n

$$\left(S_n + \lambda B_{x^k}^n\right)x^{k+1} = y^n$$

where $y^n(t) = \sum_{i=1}^N y_i^n e_i(t)$ with $y_i^n = \langle y, e_i \rangle$. To obtain a matrix representation A^n for $B_{x^k}^n$, we let

$$b_{ij}^k = \frac{1}{2}\int_0^1 \int_0^1 e_k(s)e_i(s)e_j(t)\frac{s}{s+t}dtds \quad (i, j, k = 1, \ldots, N) \tag{4.226}$$

so that the entries of A^n are given by

$$a_{pq} = \sum_{i=1}^{N} b_{iq}^p x_i^k \quad \text{where} \quad x^k(t) = \sum_{i=1}^{N} x_i^k e_i(t) .$$

We start with $x^1 := y$ and then iterate by repeatedly solving the system

$$\begin{bmatrix} y_1^n \\ \vdots \\ y_N^n \end{bmatrix} = (I_n + \lambda A^n) \begin{bmatrix} x_1^{k+1} \\ \vdots \\ x_N^{k+1} \end{bmatrix} ,$$

where the $n \times n$ identity matrix I_n is the matrix representation of S_n. Note that the b_{ij}^ks must only be computed once in this scheme and that b_{ij}^k is symmetric in i and k. In addition, if the basis functions have compact support within $[0, 1]$, then it is possible to *a priori* assign certain $b_{ij}^k = 0$. Certain wavelet bases have this property.

Here $P_3([0, 1])$ is the space of polynomials of degree three or less spanned by the classical Legendre polynomials. $S_1^2([0, 1])$ is the space of piecewise continuously differentiable quadratic polynomials with possible breakpoints at $0, \frac{1}{2}$, and 1 spanned by orthonormalized B-splines. Finally $S_0^1([0, 1])$ is the space of continuous linear polynomials with possible breakpoints at $0, \frac{1}{4}, \frac{1}{2}, \frac{3}{4}$, and 1, also spanned by orthonormalized B-splines.

The solution to the Chandrasekhar equation (4.222) is usually approximated in $C[0, 1]$ rather than $L^2[0, 1]$. Theorem 4.37 is general enough to be applied to $C[0, 1]$ since only a Schauder basis is assumed. A wavelet basis for a dense subspace of $C[0, 1]$ has been reported by Wang [196], so that the benefits of wavelets can be retained if an approximate solution in $C[0, 1]$ is desired.

Example 4.11
Solve for $x \in L^2[a, b]$ the *Hammerstein equation*

$$y(s) = x(s) + \lambda \int_a^b k(s, t)(x(t))^2 dt . \tag{4.227}$$

This equation can be expressed as a bilinear one in $L^2[a, b]$ where

$$K(f_1, f_2)(s) = \int_a^b k(s, t) f_1(t) f_2(t) dt$$

defines a bilinear operator $K: L^2[a, b] \times L^2[a, b] \mapsto L^2[a, b]$. Using the Cauchy-Schwarz inequality, the map K can be shown to be bounded if

$$k^*(s) = \sup_{a \le t \le b} |k(s, t)| \in L^2[a, b] . \tag{4.228}$$

The linear operator

$$K(f_1, \cdot)(s) = \int_a^b k(s, t) f_1(t)(\cdot) dt$$

is compact for each $f_1 \in L^2[a, b]$ when (4.228) holds. Thus, Theorem 4.37 (a) applies. Actually, we can claim that Theorem 4.37 (b) applies because K is actually compact.

We have used the iterative scheme (4.218). This method is quite similar to that used in Example 4.10 only now each successive iteration takes place in a larger subspace. In the case when $X = L^2(\mathbf{R})$, it is quite convenient to employ the ladder of subspaces given in Daubechies [88].

Example 4.12
Consider the Hammerstein integral equation

$$y(s) = x(s) + \lambda \int_a^b k(s, t_1, t_2, \ldots, t_n) x(t_1) \cdots x(t_n) dt_1 \cdots dt_n . \tag{4.229}$$

As in Example 4.11, (4.229) can also be viewed as an n-linear equation in $L^2[a, b]$

$$y = x + \lambda L(x, \ldots, x)$$

where the n-linear operator

$$L: L^2[a, b] \times L^2[a, b] \times \cdots \times L^2[a, b] \mapsto L^2[a, b]$$

is defined by

$$L(f_1, f_2, \ldots, f_n)(s) = \int_a^b k(s, t_1, t_2, \ldots, t_n)$$

$$f_1(t_1) f_2(t_2) \cdots f_n(t_n) dt_1 dt_2 \cdots dt_n .$$

It can be shown (see Section 5.6) that L is in fact compact when the kernel function is in $L^2([a, b]^{n+1})$, so Theorem 4.37 (b) and the iterative scheme (4.218) apply.

4.7 Uniformly Contractive Systems and Quadratic Equations in Banach Space

Here we examine the solvability of Equation (1.19) using uniformly contractive systems. We then apply the results to solve Equation (1.22) [173].

(V) Suppose that X has a sequence of proper subspaces $\{V_n\}$ and linear projections $P_n: X \to V_n$ for which

$$\lim P_n x = x$$

for each $x \in X$.

We make the following observations.

- $\mu = \sup \|P_n\| < \infty$ since X is complete.

- The subspaces need not be nested, so finite element methods may be applied.

- Any space X with a Schauder basis satisfies condition (V).

The spaces $V_n = P_n(X)$ are usually taken to be finite dimensional, and Equation (1.19) is replaced by

$$x = P_n B(x, x) + P_n y \tag{4.230}$$

which is solved in V_n using the iterative method

$$x_{k+1} = P_n B(x_k, x_k) + P_n y . \tag{4.231}$$

Uniformly contractive systems will be used to show that $z_n \to z_s$, where z_s solves (1.19) and the z_n solve (4.230). These results require the map B only to be bounded and bilinear so the finite rank operators $P_n B$ need only converge pointwise to B. If B is compact, a routine that avoids solving for any of the z_n will be shown to converge to z_s.

Recall that a bilinear operator $B: X \times X \to X$ is *compact* if for any bounded set $S \subset X$, the set $B(S, S)$ is relatively compact.

PROPOSITION 4.11
Suppose that X satisfies condition (V). If B is compact, then

$$\lim \|P_n B - B\| = 0 .$$

PROOF   It is clear that P_n converges uniformly to the identity map I on relatively compact sets. Since B is a compact map, for any bounded set $S \subset X$, the set $B(S)$ is relatively compact. Hence

$$\|(P_n B - B)(S)\| = \|(P_n - I)B(s)\| \to 0 . \quad \blacksquare \qquad (4.232)$$

We conclude the case of approximating solutions using a continuation method similar to that given in Chapters 1 and 2. This method is illustrated on Chandrasekhar's equation (1.22) and increases the range of positive values of λ for which (1.22) can be solved from 0.424059 given in Chapter 1 to 0.473571.

We begin by defining and giving relevant theorems for a uniformly contractive system (UCS). The notion of a UCS was developed and used in Section 4.6 to provide a general framework for obtaining iterative solutions to a class of multipower equations. We shall use the concept of the UCS in conjunction with the scheme discussed in the introduction to construct approximate solutions to Equation (1.39).

DEFINITION 4.11   *Let X be a Banach space, $\{V_n\}$ a sequence of subspaces of X such that*

$$\lim_{n \to \infty} \text{dist}(V_n, x) = 0 \qquad (4.233)$$

for each $x \in X$. Let U be a closed set in X and define the sets $U_n = V_n \cap U$ and the operators $Q_n: X \to V_n$. We say that $\{U_n, Q_n\}$ is a uniformly contractive system (UCS) if conditions (1) and (2) below hold.

1. *There exists a $c \in R$, $0 < c < 1$, and an $N \in \mathbf{N}$ such that if $n \geq N$ and $x, y \in U$, then $Q_n(U) \subset U_n$ and $\|Q_n(x) - Q_n(y)\| \leq c\|x - y\|$.*

2. *For any $x, y \in U$ and $\varepsilon > 0$, there exists an $N \in \mathbf{N}$ such that if $k \geq j \geq N$ then $\|Q_k(x) - Q_j(y)\| \leq c\|x - y\| + \varepsilon$.*

A space X satisfying condition (V) will satisfy (4.233).

Note that no solution in any individual V_n space needs to be found for this iterative routine to converge.

In order to apply these UCS results to solve the quadratic equation (1.19) in a space X satisfying condition (V), we define $Q: X \to X$ by

$$Q(x) = B(x, x) + y \qquad (4.234)$$

and $Q_n: X \to V_n$ by

$$Q_n(x) = P_n B(x, x) + P_n y . \qquad (4.235)$$

We now give sufficient conditions on B under which the hypotheses for Theorem 4.35 hold.

PROPOSITION 4.12
Suppose that X satisfies condition (V). If $B: X \times X \to X$ is compact then $\{Q_n\}$ converges uniformly to Q on any bounded set.

PROOF Apply Proposition 4.11. ∎

Recall that the Fréchet derivative B' of a bilinear operator B is defined by

$$B'(x)(u) = B(x, u) + B(u, x) .$$

Note that the maps B' and $B'(x)$ are both linear. We shall make use of the following identities in the sequel.

$$B(u, u) - B(v, v) = B' \left(\frac{u + v}{2} \right) (u - v) . \tag{4.236}$$

In the case where $v = 0$, this simplifies to

$$B(u, u) = B' \left(\frac{u}{2} \right) (u) . \tag{4.237}$$

In order to prove our uniqueness claim below we require the following theorem, which is a variation on a result due to Rall [167].

THEOREM 4.38
Any solution $z \in C = \{x: \|B'(x)\| < 1\}$ to Equation (1.19) is unique in C.

PROOF If $z_1, z_2 \in C$ are solutions to (1.19), then $z_1 - z_2 = B(z_1, z_1) - B(z_2, z_2)$. By the identity (4.54) we have

$$\|z_1 - z_2\| \leq \left\| B' \left(\frac{z_1 + z_2}{2} \right) \right\| \cdot \|z_1 - z_2\| . \tag{4.238}$$

Note that C is convex by the linearity of B', so $(z_1 + z_2)/z \in C$. By hypothesis, inequality (4.238) can only be true if $z_1 = z_2$. ∎

We can now prove the main result of this section.

THEOREM 4.39

Suppose that X satisfies condition (V). Let $B: X \times X \to X$ be bounded and bilinear, with $y, z \in X$. Define $Q: X \to X$ by

$$Q(x) = B(x, x) + y .$$

Suppose that

$$a = \frac{1 - \mu \| B'z \|}{\mu \| B' \|} > \sqrt{\frac{2 \| Q(z) - z \|}{\| B' \|}} . \tag{4.239}$$

Then

(1) *$Q(x)$ has a unique fixed point z_s in the set $C = \{x : \mu \| B'x \| < 1\}$.*

(2) *This fixed point z_s lies in $\bar{S}(z, b)$, where*

$$b = a - \sqrt{a^2 - \frac{2 \| Q(z) - z \|}{\| B' \|}} . \tag{4.240}$$

(3) *The equation $Q_n(x) = x$ has solutions z_n for sufficiently large n, and these solutions converge to the fixed point z_s of $Q(x)$. These solutions are unique in C, and lie in $\bar{S}(z, b_n)$, where*

$$b_n = a - \sqrt{a^2 - \frac{\mu^2 \| Q(z) - z \| + \| P_n z - z \|}{\mu \| B' \|}} .$$

(4) *If B is compact, then the iterative scheme given in Theorem 4.35 converges to the solution z_s of (1.19).*

PROOF   Choose $N \in \mathbf{N}$ so that

$$a^2 > \frac{2 \left(\| P_n z - z \| + \mu \| Q(z) - z \| \right)}{\mu \| B' \|}$$

for all $n \geq N$. Choose $r \in [b, a)$. For $x, w \in \bar{S}(z, r)$ let $\delta = (x + w)/2 - z$. We have

$$\| Q_n(x) - Q_n(w) \| = \left\| B_n' \left(\frac{x + w}{2} \right) (x - w) \right\|$$

$$\leq \mu \left\| B'(z + \delta) \right\| \cdot \| (x - w) \| \tag{4.241}$$

$$\leq \mu \left(\left\| B'z \right\| + \left\| B' \right\| r \right) \left\| x - w \right\| .$$

Put

$$c = \mu \left(\left\| B'z \right\| + \left\| B' \right\| r \right) . \tag{4.242}$$

Now $c < 1$ by the definition of r and a, so Q_n is a contraction on $\bar{S}(z, r)$ for $n \geq N$.

To see that $Q_n(\bar{S}(z, r)) \subset \bar{S}(z, r)$, let $x \in \bar{S}(z, r)$ and set

$$\gamma_n = \mu \| Q(z) - z \| + \| P_n z - z \| .$$

Then

$$\| Q_n(x) - z \| \leq \| Q_n(x) - Q_n(z) \| + \| Q_n(z) - P_n z \| + \| P_n z - z \|$$

$$\leq \left\| B'_n \left(\frac{x+z}{2} \right) (x - z) \right\| + \mu \| Q(z) - z \| + \| P_n z - z \|$$

$$\leq \mu \left(\left\| B'z + B' \left(\frac{x-z}{2} \right) \right\| \right) \| x - z \| + \gamma_n$$

$$\leq \mu \| B'z \| r + \mu \frac{\| B' \|}{2} r^2 + \gamma_n . \tag{4.243}$$

Thus, $\| Q_n(x) - z \| \leq r$ if

$$\frac{\mu \| B' \|}{2} r^2 + \left(\mu \| B'z \| - 1 \right) r + \gamma_n \leq 0 . \tag{4.244}$$

This quadratic inequality in r is satisfied for $r \in [b_n, a)$, so $Q_n(\bar{S}(z, r)) \subset \bar{S}(z, r)$. Applying the contraction mapping principal, Q_n has a unique fixed point z_n in $\bar{S}(z, a)$, and in fact $z_n \in \bar{S}(z, b_n)$.

Note that all the contractions Q_n have the same contraction factor c defined in (4.242). Since Q_n converges pointwise to Q by condition (V), Q is clearly a contraction and $Q(S(z, r)) \subset \bar{S}(z, r)$. If we put $U_n = V_n \cap \bar{S}(z, r)$, then by Theorem 1.3 $\{U_n, Q_n\}$ is a UCS. Theorem 4.34 yields conclusion (3). Since each $z_n \in \bar{S}(z, b_n)$ and $\lim b_n = b$, conclusion (2) is proved. If B is compact, then Q_n converges uniformly to Q, so applying Theorem 4.35 yields conclusion (4).

To prove (1), we note that the contraction mapping theorem guarantees uniqueness in $S(z, a)$. Next consider $x \in S(z, a)$ and write $x = z + \delta$ for some δ with

$\|\delta\| < a$. By the linearity of B' we have

$$\|B'x\| \le \|B'z\| + \|B'\delta\| < \|B'z\| + \|B'\| a = 1/\mu \le 1 . \qquad (4.245)$$

This guarantees that

$$S(z, a) \subset C = \{x : \|B'x\| < 1\} . \qquad (4.246)$$

The set C contains a solution to (1.19), so by Theorem 4.38 we have uniqueness in C. ∎

REMARK 4.10   If we set $V_n = X$, $P_n = I$ for all n so $\mu = 1$, the proof is unchanged for parts (1) and (2). Thus, parts (1) and (2) of the theorem hold for any Banach space X with $\mu = 1$. ∎

REMARK 4.11   We note that since $U(z, a) \subset C = \{x : \|B'x\| < 1\}$, Theorem 4.39 yields greater uniqueness information than Theorem 1.4. Also note that since $\|B'z\| \le 2\|B\| \cdot \|z\|$, Theorem 4.39 gives greater flexibility than Theorem 1.4 in searching for approximate solutions z for which the hypotheses hold. This advantage will be used later.

It should also be noted in Theorem 4.39 that the operators $P_n B$ need only converge pointwise to B for part (3) to hold—B need not be compact. Compare also with Theorem 1.12.

We also observe the following necessary condition on solutions of (4.29). ∎

COROLLARY 4.9
If Equation (4.29) has a solution z_s with

$$\|B'z_s\| < 1 , \qquad (4.247)$$

then there is an open ball S about z_s such that for any initial estimate $x_0 \in S$, the iterative scheme (1.44) converges to the solution z_s.

PROOF   In the proof of Theorem 4.39, let each $V_n = X$ so $P_n = I$, $\mu = 1$, and choose $z = z_s$. Then (4.239) is satisfied, and by the contraction mapping theorem the iterative routine (1.44) will converge. ∎

The following will be useful.

PROPOSITION 4.13

Let $B: X \times X \to X$ be bounded and bilinear, $y \in X$. If

$$\|B'y\| < \frac{1}{2}$$

then

$$x = y + B(x, x)$$

has a unique solution in $C = \{x \in X: \|B'x\| 1\}$.

PROOF   Let $D = \{x: \|B'x\| \leq \delta\}$, where $\delta = \|B'y\| + 1/2 < 1$. We shall show that $Q(x) = B(x, x) + y$ has a fixed point in D. Let $u, v \in D$. Then

$$\|Q(u) - Q(v)\| = \|B(u, u) - B(v, v)\| = \left\| B'\left(\frac{u+v}{2}\right)(u - v) \right\|$$

$$\leq \frac{1}{2} \|B'u + B'v\| \cdot \|u - v\| \leq \delta\|u - v\| ,$$

so Q is a contraction on D. Now if $x \in D$, then

$$B'(Q(x)) = B'(B(x, x)) + B'y = B'\left(\frac{B'x}{2}\right) + B'y$$

by identity (4.237). Hence

$$\|B'(Q(x))\| \leq \frac{\|B'x\|^2}{2} + \|B'y\| < \delta ,$$

so $Q(D) \subset D$. Since D is closed, Q has a fixed point in D by the contraction mapping principle. The uniqueness follows from Theorem 4.38. ∎

REMARK 4.12   Corollary 1.1 was proved with the hypothesis "$\|B'y\| < 1/2$" replaced by "$\|B\| \cdot \|y\| < 1/4$". The latter is a stronger assumption since $\|B'\| \leq 2\|B\|$.

It is often the case that we seek a solution to Equation (1.37) for large λ, but finding an approximate solution $z \in X$ for which Theorem 4.39 applies may be difficult or impractical. One way to handle this problem is the *continuation* technique, whereby (1.37) is solved for small enough λ so that an initial guess z can be easily found for which Theorem 4.39 applies. An approximate solution z_n is found in some V_n space, and λ is increased with z_n used as an initial guess for the new equation (1.37) with larger λ. This process is repeated until the desired large λ is reached and a satisfactory approximation obtained. In this section we present an algorithm and a theorem that make this precise for our problem of solving quadratic equations in a space that satisfies condition (V). In particular, we give conditions under which the desired large λ can be reached in a finite number of repetitions of the continuation process. This scheme is then illustrated on Chandrasekhar's equation, extending the range of λ values from 0.424059 given in Chapter 1 to 0.473571. ∎

Continuation Algorithm

1. Choose λ_0 small enough so that (1.37) is guaranteed to have a solution z_0 for which $\mu\lambda_0 \| B'z_0 \| < 1$. We note that this is always possible (for example, $\lambda_0 = 0$).

2. Choose n sufficiently large so that

$$x = P_n y + \lambda_0 P_n B(x, x)$$

has a solution z_n in V_n that satisfies

$$1 - \mu\lambda_0 \left\| B'z_n \right\| > \mu\sqrt{2\lambda_0} \|B'\|\sqrt{\|E_{0,n}\|} \qquad (4.248)$$

where the "error" for z_n is

$$E_{0,n} = \lambda_0 B(z_n, z_n) + y - z_n \ .$$

That such an n exists follows from Theorem 4.39 (3), for

$$\lim_{n\to\infty} z_n = z_0 \quad \text{and} \quad \lim_{n\to\infty} E_{0,n} = 0 \ .$$

3. Solve

$$1 - \mu\lambda_1 \left\| B'z_n \right\| = \mu\sqrt{2\lambda_1} \|B'\|\sqrt{(\lambda_1 - \lambda_0) \|B(z_n, z_n)\| + \|E_{0,n}\|} \qquad (4.249)$$

for λ_1.

CLAIM 4.6

For each λ satisfying $\lambda_0 \leq \lambda < \lambda_1$, Equation (1.37) has a solution z_λ for which $\mu\lambda\|B'z_\lambda\| < 1$.

PROOF   It is clear that replacing λ_1 by λ in (4.249) will yield

$$1 - \mu\lambda \left\| B'z_n \right\| > \mu\sqrt{2\lambda} \left\| B' \right\| \sqrt{(\lambda - \lambda_0) \left\| B(z_n, z_n) \right\| + \left\| E_{0,n} \right\|} .$$

Define Q_λ by $Q_\lambda(x) = \lambda B(x, x) + y$. Then

$$\left\| Q_\lambda(z_n) - z_n \right\| \leq (\lambda - \lambda_0) \left\| B(z_n, z_n) \right\| + \left\| E_{0,n} \right\| ,$$

so

$$\frac{1 - \mu\lambda \left\| B'z_n \right\|}{\mu\lambda \left\| B' \right\|} > \sqrt{\frac{2 \left\| Q_\lambda(z_n) - z_n \right\|}{\lambda \left\| B' \right\|}} . \tag{4.250}$$

The Claim then follows from Theorem 4.39.   ∎

4. If λ_1 is not large enough, return to step (1) with λ_0 replaced by $\lambda_1 - \varepsilon$ for small $\varepsilon > 0$.

REMARK 4.13   Observe that inequality (4.248) is satisfied if it holds when upper bounds for $\| B'z_n \|$ and $\| B' \|$ are used in place of $\| B'z_n \|$ and $\| B' \|$, respectively.

Information on the location and uniqueness of each "intermediate" solution z_n in this algorithm can be obtained from Theorem 4.39.

There is some question about whether this algorithm will eventually reach the desired large λ value. The next result gives conditions that guarantee this convergence—in a finite number of steps.   ∎

THEOREM 4.40
Suppose that $\lambda_E > 0$ and Equation (1.37) has a solution z_λ with $\mu\|B'z_\lambda\| < 1$ for all λ, $0 \leq \lambda \leq \lambda_E$. Then after a finite number of iterations, the algorithm given above will obtain a λ_1 for which $\lambda_E \leq \lambda_1$.

PROOF   For each λ, $0 \leq \lambda \leq \lambda_E$, the inequality (4.250) and a continuity argument on λ guarantee some $\delta_\lambda^n > 0$ for which $t \in (\lambda - \delta_\lambda^n, \lambda + \delta_\lambda^n) \Rightarrow x = y + tB(x, x)$ has a solution z_t with $\mu\|B'z_t\| < 1$. The open sets $(\lambda - \delta_\lambda^n, \lambda + \delta_\lambda^n)$ form an open cover of the compact set $[0, \lambda_E]$, so there exists some δ such that $0 < \delta < \delta_\lambda^n$ for all $\lambda \in [0, \lambda_E]$. Therefore, each iteration of the algorithm increases λ by at least δ.   ∎

The final result of the algorithm given is an estimate $z_n \in V_n$. This is less than satisfactory because the true solution to (1.37) must be of the form $y + f$, where

$f \in$ Range (B). To get an approximation of this form, one approach is to find

$$\hat{z} = y + B\,(z_n, z_n)\ .$$

While calculating $y + B(z_n, z_n)$ is more expensive than an iteration in V_n, the following result shows that $\hat{z}$ must be an improvement on z_n. Numerical experiments suggest that one such calculation is worth the price in many situations.

PROPOSITION 4.14
Let $B, F\colon X \times X \to X$ be bounded and bilinear. Suppose that the equations

$$x = \tilde{y} + F(x, x) \tag{4.251}$$

and

$$x = y + B(x, x)$$

have solutions z_n and z_s, respectively, with

$$\|B'z_n\|\,,\ \|B'z_s\| < 1\ .$$

Then

$$\|z_s - \hat{z}\| < \|z_s - z_n\|\ .$$

PROOF   We calculate

$$\|z_s - \hat{z}\| = \|B\,(z_s, z_s) - B\,(z_n, z_n)\| = \left\|B'\left(\frac{z_s + z_n}{2}\right)(z_s - z_n)\right\|$$

$$\leq \frac{\|B'\,(z_n)\| + \|B'\,(z_s)\|}{2}\,\|z_s - z_n\| < \|z_s - z_n\|\ . \quad \blacksquare$$

We now illustrate the continuation algorithm to approximate solutions z_λ to the Chandrasekhar equation (1.22).

Example 4.13
 Equation (1.22) is usually solved in $C[0, 1]$ for physical reasons. We shall first seek solutions in $L^2[0, 1]$, taking advantage of certain properties of this space, and then show that such solutions lie in $C[0, 1]$. For these reasons, let $X = L^2[0, 1]$ and let V_n be the span of the first n Legendre polynomials $P_0, \ldots, P_{n-1}$. Observe

that $\mu = 1$ in a Hilbert space. Define $B: X \times X \to X$ by

$$B(f, g)(s) = f(s) \int_0^1 \frac{s}{s+t} g(t)dt .$$

We seek the maximal λ value for which our algorithm applies. A first estimate of $y = 1$ is natural. From Proposition 4.13, any $\lambda_0 < 1/(2\|B'y\|)$ will satisfy step 1 of the algorithm. We bound $\|B'y\|$ as follows. For $f \in X$, $\|f\| \leq 1$, we have

$$\|B'y(f)\| \leq \left\| \int_0^1 \frac{s}{s+t} f(t)dt \right\| + \left\| f(s) \int_0^1 \frac{s}{s+t} dt \right\| .$$

By Cauchy-Schwartz, we obtain

$$\left\| \int_0^1 \frac{s}{s+t} f(t)dt \right\|^2 \leq \|f\|^2 \cdot \int_0^1 \int_0^1 \left(\frac{s}{s+t} \right)^2 dt ds \leq 1 - \ln 2$$

and

$$\left\| f(s) \int_0^1 \frac{s}{s+t} dt \right\| \leq \|f\| \cdot \sup_s \int_0^1 \frac{s}{s+t} dt \leq \ln 2 .$$

Therefore,

$$\|B'y\| \leq \sqrt{1 - \ln 2} + \ln 2 ,$$

so we set $\lambda_0 = 0.40 < 1/(2\|B'y\|)$. Similar arguments yield

$$\|B'\| \leq 2\|B\| \leq 2 \sup_s \sqrt{\int_0^1 \left(\frac{s}{s+t} \right)^2 dt} = \frac{1}{\sqrt{2}} .$$

A choice of $n = 1$ is not large enough to satisfy step 2 of the continuation algorithm, so we begin with $n = 2$ for our initial V_n space. The following table gives the results of the algorithm. At each step, the iterative routine (4.231) was carried out in V_n space until consecutive iterates differed by less than 10^{-12} in the L^2 norm. Each column represents one iteration of the algorithm. For each n value, the λ values increase until reaching a limiting value. At this point, we increase n as directed by step 2 of the algorithm and continue.

We now show that the solutions guaranteed are not only in $X = L^2[0, 1]$, but are in fact continuous.

Table 4.4   Numerical solution of Chandrasekhar's equation VI

n	2	2	$\cdots$	2	6	$\cdots$	6	9	$\cdots$	9
λ_0	.4000	.4102	$\cdots$	.4371	.4371	$\cdots$	.4708	.4708	$\cdots$	.473571
λ_1	.4102	.4198	$\cdots$	.4371	.4428	$\cdots$	.4708	.4712	$\cdots$	.473571
$\lambda_1\|B'z_n\|$	.7123	.7531	$\cdots$	.8342	.8521	$\cdots$	.9645	.9651	$\cdots$	.979623

LEMMA 4.9
If f is an $L^2[0, 1]$ solution to the Chandrasekhar equation (1.22), then $f \in C[0, 1]$.

PROOF   Suppose that f solves (1.22), and define F_f by

$$F_f(s) = \int_0^1 \frac{s}{s+t} f(t)dt .$$

Obviously $F_f(0) = 0$, and $\lim_{s \to 0} F_f(s) = 0$ since

$$|F_f(s)| \le \|f\| \sqrt{\int_0^1 \left(\frac{s}{s+t}\right)^2 dt} .$$

Hence, F_f is continuous at 0. Now F_f is clearly continuous for $s > 0$, so $F_f(s)$ is continuous on $[0, 1]$. By hypothesis $f(s) = 1 + f(s)F_f(s)$, so $F_f(s) \ne 1$ for $s \in [0, 1]$, and thus

$$f(s) = \frac{1}{1 - F_f(s)} .$$

We conclude that $f \in C[0, 1]$.   ∎

REMARK 4.14   The Legendre polynomials were used solely for simplicity; wavelet bases have been shown to be superior in many aspects (see the previous section). Certainly, a Banach space with a multiresolution analysis satisfies condition (V).

All computations for the example were done using *Maple V* on a 486 PC. No Pentium chips were involved in this work.

With $n = 9$, we obtain $\lambda \approx 0.473571$. By increasing n this λ value can be increased. It has been shown elsewhere [77] that the maximum value for λ is 0.5— no solution exists for larger λ values. The next result shows that this maximum value cannot be achieved by our algorithm.   ∎

THEOREM 4.41
Let $BL(X \times X, X)$ denote the bounded bilinear operators on $X \times X$ into X, and

suppose that $y \in X$. *Then the set* $\mathcal{O}$ *of all* $B \in BL(X \times X, X)$ *such that* (1.19) *has a solution* z *with* $\|B'z\| < 1$ *is open (in the operator topology) in* $BL(X \times X, X)$.

PROOF   Suppose that $F \in \mathcal{O}$ has solution z with $\|F'z\| < 1$. It is easy to verify that $\|B - F\| < \varepsilon \Rightarrow \|B' - F'\| < 2\varepsilon$ when $B \in BL(X \times X, X)$. Thus, we can choose $\varepsilon > 0$ sufficiently small so that

$$\frac{1 - \|B'z\|}{\|B'\|} > \sqrt{\frac{2\|B(z,z) + y - z\|}{\|B'\|}} = \sqrt{\frac{2\|B(z,z) - F(z,z)\|}{\|B'\|}}$$

holds. Then by Theorem 4.39, Equation (1.19) has a solution s with $\|B's\| < 1$. We conclude that there is an open ball of radius ε about F that is contained in set $\mathcal{O}$.   ∎

Now if Chandrasekhar's equation (1.22) with $\lambda = 0.5$ had a solution z with $\|0.5B'z\| < 1$, then by Theorem 4.41 Equation (1.19) would have a solution for some $\lambda > 0.5$, which is false as noted in the remarks before Theorem 4.41.

Exercises

1.   Assuming that L_0^{-1} exists, show that the operator $T = L_0 + cP$ has an inverse for

$$|c| < \frac{1}{\left\|L_0^{-1}\right\| \cdot \|P\|} .$$

Write the Neumann series for T^{-1}.

2.   Prove that if L is a bounded linear operator from a Banach space X into a Banach space Y, and there exists a nonsingular bounded linear operator K from Y onto X, such that $\|I - KL\| < 1$, where I is the identity operator in X, then the equation $Lx = y$ has the unique solution $x = \sum_{n=0}^{\infty}(I - KL)^n Ky$, for each $y \in Y$. Use this result to invert the linear differential operator $\lambda I - D$, $D = \frac{d}{ds}$, which maps $X = C'[0, 1]$ into $Y = C[0, 1]$, for $|\lambda| < 1$.

3.   Calculate the derivative of the Uryson integral operator $U(x) = \int_0^1 f(s, t, x(t))dt$, in $C[0, 1]$ at $x_0 = x_0(s)$.

4. Calculate the derivative of the Ricatti differential operator

$$R(y) = \frac{dy}{ds} + p(s)y^2 + q(s)y + r(s)$$

from $C'[0, a]$ into $C[0, a]$ at $y_0 = y_0(s)$ in $C'[0, a]$.

5. Use Newton's method to solve the real quadratic equation

$$P(x) = ax^2 + bx + c = 0, \quad \text{for real } x_0 .$$

After the calculation of x_1 and $P(x_1)$, show that the error estimate

$$\left\| x^* - x_1 \right\| \leq \frac{2}{1 - h_0 + \sqrt{1 - 2h_0}} B_0 \left\| P(x_1) \right\|$$

holds if $h_0 = B_0 \eta_0 K \leq \frac{1}{2}$, where

$$\left\| P'(x_0)^{-1} \right\| \leq B_0, \ \|x_1 - x_0\| \leq \eta_0 \quad \text{and} \quad \left\| P'(x) - P'(y) \right\| \leq K \|x - y\|$$

for all $x, y \in U(x_0, \|x_0 - x_1\|)$.

6. Verify the claims made about the functions f_1 and f_2 after Definition 4.2.

7. For $\bar{B}$ as defined after Theorem 4.2 show that $\bar{B}(v, w) \in C[s, 1]$.

8. Show a result similar to Proposition 4.2 for negative λ values.

9. Show that R, F are isotone and antitone respectively in Theorem 4.4.

10. Show that f is antitone, and $F(M)$ bounded and equicontinuous in Theorem 4.5.

11. Verify the claims made in Remark 4.3.

12. Verify the claims made in Example 4.2.

13. Verify the claims made in Example 4.3.

14. Prove Corollary 4.2.

15. Prove the claim made at the end of Proposition 4.7.

16. Prove Corollary 4.3.

17. Prove Corollary 4.4.

18. Complete the details in the proof of Theorem 4.25.

19. Prove Theorem 4.27.

20. Prove the claims in Remark 4.5.

21. Let $X = C[0, 1]$ denote the Banach space of all real valued continuous functions f on $[0, 1]$ with $\|f\| = \sup_{x \in [0, 1]} |f(x)|$. For $0 < A < B < \infty$, let

$$G_A^B = \left\{ f \in X \mid \inf_{x \in [0, 1]} f(x) > A \text{ and } \|f\| < B \right\}$$

and

$$G^B = \bigcup_{0 < A < B} G_A^B \quad (\text{for } B > 0) .$$

(a) Let K be a real valued continuous function on

$$[0, 1] \times [0, 1] \setminus \{(x, x) \mid x \in [0, 1]\} .$$

Then if $\exists m > 0$ such that $\|K(x, y)\| < c\|x - y\|^{m-1}$ for all $x, y \in [0, 1]$, $x \neq y$, then show that the integral operator

$$(Kf)(x) = \int_0^1 K(x, y) f(y) dy$$

is compact from X into X with

$$\|K\| \leq \frac{c 2^{1-m}}{m} .$$

(b) Suppose that

$$0 \leq K(x, y) \quad \text{for all } x, y \in [0, 1] .$$

Given $\eta \in (0, 1)$ let $A = 1 - \eta$ and $B > 1 + M A^{-1}$ with $M = \|K\|$. Then show that for each $\lambda \in [0, 1]$ $\exists f_\lambda \in G_A^B$ which satisfies

$$g(x) = 1 + \lambda \int_0^1 K(x, y) g(y)^{-1} dy \qquad (4.252)$$

and $f_\lambda \geq 1$.

(c) Suppose that $A = 0$, $B = 1$ and

(c_1) R: is a continuous real-valued function on $[0, 1] \times [0, 1]$ such that

$$R(x, y) > 0 \quad \text{if } x > y$$

$$R(x, y) < 0 \quad \text{if } x < y .$$

(c_2) $\exists m > 0$ such that $\|R(x, y)\| < c\|x - y\|^m \|x + y\|, x \neq y$. Then show that (4.252) has at least one solution $0 < f \leq 1$. Further show that if R satisfies (c_1), (c_2), and

(c_3) $c2^{1-m} m^{-1} \leq 1$, then (4.252) has one and only one positive solution f, and $0 < f(x) \leq 1$ for all $x \in [0, 1]$.

22. Let us define the quadratic operator

$$T(u) = \int_x^1 u(y)u(y - x)dy$$

in the space $X = C[0, 1]$. Show that the operator T maps X into itself. Use this result to find sufficient conditions for the existence of a solution of the equation

$$u(x) = 1 + \lambda \int_x^1 u(y)u(y - x)dx = 1 + \lambda T(u), \quad (0 \leq x \leq 1) .$$

23. Consider the (1.22) equation in the form

$$H(s) = 1 + \frac{1}{2}sH(s) \int_0^1 \frac{H(t)}{s + t}dt .$$

Show that the following two expressions satisfy the above equation

$$H(s) = (1 + s) \exp \left\{ \frac{s}{\pi} \int_0^{\pi/2} \frac{\ln\left[\sin^2\theta/(1 - \theta \cot\theta)\right]}{\cos^2\theta + s^2\sin^2\theta} d\theta \right\}$$

and

$$H(s) = \frac{1}{(1 + s)^{1/2}} \exp \left\{ \frac{1}{\pi} \int_0^{\pi/2} \frac{\theta \tan^{-1}(s \tan\theta)}{1 - \theta \cot\theta} d\theta \right\} .$$

24. Let us consider the (1.22) equation in the form

$$H(\mu, c) = 1 + cH(\mu, c) \int_0^1 \frac{v}{\mu + v} H(v, c)g(v)dv . \qquad (4.253)$$

Here $c \in \mathbf{C}, g \in L^1[0, 1]$ is given and real valued. Let us assume that

$$\int_0^1 g(t)dt = \frac{1}{2} . \qquad (4.254)$$

(a) Assume that $g \in L^1[0, 1]$ is real valued and satisfies (4.254). Let $H(c_0)$ be any solution to (4.253) with $c = c_0$. Then show that $F'(H(c_0), c_0)$ is an invertible operator on $C[0, 1]$ with bounded inverse if $c_0 \neq 1$. If $c_0 = 1$ show that $F'(H(1), 1)$ has a one-dimensional null space spanned by the function $f(t) = tH(t, 1)$. The range $F'(H(1), 1)$ is

$$\left\{ F \in C[0, 1] \Bigg/ \int_0^1 F(t)g(t)dt \right\} .$$

(b) Let $H_0(t)$ be any solution to $C(K)$ for $c = c_0 \neq 1$. Then show there exist $\varepsilon > 0$ and a function $H(t, c)$ that is analytic and $C[0, 1]$ valued for $c \in \{d \mid |d - c_0| < \varepsilon\}$ so that $F'(H(c), c) = 0$ and $H(t, c_0) = H_0(t)$. Hence, show that the set $\{c \mid$ Equation (4.254) has a solution, $c \neq 1\}$ is open.

(c) Let $c \neq 1$ be such that a solution $H(t, c)$ to (4.253) exists. Then show there is $\varepsilon_0 > 0$ so that if $\| H_0 - H(c)\| < \varepsilon_0$, Newton's method converges and

$$\| H_{n+1} - H(c)\| \leq 3|c| \left\| F'(H(c), c)^{-1} \right\| \cdot \| H_n - H(c)\|^2 \quad (n \geq 0) .$$

25. Let us consider the (1.22) equation in the form

$$H(t) = 1 + H(t) \int_0^1 \frac{t}{t + s} g(t)H(s)ds$$

$$+ \int_0^1 P(t, s, H(t), H(s))ds . \qquad (4.255)$$

Here we work in $C[0, t]$, P is the perturbation of the (1.22) equation. Denote by $C_+[0, 1]$ the cone of nonnegative functions in $C[0, 1]$, $C^\alpha[0, 1]$,

$0 < \alpha < 1$ denotes the Banach space of all real continuous functions on $[0, 1]$ such that

$$\sup_{t,s\in[0,1]} \frac{|x(t) - x(s)|}{|t - s|^\alpha} < \infty ,$$

with the norm

$$\|x\|_{C^\alpha} = \max_{0 \le t \le 1} |x(t)| + \sup_{t,s\in[0,1]} \frac{|x(t) - x(s)|}{|t - s|^\alpha}$$

and $C_+^\alpha[0, 1]$ the cone of nonnegative functions in $C^\alpha[0, 1]$. Moreover for $\delta > 0$ and $R > 0$, let

$$D_\delta = \left\{ H(t) \in C_+[0, 1] \,\middle/\, 1 - \int_0^1 \frac{t}{t+s} g(s)H(s)ds \ge \delta \right\} ,$$

$$D_\delta^R = \{H(t) \in D_\delta \mid \|H\|_C \le R\} ,$$

$$D_\delta^\alpha = \left\{ H(t) \in C_+^\alpha[0, 1] \,\middle/\, 1 - \int_0^1 \frac{t}{t+s} g(s)H(s)ds \ge \delta \right\} ,$$

$$D_\delta^{\alpha R} = \left\{ H(t) \in D_\delta^\alpha \mid \|H\|_{C^\alpha} \le R \right\} .$$

(a)   Suppose that g and P satisfy the condition

     (i): $|P(t, s, u_1, v_1) - P(t, s, u_2, v_2)| \le K_1|u_1 - u_2| + K_2|v_1 - v_2|$
        $(u_i, v_i \in \mathbf{R}^+, i = 1, 2)$;

     (ii): $|P(t, s, u, v)| \le \varepsilon$;

     (iii): $b = \sup_{0 \le t \le 1} \int_0^1 \frac{tg(s)}{t+s} ds \le \dfrac{1}{4(1+\varepsilon)} = \dfrac{1}{4\mu}$.

     Then for all $\delta > 0$ such that

     (iv) $\frac{1}{\delta}(K_1 + K_2) < 1$ and

     (v) $\delta_1 \le \delta \le \delta_2$ where

$$\delta_1 = \frac{1 - \sqrt{1 - 4\mu b}}{2} , \qquad \delta_2 = \frac{1 + \sqrt{1 - 4\mu b}}{2}$$

and for all $R > 0$ such that $R_1 \le R \le R_2$, where

$$R_1 = \frac{1 - \sqrt{1 - 4\mu b}}{2b} , \qquad R_2 = \frac{1 + \sqrt{1 - 4\mu b}}{2b}$$

show there exist a solution to (4.255) which belongs to D_δ^R.

(b) Assume that (i), (ii), (iii), (v) are satisfied. Moreover, assume that the second order partial derivatives $\frac{\partial^2 P}{\partial t \partial u}$, $\frac{\partial^2 P}{\partial t \partial u}$, $\frac{\partial^2 P}{\partial u \partial v}$, $\frac{\partial^2 P}{\partial v^2}$ exist and

$$K_3 = \sup \left| \frac{\partial^2 P}{\partial t \partial u} \right|, \quad K_4 = \sup \left| \frac{\partial^2 P}{\partial t \partial v} \right|,$$

$$K_5 = \sup \left| \frac{\partial^2 P}{\partial u \partial v} \right|, \quad K_6 = \sup \left| \frac{\partial^2 P}{\partial v^2} \right|,$$

are finite, and that

(vi) $\sup_{H \in D_\delta^{\alpha R}} \| K H \|_{C^\alpha} \Gamma < 1,$

where

$$KH(t) = \left[1 - \int_0^1 \frac{t g(s)}{t + s} H(s) ds \right]^{-1},$$

and

$$\Gamma = 2K_1 + K_2 + K_3 + H_4 + R K_5 + R K_6 .$$

Then show there exists a solution to (4.255) which belongs to $D_\delta^{\alpha R}$, with $R_1 \leq R \leq R_2$, δ as in (v).

$$R_1 = \frac{1 - (ac + db) - \sqrt{1 + (ac + bd)^2 - 2ac - 2bd}}{2ad},$$

$$= \frac{1 - ac - bd - \sqrt{\Delta}}{2ad}, \quad R_2 = \frac{1 - ac - bd + \sqrt{\Delta}}{2ad}$$

$$a = \frac{m}{\delta^2}, m = \sup_{t_1, t_2 \in [0,1]} \int_0^1 \frac{s g(s)}{(t_1 + s)(t_2 + s)} ds ,$$

$$b = \frac{1}{\delta}, \ c = \ell + \varepsilon + K_0, d = K_1 \text{ and } K_0 = \sup \left| \frac{\partial P}{\partial t} \right| .$$

(c) Suppose that (i) through (v) are satisfied and that $P(t, s, u, v)$ is a nondecreasing function with respect to u and v and

$$\int_0^1 \left[\frac{t}{t + s} g(s) + P(t, s, 1, 1) \right] ds \geq 0 \quad \text{for } 0 \leq t \leq 1 .$$

Let $H_0 = 1$ and define $H_{n+1} = T H_n = K H_n L H_n$ $(n \geq 0)$ where

$$LH_n(t) = 1 + \int_0^1 P(t, s, H_n(t), H_n(s)) \, ds .$$

Then show the sequence $\{H_n\}$ $(n \geq 0)$ converges uniformly to a solution of (4.255) which belongs to D_δ^R.

26. Show that H in (4.87) is analytic in $|c|$ for $|c| < c_0 = \frac{1}{\lambda_0}$, where λ_0 is the spectral radius of K given by (4.90).

27. Show that the iteration $H_0 = 1$, $H_{n+1} = 1 + cB(H_n, H_n)$ related with Equation (4.95) converges to H in an appropriate Banach space sense for $|c| < c_0$.

28. Show Lemma 4.2.

29. Show estimates (4.118) and (4.119).

30. Show Theorem 4.30.

31. Show estimate (4.141).

32. Show estimate (4.144).

33. Show estimates (4.163) and (4.164).

34. Verify estimates (4.181) and (4.182).

35. Verify estimates (4.185) and (4.186).

36. Consider the Chandrasekhar's equation (1.22) in the form

$$H(s) = 1 + \lambda H(s) \int_0^1 K(s, t) H(t) \psi(t) dt . \qquad (4.256)$$

In Equation (4.256), λ is a real (or complex) parameter, $\psi \in X = L'[0, 1]$ is given and real valued, the kernel $K(s, t)$ on $X \times X$ satisfies

$$0 < K(s, t) < 1$$

and
$$K(s, t) + K(t, s) = 1 \quad \text{for all } s, t \in X .$$

The question of finding a solution of Equation (4.256) has been answered when $|\lambda| \leq 1$ and ψ satisfies certain positivity assumptions. We suggest an iteration that converges for any λ for which a solution to Equation (4.256) exists and do not require positivity assumptions on ψ and restrictions on λ.

We consider the space $C[0, 1]$ with the sup-norm. We define a bounded bilinear operator on $C[0, 1]$ by

$$B(w, z)(s) = w(s)(Lz)(s) + z(s)(Lw(s))$$

where

$$(Lw)(s) = \int_0^1 K(s, t)\psi(t)w(t)dt \ .$$

Show:

(i) Let $H = H(\lambda_0)$ be any solution to Equation (4.256) with $\lambda = \lambda_0$. If $\int_0^1 \psi(t)dt \neq 0$ and $\text{Kern}(L) = \{0\}$, then $B'(H(\lambda_0))$ is an invertible operator on $C[0, 1]$ with bounded inverse for any λ.

(ii) Introduce the iteration

$$H_n = H_{n-1} + B(H_{n-1})^{-1}$$

$$(H_{n-1} - 1 - B(H_{n-1}, H_{n-1})) \qquad (4.257)$$

and let $\varepsilon_0 > 0$ be such that

$$\|H_0 - H(\lambda_0)\| < \varepsilon_0$$

where $H(\lambda_0)$ is as in (i).

Then iteration (4.257) converges to $H(\lambda_0)$ and

$$\|H_{n+1} - H(\lambda_0)\| \leq \left\|I - B(H(\lambda_0))^{-1}\right\| \cdot \|H_n - H(\lambda_0)\| \ (n \geq 0)$$

(see also Case 2.2).

37. Prove Lemma 4.5.

38. Show that an L^2-solution to Equation (4.222) is in $C[0, 1]$.

39. Show that Equation (4.222) has a solution for $0 \leq \lambda \leq 1$.

40. Show that operator L in Example 4.12 is compact when the kernel function is on $L^2([a, b]^{n+1})$.

Chapter 5

Polynomial Operators in Linear Spaces

In this chapter we cover the following: A Weierstrass theorem for real normed linear spaces, matrix representations of polynomial operators, Lagrange and Hermite interpolation in Banach spaces, upper and lower bounds for the number of solutions of a polynomial. These sections must constitute part of every advanced course on polynomial operators and they are attributed to the excellent works of Prenter, Cronin, and Ruch [174] (adjusted appropriately for our purposes) (see, e.g., [159, 169]). Case 5.4.6, however, constitutes our work.

5.1 A Weierstrass Theorem

Let X and Y be real normed linear spaces and let K be a compact subset of X. Let $C(K, Y)$ denote the family of continuous functions from K to Y and let $\tilde{C}(K, Y)$ denote the family of continuous functions from X to Y restricted to K. Both $C(K, Y)$ and $\tilde{C}(K, Y)$ carry the uniform norm topology given by

$$\|f - g\|_u = \sup_{x \in K} \|f(x) - g(x)\| .$$

In the event $X = Y = R$, where R denotes the real numbers, the classical Weierstrass theorem states that the family of real-valued polynomials on X is dense in $C(K, Y)$ where $K = [a, b]$. If $X = R^n$ and $Y = R$, an application of the Stone–Weierstrass theorem states that the family of polynomials in n real variables, $n = 1, 2, \ldots$, is dense in $C(K, Y)$. Various generalizations of the Stone–Weierstrass theorem have been given.

In the present section we extend the Weierstrass theory to a normed linear space setting.

If K is a compact subset of X, we let $P(K, Y)$ denote the family of continuous polynomial operators from K to Y and we let $\tilde{P}(K, Y)$ denote the family of continuous polynomial operators from X to Y restricted to K. Both $P(K, Y)$

and $\tilde{P}(K, Y)$ carry the uniform norm topology. Clearly, not all polynomials are continuous since any unbounded, linear operator is a polynomial of degree one.

Example 5.1
Polynomials occur frequently, the Navier–Stokes equations being an impressive example.

(a) Consider the expression

$$a_2(t)x''(t) + a_1(t)x'(t) + x^3(t)$$

where a_1, a_2, and a_3 are continuous functions. If we let $X = C^2[0, 1]$ and $Y = C[0, 1]$, then this is a polynomial of degree three (a cubic equation) from X to Y.

(b) Moreover let

$$x(s) - \int_0^1 \int_0^1 k(s, t_1, t_2)x(t_1)x(t_2)dt_1dt_2 \, .$$

Similarly then, if we let $X = L^2[0, 1]$, the Lebesgue square integrable functions in $[0, 1]$, and insist that

$$\int_0^1 \int_0^1 \int_0^1 |K(s, t_1, t_2)| \, dt_1 dt_2 ds < \infty \, ,$$

then this is a polynomial of degree two (a quadratic) from X into X.

(c) Perhaps the simplest n-linear operators are the p-way matrices $(a_{i, j_1, j_2, \ldots, j_p})$. For example, let $X = \mathbf{R}^n = \{(x_1, x_2, \ldots, x_n) : x_i \in R, \ i = 1, 2, \ldots, n\}$. Let $M = (m_{i, j_1, j_2, \ldots, j_p})$, where $j_1, j_2, \ldots, j_p = 1, 2, \ldots, n$ and $i = 1, 2, \ldots, k$, be a p-way matrix of real numbers. For each p vectors

$$x^1 = \left(x_1^1, x_2^1, \ldots, x_n^1\right)$$

$$x^2 = \left(x_1^2, x_2^2, \ldots, x_n^2\right)$$

$$\vdots$$

$$x^p = \left(x_1^p, x_2^p, \ldots, x_n^p\right)$$

in X and each $i = 1, 2, \ldots, k$, define

$$a_i = \left(M\left(x^1, x^2, \ldots, x^p\right)\right)_i$$

$$= \sum_{j_1, j_2, \ldots, j_p = 1} m_{i, j_1, j_2, \ldots, j_p} x_{j_p}^1 x_{j_{p-1}}^2 \cdots x_{j_1}^p \, .$$

Clearly $(a_1, a_2, \ldots, a_k) \in R^k$ and M defines a p-linear operator on X^p to R^k.

Let X and Y be finite dimensional, real (or complex) linear spaces of dimensions s and t, respectively. Let $\varphi = \{\varphi_1, \varphi_2, \ldots, \varphi_s\}$ be a basis for X and let $\psi = \{\psi_1, \psi_2, \ldots, \psi_t\}$ be a basis for Y. Let M be an n-linear operator on X^n to Y. It is then possible to discuss the matrix representation of M with respect to the bases φ and ψ. That is, we define $(m_{i, j_1, j_2, \ldots, j_n})$, $i = 1, 2, \ldots, t$ and $j_1, j_2, \ldots, j_n = 1, 2, \ldots, s$ to be a *matrix representation of M with respect to the bases φ and ψ* if, for each $x^1, x^2, \ldots, x^n$ in X,

$$y_i = \sum_{j_1, j_2, \ldots, j_n = 1}^{s} m_{i, j_1, j_2, \ldots, j_n} x_{j_n}^1 x_{j_{n-1}}^2 \cdots x_{j_1}^n$$

where

$$y = M\left(x^1, x^2, \ldots, x^n\right)$$

$$x^k = \sum_{i=1}^{s} x_i^k \varphi_i$$

and

$$y = \sum_{i=1}^{t} y_i \psi_i \, .$$

In fact, it is a simple matter to prove:

THEOREM 5.1

Let X and Y be finite dimensional, real (or complex) linear spaces. Let $\varphi = \{\varphi_1, \varphi_2, \ldots, \varphi_s\}$ be a basis for X and let $\psi = \{\psi_1, \psi_2, \ldots, \psi_t\}$ be a basis for Y. Then

a) *All k-linear operators, $k = 0, 1, 2, \ldots$ are continuous.*

b)   Each n-linear operator, M, $n = 1, 2, \ldots$ has a unique matrix representation $(m_{i,j_1,j_2,\ldots,j_n})$ with respect to φ and ψ where $i = 1, 2, \ldots, t$, $j_1, j_2, \ldots, j_n = 1, 2, \ldots, s$ and $m_{i,j_1,j_2,\ldots,j_n}$ is the ith component of $M(\varphi_{j_n}, \varphi_{j_{n-1}}, \ldots, \varphi_{j_1})$.

c)   Every matrix $(m_{i,j_1,j_2,\ldots,j_n})$, $i = 1, 2, \ldots, t$ and $j_1, j_2, \ldots, j_n = 1, 2, \ldots, s$ is a matrix representation of an n-linear operator.

The proof, being straightforward, is left as an exercise.

At this point, we note that a 0-linear operator L_0 has no matrix representation as we have defined matrix representations. However, if $y = \sum_{i=1}^{t} c_i \varphi_i$ ($c_1, c_2, \ldots, c_t$ all constants) is that fixed member of Y for which

$$L_0 = L_0 x$$

for all $x \in X$ where L_0 is a 0-linear operator, we can associate L_0 with the "one-dimensional matrix" (c_i), $i = 1, 2, \ldots, t$. We shall call (c_i) a matrix representation of the 0-linear operator L_0 with respect to the bases φ and ψ. Conversely, with each t-tuple (c_i), $i = 1, 2, \ldots, t$, of constants we can associate a 0-linear operator L_0 given by

$$L_0 = L_0 x = y$$

for all $x \in X$, where $y = \sum_{i=1}^{t} c_i \psi_i$.

We have restricted ourselves to matrix representations of polynomials in the event X and Y are finite dimensional. The problem of matrix representations of polynomial operators on a separable Hilbert space X into X is discussed in the next section.

Let X and Y be real, finite dimensional, normed linear spaces of dimensions s and t, respectively. Then there exists homeomorphisms h and k, on X onto R^s and Y onto R^t where $R^s = h[X]$ and $R^t = k[Y]$. Let K be a compact subset of X and let $F \in C(K, Y)$. Then the mapping F induces a mapping kFh^{-1} of a subset of R^s into R^t. That is, if $\{\varphi_1, \varphi_2, \ldots, \varphi_s\}$ is a basis for X and $\{\psi_1, \psi_2, \ldots, \psi_t\}$ is a basis for Y, then $y = F(x)$ is equivalent to the system of equations

$$y_i = F_i(x_1, x_2, \ldots, x_s)$$

where $i = 1, 2, \ldots, t$, $x = \sum_{j=1}^{s} x_j \varphi_j$ and $y = \sum_{j=1}^{t} y_j \varphi_j$. This line of reasoning enables us to prove:

THEOREM 5.2

Let X and Y be finite dimensional, real, normed linear spaces. Let K be a compact subset of X. Then $P(K, Y)$ is dense in $C(K, Y)$.

Fig 2-5   Wild saffron *(Crocus cartwrghtianus)*

Fig. 2-6   Joe quasem saffron
(*Crocus pallasii* Subsp.
Haussknechtii)

Fig. 2-7   Almeh saffron (*C. almehensis*)

Corms about 10 to 15 mm but less than 20 mm diameter, spherical but solid and flat at basal position. Corm tunics delicately fibrous, reticulate and extended toward perigone about 2 to 3 and sometimes 4.5 cm. Cataphylls 3 to 5 white and membraneous. Leaves 7 to 12, synanthus, about the same size as flowers, green and scattered around the stem, 1.5 to 2.5 mm width, glabrous or pubescent. Fall flowering, 1 to 5, pale lilac or purple. Perianth white mostly with reticulate orientation, perianth tube glabrous, white or lilac with pronounced spathe. Bract and bracteole are present and unequal, white and membraneous, gradually tapering end. Perigone tube 3 to 5 and sometimes up to 7 cm long. Perianth segments open somehow unequal, 1.4 to 3.2 × 0.7 to 1.2 cm Width, lanceolate to obovate. Filament 3 to 7 mm and yellow. Style with 3 flat clubs shaped red stigmatic arms, each arm between 10 and 27 mm length, exceeding the anther or about the same size as the anther. At least half of the petal segments located under the anther. Fruit is an oval capsule, 1.5 to 2.5 cm long and 0.6 to 0.7 cm wide, located on a short peduncle above the ground level at maturity. Seeds brownish red irregularly round, 3 to 4 mm diameter. Raphe has an irregular ridge extended along the seed and terminated into a tiny sharp caruncle (<1 mm), testa is covered with long and dense papilla.

This is a diploid species with $2n = 16$ chromosomes. Flowering time is late October to early December (16) (see also figure 2-5).

The habitat of this saffron is open and rocky hillsides, and sometimes on lawn or pine, schist, shale and calcareous areas at 1000 m altitude.

Wild saffron is mostly distributed in Greece in Athica and cyclades region. The geographical distribution of this species differs from cultivated species, However, most specialists believe that this wild saffron species is most probably the ancestor of cultivated saffron (18). The type specimens of this plant have been observed in Greek islands.

Fig. 2-8   Violet saffron (*C. michelsonii*)

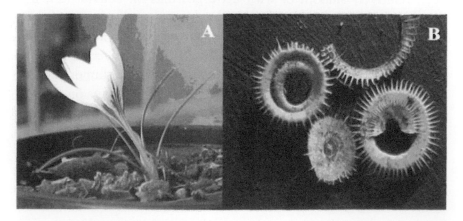

Fig. 2-9   A–Two flowered saffron (*C. biflorus*) B–tunic basal rings

during fruit formation the leaves are about 10-30 cm (20, 13), grayish green. Cataphylls 3. Spathe absent. Bract and bracteole exserted from cataphylls, 4-6.5 cm long, bracteole slender than bracts (13, 15, 20, 23, 25). Flowers 1 or 2, usually less than 3, perigone tube 4 to 8 cm, flower throat pubescent, perianth segments 2.5-4 cm, obovate and equal. Anther 1.3 cm with yellow pollens, filaments 0.6 cm. Style at the same level of anther, white and terminated in 3 stigmatic arms (Figure 2-7). Fruit an elliptical capsule, located on a peduncle which makes it adjacent to soil surface at maturity, 0.5-1 cm long. Seeds black, 0.3-1cm (15, 23).

This saffron is located in north and northeastern Khorasan, in Quchan, Kopet daugh Mountains, Sisab in Bojnourd and in dry slopes of north Khorasan Steppe flora at 1200-2300 m altitude (15, 20, 21, 22, 23, 25). Flowering time late February and March. Diploid and 2n= 20 chromosomes.

2-9 TWO FLOWERED SAFFRON (*C. BIFLORUS*)

Corm tunics coriaceous with hard sharp pointed fibrous (19, 25), tunica concentric with tough annular and ± smooth structures near the base but serrated at edges. Leaves 3-9, synanthus, 0.1-0.3 cm wide, green to gray green, one or more nerves on either side of the keel, relatively smooth or with little postules on abaxial (23, 25). Spathe absent, bract and bracteole pronouncedly exserted from cataphylls. Spring flowering with 1-4 flowers; perigone tube 5-9 cm long; perianth segments obovate, obtuse to subacute, ± equal, 0.7 - 1.3 × 2-3.5 cm, smooth or slightly pubescent where stamens adnate to perianth, white with violet or purple reticulated veins on abaxial, sometimes 3 outer perianths are purple to violet blue on abaxial, and white to yellow on adaxial. Anther yellow, 0.7-1.2 cm filament 0.3 to 0.4 and sometimes up to 0.9 cm, yellow or white. Style at the same level as stamen, usually yellow and terminated to 3 frilled stigmatic branches, yellow or orange and rarely white (figure 2-9). Fruit a cylindrical capsule, 1-1.5 cm located on a small peduncle at maturity. Seeds ± spherical with broad caruncle at the base, reddish brown, 0.2 - 0.25 cm long.

PROOF   Let $F \in C(K, Y)$ so that F is a continuous function from K to Y. Let $\{\varphi_1, \varphi_2, \ldots, \varphi_s\}$ be a basis for X and let $\{\psi_1, \psi_2, \ldots, \psi_n\}$ be a basis for Y. Then using the homeomorphisms h and k, $y = F(x)$ is equivalent to the system of equations

$$y_i = F_i(x_1, x_2, \ldots, x_s)$$

where $i = 1, 2, \ldots, t$, $x = \sum_{j=1}^{s} x_j \varphi_j$ and $y = \sum_{i=1}^{t} y_i \psi_i$. Since F is continuous on K, each of the functions F_i, $i = 1, 2, \ldots, t$, is a continuous function of the s real variables $x_1, x_2, \ldots, x_s$. By the Weierstrass theorem for n dimensions, for each $i = 1, 2, \ldots, t$, given $\varepsilon > 0$, there exists a polynomial $P_i(x_1, x_2, \ldots, x_s)$ which approximates F_i uniformly on the compact set $\bar{K} = h[K]$. That is, we can find polynomials P_i for which

$$\sup_{(x_1, x_2, \ldots, x_s) \in \bar{K}} |F_i(x_1, x_2, \ldots, x_s) - P_i(x_1, x_2, \ldots, x_s)| < \frac{\varepsilon}{t}.$$

Let n_i be the degree of P_i. Without loss of generality, we can find n such that $n_i = n$ for all $i = 1, 2, \ldots, t$ by simply adding higher order terms with zero coefficients to each of the P_is. But, for each i there exist real numbers $a_{i, j_1, j_2, \ldots, j_k}$ and a_i such that

$$P_i(x_1, x_2, \ldots, x_s) = a_i + \sum_{k=1}^{} \sum_{j_1, j_2, \ldots, j_k = 1} a_{i, j_1, j_2, \ldots, j_k} x_{j_k} \cdots x_{j_2} x_{j_1}.$$

For example, if P_i is

linear: $P_i(x_1, x_2, \ldots, x_s) = a_i + \sum_{j=1}^{s} a_{ij} x_j$,

quadratic: $P_i(x_1, x_2, \ldots, x_s) = a_i + \sum_{j=1}^{s} a_{ij} x_j + \sum_{k,j=1}^{s} a_{i,j,k} x_k x_j$,

and so forth.

However, the one-dimensional matrix (a_i), $i = 1, 2, \ldots, t$ corresponds to a 0-linear operator L_0 on X to Y and for each $k = 1, 2, \ldots, n$, the matrix $(a_{i, j_1, j_2, \ldots, j_k})$ where $j_1, j_2, \ldots, j_k = 1, 2, \ldots, s$ corresponds to a k-linear operator L_k on X to Y. Thus, the polynomial

$$Px = L_n x^n + L_{n-1} x^{n-1} + \cdots + L_1 x + L_0$$

is that polynomial operator corresponding to $\sum_{i=1}^{t} P_i(x_1, x_2, \ldots, x_s)\psi_i$. For this reason

$$\|F - P\|_u = \sup_{x \in K} \|Fx - Px\|$$

$$= \sup_{h(x) \in \tilde{K}} \left\| \sum_{i=1}^{t} [F_i(x_1, x_2, \ldots, x_s) - P_i(x_1, x_2, \ldots, x_s)] \psi_i \right\|$$

$$\leq \sum_{i=1}^{t} \sup_{(x_1, x_2, \ldots, x_s) \in \tilde{K}} |F_i(x_1, x_2, \ldots, x_s) - P_i(x_1, x_2, \ldots, x_s)|$$

$$< \varepsilon .$$

where $\|\psi_i\| = 1$, $i = 1, 2, \ldots, t$. This completes the proof of the theorem. ∎

Let X and Y be real, normed linear spaces and let K be a compact subset of X. The following theorem for completely continuous (linear or nonlinear) operators on a bounded set is useful:

THEOREM 5.3

Let X and Y be real normed linear spaces and let $K \subset X$ be compact. Let F be a continuous function from X to Y. Then for each $\varepsilon > 0$ there exists an open set D containing K and a function $\tilde{F}$ continuously defined on D which satisfies the following two conditions:

a)   *There exists a finite dimensional subspace $Y^{(m)}$ of Y such that $\tilde{F}(x) \in Y^{(m)}$ for all $x \in D$.*

b) $\|F(x) - \tilde{F}(x)\| < 2\varepsilon$ *for all $x \in D$.*

PROOF   Since $F[K]$ is compact, given $\varepsilon > 0$, there exists a finite ε-net $\{y_1, y_2, \ldots, y_m\} \subset F[K]$ such that for any $x \in K$ there is a y_i which satisfies the inequality $\|F(x) - y_i\| < \varepsilon$. The span

$$\left\{ \sum_{i=1}^{m} a_i y_i : a_1, a_2, \ldots, a_m \text{ are real} \right\}$$

is a subspace $Y^{(m)}$ of Y of dimension $k \leq m$. For each $i = 1, 2, \ldots, m$ define the function $m_i(x)$ by

$$m_i(x) = \begin{cases} 2\varepsilon - \|F(x) - y_i\| & \text{if } \|F(x) - y_i\| < 2\varepsilon \\ 0 & \text{if } \|F(x) - y_i\| \ge 2\varepsilon. \end{cases}$$

Clearly each m_i is continuous since

$$m_i(x) = \sup_{i=1,2} \{g_i(x), h_i(x)\}$$

where the continuous functions g_i and h_i are given by

$$g_i(x) = 2\varepsilon - \|F(x) - y_i\|$$

$$h_i(x) \equiv 0.$$

Let

$$D = \{x \in X : \|F(x) - y_i\| < 2\varepsilon \text{ for some } i = 1, 2, \ldots, m\}.$$

Then $K \subset D$ and $\sum_{i=1}^{m} m_i(x) > 0$ for all $x \in D$. Define $\tilde{F}: D \to Y^{(m)}$ by

$$\tilde{F}(x) = \frac{\sum_{i=1}^{m} m_i(x) y_i}{\sum_{i=1}^{m} m_i(x)}.$$

It follows from the nonvanishing of the denominator of $\tilde{F}$ on D that $\tilde{F}$ is continuous on D. Furthermore

$$\|F(x) - \tilde{F}(x)\| = \left\| \frac{\sum_{i=1}^{m} m_i(x)}{\sum_{i=1}^{m} m_i(x)} F(x) - \frac{\sum_{1}^{m} m_i(x) y_i}{\sum_{1}^{m} m_i(x)} \right\|$$

$$= \left\| \frac{\sum_{1}^{m} m_i(x) [F(x) - y_i]}{\sum_{1}^{m} m_i(x)} \right\| = \left\| \sum_{1}^{m} \left[\frac{m_i(x)}{\sum m_i(x)} \right] [F(x) - y_i] \right\|$$

$$\le \sum_{1}^{m} \left[\frac{m_i(x)}{\sum_{1}^{m} m_i(x)} \right] \|F(x) - y_i\| \quad \text{since } m_i(x) \ge 0.$$

Thus,

$$\|F(x) - \tilde{F}(x)\| \le \left\{ \sum_{1}^{m} \left[\frac{m_i(x)}{\sum_{1}^{m} m_i(x)} \right] \right\} 2\varepsilon = 2\varepsilon$$

for all $x \in D$. This completes the proof of the theorem. ∎

In order to prove a Weierstrass theorem for arbitrary real normed linear spaces X and Y we shall need some restrictions either on the compact set K involved in the theorem or else some restrictions on boundedness of projections. We choose the former course and give:

DEFINITION 5.1   *A compact set K has property M, $M > 0$, if for each ε-net, $\varepsilon > 0$, $\{x_1, x_2, \ldots, x_p\} = A$ in K there exists a continuous projection S of X onto span A such that*

$$\|S(x - x_i)\| \leq M \|x - x_i\|$$

for each $x \in K$ and each $x_i \in A$.

Now let $F \in C(K, Y)$. We signal out uniform continuity of F in K in the following form.

THEOREM 5.4
Let X and Y be normed linear spaces and let $F \in \tilde{C}(K, Y)$ where $K \subset Y$ is compact. Given $\varepsilon > 0$ there exists a $\delta > 0$ such that $\|F(x) - F(z)\| < \varepsilon$ whenever $\|x - z\| < \delta$, $x \in K$ and $z \in X$.

The important thing to notice is that z need not belong to K. We are now able to prove:

THEOREM 5.5
(A Weierstrass theorem for normed linear spaces). Let X and Y be real normed linear spaces. Let K be a compact subset of X which has property M, $M > 0$. Then $P(K, Y)$ is dense in $\tilde{C}(K, Y)$ where $\tilde{C}(K, Y)$ carries the uniform norm topology.

PROOF   Let $F \in \tilde{C}(K, Y)$. Since F is uniformly continuous on K, given $\varepsilon > 0$, $\exists \delta > 0$ such that $\|F(x) - F(y)\| < \varepsilon$ whenever $x \in K$, $y \in X$ and $\|x - y\| < \delta$. Let $A = \{x_1, x_2, \ldots, x_m\}$ be a $\frac{\delta}{M+1}$ net for K. Then $F[A] = \{y_1, y_2, \ldots, y_m\}$ where $y_i = F(x_i)$ is an ε-net for $\bar{F}[K]$. Let $D = \{x : \|F(x) - y_i\| < 2\varepsilon$ for some $i = 1, 2, \ldots, n\}$ and let $\tilde{F}$ map D continuously into $Y^{(m)} = $ span $F(A)$ according to the construction in Theorem 5.1. Clearly for all $x \in K$ and $y \in X$ for which $\|x - y\| < \delta$, $\|F(x) - F(y)\| < \varepsilon$. But there exists $x_i \in K$ such that $\|x - x_i\| < \frac{\delta}{M+1}$. Thus,

$$\|y_i - F(y)\| = \|F(x_i) - F(y)\|$$

$$\leq \|F(x_i) - F(x)\| + \|F(x) - F(y)\| < 2\varepsilon .$$

This implies that $y \in D$ whenever $\|x - y\| < \delta$ for some $x \in K$. That is

$$D \supset U \{S_\delta(x) : x \in K\} . \tag{5.1}$$

Now let S be a projection of X onto span A for which $\|S(x - x_i)\| \le M\|x - x_i\|$ for all $sx \in K$ and for all $x_i \in A$. This exists since K has property M, $M > 0$. Let $x \in K$. Then there exists $x_i \in A \subset K$ such that $\|x - x_i\| < \frac{\delta}{M+1}$. Since $x_i \in A$, $Sx_i = x_i$. Thus,

$$\|x_i - Sx\| = \|Sx_i - Sx\| = \|S(x_i - x)\|$$

$$\le M \cdot \|x_i - x\| < \delta .$$

But $Sx \in X$, $x_i \in K$ and $\|x_i - Sx\| < \delta$. Thus by (5.1) $Sx \in D$ and thus $\tilde{F}(Sx)$ is defined. Recall that $\|F(x) - \tilde{F}(x)\| < 2\varepsilon$ for all $x \in D$. Thus, for all $x \in K$

$$\|\tilde{F}(x) - \tilde{F}(Sx)\| \le \|\tilde{F}(x) - F(x)\| + \|F(x) - F(Sx)\| + \|F(Sx) - \tilde{F}(Sx)\|$$

$$\le 2\varepsilon + 2\varepsilon + \|F(x) - F(x_i)\| + \|F(x_i) - F(Sx)\| \tag{5.2}$$

where $x_i \in A$ and

$$\|x - x_i\| < \frac{\delta}{M+1} = 2\varepsilon + 2\varepsilon + \|F(x) - F(x_i)\| + \|F(Sx_i) - F(Sx)\| < 6\varepsilon$$

since

$$\|x_i - Sx\| = \|Sx_i - Sx\| < \delta .$$

But $S[K] = \tilde{K}$ is a compact subset of $X^{(n)} = $ span A where dim $X^{(n)} = n \le m$ and $\tilde{F}$ is a continuous function from $\tilde{K}$ into $Y^{(m)}$. Thus, by Theorem 5.2 there exists a continuous polynomial operator $\tilde{P}$ from $\tilde{K}$ to $Y^{(m)}$ with the property that

$$\sup_{x \in \tilde{K}} \|\tilde{F}(x) - \tilde{P}(x)\| < \varepsilon .$$

Define the operator P from K to Y by

$$P(x) = \tilde{P}(Sx) .$$

Since S is continuous and linear, it follows that P is a continuous polynomial operator. That is

$$\tilde{P}(x) = \tilde{L}_n x^n + \tilde{L}_{n-1} x^{n-1} + \cdots + \tilde{L}_0$$

where each $\tilde{L}_k$ is a continuous, k-linear operator and $x \in S[X]$. Then, for each $x \in K$

$$P(x) = L_n x^n + L_{n-1} x^{n-1} + \cdots + L_0$$

where $L_k(x_1, x_2, \ldots, x_k) = \tilde{L}_k(Sx_1, Sx_2, \ldots, Sx_n)$ for each $(x_1, x_2, \ldots, x_k) \in X^k$. The k-linearity of L_k follows from the linearity of S.

Observe that

$$\sup_{x \in K} \| \tilde{F}(Sx) - P(x) \| = \sup_{x \in K} \| \tilde{F}(Sx) - \tilde{P}(Sx) \|$$

$$= \sup_{x \in K} \| \tilde{F}(x) - \tilde{P}(x) \| < \varepsilon \,.$$

Thus

$$\| \tilde{F} S - P \|_u < \varepsilon \,. \tag{5.3}$$

Combining (5.2) and (5.3) we have

$$\| F - P \|_u \leq \| F - \tilde{F} \|_u + \| \tilde{F} - \tilde{F} S \|_u + \| \tilde{F} S - P \|_u < 8\varepsilon \,.$$

This completes the proof of the theorem. ∎

The extension of $\tilde{P}$ to P depended directly on the linearity of S; whereas the uniform approximation of $\tilde{F}$ by $\tilde{F} S$ depended on the fact that K had property M. Even though K may not have property M, we can, nonetheless, always approximate F uniformly by a polynomial in the following sense:

THEOREM 5.6

Let X and Y be real normed linear spaces and let K be a compact subset of X. Then, for each $\varepsilon > 0$ and for each $F \in \tilde{C}(K, Y)$ there exists a finite dimensional subspace $X^{(n)}$ of X, a continuous function S from X to $X^{(n)}$ and a continuous polynomial operator P from $S[K]$ into a finite dimensional subspace $Y^{(m)}$ of Y which satisfies the following condition:

a) $\| F(x) - PS(x) \| < \varepsilon$ *for all* $x \in K$.

PROOF   Given $\varepsilon > 0$, construct $\tilde{F}$ and $Y^{(m)}$ according to Theorem 5.3 so that $\| \tilde{F}(x) - F(x) \| < \varepsilon/3$ for all $x \in K$. Find a $\delta > 0$ for which $\| \tilde{F}(x) - \tilde{F}(x') \| < \varepsilon/3$ whenever $\| x - x' \| < \delta$, $x \in K$ and $x' \in X$. Let $\{\varphi_1, \varphi_2, \ldots, \varphi_s\}$ be a δ-net for K. Let $X^{(n)} = \text{span}\{\varphi_1, \varphi_2, \ldots, \varphi_s\}$. Construct $S: X \to X^{(n)}$ from Theorem 5.1 so that $\| Sx - x \| < \delta$ for all $x \in K$. Although S is nonlinear, $\| \tilde{F}(x) - \tilde{F}(Sx) \| < \varepsilon/3$ for all $x \in K$. But $\tilde{F}$ maps $S[K]$ into $Y^{(m)}$; so that, by Theorem 5.2, there exists a continuous polynomial operator P for which $\|P(Sx) - \tilde{F}(Sx)\| < \varepsilon/3$ for all $x \in K$. ∎

THEOREM 5.7
Let X be a finite dimensional linear space of dimension n. There exists a homeomorphism $h: X \to \mathbf{R}^n$ where $\mathbf{R}$ denote the real numbers and $\mathbf{R}^n$ carries the usual topology.

PROOF Let $\{\varphi_1, \varphi_2, \ldots, \varphi_n\}$ be a basis for X. Let $f_1, f_2, \ldots, f_n$ be bounded, linear functionals on X for which $f_i(\varphi_j) = \delta_{ij}$. Define $h(x) = (f_1(x), f_2(x), \ldots, f_n(x))$. Clearly h is a continuous bijection of X onto R^n since each f_i is continuous and $\{\varphi_1, \varphi_2, \ldots, \varphi_n\}$ is a basis. Furthermore, h is open. To see this let $U = S_\varepsilon(x)$ be open in X. Let $V = \times_{i=1}^n S_{\delta_i}(f_i(x))$ where $\delta_i = \varepsilon/n\|\varphi_i\|$. Then $x \in h^{-1}[V]$ and for each $y \in h^{-1}[V]$

$$\|x - y\| = \left\| \sum_{i=1}^n (x_i - y_i)\, \varphi_i \right\|$$

$$= \left\| \sum_{i=1}^n (f_i(x) - f_i(y))\, \varphi_i \right\|$$

$$\leq \sum_{i=1}^n |f_i(x) - f_i(y)|\, \|\varphi_i\|$$

$$< \sum_{i=1}^n \frac{\varepsilon}{n} = \varepsilon .$$

Thus $h^{-1}[V] \subset U$ and h^{-1} is continuous. ∎

5.2 Matrix Representations

It is well known that if H is a separable Hilbert space with a Schauder basis $\{\varphi_i\}_{i=1}^\infty$ and if A is a continuous, linear operator on H into H, then A has a matrix representation (a_{ij}) with respect to $\{\varphi_i\}$ given by $(a_{ij}) = (A\varphi_j, \varphi_i)$. Furthermore, every linear operator A having a matrix representation (a_{ij}) must be a continuous, linear operator.

The purpose of this section is to show that these results extend to a much larger class of operators, the polynomial operators, on H into H, of which the linear operators are a proper subset. That is, we shall prove that each continuous polynomial operator P on H into H has a matrix representation and that each polynomial endowed with a matrix representation is continuous.

THEOREM 5.8

For each $n = 0, 1, 2, \ldots,$ let T be an n-linear operator. Then

    a) T is bounded iff T is continuous.

    b) T is bounded iff T is continuous at $\tilde{0}$.

    c) T is continuous iff T is continuous at $\tilde{0}$.

A function f from a topological space Z into the real numbers R (complex numbers C) is said to be *bounded at a point* x_0 if there exists an open neighborhood $U(x_0)$ and a positive constant K such that $\|f(x)\| \leq K$ for all $x \in U(x_0)$.

THEOREM 5.9

Let T be an n-linear operator. Then for all $n \geq 1$

    a) T is bounded at $z \in X^n$ iff T is bounded at $\tilde{0}$.

    b) T is bounded at $z \in X^n$ iff T is a bounded operator.

We shall actually need a stronger theorem than Theorem 5.9(b), which, worded in a somewhat different manner, says that any functional in the family

$$\{P: P(x) = \|Tx\| \text{ for } T \in \mathcal{B}_n(X, Y)\}$$

is bounded at $z \in X^n$ iff it is bounded at $\tilde{0}$. We shall prove that a positive, n-convex, n-homogeneous functional also has this property. A functional P from X^n into the reals is said to be *n-convex* if for each $k = 1, 2, \ldots, n$

$$P\left(x^1, x^2, \ldots, x^k + \bar{x}^k, \ldots, x^n\right)$$

$$\leq P\left(x^1, \ldots, x^k, \ldots, x^n\right) + P\left(x^1, \ldots, \bar{x}^k, \ldots, x^n\right).$$

The functional P is said to be *n*-homogeneous if

$$P\left(a_1 x^1, a_2 x^2, \ldots, a_n x^n\right) = |a_1|\, |a_2| \ldots |a_n|\, P\left(x^1, x^2, \ldots, x^n\right)$$

for all $z \in X^n$.

Let A be an operator on a separable Hilbert space H and let $\{\varphi_i\}_1^\infty$ be a complete, orthonormal basis for H. Then A is said to have a matrix representation (a_{ij}) with respect to $\{\varphi_i\}_1^\infty$ if for each $x \in H$

$$Ax = \sum_{i=1}^\infty d_i \varphi_i$$

where

$$d_i = \sum_{j=1}^\infty a_{ij} x_j$$

and

$$x = \sum_{j=1}^\infty x_j \varphi_j \, .$$

It is well known that every continuous, linear operator has a matrix representation and that any linear operator having a matrix representation is continuous.

Now let A be an operator on H^k to H where $k \geq 1$. Then A is said to have a *matrix representation* $(a_{j,i_1,i_2,\ldots,i_k})$ *with respect to* $\{\varphi_j\}_{j=1}^\infty$ if, for each k elements $x, y, \ldots, z$ in H,

$$A(x, y, \ldots, z) = \sum_{j=1}^\infty d_j \varphi_j$$

where

$$d_j = \sum \left\{ a_{j,i_1,i_2,\ldots,i_k} x_{i_k} y_{i_{k-1}} \cdots z_{i_1} : i_1, i_2, \ldots, i_n = 1, 2, \ldots \right\}$$

and

$$x = \sum_{i=1}^\infty x_i \varphi_i$$

$$y = \sum_{i=1}^\infty y_i \varphi_i$$

$$\vdots$$

$$z = \sum_{i=1}^{\infty} z_i \varphi_i \ .$$

Then it is a simple matter to prove:

THEOREM 5.10
Let H be a separable Hilbert space with a complete, orthonormal basis $\{\varphi_i\}_{i=1}^{\infty}$. Then

a) *every continuous k-linear operator A, $k = 1, 2, \ldots$ has a unique matrix representation $(a_{j,i_1,i_2,\ldots,i_k})$, with respect to $\{\varphi_i\}_{i=1}^{\infty}$ where j, i_1, $i_2, \ldots, i_k = 1, 2, \ldots$ and*

b) *$(a_{j,i_1,i_2,\ldots,i_k}) = (A\varphi_{i_k}\varphi_{i_{k-1}} \cdots \varphi_{i_1}, \varphi_j)$ where $(\ ,\ )$ denotes inner product.*

PROOF   Let $x^1, x^2, \ldots, x^k$ be any k elements of H. Let $x_N^i = \sum_{j=1}^{N}(x^i, \varphi_j)\varphi_j$ for each $i = 1, 2, \ldots, k$ and let

$$a_{j,i_1,i_2,\ldots,i_k} = \left(A\varphi_{i_k}\varphi_{i_{k-1}} \cdots \varphi_{i_1}, \varphi_j\right) \ .$$

Then

$$Ax_N^1 x_N^2 \cdots x_N^k = \sum_{i_1,i_2,\ldots,i_k=1}^{N} x_{i_1}^1 x_{i_2}^2 \cdots x_{i_k}^k A\varphi_{i_1}\varphi_{i_2}\cdots\varphi_{i_k} \ .$$

It follows that

$$\left(Ax_N^1 x_N^2 \cdots x_N^k, \varphi_j\right) = \sum_{i_1,i_2,\ldots,i_k=1}^{N} a_{j,i_k,i_{k-1},\ldots,i_1} x_{i_1}^1 x_{i_2}^2 \cdots x_{i_k}^k \ .$$

Since A is continuous

$$\lim_{N\to\infty} Ax_N^1 x_N^2 \cdots x_N^k = Ax^1 x^2 \cdots x^k$$

and

$$\left(Ax^1 x^2 \ldots x^k, \varphi_j\right) = \lim_{N\to\infty} \sum_{i_1,i_2,\ldots,i_k=1}^{N} a_{j,i_k,i_{k-1},\ldots,i_1} x_{i_1}^1 x_{i_2}^2 \cdots x_{i_k}^k$$

$$= \sum_{i_1,i_2,...,i_k=1}^{\infty} a_{j,i_k,i_{k-1},...,i_1} x_{i_1}^1 x_{i_2}^2 \cdots x_{i_k}^k ,$$

converges for all $x^1, x^2, \ldots, x^k$ in H. Letting $d_j = (Ax^1x^2 \cdots x^k, \varphi_j)$ we see that A has a matrix representation with respect to $\{\varphi_i\}_1^{\infty}$. Uniqueness follows simply from the uniqueness of $(A\varphi_{i_1}\varphi_{i_2} \cdots \varphi_{i_k}, \varphi_j)$. This completes the proof of the theorem. ∎

Every continuous k-linear operator has a matrix representation. Does every infinite matrix $(a_{j,i_1,i_2,...,i_k})$ represent a continuous k-linear operator on H^n to H or even a noncontinuous k-linear operator on H^k to H? The answer to this question is clearly "no" since counterexamples are easily constructed.

A somewhat more subtle conjecture which does have an affirmative answer asks whether every operator on H^k to H, $k \geq 1$, having a matrix representation is a continuous k-linear operator. Well-known proofs of this theorem in the case $k = 1$ are based on either Landau's Theorem combined with the closed graph theorem, or on properties of the adjoint of a linear operator, or on convexity arguments. It is a simple matter to adapt the theory for the linear case to weak linear case to prove:

THEOREM 5.11
Let H be a complete, separable Hilbert space. Every operator L on H^k to H possessing a matrix representation $(a_{j,i_1,i_2,...,i_n})$ is a continuous k-linear operator.

We shall prove this theorem through a sequence of lemmas using mathematical induction. First recall that if $\sum_{k=1}^{\infty} a_k b_k$ converges for all $(b_k) \in \ell^p$, $p > 1$, then Landau's Theorem states that $(a_k) \in \ell^q$ where $\frac{1}{p} + \frac{1}{q} = 1$. If L has a matrix representation $(a_j, i_1, i_2, \ldots, i_k)$, then for each $x = (x^1, x^2, \ldots, x^k) \in H^k$

$$Lx = \sum_{j=1}^{\infty} \Omega_j(x)\varphi_j . \tag{5.4}$$

If $x^n = \sum_{i=1}^{\infty} x_i^n \varphi_i$, $n = 1, 2, \ldots, k$ then

$$\Omega_j(x) = \sum_{i_1,i_2,...,i_k=1}^{\infty} a_{j,i_1,i_2,...,i_k} x_{i_k}^1 x_{i_{k-1}}^2 \cdots x_{i_1}^k . \tag{5.5}$$

It is well known that Theorem 5.10 is true when $n = 1$. We now make the *induction hypothesis* that Theorem 5.11 is true when $n = k - 1$.

With this assumption we are able to prove:

LEMMA 5.1

Let $(a_{j,i_1,i_2,\ldots,i_k})$ be a matrix representation of an operator L from H^k to H where $Lx = \sum_{j=1}^{\infty} \Omega_j(x)\varphi_j$ and each Ω_j is given by Equation (5.5). Then each Ω_j is a continuous k-linear functional.

PROOF   It is obvious that each Ω_j is k-linear. For fixed j let $c = (c_{i_1}:$ $i_1 = 1, 2, \ldots)$ be given by

$$c_{i_1} = \sum_{i_2,i_3,\ldots,i_k=1} a_{j,i_1,i_2,\ldots,i_k} x_{i_k}^1 x_{i_{k-1}}^2 \cdots x_{i_2}^{k-1} .$$

It is clear from Landau's theorem that $c \in \ell^2$ since $\sum c_{i_1} x_{i_1}$ converges for all $(x_{i_1}) \in \ell^2$. We now use the induction hypothesis. Since for each

$$j = 1, 2, \ldots, \sum_{i_2,\ldots,i_k} a_{j,i_1,i_2,\ldots,i_k} x_{i_k}^1 x_{i_{k-1}}^2 \cdots x_{i_2}^{k-1}$$

converges for all $(x^1, x^2, \ldots, x^{k-1}) \in H^{k-1}$ it follows that, for each j, $(a_{j,i_1,i_2,\ldots,i_k})$ is a matrix representation of a continuous $(k - 1)$-linear operator A_j. Furthermore, $c = A_j(x^1, x^2, \ldots, x^{k-1})$. Now simply note that

$$\Omega_j(x^1, x^2, \ldots, x^k) = (c, x^k)$$

$$= \left(A_j\left(x^1, x^2, \ldots, x^{k-1} \right), x^k \right) .$$

Thus, using Schwartz's inequality

$$\left| \Omega_j\left(x^1, x^2, \ldots, x^k \right) \right| = \left| \left(A_j\left(x^1, x^2, \ldots, x^{k-1} \right), x^k \right) \right|$$

$$\leq \left\| A_j\left(x^1, x^2, \ldots, x^{k-1} \right) \right\| \cdot \left\| x^k \right\|$$

$$\leq \left\| A_j \right\| \cdot \left\| x^1 \right\| \cdot \left\| x^2 \right\| \cdots \left\| x^k \right\| .$$

Thus, Ω_j is continuous as was to be proved.   ∎

LEMMA 5.2

Let

$$Lx = \sum_{j=1}^{\infty} \Omega_j(x)\varphi_j \ .$$

If each Ω_j is continuous, then L is closed.

PROOF   Let $(x_n) = (x_n^1, x_n^2, \ldots, x_n^k)$ be a sequence in H^k which converges to $y = (y^1, y^2, \ldots, y^k)$ in H^k. Suppose Lx_n converges to z in H. We must prove that $Ly = z$. But $(Lx_n, \varphi_j) = \Omega_j(x_n)$ which converges to $\Omega_j(y)$ since Ω_j is continuous. Since $(Lx_n, \varphi_j) \to (z, \varphi_j)$, it follows that $\Omega_j(y) = (z, \varphi_j)$ for each $j = 1, 2, \ldots$. But then

$$\|z - Ly\|^2 = \sum_{j=1}^{\infty} |(z - Ly, \varphi_j)|^2$$

$$= \sum_{j} |(z, \varphi_j) - (Ly, \varphi_j)|^2$$

$$= \sum_{j} |\Omega_j(y) - (Ly, \varphi_j)|^2 = 0 \ .$$

Thus, $Ly = z$ and L is closed.   ∎

Now let L be given as in Lemma 5.2. Observe that if $Tx = Lx$ where $x \in H$, then it follows from the induction hypotheses that $Tx \in B_{k-1}[H, H]$, the family of bounded, $(k-1)$-linear operators from H^{k-1} to H. We then have the following:

LEMMA 5.3

Let L be an operator from H^k to H having a matrix representation $(a_{j,i_1,i_2,\ldots,i_k})$. If L is closed and $T: H \to B_{k-1}[H, H]$ is given by $Tx = Lx$ where $x \in H$, then T is closed.

PROOF   Let (x_n) be a sequence in H which converges to x in H and suppose Tx_n converges to A in $B_{k-1}[H, H]$. We must prove that $Tx = A$. For each $y = (y^1, y^2, \ldots, y^{k-1})$ in H^{k-1}, $(Tx_n)(y)$ converges pointwise to Ay. However, L is closed. Thus, for all y in $H^{k-1}(x_n, y^1, y^2, \ldots, y^{k-1})$ converges to $(x, y^1, y^2, \ldots, y^{k-1})$, $(Tx_n)(y) = L(x_n, y^1, y^2, \ldots, y^{k-1})$ converges to Ay and thus by the closedness of L

$$L\left(x, y^1, y^2, \ldots, y^{k-1}\right) = (Tx)(y) = Ay$$

for all $y \in H^{k-1}$. Thus, $(A - Tx)$ is a $(k - 1)$-linear operator which vanishes as all $y \in H^{k-1}$. Thus, $A = Tx$ which was to be proved. ∎

LEMMA 5.4
Let $T: H \to B_{k-1}[H, H]$ linearly. If T is closed, then T is continuous.

PROOF   The proof is a simple adaptation of a canonical proof for the linear case. Let graph $T = \{(x, Tx): x \in H\}$. This set is closed in the product topology on $H \times B_{k-1}[H, H]$. Observe that graph T is a subspace when addition on $H \times B_{k-1}[H, H]$ is defined by

$$(x, A) + (y, B) = (x + y, A + B)$$

where $x, y \in H$ and $A, B \in B_{k-1}[H, H]$. Let $H \oplus B_{k-1}[H, H]$ be the space $H \times B_{k-1}[H, H]$ topologized by the norm

$$\|(x, B)\| = \|x\| + \|B\|$$

where $B \in B_{k-1}[H, H]$. This topology is equivalent to the product topology on $H \times B_{k-1}[H, H]$. Since H and $B_{k-1}[H, H]$ are complete, $H \oplus B_{k-1}[H, H]$ is complete. Observe that graph T is a closed subspace of a complete space and thus is itself complete. Define a function A from graph T to H by

$$A(x, Tx) = x \ .$$

Then

$$\|A(x, Tx)\| = \|x\| \leq \|x\| + \|Tx\| = \|(x, Tx)\| \ ;$$

so that A is bounded. Furthermore, A is linear since

$$A((x, Tx) + (y, Ty)) = A(x + y, T(x + y))$$

$$= A(x + y, Tx + Ty)$$

$$= x + y = A(x, Tx) + A(y, Ty) \ .$$

Thus, A is a bounded linear transformation on graph T *onto* H. Also A is $1 - 1$. Thus, A^{-1} exists and is continuous since graph T and H are both complete. It follows that if the sequence (x_n) in H converges to x in H, then $A^{-1}(x_n)$ converges to $A^{-1}x$. That is, (x_n, Tx_n) converges to (x, Tx) in $H \oplus B_{k-1}[H, H]$. But then Tx_n converges to Tx since

$$\|(x_n, Tx_n) - (x, Tx)\| = \|x - x_n\| + \|Tx_n - Tx\| .$$

Thus, T is continuous as was to be proved. ∎

We can now prove:

THEOREM 5.12
Let H be a complete, separable Hilbert space. Every operator L on H^k to H possessing a matrix representation $(a_{j, i_1, i_2, \dots, i_k})$ is a continuous k-linear operator.

PROOF   The theorem is clearly true when $n = 1$. Assume the theorem is true when $n = k - 1$ (induction hypothesis) and let $n = k$. Define T from H to $B_{k-1}[H, H]$ by $Tx = Lx$. Invoking Lemmas 5.1 through 5.4 we see that L is closed which implies T is closed which implies T is continuous. Let $x = (x^1, x^2, \dots, x^k) \in H^k$. Then

$$\left\| L\left(x^1, x^2, \dots, x^k\right) \right\| = \left\| \left(Tx^1\right)\left(x^2, x^3, \dots, x^k\right) \right\|$$

$$\leq \left\| Tx^1 \right\| \left\| x^2 \right\| \left\| x^2 \right\| \cdots \left\| x^k \right\|$$

$$\leq \|T\| \left\| x^1 \right\| \left\| x^2 \right\| \cdots \left\| x^k \right\|$$

and L is continuous. This completes the proof of the theorem. ∎

We have given a proof of Theorem 5.12 by arguments involving closed graph theory. There is a second and somewhat more complicated proof involving n-convexity notions for n-linear functionals. We shall, however, omit this proof.

The preceding theory pertains to n-linear operators where $n \geq 1$. When $n = 0$ we can define a *matrix representation of a 0-linear operator*. Let L_0 be a constant function on H to H. Then there exists a fixed $y = L_0 \in H$ such that $L_0 x = y$ for all $x \in H$. But $y = \sum_{i=1}^{\infty} y_i \varphi_i$. The one-way matrix $(y_i : i = 1, 2, \dots)$ is said to be a matrix representation of the operator L_0. It is clear that to each one-way matrix $(c_i)_{i=1}^{\infty}$ for which $\sum_{i=1}^{\infty} |c_i|^2 < \infty$ there corresponds a 0-linear operator L_0 defined by $L_0 x = \sum_{i=1}^{\infty} c_i \varphi_i$ for all $x \in H$.

REMARK 5.1   We have proved that every continuous n-linear operator (and hence every continuous polynomial operator) has a unique matrix representation with respect to each complete, orthonormal system $\{\varphi_i\}_1^\infty$ in a separable Hilbert space. However, not every matrix $(a_{j,i_1,i_2,\ldots,i_n})$, $j, i_1, i_2, \ldots, i_n = 1, 2, \ldots$, where

$$A\left(x^1, x^2, \ldots, x^n\right) = \sum_{j=1}^\infty d_j \varphi_j \,,$$

$$d_j = \sum_{i_1, i_2, \ldots, i_n} a_{j,i_1,i_2,\ldots,i_n} x_{i_n}^1 \cdots x_{i_1}^n$$

and

$$x^k = \sum_{j=1}^n x_j^k \varphi_j$$

corresponds to an n-linear operator on H^n to H. For example, let $n = 2$ and define $a_{j,k,\ell} = j \cdot \delta_{k\ell}$ where $\delta_{k\ell}$ is the Kronecker delta. Then for each j

$$a_{jk\ell} = \begin{bmatrix} j & 0 & 0 & \cdots \\ 0 & j & 0 & \cdots \\ 0 & 0 & j & \cdots \\ \vdots & & & \end{bmatrix}.$$

If $x = \varphi_1$

$$Ax = \begin{bmatrix} 1 & 0 & 0 & \cdots \\ 2 & 0 & 0 & \cdots \\ \vdots & & & \\ n & 0 & 0 & \cdots \\ \vdots & & & \end{bmatrix}$$

and

$$Ax^2 = \sum_{n=1}^\infty n\varphi_n \notin H \,.$$

Necessary conditions upon the terms of the matrix $(a_{j,i_1,i_2,\ldots,i_n})$ to guarantee it to be an operator on H^n to H are rather easy to come by. However, such conditions are usually not sufficient conditions. In the event $n = 1$, Schur's Lemma gives a very stringent set of sufficient conditions on (a_{jk}) to guarantee it to be a matrix representation of a linear operator. Can Schur's Lemma be extended to the cases $n \geq 2$? Also conditions exist which guarantee that the matrix (a_{ij}) represents a compact linear operator on H to H. Certainly, these conditions can also be extended to $n \geq 2$.   ∎

All of the foregoing theory reduces to the case $H = \ell^2(n)$ where $n = 1, 2, \ldots$. A good part of it should carry over to the spaces $H = \ell^p(n)$ where $p > 1$ or $p = 1$. At least, one can certainly ask the same questions in any separable Banach space.

Finally the applicability of the foregoing theory to approximate solutions of polynomial equations $Px = y$ in separable Hilbert (Banach) spaces may prove fruitful.

5.3 Lagrange and Hermite Interpolation

Let X and Y be Banach spaces and let f be a function mapping X into Y. If X and Y are the real line R^1, the classical Lagrange and Hermite interpolation problems are, respectively, to find a polynomial $y(x)$ of degree $(n - 1)$ which interpolates f at n given distinct points $x_1, x_2, \ldots, x_n$, and to find a polynomial $\bar{y}(x)$ of degree $(2n - 1)$ which interpolates f at $x_1, x_2, \ldots, x_n$ while $\bar{y}'$ interpolates f' at these points. In this work we solve the Banach space analogs of these two problems, using polynomial operators. That is, we exhibit polynomials $y(x)$ and $\bar{y}(x)$, of degrees $n - 1$ and $2n - 1$, such that y interpolates f at the n given distinct points $x_1, x_2, \ldots, x_n$, $\bar{y}$ interpolates f, and $\bar{y}'$ interpolates f' at these points. In particular, we show that $y(x)$ has a Lagrange representation

$$y(x) = \sum_{i=1}^{n} \left[w_i'(x_i) \right]^{-1} \frac{w_i(x)}{(x - x_i)} f(x_i) ,$$

where $w_i'(x_i)$ is the Fréchet derivative of w_i at x_i, and $w_i(x)$ is the n-th degree polynomial

$$w_i(x) = L_i(x - x_1, x - x_2, \ldots, x - x_n) ,$$

L_i being an appropriately chosen n-linear operator. In the event X is a Hilbert space, the polynomials $y(x)$ and $\bar{y}(x)$ are shown to have simple representations in terms of inner products.

If L is n-linear $(n > 1)$, we shall let $\partial_i L$ denote the $(n - 1)$-linear operator on X into $\mathcal{L}_1[X, Y]$ defined by

$$\partial_i L(x_1, x_2, \ldots, x_{i-1}, x_{i+1}, \ldots, x_n) = L(x_1, x_2, \ldots, x_{i-1}\cdot, x_{i+1}, \ldots, x_n) ,$$

where

$$(L(x_1, x_2, \ldots, x_{i-1}, \cdot, x_{i+1}, \ldots, x_n))(x) = L(x_1, x_2, \ldots, x_{i-1}, x, x_{i+1}, \ldots, x_n) .$$

In general, n-linear operators are not *symmetric*. That is, it need not be true that

$$L\left(x_1, x_2, \ldots, x_n\right) = L\left(x_{i_1}, x_{i_2}, \ldots, x_{i_n}\right)$$

for all permutations $(i_1, i_2, \ldots, i_n)$ of $(1, 2, \ldots, n)$. For this reason, in general,

$$\partial_i L\left(x_1, x_2, \ldots, x_{i-1}, x_{i+1}, \ldots, x_n\right) \neq L\left(x_1, x_2, \ldots, x_{i-1}, x_{i+1}, \ldots, x_n\right) .$$

Now let L be any n-linear operator and let $x_1, x_2, \ldots, x_n$ be any points of X. We define a function w on X into Y by

$$w(x) = L\left(x - x_1, x - x_2, \ldots, x - x_n\right) .$$

Clearly, $w(x)$ is a polynomial

$$L_n x^n + L_{n-1} x^{n-1} + \cdots + L_1 x + L_0$$

of degree n on X, where $L_n = L$, and $L_0 = (-1)^n L(x_1, x_2, \ldots, x_n)$. For example, if L is bilinear,

$$L\left(x - x_1, x - x_2\right) = Lx^2 - L\left(x_1, x\right) - L\left(x, x_1\right) + L\left(x_1, x_2\right) .$$

Thus $L_2 = L$, $L_0 = L(x_1, x_2)$, and $L_1 = -L(x_1, \cdot) - L(\cdot, x_2)$.

We shall need the derivative of w. Let L be n-linear and let $x_1, x_2, \ldots, x_n$ be points of X. We let $\partial_i w$ or $w/(x - x_i)$ denote the operator on X into $\mathcal{L}_i[X, Y]$ defined by

$$\partial_i w(z) = L\left(z - x_1, z - x_2, \ldots, z - x_{i-1}, \cdot, z - x_{i+1}, \ldots, z - x_n\right) .$$

We set

$$\partial_i w(z) = (w/(x - x_i))(z) = w(z)/(x - x_i) .$$

It should be noted that the operator $w/(x - x_i)$ is completely independent of the x in the denominator; the denominator $(x - x_i)$ is purely symbolic.

THEOREM 5.13

Let L be a bounded, n-linear operator. Let $x_1, x_2, \ldots, x_n \in X$, and set

$$w(x) = L\left(x - x_1, x - x_2, \ldots, x - x_n\right) .$$

Then $w'(x_0) = \sum_{i=1}^{n} w(x_0)/(x - x_i)$ and, in particular, $w'(x_i) = w(x_i)/(x - x_i) = \partial_i w(x_i)$.

PROOF Let x_0 be a fixed point of X. Then, using the multilinearity and boundedness of L,

$$\left\| w(x_0 + \Delta x) - w(x_0) - \sum_{i=1}^{n} \frac{w(x_0)}{(x - x_i)}(\Delta w) \right\|$$

$$= \left\| L(x_0 - x_1 + \Delta x, x_0 - x_2 + \Delta x, \ldots, x_0 - x_n + \Delta x) \right.$$

$$- L(x_0 - x_1, \ldots, x_0 - x_n)$$

$$\left. - \sum_{i=1}^{n} L(x_0 - x_1, \ldots, x_0 - x_{i-1}, \Delta x, x_0 - x_{i+1}, \ldots, x_0 - x_n) \right\|$$

$$\leq \sum_{k=2}^{n} M_k \|\Delta x\|^k = o(\|\Delta x\|),$$

where each M_k is a positive constant arising from $\|L\|$ and from the norms $\|x_0 - x_i\|$, $i = 1, 2, \ldots, n$. ∎

One can speak also of higher order Fréchet derivatives. If $f : X \to Y$ and if f' exists on an open neighborhood V of x_0 in X, then $f''(x_0) = (f')'(x_0)$ is a linear operator on X into $\mathcal{L}_1[X, Y]$ for which

$$\left\| f'(x_0 + \Delta x) - f'(x_0) - f''(x_0)(\Delta x) \right\| = o(\|\Delta x\|).$$

Thus, $f''(x_0) \in \mathcal{L}_1[X, \mathcal{L}_2[X, Y]]$ and, since $\mathcal{L}_1[X, \mathcal{L}_1[X, Y]]$ is isometric to $\mathcal{L}_2[X, Y]$, it follows that $f''(x_0)$ can be considered a bilinear operator on X into Y which is usually not symmetric. In general, $f^{(n)}(x_0)$, the nth Fréchet derivative of f at x_0, is a linear operator on X into $\mathcal{L}_{n-1}[X, Y]$; so that $f^{(n)}(x_0)$ can be considered as belonging to $\mathcal{L}_n[X, Y]$.

Some examples of Fréchet derivatives are instructive. Let $X = Y = R^n$, the (real) Euclidean n-space. Then, if $f : X \to Y$ and for $(x_1, x_2, \ldots, x_n) \in X$, $f(x_1, x_2, \ldots, x_n) = (y_1, y_2, \ldots, y_n)$, where $y_i = f_i(x_1, x_2, \ldots, x_n)$, $i = 1, 2, \ldots, n$, each f_i is a real-valued function of n real variables. It can then be shown that if each f_i has continuous first partial derivatives on some open set V

in X, then $f'(x)$ exists on V and is given by the matrix

$$f'(x) = \begin{bmatrix} \frac{\partial f_1}{\partial x_1}(x) & \frac{\partial f}{\partial x_2} & \cdots & \frac{\partial f_1}{\partial x_n}(x) \\ \frac{\partial f_2}{\partial x_1}(x) & \frac{\partial f_2}{\partial x_2}(x) & \cdots & \frac{\partial f_2}{\partial x_n}(x) \\ \vdots & & & \\ \frac{\partial f_n}{\partial x_1}(x) & & \cdots & \frac{\partial f_n}{\partial x_n}(x) \end{bmatrix}$$

$$= \left(\frac{\partial f_i}{\partial x_j}(x) \right), \qquad i, j = 1, 2, \ldots, n \, ,$$

which is the gradient of f at x. Analogously, if each f_i has continuous second partials on some neighborhood U of x, then $f''(x)$ is given by the three-way matrix

$$f''(x) = \frac{\partial^2 f_i(x)}{\partial x_j \partial x_k}, \quad i, j, k = 1, 2, \ldots, n \, ,$$

which is the Hessian of f at x.

More generally, one can show that if L is a bounded, n-linear operator on X, then $L^{(k)}(x)$ is a bounded, k-linear operator on X.

Let $c_1, c_2, \ldots, c_n$ be points of a Banach space X. The *interpolation problem* is that of finding, for each sequence $\{x_1, x_2, \ldots, x_n\}$ of distinct points of X, a polynomial operator p which interpolates $\{c_1, c_2, \ldots, c_n\}$ at $\{x_1, x_2, \ldots, x_n\}$, so that $p(x_i) = c_i$. We shall prove that there always exists a polynomial of degree $(n - 1)$ which solves the interpolation problem.

To this end, let L be a bounded n-linear operator in $\mathcal{L}_n[X]$; let $x_1, x_2, \ldots, x_n$ be distinct points of X and let $w(x) = L(x - x_1, x - x_2, \ldots, x - x_n)$. Then w is a polynomial of degree n mapping X into X, and

$$\frac{w(x)}{(x - x_i)} = \partial_i w(x) = L(x - x_1, x - x_2, \ldots, x - x_{i-1}, x - x_{i+1}, \ldots, x - x_n) \, ,$$

is a polynomial of degree $(n - 1)$ which maps X into $\mathcal{L}_1[X]$. We have shown that $w'(x) = \sum_{i=1}^{n} w(x)/(x - x_i)$, so that $w'(x_i) = w(x_i)/(x - x_i) = \partial_i w(x_i)$ is a linear operator. Thus, should $w'(x_i)$ be nonsingular for $i = 1, 2, \ldots, n$, then since $l_i(x) = [w'(x_i)]^{-1} w(x)/(x - x_j)$, l_i would be a linear and operator-valued function having the property

$$l_i(x_j) = \delta_{ij} I.$$

Furthermore, for each $x_0 \in X$, it is easily seen that $[l_i(x)](x_0) = l_i(x)x_0$ is a polynomial of degree $(n - 1)$. That is, we have proved:

THEOREM 5.14

If there exists an n-linear operator L such that $[w'(x_i)]^{-1}$ exists for each $i = 1, 2, \ldots, n$, where

$$w(x) = L(x - x_1, x - x_2, \ldots, x - x_n),$$

then the Lagrange polynomial $y(x)$ of degree $(n - 1)$ given by

$$y(x) = \sum_{i=1}^{n} l_i(x) c_i \left(= \sum_{i=1}^{n} l_i(x) f(x_i) \right),$$

where $l_i(x) = [w'(x_i)]^{-1} w(x)/(x - x_i) = [w'(x_i)]^{-1} \partial_i w(x)$, solves the interpolation problem (interpolates the function f at the n distinct points $x_1, x_2, \ldots, x_n$ of X).

Thus, to solve the interpolation problem, it is enough to prove that such an n-linear operator exists. It would actually suffice to prove the existence of a family $\{L_1, L_2, \ldots, L_n\}$ of n-linear operators having the property that $[w_i'(x_i)]^{-1}$ exists for $i = 1, 2, \ldots, n$, where $w_i(x) = L_i(x - x_1, x - x_2, \ldots, x - x_n)$. If this were the case, we could take

$$y(x) = \sum_{i=1}^{n} [w_i'(x_i)]^{-1} \frac{w_i(x)}{(x - x_i)} (c_i)$$

as our interpolating polynomial. We shall prove the existence of such a family of L_is.

THEOREM 5.15

Let $x_1, x_2, \ldots, x_n$ be distinct points of a Banach space X. Then for each $i = 1, 2, \ldots, n$ there exists an n-linear operator L_i for which $[w_i'(x_i)]^{-1}$ exists, where

$$w_i(x) = L_i(x - x_1, x - x_2, \ldots, x - x_n).$$

Furthermore, the L_is can be chosen so that $w_i'(x_i) = I$, where I is the identity operator in $\mathcal{L}_1[X]$.

PROOF We start with $i = 1$. We must produce an n-linear operator L_1 for which $w_1'(x_1)$ exists and is nonsingular, where

$$w_1(x) = L_1(x - x_1, x - x_2, \ldots, x - x_n).$$

Recall that if such an L_1 exists, then

$$w_1'(x_1) = \frac{w_1(x_1)}{(x - x_1)} = \partial_1 w_1(x_1)$$

$$= L_1(\cdot, x_1 - x_2, x_1 - x_3, \ldots, x_1 - x_n) ,$$

which belongs to $\mathcal{L}_1[X]$. Also, $L_1 : X^{n-1} \to \mathcal{L}_1[X]$. With this in mind, let $X_{1j} = \operatorname{span}\{x_1 - x_j\}$. Since each X_{1j} $(j = 2, 3, \ldots, n)$ is one-dimensional, there exist continuous projections P_{1j} of X onto X_{1j}. Define

$$\tilde{T}_1 : X_{12} \times X_{13} \times \cdots \times X_{1n} \to \mathcal{L}_1[X]$$

by linearity, through the equation

$$\tilde{T}_1(x_1 - x_2, x_1 - x_3, \ldots, x_1 - x_n) = I .$$

Then $\tilde{T}_1$ is a bounded (continuous), $(n-1)$-linear operator in

$$\mathcal{L}_1[X_{12} \times X_{13} \times \cdots \times X_{1n}, Y] .$$

That is,

$$\left\| \tilde{T}_1\left(a_2(x_1 - x_2), a_3(x_1 - x_3), \ldots, a_n(x_1 - x_n)\right) \right\|$$

$$= \left\| a_2 a_3 \cdots a_n \tilde{T}_1(x_1 - x_2, x_1 - x_3, \ldots, x_1 - x_n) \right\|$$

$$= |a_2 a_3 \cdots a_n| \cdot \|I\|$$

$$= \frac{1}{\|x_1 - x_2\| \; \|x_1 - x_3\| \cdots \|x_1 - x_n\|}$$

$$\|a_1(x_1 - x_2)\| \cdot \|a_n(x_1 - x_n)\| ,$$

so that $\|\tilde{T}_1\| = 1/\|x_1 - x_2\| \; \|x_1 - x_3\| \cdots \|x_1 - x_n\|$.

We extend $\tilde{T}_1$ to a continuous, $(n-1)$-linear operator $T_1 : X^{n-1} \to \mathcal{L}_1[X]$ through the projections P_{1j}. That is, we define

$$T_1(y_1, y_2, \ldots, y_{n-1}) = \tilde{T}_1(P_{12} y_1, P_{13} y_2, \ldots, P_{1n} y_{n-1}) .$$

Since the projections P_{1j} are linear and continuous, it follows that T_1 is $(n-1)$-linear and continuous. In particular, the map P,

$$P : X^{n-1} \to X_{12} \times X_{13} \times \cdots \times X_{1n}$$

given by $P(y_2, y_3, \ldots, y_n) = (P_{12}y_2, P_{13}y_3, \ldots, P_{1n}y_n)$ is continuous, so that the decomposition $\tilde{T}_1 \circ P = T_1$, is continuous.

Now define the n-linear operator L_1 by

$$L_1(y_1, y_2, \ldots, y_n) = [T_1(y_2, y_3, \ldots, y_n)](y_1) \ .$$

The n-linearity of L_1 follows directly from the $(n-1)$-linearity of T_1 and the fact that T_1 is linear and operator-valued. The boundedness of T_1 is also apparent. If $P_{1k}y_k = a_k[(x_1 - x_k)/\|x_1 - x_k\|]$, then $\|P_{1k}y_k\| = |a_k|$. Thus,

$$L_1(y_1, y_2, \ldots, y_{n-1}, y_n)$$

$$= [T_1(y_2, y_3, \ldots, y_n)](y_1)$$

$$= \left[\tilde{T}_1(P_{12}y_2, P_{13}y_3, \ldots, P_{1n}y_n)\right](y_1)$$

$$= \frac{a_2 \cdot a_3 \cdots a_n}{\|x_1 - x_2\| \cdot \|x_1 - x_3\| \cdots \|x_1 - x_n\|}$$

$$\left[\tilde{T}_1(x_1 - x_2, \ldots, x_1 - x_n)\right](y_1)$$

$$= \frac{a_1 \cdot a_3 \cdots a_n}{\|x_1 - x_2\| \cdot \|x_1 - x_3\| \cdots \|x_1 - x_n\|}y_1 \ .$$

Therefore, if $K = 1/\|x_1 - x_2\| \cdot \|x_1 - x_3\| \cdots \|x_1 - x_n\|$, then

$$\|L_1(y_1, y_2, \ldots, y_n)\| = K\,|a_1| \cdot |a_2| \cdots |a_n|\,\|y_1\|$$

$$= K \cdot \|P_{12}y_2\| \cdot \|P_{13}y_3\| \cdots \|P_{1n}y_n\| \cdot \|y_1\|$$

$$\leq \bar{K}\,\|y_1\| \cdot \|y_2\| \cdots \|y_n\| \ ,$$

since each P_{1k} is a projection and $\|P_{1k}y\| = \|P_{1k}\| \cdot \|y\|$.

Now let $w_1(x) = L_1(x - x_1, x - x_2, \ldots, x - x_n)$. Since L_1 is a bounded, n-linear operator, $w_1(x)$ is differentiable and

$$w_1'(x_1) = \frac{w(x_1)}{(x - x_1)}$$

$$= L_1(\cdot, x_1 - x_2, x_1 - x_3, \ldots, x_1 - x_n)$$

$$= \tilde{T}_1(x_1 - x_2, x_1 - x_3, \ldots, x_1 - x_n)$$

$$= I.$$

Thus $w_1'(x_1)$ is a nonsingular, linear operator.

A similar line of argument proves the existence, for each $i = 1, 2, \ldots, n$, of an n-linear operator L_i for which $w_i'(x_i) = I$, where

$$w_i(x) = L_i(x - x_1, x - x_2, \ldots, x - x_n) .$$

This completes the proof of the theorem.   ∎

As a direct result of Theorem 5.15 we have:

THEOREM 5.16

The interpolation problem can always be solved by a polynomial $y(x)$ of degree $(n - 1)$ having a Lagrange representation

$$y(x) = \sum_{i=1}^{n} l_i(x)c_i ,$$

where $l_i(x) = [w_i'(x_i)]^{-1}/(x - x_i) = [w_i'(x_i)]^{-1}\partial_i w_i(x)$ and $w_i(x) = L_i(x - x_1, x - x_2, \ldots, x - x_n)$ for appropriately chosen n-linear operators $L_1, L_2, \ldots, L_n$.

In the event X is a Hilbert space with inner product (x, y), Theorem 5.15 also yields a representation theorem. Consider the projection P_{1j} of X onto X_{1j} given in the proof of Theorem 5.15. If X is a Hilbert space, then

$$P_{1j}y_j = \left(y_j, \frac{x_1 - x_j}{\|x_1 - x_j\|}\right)\frac{x_1 - x_j}{\|x_1 - x_j\|} .$$

Thus

$$L_1(y_1, y_2, \ldots, y_n) = \frac{(y_2, x_1 - x_2) \cdot (y_3, x_1 - x_3) \cdots (y_n, x_1 - x_n)}{\|x_1 - x_2\|^2 \cdot \|x_1 - x_3\|^2 \cdots \|x_1 - x_n\|^2}.$$

In particular, since $w_1'(x_1) = I$,

$$l_1(x) = I \circ \frac{w_1(x)}{(x - x_1)}$$

$$= L_1(\cdot, x - x_2, x - x_3, \ldots, x - x_n)$$

$$= \frac{(x - x_2, x_1 - x_2) \cdot (x - x_3, x_1 - x_3) \cdots (x - x_n, x_1 - x_n)}{\|x_1 - x_2\|^2 \cdot \|x_1 - x_3\|^2 \cdots \|x_1 - x_n\|^2}.$$

Analogously, one can prove that

$$l_j(x) = \left[\prod_{\substack{k=1 \\ k \neq j}}^{n} (x - x_k, x_j - x_k) \right] \left[\prod_{\substack{k=1 \\ k \neq j}}^{n} \|x_j - x_k\| \right]^{-1} I.$$

Thus we arrive at:

THEOREM 5.17
Let X be a Hilbert space with inner product (x, y) and let $c_1, c_2, \ldots, c_n$ be points of X. Then, for any distinct points $x_1, x_2, \ldots, x_n$ of X, the polynomial $y(x)$ of degree $n - 1$, given by

$$y(x) = \sum_{i=1}^{n} \frac{\pi_i(x)}{\pi_i(x_i)} c_i,$$

where

$$\pi_i(x) = \prod_{\substack{k=1 \\ k \neq i}}^{n} (x - x_k, x_i - x_k),$$

satisfies $y(x_i) = c_i$, $i = 1, 2, \ldots, n$.

This theorem is evident by inspection; however, it is interesting to note how it followed naturally from the theory of Theorems 5.15 and 5.16.

Recall the classical Hermite polynomial $y(x)$ of degree $(2n - 1)$ which interpolates a real-valued function f of a real variable at the n distinct points x_1,

$x_2, \ldots, x_n$ and for which $y'(x)$ interpolates f' at these points. This $y(x)$ is given by the formula

$$y(x) = \sum_{i=1}^{n} \left\{ H_i(x)f(x_i) + \bar{H}_i(x)f'(x_i) \right\} ,$$

where $H_i(x) = [1 - 2l_i'(x_i)(x - x_i)]l_i^2(x)$, and $\bar{H}_i(x) = (x - x_i)/l_i^2(x)$. Here $l_i(x)$ is the polynomial $w(x)/w'(x_i)(x - x_i)$ occurring in the classical Lagrange formula, and $w(x) = (x - x_1)(x - x_2) \cdots (x - x_n)$. It follows that

$$H_i\left(x_j\right) = \delta_{ij} = \bar{H}_i'\left(x_j\right) ,$$

and

$$H_i'\left(x_j\right) = 0 = \bar{H}_i\left(x_j\right), \quad \text{for } i, j = 1, 2, \ldots, n .$$

Now suppose X is a Banach space and f is a function from X into X which has a continuous Fréchet derivative at n distinct points $x_1, x_2, \ldots, x_n$ of X. Referring to Theorem 5.15, let $l_i(x) = [w_i'(x_i)]^{-1}w_i(x)/(x - x_i) = [w_i'(x_i)]^{-1}\partial_i w_i(x)$. Since $l_i(x)$ is linear and operator-valued, $l_1^2(x) = l_i(x) \circ l_i(x)$, being the composition of two linear operators, is itself linear and operator-valued. Furthermore, $l_i' : X \to \mathcal{L}_1[X, \mathcal{L}_1[X]]$ so that $[l_i'(x)](y)$ is linear and operator-valued. It is thus obvious that, for each $x \in X, l_i'(x_i)(x - x_i)$ is linear and operator-valued. We now define the Banach space analog of the above function $H_i(x)$ to be the linear operator-valued function on X:

$$H_i(x) = \left[I - 2l_i'(x_i)(x - x_i) \right] l_i^2(x) ,$$

where I is the identity in $\mathcal{L}_1[X]$. Since $l_i(x_j) = \delta_{ij}I$, it is evident that

$$H_i\left(x_j\right) = \delta_{ij}I .$$

Furthermore, we can show that $H_i'(x_j) = 0$, the zero linear operator from X to $\mathcal{L}_1[X]$, for $i, j = 1, 2, \ldots, n$. A proof of this requires some basic facts about Fréchet derivatives (see Chapter 1).

If A is a linear operator from X into Y, then $A'(x) = A$ for all $x \in X$. If $F : X \to Y$ and $F(x) = L_0$, a constant, for all $x \in X$, then $F'(x) = 0 \in \mathcal{L}_1[X, Y]$ for all $x \in X$. Let X, Y and Z be Banach spaces, and let $F : X \to Y$ and $G : Y \to Z$ be functions such that F is differentiable at x_0 and G is differentiable at $y_0 = F(x_0)$. Then GF is differentiable at x_0, and $(GF)'(x_0) = G'(y_0)F'(x_0)$. In particular, if G is linear, $(GF)'(x_0) = GF'(x_0)$. Finally,

LEMMA 5.5

Let A and B be functions from X into $\mathcal{L}_1[X]$ which are bounded, linear and operator-valued. If both A and B are differentiable at x_0 and if $F(x) = A(x)B(x)$,

then

$$F'(x_0)(x) = A(x_0) B'(x_0)(x) + A'(x_0)(x)B(x_0).$$

PROOF The proof follows directly from the continuity of A and B at x_0 and the definition of the Fréchet derivative. ∎

Now let $A_i(x) = I - 2l_i'(x_i)(x - x_i)$, $B(x) = l_i^2(x)$. Then $A_i'(x_0) = -2l_i'(x_i)$ since I and $-2l_i'(x_i)(x_i)$ are constant and $l_i'(x_i)$ is a linear operator. Using Lemma 5.5 we see that

$$B_i'(x_j)(x) = l_i(x_j) l_i'(x_j)(x) + l_i'(x_j)(x)l_i(x_j)$$

so that

$$B_i'(x_j)(x) = \begin{cases} 0 \in \mathcal{L}_1[X] & \text{if } j \neq i, \\ -2l_i'(x_i)(x) & \text{if } j = i. \end{cases}$$

But $H_i(x) = A_i(x)B_i(x)$ so that, invoking again Lemma 5.5,

$$H_i'(x_j)(x) = A_i'(x_j)(x)B_i(x_j) + A_i(x_j) B_i'(x_j)(x)$$

$$= -2l_i'(x_i)(x)l_i^2(x_j) + \left[I - 2l_i'(x_i)(x_j - x_i) \right] B_i'(x_j)(x)$$

$$= -2l_i'(x_j)(x)\delta_{ij} I$$

$$\qquad + \left[I - 2l_i'(x_i)(x_i - x_j) \right] \left[(\delta_{ij} I) l_i'(x_i)(x) + l_i'(x_j)(x) (\delta_{ij} I) \right]$$

$$= \begin{cases} 0 \in \mathcal{L}_1[X] & \text{if } j \neq i, \\ 2l_i'(x_i)(x) - 2l_i'(x_i)(x) = 0 \in \mathcal{L}_1[X] & \text{if } j = i. \end{cases}$$

That is, $H_i'(x_j) = 0 \in \mathcal{L}_1[X]$ for all $i, j = 1, 2, \ldots, n$.

If $\bar{H}_i(x)$ were a polynomial of degree $2n - 1$ from X into X for which $\bar{H}_i(x_j) = 0$ for all $i, j = 1, 2, \ldots, n$, and for which $\bar{H}_i'(x_j) = \delta_{ij} I$, then

$$y(x) = \sum_{i=1}^{n} \left\{ H_i(x)f(x_i) + f'(x_i) \bar{H}_i(x) \right\} \tag{5.6}$$

would be a polynomial of degree $2n - 1$ interpolating f at $x_1, x_2, \ldots, x_n$, with y' interpolating f' at these points. This follows directly from

$$y'(x) = \sum_{i=1}^{n} H_i'(x)f(x_i) + f'(x_i) \bar{H}_i'(x).$$

Note that since $H_i(x) \in X$ and $f'(x_i) \in \mathcal{L}_1[X]$, $f'(x_i)$ must precede $H_i(x)$ in formula (5.6). Looking at the proof of Theorem 5.15, we find it can be readily adapted to produce a $(2n - 1)$-linear operator L_i, for each $i = 1, 2, \ldots, n$, for which $[w_i'(x_i)]^{-1}$ exists and equals I, where

$$w_i(x) = L_i\,(x - x_1, x - x_1, x - x_2, x - x_2, \ldots, x - x_i, \ldots, x - x_n, x - x_n)$$

$$= L_i\left((x - x_1)^2, (x - x_2)^2, \ldots, (x - x_i), \ldots, (x - x_n)^2\right) .$$

It follows easily that

$$\bar{H}_i(x) = \left[w_i'\,(x_i)\right]^{-1} w_i(x) = w_i(x)$$

obeys the following relations:

$$\bar{H}_i\left(x_j\right) = 0 \qquad \text{for all } i, j = 1, 2, \ldots, n ,$$

$$\bar{H}_i'\left(x_j\right) = \delta_{ij} I .$$

Thus we arrive at:

THEOREM 5.18
Let $x_1, x_2, \ldots, x_n$ be distinct points of a Banach space X and let $f : X \to X$ be differentiable at $x_1, x_2, \ldots, x_n$. Then there exists a polynomial y of degree $(2n - 1)$,

$$y(x) = \sum_{i=1}^{n} \left\{H_i(x)f\,(x_i) + f'\,(x_i)\,\bar{H}_i(x)\right\} ,$$

which interpolates f at $x_1, x_2, \ldots, x_n$, with $y'(x)$ interpolating f' at these points. Furthermore,

$$H_i(x) = \left[I - 2l_i'\,(x_i)\,(x - x_i)\right]l_i^2(x) , \quad \text{and} \quad \bar{H}_i(x) = \left[w_i'\,(x_i)\right]^{-1} w_i(x) ,$$

where $w_i(x) = L_i((x - x_1)^2, (x - x_2)^2, \ldots, (x - x_i), \ldots, (x - x_n)^2)$, L_i being an appropriately chosen $(2n - 1)$-linear operator. I is the identity in $\mathcal{L}_1[X]$.

In the event X is a Hilbert space, we can obtain a simple representation of $y(x)$ in terms of inner products. First, one can show

$$\bar{H}_i(x) = \frac{\pi_i^2(x)}{\pi_i^2\,(x_i)}\,(x - x_i) ,$$

where

$$\pi_i(x) = \prod_{\substack{k=1 \\ k \neq i}}^{n} (x - x_k, x_i - x_k)$$

and (,) denotes inner product. Then, since $l_i(x) = \pi_i(x)/\pi_i(x_i)I$, it follows upon differentiation that

$$l_i'(x)(y) = \sum_{\substack{j=1 \\ j \neq i}}^{n} \frac{\pi_i(x) \cdot (y, x - x_j)}{(x - x_j, x_i - x_j) \cdot \pi_i(x_i)} I .$$

Thus

$$l_i'(x_i)(x - x_i) = \sum_{\substack{j=1 \\ j \neq i}}^{n} \frac{(x - x_i, x_i - x_j)}{(x - x_j, x_i - x_j)} I$$

and

$$H_i(x) = \left[1 - \sum_{\substack{j=1 \\ j \neq i}}^{n} \frac{(x - x_i, x_i - x_j)}{(x - x_j, x_i - x_j)} \right] \frac{\pi_i^2(x)}{\pi_i^2(x_i)} I .$$

Therefore, we arrive at:

THEOREM 5.19

Let X be a Hilbert space with inner product (x, y) and let $x_1, x_2, \ldots, x_n$ be distinct points of X. Then the polynomial of degree $2n - 1$ given by

$$y(x) = \sum_{i=1}^{n} \left\{ H_i(x) f(x_i) + f'(x_i) \bar{H}_i(x) \right\} ,$$

where

$$H_i(x) = \left[1 - \sum_{\substack{j=1 \\ j \neq i}}^{n} \frac{(x - x_i, x_i - x_j)}{(x - x_j, x_i - x_j)} \right] \frac{\pi_i^2(x)}{\pi_i^2(x_i)} I$$

and

$$\bar{H}_i(x) = \left[\frac{\pi_i^2(x)}{\pi_i^2(x_i)} \right] (x - x_i) ,$$

interpolates the function $f : X \to X$, while y' interpolates f' at $x_1, x_2, \ldots, x_n$.

5.4 Bounds of Polynomial Equations

In this section we find bounds for the solutions of polynomials by considering two cases [85].

CASE 5.1

In this we find upper and lower bounds for the number of solutions of polynomial equations using degree theory.

The operators studied are of the form $I + P$ where I is the identity operator and P is a completely continuous (i.e., compact) polynomial operator. One further hypothesis is imposed: in defining the Leray–Schauder degree of a map $I + F$, the usual procedure is to use the compactness of F to get a finite-dimensional approximation of $I + F$. More precisely, the degree of $I + F$ at 0 and relative to the closure $\bar{W}$ of a bounded open set W is defined to be the Brouwer degree of an approximation to $I + F$ in the finite-dimensional space which contains a certain ε-net in the compact set $F(\bar{W})$. The existence of the ε-net follows directly from the compactness, but the ε-net is not obtained constructively. Consequently, the finite-dimensional approximation of $I + F$ is not described explicitly and there is little hope, in most cases, of computing the degree. We impose the stronger hypothesis (see Assumptions 3 and 5) that there exists a finite-dimensional linear map from which the finite-dimensional approximation can be obtained. This extra hypothesis is esthetically unattractive, but it is satisfied in applications to integral and elliptic differential equations and it makes possible the explicit computation of the Leray–Schauder degree in many cases.

We obtain an upper bound for the number of solutions of a functional equation in a complex Banach space. This result is a fairly strict analog of the statement that a polynomial of degree n has n roots. Real solutions of a functional equation are also studied. We require (Assumption 4) that an analog of the condition that the polynomial have real coefficients be satisfied and we obtain upper and lower bounds for the number of real solutions. These bounds contain generalizations of the statement that the number of real roots of a polynomial with real coefficients is less than or equal to the degree of the polynomial and is equal (mod 2) to the degree. We show how our results can be applied to nonlinear elliptic differential equations.

Let B_0 be a Banach space over the complex numbers which is the complexification of a real Banach space B and let P_m $(m = 1, \ldots, q)$ be a continuous homogeneous polynomial of degree m from B_0 into B_0. Let P_0 denote the constant function defined by: for each $z \in B$, $P_0(z) = z_0$ where z_0 is a fixed element of B.

Assumption 1. The operator $P = \sum_{m=1}^{q} P_m$ is compact, i.e., if $\mathcal{E}$ is a bounded set in B_0, then $P(\mathcal{E})$ is compact in the space (each infinite subset of $P(\mathcal{E})$ has a limit point in B_0).

Assumption 2. There exists a constant $r_0 > 0$ such that if $\|z\| = r_0$ then

$$\|(I - P)z\| \neq 0 \,,$$

where I is the identity map on B_0.

REMARK 5.2 If Assumption 2 is not satisfied, then the equation

$$(I - P)z = 0$$

is such that for each positive number r, the equation has a solution $z = z_1$, such that $\|z_1\| = r$. ∎

LEMMA 5.6

If $\mathcal{W}$ is a bounded open set in B such that $z \in \bar{\mathcal{W}} - \mathcal{W}$ implies $(I - P)z \neq 0$, then there exists a constant $b > 0$ such that $z \subset \bar{\mathcal{W}} - \mathcal{W}$ implies

$$\|(I - P)z\| > b \,. \tag{5.7}$$

PROOF Since P is compact, then $(I - P)$ is a closed map. ∎

Assumption 3. Suppose $\mathcal{W}$ is a bounded open set in B_0 such that $z \in \bar{\mathcal{W}} - \mathcal{W}$ implies $(I - P)z \neq 0$. Let b be the constant given by Lemma 5.6. There exists a finite-dimensional (say of dimension n) subspace $\mathcal{N}$ of B_0 and a linear (continuous) map $A : B_0 \to B_0$ such that at $z \in \bar{\mathcal{W}}$ implies $AP(z) \in N$ and

$$\|AP(z) - P(z)\| < b/2 \,.$$

In order to discuss the topological degree of a map, we must consider the corresponding real linear space $B \times B$. First we obtain, corresponding to each P_m ($m = 1, \ldots, q$), a continuous homogeneous polynomial $\tilde{P}_m$ of degree m from $B \times B$ into $B \times B$. Let $\mathbf{P}(z_1, \ldots, z_m)$ denote the continuous symmetric m-linear form which is the polar form of P_m. If $z \in B_0$, then z can be represented as (x, y) where $x, y \in B$ and we have

$$P_m(z) = (\mathbf{P}(x, y), \ldots, (x, y)) \,.$$

If $(u, v) \in B \times B$, we define the projections:

$$\pi_1 : (u, v) \to (u, 0) \quad \pi_2 : (u, v) \to (0, v) \,.$$

Then it follows that the map

$$\tilde{\mathbf{P}}: ((x_1, y_1), \ldots, (x_m, y_m)) \to \pi_1 \mathbf{P} ((x_1, y_1), \ldots, (x_m, y_m))$$

$$+ \pi_2 \mathbf{P} ((x_1, y_1), \ldots, (x_m, y_m))$$

is a continuous symmetric m-linear form.

DEFINITION 5.2   *Let $\tilde{P}_m$ be the homogeneous polynomial of degree m whose symmetric m-linear form is $\tilde{\mathbf{P}}$. Let $\tilde{P}_0$ be the constant map from $B \times B$ into (x_0, y_0) where $z_0 = (x_0, y_0)$.*

Corresponding to map A, we have a map $\tilde{A} : B \times B \to B \times B$ defined as follows:

DEFINITION 5.3   *If $z = (x, y)$, then*

$$\tilde{A} : (x, y) \to \pi_1 A_z + \pi_2 A_z .$$

Now let $S = \{z \in B \mid \|z\| \le r_0\}$ where r_0 is the constant in Assumption 2. Since the natural isomorphism between B_0 and $B \times B$, i.e., $z \leftrightarrow (x, y)$ is norm-preserving, then from Assumption 2 and Assumption 3 with $\tilde{W} = S$, it follows that the Leray–Schauder degree

$$d \left(I - \sum_{m=0}^{q} \tilde{P}_m, \tilde{S}, 0 \right) , \tag{5.8}$$

where I is the identity map on $B \times B$ and $\tilde{S}$ is the sphere in $B \times B$ with center 0 and radius r_0, is defined and is equal to

$$d \left(I - \tilde{A} \sum_{m=0}^{q} \tilde{P}_m, \tilde{S} \cap \tilde{N}, 0 \right) \tag{5.9}$$

where $\tilde{N}$ is the $(2n)$-dimensional real Euclidean n-space underlying $\mathcal{N}$. Because it is more convenient to work with the maps in the complex spaces, we write (5.8) and (5.9) as $d(I - P, S, 0)$ and $d(I - AP, S \cap \mathcal{N}, 0)$, respectively.

Now we compute the explicit form of

$$(I - AP)/S \cap \mathcal{N} .$$

Let $z_1, \ldots, z_n$ be a fixed basis in $\mathcal{N}$. Then $z \in \mathcal{N}$ implies $z = a_1 z_1 + \cdots a_n z_n$. The constant $A z_0$ is given, say, by $A z_0 = a_1^0 a_1 + \cdots + a_n^0 z_n$. Hence,

$$(I - AP)(a_1 z_1 + \cdots + a_n z_n) = a_1 z_1 + \cdots + a_n z_n - a_1^0 z_1 - \cdots - a_n$$

$$-AP_1 (a_1 z_1 + \cdots + a_n z_n) - \cdots - AP_q (a_1 z_1 + \cdots + a_n z_n) .$$

Since P_m is a homogeneous polynomial of degree m, then by

$$P_m (a_1 z_1 + \cdots + a_n z_n)$$

$$= \sum_{\substack{k_1 + \cdots + k_n = m \\ k_1 \geq 0}} \left(a_1^k \cdots a_n^k \right) \mathcal{Q}_{k_1 \cdots k_n}^{(m)} (z_1, \ldots, z_n) \qquad (5.10)$$

where $\mathcal{Q}_{k_1 \cdots k_n}^{(m)} (z_1, \ldots, z_n)$ depends on $z_1, \ldots, z_n$ only and is homogeneous of degree k_i in z_i $(i = 1, \ldots, n)$. Also $\mathcal{Q}_{k_1 \cdots k_n}^{(m)}$ is continuous (see Chapter 1 also).
   Suppose

$$AP_1 (z_1) = b_1^{(1)} z_1 + \cdots + b_n^{(1)} z_n$$

$$\cdots \cdots \cdots \cdots$$

$$AP_1 (z_n) = b_1^{(n)} z_1 + \cdots + b_n^{(n)} z_n$$

$$A\mathcal{Q}_{k_1 \cdots k_n}^{(m)} (z_1, \ldots, z_n) = b_1^{(k_1 \cdots k_n)m} z_1 + \cdots + b_n^{(k_1 \cdots k_n)} .$$

Then

$$(I - AP)(z) = - \left(a_1^0 z_1 + \cdots + a_n^0 z_n \right) + a_1 z_1 + \cdots + a_n z_n$$

$$- a_1 \left(b_1^{(1)} z_1 + \cdots + b_n^{(1)} z_n \right)$$

$$- \cdots - a_n \left(b_1^{(n)} z_1 + \cdots + b_n^{(n)} z_n \right)$$

$$- a_1^q \left[b_1^{(k_1 0 \cdots 0)q} z_1 + \cdots + b_n^{(k_1 0 \cdots 0)q} z_n \right]$$

$$- \cdots .$$

Thus if the map $(I - AP)/\mathcal{N}$ is described in terms of the coefficients of $z_1, \ldots, z_n$, it has the form:

$$a_i = a_i - a_i^0 - a_1 b_1^{(i)} - \cdots - a_n b_n^{(i)} - \cdots - a_1^q b_i^{(k_1 0 \cdots 0)q} - \cdots -$$

$$= p_i (a_1, \ldots, a_n)$$

where p_i is a polynomial of degree q in $a_1, \ldots, a_n$ $(i = 1, \ldots, n)$. Let $p_i^{(q)}$ denote the homogeneous polynomial which is the sum of the terms of degree q in p_i.

THEOREM 5.20
If Assumptions 1, 2, 3 are satisfied, then $d(I - P, S, 0)$ is defined and

$$0 \le d(I - P, S, 0) \le q^n .$$

If the resultant of $p_1^{(q)}, \ldots, p_n^{(q)}$ is nonzero and if the radius of S is sufficiently large, then

$$d(I - P, S, 0) = q^n .$$

PROOF   The proof is straightforward because P is differentiable, and is left as an exercise.   ∎

REMARK 5.3   Notice that q is independent of the radius r_0 of S and of the number b. The number n depends on b in this sense: if b is very small, then the map A must in general be "fine", i.e., the dimension n of subspace $\mathcal{N}$ must be large.   ∎

THEOREM 5.21
Let C be the open set which is the component (of connectedness) of $B_0 - (I - P)(\dot{S})$ (where $\dot{S}$ is the point set boundary of S) which contains 0. Then there is a first category set $K \subset C$ such that if $v \in C - K$, then the number of distinct solutions of

$$(I - P)u = v \tag{5.11}$$

in the interior of S is equal to $d(I - P, S, 0)$. If $v \in K$, then the number of distinct solutions of (5.11) in the interior of S is infinite or is less than or equal to $d(I - P, S, 0)$.

PROOF   From the basic properties of topological degree, if $w \in C$, then

$$d(I - P, S, w) = d(I - P, S, 0) .$$

Let

$$\mathcal{H} = \{w \in C \mid w \text{ is a singular value of } I - P\}.$$

Since $I - P$ is differentiable and P is compact, then $I - P$ is a Fredholm map of index zero and by the Smale–Sard Theorem, the set $\mathcal{K}$ is first category. If $v \in C - \mathcal{K}$ and if $u_0 \in S \cap (I - P)^{-1}v$, then from the definition of regular value it follows that the differential of $I - P$ at u_0 is $1 - 1$. Hence, the topological index of $I - P$ at u_0 is $+1$. From this fact and the fact that P is compact, it follows that $S \cap (I - P)^{-1}u_0$ is finite. Thus, $d(I - P, S, 0)$ is the sum of the topological indices of the points in $S \cap (I - P)^{-1}u_0$. Since each index is $+1$, this yields the first conclusion of the theorem.

Now suppose $v \in \mathcal{K}$ and suppose $S \cap (I - P)^{-1}v$ is a finite set with more than $d(I - P, S, 0)$ elements. Call these elements $u_1, \ldots, u_t$ where $t > d(I - P, S, 0)$. Since $\{u_1, \ldots, u_t\}$ is a finite set, then each u_i $(i = 1, \ldots, t)$ is an isolated v-point and hence the topological index of $I - P$ at u_i is defined and is positive. The sum of the topological indices is greater than or equal to t and is equal to $d(I - P, S, 0)$. Contradiction. ∎

COROLLARY 5.1

Let $F: S \to B$ be a completely continuous operator such that $z \in \dot{S}$ implies

$$\|(I - \mathcal{F})z - (I-)z\| \leq \max\{\|(\mathcal{I} - P)z\|, \|(I - \mathcal{F})z\|\}.$$

Then $d(I - \mathcal{F}, S, 0)$ is defined and

$$d(I - \mathcal{F}, S, 0) = d(I - P, S, 0),$$

and if $\mathcal{F}$ is differentiable, the conclusions of Theorem 5.21 hold for $I - \mathcal{F}$.

PROOF That $d(I - \mathcal{F}, S, 0) = d(I - P, S, 0)$ follows from the invariance under homotopy of the degree. The argument in the proof of Theorem 5.21 is applicable because $\mathcal{F}$ is differentiable. ∎

Besides Assumptions 1, 2, 3, we want now to impose an assumption that corresponds to the condition that the coefficients of a polynomial are real. We have for each positive integer n,

$$P_m (a_1 z_1 + \cdots + a_n z_n) = \sum_{k_1 + \cdots + k_n = m} \left(a_1^{k_1} \cdots a_n^{k_n}\right) \mathcal{Q}_{k_1 \cdots k_n}^{(m)} (z_1, \ldots, z_n)$$

$$k_i \geq 0 \quad (i = 1, \ldots, n)$$

where $Q^{(m)}_{k_1 \cdots k_n}(z_1, \ldots, z_n)$ is a continuous homogeneous polynomial of degree k_i in z_i. If $z \in B$, then z can be represented as (x, y) where $x, y \in B$. Let z^* denote $(x, -y)$.

Assumption 4. For $m = 1, \ldots, q$ and for all $k_1, \ldots, k_n$ such that $k_1 + \cdots + k_n = m$ and for all $z_1, \ldots, z_n \in B$,

$$Q^{(m)}_{k_1 \cdots k_n}(z_1^*, \ldots, z_n^*) = \left[Q^{(m)}_{k_1 \cdots k_n}(z_1, \ldots, z_n) \right]^* \tag{5.12}$$

for all $z \in B[P_0(z)]^* = P_0(z)$, i.e., $z_0^* = z_0$.

Assumption 4 implies that $\mathbf{P}$, the polar form of P_m, is such that

$$\mathbf{P}((x, 0), \ldots, (x, 0)) \in B \times \{0\}$$

because

$$[\mathbf{P}(x, 0), \ldots, (x, 0)]^* = [P_m(x, 0)]^*$$

$$= P_m \left([(x, 0)]^* \right)$$

$$= P_m((x, 0))$$

$$= \mathbf{P}((x, 0), \ldots, (x, 0)) \, .$$

Then if j is the standard map from $B \times \{0\}$ into B, the function defined by $j\mathbf{P}((x, 0), \ldots, (x, 0))$ is a symmetric m-linear form on B.

DEFINITION 5.4   *Let P_m denote the continuous homogeneous polynomial of degree m defined by: if $x \in B$, then*

$$P_m(x) = j\mathbf{P}((x, 0), \ldots, (x, 0)) \, .$$

Let $S = j[S \cap (B \times \{0\})]$ and let $P = \sum_{m=0}^{q} P_m$. Since we will be concerned with $d(I - P, S, 0)$, we need an approximation map in the space B.

Assumption 5. There is a finite-dimensional subspace N of B and a linear continuous map $A \colon B \to B$ such that if $x \in S$, then $AP(x) \in N$ and

$$\|AP(x) - P(x)\| < b/2 \, .$$

Also $\mathcal{N}$ (the linear subspace of B_0 described in Assumption 3 which corresponds to $\tilde{W} = S$) is the complexification of N and if $z \in S$ and $z = (x, y)$, then

$$A(z) = A(x, y) = (Ax, Ay) .$$

THEOREM 5.22

If Assumptions 1, 2, 3, 4, 5 hold and if $C = j[C \cap (B \times \{0\})]$ (where C is the open set in B which is the component of connectedness of $B_0 - (I - P)(\dot{S})$ that contains 0), then there exists a first category set $K \subset C$ such that the number t of distinct solutions of

$$(I - P)x = y , \tag{5.13}$$

where $y \in C - K$, in the interior of S is such that

$$t \equiv d(I - P, S, 0) \ (mod \ 2) \tag{5.14}$$

and

$$|d(I - P, S, 0)| \le t \le d(I - P, S, 0) . \tag{5.15}$$

If $y \in K$, the set of solutions in the interior of S of the equation

$$(I - P)(u, v) = (y, 0)$$

is infinite or has q elements where $q \le d(I - P, S, 0)$.

PROOF   Since $I - P$ is a Fredholm map of index zero and the Leray–Schauder degree $d(I - P, S, 0)$ is defined (this latter follows from Assumption 2 and the definition of P) then, by the Smale–Sard Theorem and the invariance under homotopy of the degree, it follows that there is a first category set $K \subset C$ such that if $u \in C - K$, then

(i) $d(I - P, S, 0) = d(I - P, S, u)$;

(ii) $d(I - P, S, 0) = d(I - P, S, (u, 0))$;

(iii)   the set $[(I - P)^{-1}u]S$ is finite and the differential of $I - P$ at each point of this set has an inverse.   ∎

(From now on, we work entirely within S or S. So we omit "$\cap S$" or "$\cap S$" in the remainder of this discussion.) Let $w_1, \ldots, w_t = [(I - P)^{-1}u]$ and denote the differential of $(I - P)$ at w by $d(I - P)_w$. Since $d(I - P)_{w_i}$ is 1–1 $(i = 1, \ldots, t)$, then if $d(I - P)_{(w_i, 0)}$ denotes the differential of $(I - P)$ at $(w_i, 0)$, the Leray–Schauder index of $d(I - P)_{(w_i, 0)}$ is $+1$.

LEMMA 5.7

There exists an open set $\mathcal{U} \subset S$ such that if $U = j[\mathcal{U} \cap (B \times \{0\})]$ then

$$j\left\{(\bar{U} \times \{0\})\right\} \cap \left[(I - P)^{-1}(u, 0)\right] = \{w_1, \ldots, w_t\} .$$

PROOF   Since $d(I - P)_{(w_i, 0)}$ is 1–1 in a neighborhood of $(w_i, 0)$, there is an open sphere, σ_i in B_0 with center $(w_i, 0)$ such that

$$\bar{\sigma}_i \cap \left[(I - P)^{-1}(u, 0)\right] = (w_i, 0)$$

and

$$d(I - P, \bar{\sigma}_i, (u, 0)) = +1 . \tag{5.16}$$

Also the σ_i can be chosen so that $\bar{\sigma}_1, \ldots, \bar{\sigma}_t$ are pairwise disjoint. Let $\mathcal{U} = \bigcup_{i=1}^{t} \sigma_i$. Now let $\mathcal{V}$ denote the open set Int $S - \bar{\mathcal{U}}$. Then $S = \bar{\mathcal{V}} \cap \bar{\mathcal{U}}$, and $\bar{\mathcal{V}} \cap \bar{\mathcal{U}}$ contains only boundary points of $\mathcal{U}$ and $\mathcal{V}$. Therefore

$$d(I - P, S, 0) = d(I - P, S, (u, 0))$$

$$= d\left(I - P, \bar{\mathcal{V}}, (u, 0)\right) + d\left(I - P, \bar{\mathcal{U}}, (u, 0)\right) . \tag{5.17}$$

From the definition of $\mathcal{U}$ and (5.16), we have

$$d\left(I - P, \bar{\mathcal{U}}, (u, 0)\right) = t . \quad \blacksquare \tag{5.18}$$

LEMMA 5.8

$d(I - P, \bar{\mathcal{V}}, 0)$ *is a nonnegative even number.*

PROOF   It follows from Assumption 3 and the Reduction Theorem that there is a map $I - AP$ from a complex finite-dimensional Euclidean space $\mathcal{N}$ into itself such that $(u, 0) \in \mathcal{N}$ and

$$d(I - P, \bar{\mathcal{V}}, 0) = d(I - AP, \bar{\mathcal{V}} \cap \mathcal{N}, 0) = d(I - AP, \bar{\mathcal{V}} \cap \mathcal{N}, (u, 0)) .$$

By Assumption 5, $\mathcal{N}$ is the complexification of a finite-dimensional subspace N of B and $u \in N$. Let $x_1, \ldots, x_n$ be a basis for N. Since $\mathcal{N}$ is the complexification of N, then $(x_1, 0), \ldots, (x_n, 0)$ is a basis for $\mathcal{N}$. Let $z_i = (x_i, 0)$. Then $z_i = z_i^*$ and by Assumption 4,

$$Q_{k_1 \cdots k_n}^{(m)}(z_1, \ldots, z_n) = Q_{k_1 \cdots k_n}^{(m)}\left(z_1^*, \ldots, z_n^*\right)$$

$$= \left[\mathcal{Q}^{(m)}_{k_1 \cdots k_n} (z_1, \ldots, z_n) \right]^* .$$

Thus $\mathcal{Q}^{(m)}_{k_1 \cdots k_n} (z_1, \ldots, z_n)$ has the form $(v^{(m)}_{k_i}, 0)$ where $v^{(m)}_{k_i}$ denotes an element of B which depends on $m, k_1, \ldots, k_n$. Since $v^{(m)}_{k_i}$ can be expressed uniquely as

$$v^{(m)}_{k_i} = \gamma^{(m)}_1 x_1 + \cdots + \gamma^{(m)}_n x_n ,$$

where $\gamma^{(m)}_1, \ldots, \gamma^{(m)}_n$ are real, then Equation (5.10) becomes

$$P_m (a_1 z_1 + \cdots + a_n z_n) = \sum_{\substack{k_1 + \cdots + k_n = m \\ k_i \geq 0}} \left(a^{k_1}_1 \cdots a^{k_n}_n \right) \left(\gamma^{(m)}_1 x_1 + \cdots + \gamma^{(m)}_n x_n, 0 \right) .$$

Hence the map

$$a_i \rightarrow p_i (a_1, \ldots, a_n)$$

is such that each polynomial p_i has real coefficients. Since $(u, 0) \in \mathcal{N}$, then there exist unique real numbers $\eta_1, \ldots, \eta_n$ such that $u = \eta_1 x_1 + \cdots + \eta_n x_n$. Thus, $d(I - AP, \bar{\mathcal{V}} \cap \mathcal{N}, (u, 0))$ is the degree of 0 and relative to $\bar{\mathcal{V}} \cap \mathcal{N}$ of the map defined by

$$a_i \rightarrow p_i (a_1, \ldots, a_n) - \eta_i \quad (i = 1, \ldots, n)$$

which is a map described by polynomials in $a_1, \ldots, a_n$ with real coefficients. From the definition of $\bar{\mathcal{V}}$, it follows that the system of equations

$$p_i (a_1, \ldots, a_n) - \eta_i = 0 \quad (i = 1, \ldots, n) \tag{5.19}$$

has no real solutions [i.e., n-tuples $(a_1, \ldots, a_n)$ such that $a_1, \ldots, a_n$ are real] in $\bar{\mathcal{V}} \cap \mathcal{N}$. We use this fact to prove that $d(I - AP, \bar{\mathcal{V}} \cap \mathcal{N}, (u, 0))$ or $d(I - AP - (u, 0), \mathcal{V} \cap \mathcal{N}, (u, 0))$ is an even number. (The underlying idea of the proof is the same as the idea in the proof of the Index Lemma.) Let F denote the map $(I - AP - (u, 0) / \bar{\mathcal{V}} \cap \mathcal{N}$. Let

$$E = \{p \in p \in \text{Int} \{\bar{\mathcal{V}} \cap \mathcal{N}\} \text{ and } F(p) = 0\} .$$

Suppose $E_{j_1}, \ldots, j_h = \{(\xi_1, \ldots, \xi_n) \in E \mid \text{imaginary parts of } \xi_{j_1}, \ldots, \xi_{j_h} \text{ are nonnegative and the imaginary parts of the other coordinates } \xi_j \text{ are negative}\}$. Since the coefficients in the polynomials $p_i(a_1, \ldots, a_n) - \eta_i$ are real, the set

$$\left(E_{j_1, \ldots, j_h} \right)^* = \left\{ (\xi_1, \ldots, \xi_n) / \left(\xi^*_1, \ldots, \xi^*_n \right) \in E_{j_1, \ldots, j_h} \right\} ,$$

where ξ_i^* denotes the conjugate of ξ_i, is contained in E. Hence we may write

$$E = \bigcup_{i=1}^{M} (B_i \cup B_i^*)$$

where $B_1, \ldots, B_M$ are sets of the form $E_{j_1 \cdots j_h}$ and

$$B_i^* = \left\{ (\xi_1, \ldots, \xi_n) \mid (\xi_1^*, \ldots, \xi_n^*) \in B_i \right\} \tag{5.20}$$

and $B_1, \ldots, B_M, B_1^*, \ldots, B_M^*$ are pairwise disjoint. Since (5.20) has no real solutions in $\bar{V} \cap \mathcal{N}$, there exist pairwise disjoint open sets $U_1, \ldots, U_M, U_1^*, \ldots, U_M^*$ in $\bar{V} \cap \mathcal{N}$ such that

$$U_i^* = \left\{ (\xi_1, \ldots, \xi_n) \mid (\xi_1^*, \ldots, \xi_n^*) \in U_i \right\}$$

and such that $B_i \subset U_i$, $B_i^* \subset U_i^*$. By a basic property of topological degree, we have

$$d(I - AP) - ((u, 0), \bar{V} \cap \mathcal{N}, (0, 0)) = \sum_{i=1}^{M} \left[d(\bar{U}_i) + d(\bar{U}_i^*) \right]$$

where $d(\bar{U}_i) = d(F, \bar{U}_i, 0)$ and $d(\bar{U}_i^*) = d(F, \bar{U}_i^*, 0)$. To complete the proof of Lemma 5.8, it is sufficient to prove that $d(\bar{U}_i) = d(\bar{U}_i^*)$.

To prove this observe first that map F, regarded as a map from real Euclidean $(2n)$-space into itself, can be approximated on $\bigcup_{i=1}^{M} \bar{U}_i$ by a map f such that the 0-points of f are regular and such that $d(F, \bar{U}_i, 0) = d(f, \bar{U}_i, 0)$. Next extend map f to $\bigcup_{i=1}^{M} \bar{U}_i^*$ in this way: if $p^* \subset \bar{U}_i^*$, let $f(p^*) = [f(p)]^*$. That $d(f, \bar{U}_i, 0) = d(f, \bar{U}_i^*, 0)$ is a consequence of the following remarks.

Suppose $p_0, p_1, \ldots, p_{2n}$ is a set of $(2n + 1)$ points in complex Euclidean n-space. Then p_j can be written as

$$p_j = \bar{p}_j + i \bar{q}_j$$

where $\bar{p}_j, \bar{q}_j$ are points in real Euclidean n-space and if $\bar{p}_j = (x_1^{(j)}, \ldots, x_n^{(j)})$ and $\bar{q}_j = (y_1^{(j)}, \ldots, y_n^{(j)})$ then p_j may be made to correspond to the point

$$P_j = \left(x_1^{(j)}, y_1^{(j)}, \ldots, x_n^{(j)}, y_n^{(j)} \right)$$

in real Euclidean $(2n)$-space. Assume that $p_0, \ldots, p_{2n}$ are such that P_0, $P_1, \ldots, P_{2n}$ are linearly independent and let σ denote the $(2n)$-simplex determined by $P_0, P_1, \ldots, P_{2n}$. Let σ^* denote the $(2n)$-simplex determined by the points

$$\bar{P}_j = \left(x_1^{(j)}, -y_1^{(j)}, \ldots, x_n^{(j)}, -y_n^{(j)} \right) \quad (j = 0, 1, \ldots, 2n) \ .$$

(Since $P_0, \ldots, P_{2n}$ are linearly independent, it follows at once that $\bar{P}_0, \ldots, \bar{P}_{2n}$ are linearly independent.) If n is odd, σ and σ^* have opposite orientations. From the definition of f, it follows that $f(\sigma^*) = [f(\sigma)]^*$. Hence $f(\sigma)$ and $f(\sigma^*)$ also have opposite orientations. If n is even, σ and σ^* have the same orientation, and $f(\sigma)$ and $f(\sigma^*)$ also have the same orientation. This completes the proof of Lemma 5.8. ∎

Equation (5.14) and the second half of inequality (5.15) follow from Equations (5.17) and (5.18) and Lemma 5.8. The first half of inequality (5.15) follows from the facts that the index of $d(I - P)_{w_i}$ is $+1$ or -1 and that

$$d(I - P, S, 0) = \sum_{i=1}^{t} \left(\text{index of } d(I - P)_{w_i} \right) \ .$$

The proof of the last statement in Theorem 5.21 is practically the same as the proof of the corresponding statement in Theorem 5.20.

COROLLARY 5.2
If $d(I - P, S, 0)$ is odd, then the number of solutions of (5.12) is nonzero.

REMARK 5.4 If $I - \mathcal{F}$ is such that

$$I - \mathcal{F} : B \to B$$

and $\mathcal{F}$ is completely continuous and differentiable and if appropriate forms of Assumptions 4 and 5 are imposed (e.g., since $I - \mathcal{F}$ can be represented by a power series we can require that each term of the power series, which is a homogeneous polynomial P_m, be completely continuous and satisfy Assumption 4 and we can require that $F = \mathcal{F}/B$ [obtained by applying Assumption 4 to each P_m, just as P was obtained] satisfy the condition that P satisfies in Assumption 5), then Theorem 5.21 applies to $I - \mathcal{F}$. ∎

THEOREM 5.23

If $P = \sum_{1 \leq m \leq q:m \text{ odd}} P_m$ and if $d(I - P, S, 0)$ is defined where $P = \sum P_m$ and S is a sphere in B with center 0, then $d(I - P, S, 0)$ is odd. If $d(I - P, S, 0)$ is defined (and a fortiori $d(I - P, S, 0)$ is defined where $S = j\{S \cap [B \times \{0\}]\}$) there is a set K of first category such that the conclusions of Theorem 5.21 holds.

PROOF   From the hypothesis, it follows that the map $(I - AP)/N$ contains, in this case, only polynomial terms of odd degree, i.e., $(I - AP)/N$ is defined by:

$$I - AP : (x_1, \ldots, x_n) - (x'_i, \ldots, x'_n)$$

where $x_i = T_i(x_1, \ldots, x_n)$ and T_i is a polynomial in $x_1, \ldots, x_n$ with real coefficients which contains only terms of odd degree and which has no constant term. But this shows that $I - AP$ is an odd map and hence its degree is odd.   ∎

THEOREM 5.24

Suppose that P is a completely continuous operator which maps a real Banach space B into itself and is such that

$$P = \sum_{1 \leq m \leq q; m \text{ odd}} P_m$$

where each P_m is a continuous homogeneous polynomial from B into B and P satisfies the first statement in Assumption 5. Suppose there exist positive numbers M, b, r such that: if $\|x\| \geq M$, then

$$\|(I - P)x\| \geq r(\|x\|)^b .$$

Then for each $y \in B$, there exists $x \in B$ such that

$$(I - P)x = y .   \tag{5.21}$$

PROOF   Let $y \in B$ and let $\|y\| = R_1$. Then there exists $R_2 > 0$ such that if $\|x\| > R_2$, then

$$r\|x\|^q > R_1 .   \tag{5.22}$$

From (5.22) it follows that $d(I - P, S, y)$ is defined. From (5.22) and invariance under homotopy, it follows that

$$d(I - P, S, y) = d(I - P, S, 0) .$$

But $d(I - P, S, 0)$ is odd and therefore nonzero. ∎

Example 5.2

We show that the Dirichlet problem for a class of nonlinear elliptic equations can be formulated in terms of the functional equations studied earlier. For simplicity, we describe the case of two independent variables, but more general cases can be considered.

We study the equation

$$a(x, y)z_{xx} + b(x, y)z_{xy} + c(x, y)z_{yy} = p\left(x, y, z, z_x, z_y\right) + \psi(x, y) \quad (5.23)$$

where $\psi \in C_\mu(\bar{D})$ and $p(x, y, \xi, \eta, \zeta)$ is a polynomial in ξ, η, ζ with coefficients with are μ-Hölder continuous functions from R^2 into the reals and $p(x, y, 0, 0, 0) = 0$. Let ϕ be a fixed element of $C_{2+\mu}(\bar{D}')$, $\rho(x, y) \in C_\mu(\bar{D})$, and let $w(\rho, \phi)$ be the solution in $C_{2+\mu}(\bar{D})$, of the Dirichlet problem for

$$a(x, y)z_{xx} + b(x, y)z_{xy} + c(x, y)z_{yy} = \rho(x, y) .$$

Then (5.23) may be rewritten as:

$$\rho(x, y) - p\left[x, y, w(\rho, \phi), \frac{\partial}{\partial x}w(\rho, \phi), \frac{\partial}{\partial y}w(\rho, \phi)\right] = \psi(x, y) \quad (5.24)$$

and by the Schauder Theorem, the map P_ϕ described by

$$P_\phi : \rho(x, y) \rightarrow p\left[x, y, w(\rho, \phi), \frac{\partial}{\partial x}w(\rho, \phi), \frac{\partial}{\partial y}w(\rho, \phi)\right]$$

maps bounded sets in $C_\mu(\bar{D})$ into bounded sets in $C_{1+\mu}(\bar{D})$ and thus is a completely continuous map from $C_\mu(\bar{D})$ into $C_\mu(\bar{D})$.

Thus, the Dirichlet problem becomes the problem of solving the fundamental equation $(I - P_\phi)\rho = \psi$ in $C_\mu(\bar{D})$. Clearly this equation can be extended to an equation in the complexification of $C_\mu(\bar{D})$.

Suppose that Assumption 2 is satisfied and that b is the positive constant given in Lemma 5.6 for the set $\bar{W}$. It remains to show that Assumption 5 is satisfied. To show this, let $R = [a_1, b_1] \times [a_2, b_2]$ be a closed rectangle which contains $\bar{D}$. Assume that $\bar{D}$ is nice enough so that for each $f \in C_\mu(\bar{D})$ there exists $F \in C_\mu(R)$ such that $F/\bar{D} = f$ and such that $\|F\|_\mu \leq M\|f\|_\mu$ where M is a positive constant that is independent of f and that a similar condition holds for $f \in C_{1+\mu}(\bar{D})$. Function F can be approximated uniformly on R by a polynomial $U(x, y)$. The map A is defined by $A: f \rightarrow u = U/\bar{D}$. From the definition of U, it follows that

A is additive. Also the elements of $P_\phi(\bar{W})$ are a bounded set of equicontinuous functions. Hence there is an integer n such that for all $f \in P_\phi(\bar{W})$, $A(f)$ is a polynomial of degree not exceeding n, i.e., $A[P_\phi(\bar{W})]$ is contained in a finite-dimensional linear subspace. Finally it must be shown that A is bounded and hence continuous. This follows from the fact that the first derivatives of f are uniformly approximated by the first derivatives of U and the uniform approximation is independent of f for $f \in P_\phi(\bar{W})$ because $P_\phi(\bar{W})$ is contained in a bounded set in $C_{1+\mu}(\bar{D})$.

CASE 5.2
In this case we find computable bounds for the solutions of polynomial equations by refining and improving results already in the literature.

Consider the polynomial

$$P(z) = z^n + a_{n-1}z^{n-1} + \cdots + a_1 z + a_0 \tag{5.25}$$

where $a_0, a_1, \ldots, a_{n-1}$ are complex numbers. For every zero z of $P(z)$ we have

$$|z| \leq \max\{|a_0|, 1 + |a_1|, \ldots, 1 + |a_{n-1}|\}, \tag{5.26}$$

$$|z| \leq \max\{1, |a_0| + |a_1| + \cdots + |a_{n-1}|\}, \tag{5.27}$$

$$|z| \leq r, \tag{5.28}$$

where r is the unique positive zero of

$$G(z) = z^n - |a_{n-1}|z^{n-1} - \cdots - |a_1|z - |a_0|. \tag{5.29}$$

It is also known that

$$|z| \leq L_m \tag{5.30}$$

where

$$L_m = \max\{S_m, 1 + |a_{m+1}|, 1 + |a_{m+2}|, \ldots, 1 + |a_{n-1}|\},$$

$$m = 0, 1, 2, \ldots, n-1$$

and S_m is the unique positive zero of

$$G_m(z) = z^{m+1} - |a_m|z^m - |a_{m-1}|z^{m-1} - \cdots - |a_1|z - |a_0|. \tag{5.31}$$

Finally we have

$$|z| \leq \max\{1, |a_0| + \cdots + |a_m|, 1 + |a_{m+1}|, \ldots, 1 + |a_{n-1}|\} \qquad (5.32)$$

where $m \in \{0, 1, 2, \ldots, n - 1\}$.

Note that for $m = 0$ and $m = n - 1$, (5.30) yields (5.26) and (5.28), whereas for $m = 0$ and $m = n - 1$ (5.32) yields (5.26) and (5.27). These are classical results due to Cauchy and Euler.

In this case we improve further the above upper bounds. We also provide a lower bound for the absolute value of the solution z.

Finally we show how to use the above estimates to bound the solutions of polynomial equations in a normed space X.

PROPOSITION 5.1

Let $k \in \{1, 2, \ldots, n - 1\}$ be fixed. Assume:

$$0 \neq |a_k| < \bar{N}_k \qquad (5.33)$$

where

$$\bar{N}_k = \begin{cases} 1 + N_k, & k = 1, 2, \ldots, n - 2 \\ r_{n-1}, & k = n - 1 \end{cases}$$

$$N_i = \max\{|a_{i+1}|, |a_{i+2}|, \ldots, |a_{n-1}|\}, \quad i = 1, 2, \ldots, n - 2,$$

r_k is the unique positive zero of

$$P_k(z) = (q_k - 1)|a_k| z^k - |a_{k-1}| z^{k-1} - \cdots - |a_1| z - |a_0|$$

and q_k is such that

$$1 < q_k \leq \frac{\bar{N}_k}{|a_k|}, \quad k = 1, 2, \ldots, n - 1. \qquad (5.34)$$

Then every zero of the complex polynomial

$$P(z) = z^n + a_{n-1} z^{n-1} + \cdots + a_1 z + a_0$$

satisfies

$$|z| \leq M_k \qquad (5.35)$$

where

$$M_k = \begin{cases} \max\{r_k, 1+|a_{k+1}|, 1+|a_{k+2}|, \ldots, 1+|a_{n-1}|\}, & k = 1, 2, \ldots, n-2 \\ r_{n-1}, & k = n-1. \end{cases}$$

PROOF   We shall show that a complex number z satisfying $|z| > \bar{N}_k$, $|z| > r_k$, $k = 1, 2, \ldots, n-1$ cannot be a zero of P. We have

$$|P(z)| = \left| z^n + \sum_{j=0}^{n-1} a_j z^j \right| \geq |z|^n - \sum_{j=0}^{n-1} |a_j| \, |z|^j$$

$$= |z|^n - q_k \, |a_k| \cdot |z|^k + q_k \, |a_k| \, |z|^k - \sum_{j=0}^{k} |a_j| \cdot |z|^j - \sum_{j=k+1}^{n-1} |a_j| \, |z|^j$$

$$\geq \left(|z|^n - q_k \, |a_k| \, |z|^k \right) + P_k(|z|) - N_k \sum_{j=k+1}^{n-1} |z|^j$$

$$\geq \left(|z|^n - |z|^{k+1} \right) + P_k(|z|) - N_k \sum_{j=k+1}^{n-1} |z|^j$$

(by the choice of q_k)

$$= \left[|z|^n - |z|^{k+1} \right] \left[|z| - \bar{N}_k \right] / (|z| - 1) + P_k(|z|) .$$

Since $|z| > r_k$ and r_k is the unique positive zero of P_k, we have $P_k(|z|) > 0$ and thus $|P(z)| > 0$.   ∎

REMARK 5.5   Proposition 5.1 can sometimes be applied even if (5.33) is violated. Consider the real equation

$$z^3 + z^2 - 4z + 1 = 0 . \tag{5.36}$$

Here, $a_0 = a_2 = 1$, $a_1 = -4$ and for $k = 1$, $N_1 = |a_2|$ so (5.33) is violated. Set $z = x + 1$ then (5.36) is equivalent to

$$z = x + 1$$

$$x^2 + 4x^2 + x - 1 = 0 \qquad (5.37)$$

and (5.33) is now satisfied in (5.37). ∎

We now compare L_k and M_{k+1}.

PROPOSITION 5.2

Assume:

(a) $i \in \{1, 2, \ldots, n-2\}$ *is fixed;*

(b)

$$|a_{i+1}| < \tilde{N}_{i+1} \qquad (5.38)$$

and

(c)

$$\left. \begin{array}{c} \dfrac{1 + |a_{i+1}|}{|a_{i+1}|} < q_{i+1} \leq \dfrac{\tilde{N}_{i+1}}{|a_{i+1}|} \\[3mm] 1 < q_{n-1} = \dfrac{r_{n-1}}{|a_{n-1}|} \end{array} \right\} \qquad (5.39)$$

Then

$$M_{m+1} \leq L_m, \quad m = 0, 1, 2, \ldots, n-2$$

and

$$r_{n-1} = r .$$

PROOF   We have

$$G_k(z) = z^{k+1} - |a_k| z^k - |a_{k-1}| z^{k-1} - \cdots - |a_0|$$

$$P_{k+1}(z) = (q_{k+1} - 1) |a_{k+1}| z^{k+1} - |a_k| z^k - |a_{k-1}| z^{k-1} - \cdots - |a_0| .$$

Set $z = s_k$ then $G_k(s_k) = 0$ and

$$P_{k+1}(s_k) = \left[(q_{k+1} - 1) |a_{k+1}| - 1 \right] s_k^{k+1} > 0$$

by the choice of $q_{k+1} r_{m+1} < s_m \Rightarrow M_{m+1} \leq L_m, m = 0, 1, 2, \ldots, n-2$. Also $G(r_{n-1}) = 0$ by (5.38) and (5.39), so $r_{n-1} = r$. ∎

Note that a comparison between the bounds given in (5.30) and (5.32) makes sense only if the degree of the polynomials G_m is the same with the degree of P_k.

Example 5.3

(a) Consider the polynomial

$$P(z) = z^3 + 5z^2 + z + 7 .$$

Let $m = 0$ then the bound given by (5.26), (5.29), and (5.32) is 7. The bound given by (5.27) is 13, whereas the bound given by (5.35) for $k = 1$, $q_1 = 6$ is

$$M_1 = \max\{1.4, 6\} = 6 .$$

(b) Consider the polynomial

$$P(z) = z^2 + z + 1 .$$

The bound on $|z|$ given by (5.28) is

$$r = 1.6180339$$

whereas for $k = 1$, $q_1 = 1.6180339$

$$M_1 = r_1 = r .$$

REMARK 5.6   The estimates obtained above can also be used to find upper and lower bounds for the norm of the solutions of polynomial equations in a normed space X.

Let

$$B(x) = B_n x^n + B_{n-1} x^{n-1} + \cdots + B_1 x + B_0 \qquad (5.40)$$

where the B_ℓ, $\ell = 1, 2, \ldots, n$ is a bounded ℓ-linear operator on X and $B_0 \in X$ is fixed.

Set

$$C(x) = \|B_n\| \cdot \|x\|^n + \|B_{n-1}\| \, \|x\|^{n-1} + \cdots + \|B_1\| \cdot \|x\| + \|B_0\|$$

and

$$D(x) = \frac{1}{\|B_n\|} C(x) \quad \text{if } \|B_n\| \neq 0 . \quad \blacksquare$$

We can now apply Propositions 5.1 and 5.2 on $D(x)$ to obtain bounds on the norm of the solution x of (5.40).

5.5 Representations of Multilinear and Polynomial Operators on Vector Spaces

That linear functions (operators) on vector spaces are representable by matrices relative to chosen bases is familiar. By generalizing the notion of a matrix to that of a "matroid", we can extend the representation theory of linear functions to multilinear functions and polynomial functions. In Section 5.2 we worked on a Hilbert space. Here we work on general linear spaces. This is attributed to [8] and is adjusted appropriate here.

DEFINITION 5.5   *For each* $n \in \mathbf{N} = \{0, 1, 2, \ldots\}$ *denote by* $I_1, \ldots, I_n, J, K$ *any* $n + 2$ *sets. Then an* n-matroid m *of type* $(I_1, \ldots, I_n; J)$ *on* K *is a family*

$$m = \left(m_j^{i_1, \ldots, i_n} \right)_{j \in J}^{(i_1, \ldots, i_n) \in I_1 \times \cdots \times I_n} ,$$

of elements $m_j^{i_1, \ldots, i_n} \in K$, *indexed by the product set* $I_1 \times \cdots \times I_n \times J$.

Thus, an n-matroid of type $(I_1, \ldots, I_n; J)$ on K is no more than a matrix of type $(H; J)$ on K, with a product index set $H = I_1 \times \cdots I_n$. In particular, a 1-matroid of type $(I; J)$ on K is a matrix of type (I, J) on K is a matrix of type (I, J) on K. Note that we do allow n to be 0 since a 0-matroid is going to correspond to a constant ("0-linear") function; but then the index set $I_1 \times \cdots \times I_n$ is vacuous, and in effect a 0-matroid is simply a list $(m_j)_{j \in J}$ on K.

If K is a set with a distinguished ("zero") element, then a matroid m is said to be of *finite support* if the subfamily of its nonzero elements is finite; m is of *finite support over J* [alias, *of type* $(I_1, \ldots, I_n; (J))$] if for each fixed index

$$(i_1, \ldots, i_n) \in I_1 \times \cdots \times I_n$$

the subfamily $(m_j^{i_1, \ldots, i_n})_{j \in J}$ is of finite support. The *degree* n of an n-matroid m is denoted by $d(m)$. A *submatroid* of m is a restriction m' of the family m to a subdomain $I_1' \times \cdots \times I_n' \times J'$ of the index domain of m. An n-matroid of type $(I_1, \ldots, I_n; J)$ for which each set $I_1, \ldots, I_n, J$ is finite, is *finite*; it is *square* [resp. *oblong*] if $I_1, \ldots, I_n, J$ all [resp. all but J] have the same cardinal; a square matroid is *trivial* if this cardinal is 1. The set of all n-matroids of type $(I_1, \ldots, I_n; J)$ on K is denoted by

$$K^{I_1 \times \cdots \times I_n \times J} ;$$

the subset of these matroids of finite support over J is denoted by $K^{I_1 \times \cdots I_n \times (J)}$.

DEFINITION 5.6   An *n-matroid m of type* $(I, \ldots, I; J) = (I^n; J)$, *of type* $(I, \ldots, I; (J)) = (I^n, (J))$, *on K is said to be pensymmetric if for each permutation* σ *of* $\{1, 2, \ldots, n\}$ *we have* $m = \bar{\sigma} m$, *where*

$$\bar{\sigma} m = \left(m_j^{i_{\sigma 1}, \ldots, i_{\sigma n}} \right)_{j \in J}^{(i_1, \ldots, i_n) \in I^n} .$$

Thus "pensymmetric" means "almost symmetric"—specifically, "symmetric over only the upper indices"; hence every 1-matroid (matrix) is trivially pensymmetric and every 0-matroid is vacuously pensymmetric.

Now if A is a ring, and K and A-module, then we shall assume that both the sets $K^{I_1 \times \cdots \times I_n \times J}$ and $K^{I_1 \times \cdots \times I_n \times (J)}$ have been given their natural pointwise structure of an A-module. Then each n-matroid m of type $(I^n; J)$ on K uniquely determines a pensymmetric n-matroid $t_n(m)$ of similar type by the construction

$$t_n(m) = \frac{1}{n!} \Sigma_\sigma \bar{\sigma} m \quad (\text{if } n \geq 2), \; = m \; (\text{if } n = 0 \text{ or } 1), \qquad (5.41)$$

whenever the ring A contains $1/n!$ for $n \geq 2$. The matroid $t_n(m)$ is the *pensymmetric part* of m, and the operator t_n is evidently a linear projection, the *pensymmetrizer* (of degree n), on the A-module $K^{I^n \times J}$. The image $t_n(K^{I^n \times J})$ of t_n is a submodule of $K^{I^n \times J}$ called the module of pensymmetric n-matroids of type $(I^n; J)$ on K.

For the remainder of this section we shall be interested in matroids only over $K = K$ where $\mathbf{K}$ is the scalar field of reals or complexes. We shall denote by $\mathbf{M}_n(I; (J))$, or simply $\mathbf{M}_n$, the vector space of n-matroids of type $(I^n; (J))$ (of finite support over J), on $\mathbf{K}$, and by $\mathbf{Y}_n(I; (J))$, or simply $\mathbf{Y}_n$, its vector subspace $t_n(\mathbf{M}_n)$ of pensymmetric n-matroids. We have $\mathbf{M}_0 = \mathbf{Y}_0$ and $\mathbf{M}_1 = \mathbf{Y}_1$, of course.

It is well known that the vector space $L_1(E, F)$ of linear functions from one vector space E to another F is isomorphic to the vector space $\mathbf{K}^{I \times (J)}$ of matrices of type $(I, (J))$ (of finite support over J) on $\mathbf{K}$ (where card $I = \dim E$, card $J = \dim F$), relative to chosen bases $\mathbf{e} = \{\mathbf{e}^i \mid i \in I\}$ in E and $\mathbf{f} = \{\mathbf{f}^j \mid j \in J\}$ in F. Indeed, the assignment

$$\mu_1 : u \mapsto m(u) : L_1(E, F) \to \mathbf{M}_1 ,$$

where, for each $u \in L_1(E, F)$, $m(u)$ is the 1-matroid (matrix) of type $(I; (J))$ with elements $m_j^i \in \mathbf{K}$ determined uniquely by the relations

$$u(\mathbf{e}^i) = \sum_{j \in J} \mathbf{f}^j m_j^i \quad (i \in I), \qquad (5.42)$$

is the isomorphism in question.

We now consider the relation between the vector space $L_n(E, F)$ of n-linear ("multilinear, of degree n") functions $E^n \to F$ and the vector space $\mathbf{M}_n$ of n-matroids of type $(I^n; (J))$ on $\mathbf{K}$, relative to the chosen bases $\mathbf{e}$ in E and $\mathbf{f}$ in F, in the remaining cases where $n \neq 1$. First suppose that $n \geq 2$. Then it is familiar that each n-linear function $u : E^n \to F$ factors into the composition $u = 1 \circ \otimes_n$ of a unique linear function $1 : \otimes^n E \to F$ and the canonical n-linear function $\otimes_n : E^n \to \otimes^n E$, where $\otimes^n E$ is the nth-tensor power of E, spanned by its canonical basis

$$\left\{ \otimes_n \left(\mathbf{e}^{i_1}, \ldots, \mathbf{e}^{i_n} \right) \mid (i_1, \ldots, i_n) \in I^n \right\} .$$

Thus u is uniquely determined by the family of the values $1 \circ \otimes_n (\mathbf{e}^{i_1}, \ldots, \mathbf{e}^{i_n})$ of 1 on the canonical basis of $\otimes^n E$, and each such value is a linear combination

$$1 \circ \otimes_n \left(\mathbf{e}^{i_1}, \ldots, \mathbf{e}^{i_n} \right) = \sum_{j \in J} \mathbf{f}^j m_j^{i_1, \ldots, i_n} , \tag{5.43}$$

say, of the elements of the bases $\mathbf{f}$ in F. Here the coefficients $m_j^{i_1, \ldots, i_n}$ are uniquely determined scalars in $\mathbf{K}$, zero for all but finitely many values of $j \in J$, for each fixed index $(i_1, \ldots, i_n) \in I^n$. Hence these relations (5.42) assign to each n-linear function u a unique n-matroid $m(u)$ of type $(I^n; (J))$ on $\mathbf{K}$, with elements $m_j^{i_1, \ldots, i_n}$. The values of the function u at any point $(_1x, \ldots, _n x) \in E^n$ are then given by the relation (5.44) ($n \geq 2$) below. Denoting the assignment $u \mapsto m(u)$ by μ_n, it is not difficult to show that μ_n is a linear injection $L_n(E, F) \to \mathbf{M}_n$ of vector spaces. Conversely, to each n-matroid $m \in \mathbf{M}_n$ there corresponds a unique function $u : E^n \to F$ defined by setting, for each $(_1x, \ldots, _n x) \in E^n$,

$$u (_1x, \ldots, _n x) = \sum_{j \in J} \mathbf{f}^j \left[\sum_{(i_1, \ldots, i_n)} {_1x_{i_1}} \ldots {_n x_{i_n}} m_j^{i_1, \ldots, i_n} \right] , \tag{5.44}$$

where each vector $_r x = \sum_{i_r \in I} {_r x_{i_r}} \mathbf{e}^{i_r} \in E$ (the sum containing only a finite number of nonzero terms, by definition of a basis). This function u is plainly n-linear; hence the assignment μ_n is surjective, and hence an isomorphism of vector spaces.

We may dispose of the remaining case where $n = 0$ by noting that $L_0(E, F)$ is the space of constant functions $E^0 \to F$, the convention being that the vacuous Cartesian product space E^0 is the scalar field $\mathbf{K}$ itself. Then for each $u \in L_0(E, F)$ and each $k \in \mathbf{K}$ the constant value of u is a vector $u(k) \in F$ and hence

$$u(k) = \sum_{j \in J} \mathbf{f}^j m_j , \tag{5.45}$$

say, where the coefficients m_j are uniquely determined scalars in $\mathbf{K}$, zero for all but finitely many values of $j \in J$. Hence the relation (5.43) assigns to each constant function $u \in L_0(E, F)$ a unique 0-matroid $m(u) = (m_j)_{j \in J} \in \mathbf{M}_0$. The rest of the argument in this case now follows in an obviously similar way to that in the general case.

Summarizing these observations gives us the theorem:

THEOREM 5.25

For each $n \in \mathbf{N}$ the assignment $\mu_n : u \mapsto m(u)$ given by the relations (5.42) and (5.43) above, relative to chosen bases in E and F, is an isomorphism of vector spaces $\mu_n : L_n(E, F) \xrightarrow{\cong} \mathbf{M}_n$ from the vector space of n-linear functions $E^n \to F$ to the vector space of n-matroids of type $(I^n; (J))$ (of finite support over J) on $\mathbf{K}$.

As a particular instance of the linear case of this theorem, we note that the vector space $L_1(\otimes^n E, F)$ of linear functions $\otimes^n E \to F$ is isomorphic (relative to chosen bases in $\otimes^n E$ and F) to the vector space T of matrices of type $(H, (J))$ on $\mathbf{K}$, where H is a set indexing the chosen basis of $\otimes^n E$. Choosing for the latter the canonical basis (noted earlier), and for F the given basis $\mathbf{f}$, it is clear that $H = I^n$ and that T is none other than the vector space $\mathbf{M}_n$ of n-matroids of type $(I^n; (J))$ on $\mathbf{K}$. Moreover, the vector spaces $L_n(E, F)$ and $L_1(\otimes^n E, F)$ are themselves isomorphic, as is well known. Thus, the isomorphism μ_n in Theorem 5.25 is none other than the composition $L_n(E, F) \xrightarrow{\cong} L_1(\otimes^n E, F) \xrightarrow{\cong} \mathbf{M}_n$ of these two isomorphisms.

The frequently occurring set $\mathbf{e}^n = \{(\mathbf{e}^{i_1}, \ldots, \mathbf{e}^{i_n}) \mid (i_1, \ldots, i_n) \in I^n\}$ may usefully be called a *multibasis* for the product space E^n of a vector space E with a basis $\mathbf{e} = \{\mathbf{e}^i \mid i \in I\}$; if $n = 0$, the convention is that $\mathbf{e}^0 = \{k\}$, where $0 \neq k \in \mathbf{K} = E^0$. If $n \geq 2$, then a multibasis for E^n is not a basis for E^n, of course. We at once have the following corollary to Theorem 5.25.

COROLLARY 5.3

An n-linear function $u : E^n \to F$ is uniquely determined (relative to chosen bases in E and F) by specifying its values at each member of a corresponding multibasis for E^n.

Turning now to the case of a symmetric n-linear function $u : E^n \to F$, it is not difficult to see that the above discussion again holds, *mutatis mutandis*, after replacing:

(i) the vector space $L_n(E, F)$ by the vector space $S_n(E, F)$ of symmetric n-linear functions $E^n \to F$,

(ii) the nth-tensor power $\otimes^n E$ by the n-symmetric tensor power $\vee^n E \cong \otimes^n E / \mathrm{Ker}(\Sigma_n)$ whose canonical basis is the set $\{\vee_n(\mathbf{e}^{i_1}, \ldots, \mathbf{e}^{i_n}) \mid (i_1, \ldots, i_n) \in$

I^n} of symmetric tensor products, where

$$\Sigma_n = \frac{1}{n!}\Sigma_\sigma \sigma$$

is the usual symmetrizer (linear projection) operator on $\otimes^n E$, and $\vee_n = \Sigma_n \circ \otimes_n$ is the canonical symmetic n-linear function $E^n \to \vee^n E$, and

(iii) the n-matroids m and $m(u)$ by the pensymmetric n-matroids $t_n(m)$ and $t_n(m)(u)$.

Of course in the cases $n = 0$ and $n = 1$ the discussions are exactly the same because we have $S_0(E, F) = L_0(E, F)$ (vacuously), and $S_1(E, F) = L_1(E, F)$ (trivially). The details in the general case may safely be left to the reader. Writing v_n for the restriction of the assignment μ_n to $S_n(E, F) \subseteq L_n(E, F)$, we then have the following theorem.

THEOREM 5.26

For each $n \in \mathbf{N}$ the assignment $v_n : u \mapsto t_n(m)(u)$ given above, relative to chosen bases in E and F, is an isomorphism of vector spaces $v_n : S_n(E, F) \xrightarrow{\approx} \mathbf{Y}_n$ from the vector space of symmetric n-linear functions $E^n \to F$ to the vector space of pensymmetric n-matroids of type $(I^n; (J))$ on $\mathbf{K}$.

We have seen that a multibasis for E^n plays a useful role in the representation of multilinear functions. In the case of symmetric multilinear functions we need the corresponding notion of a *symmetric multibasis* for E^n. By this we mean a subset S of a multibasis $\mathbf{e}^n$ for E^n such that S is in bijection with the canonical basis of the symmetric tensor power $\vee^n E$. The two notions are the same if $n = 0$ or 1. To construct a symmetric multibasis for E^n when $n \geq 2$, simply partition the index set I^n into equivalence classes, calling two indices in I^n equivalent if and only if each is a permutation of the other. Picking out any one representative member from each equivalence class, denote the set of such representative indices by $S(I^n)$. Then the subset $S \subseteq \mathbf{e}^n$ indexed by the subset $S(I^n) \subseteq I^n$ is defined to be a symmetric multibasis for E^n. If $\dim E = d < \infty$, then {$(\mathbf{e}^{i_1}, \ldots, \mathbf{e}^{i_n}) \mid 1 \leq i_1 \leq \cdots \leq i_n \leq d$} is a symmetric multibasis for E^n, with finite cardinality equal to $\binom{d+n-1}{n}$, the dimension of $\vee^n E$. The following corollary to Theorem 5.26 is now immediate.

COROLLARY 5.4

A symmetric n-linear function $u : E^n \to F$ is uniquely determined (relative to chosen bases in E and F) by specifying its values at each member of a corresponding symmetric multibasis for E^n. The choice of the symmetric multibasis is immaterial.

DEFINITION 5.7　　*We recall that a function $p : E \to F$ from one vector space E to another F is said to be polynomial if it satisfies this condition:*

$$(\exists n \in \mathbf{N}), \ (\forall x, y \in E), (\exists z_0, z_1, \dots, z_n \in F), \ (\forall \lambda \in \mathbf{K}),$$

$$p(x + \lambda y) = \sum_{\nu=0}^{n} \lambda^{\nu} z_{\nu}. \tag{5.46}$$

Here n is the degree $d(p)$ of p if and only if $y \neq 0 \Rightarrow z_n \neq 0$, and then p is called an n-polynomial function. If p satisfies the condition: $(\exists n \in \mathbf{N}), (\forall x \in E), (\forall \lambda \in \mathbf{K}), p(\lambda x) = \lambda^n p(x)$, then p is said to be n-homogeneous; p is called an n-homogeneous polynomial function if it is both n-homogeneous and polynomial, and then $d(p) = n$. The zero function $0 : x W 0 : E \to F$ is defined to be homogeneous polynomial of arbitrary degree.

The set of all homogeneous polynomial functions $E \to F$ of fixed degree n is plainly a vector space which we shall denote by $Q_n(E, F)$. We also denote by $Q(E, F)$ the direct sum $\oplus_{n \in \mathbf{N}} Q_n(E, F)$. Again, the set of all polynomial functions $E \to F$ of every degree is plainly a vector space which we shall denote by $P(E, F)$. Then $Q(E, F)$ and $P(E, F)$ are isomorphic vector spaces; this follows from the fact that each $p \in P(E, F)$ decomposes into its unique "Fréchet sum" $p = \sum_{\nu=0}^{n} p_\nu$ of its ν-homogeneous polynomial *component* functions $p_\nu \in Q_\nu(E, F)$ $(0 \leq \nu \leq n = d(p))$ and hence may be identified with the element $(p_0, p_1, \dots, p_n, 0, 0, \dots) \in Q(E, F)$ after identifying each homogeneous component p, of p with the *homogeneous element* $(0, \dots, 0, p_\nu, 0, \dots) \in Q(E, F)$.

A crucial fact for the representation of polynomial functions is contained in the next proposition.

PROPOSITION 5.3

For each $n \geq 2$ the vector space $Q_n(E, F)$ of n-homogeneous polynomial functions $E \to F$ is naturally isomorphic to the vector space $S_n(E, F)$ of symmetric n-linear functions $E^n \to F$; if $n = 0$ or 1, then $Q_n(E, F) = S_n(E, F) = L_n(E, F)$.

The proof in the nontrivial case $n \geq 2$ may be deduced from Theorems 26.22 and 26.23 in [109]. The isomorphism in question is specified by the *polarizer operator* (of degree n) $\beta_n : Q_n(E, F) \to S_n(E, F)$. Here, for each $q \in Q_n(E, F)$ we have $\beta_n(q) = q^\vee$ where $q^\vee$ denotes the *polar* of q defined for example by the famous Mazur–Orlicz formula

$$q^\vee(_1 x, \dots, _n x) = \frac{1}{n!} \sum (-1)^{n - (z_1 + \cdots + z_n)} q\left(\varepsilon_{1\,1} x + \cdots + \varepsilon_{n\,n} x\right) \tag{5.47}$$

for all $(_1x, \ldots, _nx) \in E^n$, the summation being over all lists $(\varepsilon_1, \ldots, \varepsilon_n)$ with $\varepsilon_i = 0$ or 1. Conversely, for each $s \in S_n(E, F)$, we have $\mathcal{U}_n(s) = s'$ where s' is the *axial* of s defined by $s'(x) = s(x, \ldots, x)$ for all $x \in E$, and $\mathcal{U}_n$ is the *axializer operator* (of degree n), inverse β_n^{-1} of the polarizer operator.

Combining these facts from Proposition 5.3 with the assertions in Theorem 5.26 and Corollary 5.4 allows us to infer the following representation theorem and its corollary.

THEOREM 5.27

For each $n \in \mathbf{N}$ the composition $\eta_n = \nu_n \circ \beta_n$ of the polarizer operator β_n : $Q_n(E, F) \rightarrow S_n(E, F)$ and the assignment $\nu_n : S_n(E, F) \rightarrow \mathbf{Y}_n$ (relative to chosen basis in E and F), is an isomorphism of vector spaces $\eta_n : Q_n(E, F) \rightarrow \mathbf{Y}_n$, from the vector space of n-homogeneous polynomial functions $E \rightarrow F$ to the vector space of pensymmetric n-matroids of type $(I^n; (J))$ on $\mathbf{K}$.

COROLLARY 5.5

An n-homogeneous polynomial function $q \in Q_n(E, F)$ is uniquely determined (relative to chosen bases in E and F) by specifying the values of its polar $q^\vee = \beta_n(q) \in S_n(E, F)$ at each member of symmetric multibasis for E^n. The family of such specified values uniquely determines a pensymmetric n-matroid $m = m(q) \in \mathbf{Y}_n$ whose elements are given by the relation:

$$q^\vee \left(e^{i_1}, \ldots, e^{i_n} \right) = \sum_{j \in J} \mathbf{f}^j m_j^{i_1, \ldots, i_n} \in \mathbf{F}, \quad \left((i_1, \ldots, i_n) \in I^n \right) . \tag{5.48}$$

Conversely, if $m \in \mathbf{Y}_n$ is a pensymmetric n-matroid, then the function

$$x \mapsto \sum_{j \in J} \mathbf{f}^j \left[\sum_{(i_1, \ldots, i_n) \in I_n} x_{i_1} \ldots x_{i_n} m_j^{i_1, \ldots, i_n} \right] \tag{5.49}$$

is the n-homogeneous polynomial function $u : E \rightarrow F$ uniquely associated with m under the isomorphism $\eta_n : Q_n(E, F) \rightarrow \mathbf{Y}_n$.

Suppose we now denote by $\eta = \oplus_{n \in \mathbf{N}} \eta_n$ the isomorphism of direct sums $Q(E, F) = \oplus_{n \in \mathbf{N}} Q_n(E, F) \rightarrow \mathbf{Y} = \oplus_{n \in \mathbf{N}} \mathbf{Y}_n$ and by

$$\delta : P(E, F) \rightarrow Q(E, F)$$

the identification isomorphism $p = \sum_{\nu=0}^n p_\nu \mapsto (p_0, p_1, \ldots, p_n, 0, 0, \ldots)$ noted earlier, and by $\eta : P(E, F) \rightarrow \mathbf{Y}$ their composite isomorphism $\eta = \eta \circ \delta$. Then we may at once extend Theorem 5.27 and Corollary 5.5 to the *representation of polynomial functions* $p \in P(E, F)$ *by sequences of matroids* simply by identifying each such p with the uniquely determined, finitely nonzero sequence of pensymmetric matroids $\eta(p) = (m_0, m_1, \ldots, m_m, 0, 0, \ldots) \in \mathbf{Y}$, where $m_\nu = \eta_\nu(p_\nu)$ for each homogeneous component p_ν of p. The details are obvious and are left to the motivated reader.

Example 5.4

We illustrate Corollary 5.5 with an example of a homogeneous polynomial function of degree 2 (i.e., a quadratic operator). Illustrations involving polynomial operators of degree greater than 2 seem to be somewhat impracticable, though not impossible. We shall take $E = F = \mathbf{R}^3$, E and F both being given their natural bases in which $\mathbf{e}^1 = (1, 0, 0)$, etc. Let $q \in Q_2(\mathbf{R}^3, \mathbf{R}^3)$; then the pensymmetric 2-matroid $m(q)$ associated with q relative to the natural bases in E and F, under the isomorphism $\eta_2 : Q_2 \twoheadrightarrow \mathbf{Y}_2$ given in Theorem 5.26 above, has elements $m_j^{i,k}$ where, for each $j \in \{1, 2, 3\}$ we have $m_j^{i,k} = m_j^{k,i}$ for each $(i, k) \in \{1, 2, 3\}^2$. To determine q we specify the values of its polar $q^\vee$ on a symmetric multibasis for $\mathbf{R}^3 \times \mathbf{R}^3$, say

$$\left\{ \left(\mathbf{e}^1, \mathbf{e}^1 \right), \left(\mathbf{e}^2, \mathbf{e}^2 \right), \left(\mathbf{e}^3, \mathbf{e}^3 \right), \left(\mathbf{e}^1, \mathbf{e}^2 \right), \left(\mathbf{e}^1, \mathbf{e}^3 \right), \left(\mathbf{e}^2, \mathbf{e}^3 \right) \right\} .$$

For example we shall take the corresponding values of $q^\vee(\mathbf{e}^i, \mathbf{e}^k)$ to be $(1, 1, 0)$, $(0, 0, 1)$, $(1, 0, 1)$, $(0, 1, 0)$, $(0, 0, 1)$, $(0, 0, 0)$, resp. Then for arbitrary

$$x = a\mathbf{e}^1 + b\mathbf{e}^2 + c\mathbf{e}^3 \in \mathbf{R}^3$$

we have

$$q \left(a\mathbf{e}^1 + b\mathbf{e}^2 + c\mathbf{e}^3 \right) = q^\vee \left(a\mathbf{e}^1 + b\mathbf{e}^2 + c\mathbf{e}^3, a\mathbf{e}^1 + b\mathbf{e}^2 + c\mathbf{e}^3 \right)$$

$$= a^2 q^\vee \left(\mathbf{e}^1, \mathbf{e}^1 \right) + \cdots + 2ab q^\vee \left(\mathbf{e}^2, \mathbf{e}^3 \right)$$

$$= \left(a^2 + 2ac, a^2 + b^2, 2ab + 2ac + 2bc \right) .$$

Furthermore, writing out the *display* $[m(q)]$ of $m(q)$ as:

$$\text{dis}[m(q)] = \begin{bmatrix} m_1^{11} m_1^{12} m_1^{13} & m_2^{11} m_2^{12} m_2^{13} & m_3^{11} m_3^{12} m_3^{13} \\ m_1^{21} m_1^{22} m_1^{23} & m_2^{21} m_2^{22} m_2^{23} & m_3^{21} m_3^{22} m_3^{23} \\ m_1^{31} m_1^{32} m_1^{33} & m_2^{31} m_2^{32} m_2^{33} & m_3^{31} m_3^{32} m_3^{33} \end{bmatrix}$$

—as we may do here in this example where the degree of $m(q)$ is merely 2—and using the relation (5.48) to compute the values of the elements $m_j^{i,k}$ of $m(q)$, we have, in this example:

$$\text{dis}[m(q)] = \begin{bmatrix} 1\,0\,1 & 1\,0\,0 & 0\,1\,1 \\ 0\,0\,0 & 0\,1\,0 & 1\,0\,1 \\ 1\,0\,0 & 0\,0\,0 & 1\,1\,0 \end{bmatrix}$$

as a representation of the pensymmetric 2-matroid $m(q)$ associated with the quadratic function $q : \mathbf{R}^3 \rightarrow \mathbf{R}^3$ determined by the specifications given above. The pensymmetry of $m(q)$ is plainly apparent in this particular display because each of the three displayed submatroids (matrices) clearly exhibits the symmetry required by the definition of pensymmetry.

5.6   Completely Continuous and Related Multilinear Operators

Completely continuous multilinear operators are defined and their properties investigated. This class of operators is shown to form a closed multi-ideal. Unlike the linear case, compact multilinear operators need not be completely continuous. The completely continuous maps are shown to be the closure of a subspace of the finite rank operators. Hilbert–Schmidt operators are also considered. An application to finding error bounds for solutions of multipower equations is presented.

Let Y, Z, and X_i be real or complex Banach spaces, with U_i denoting the closed unit ball of X_i, $i = 1, \ldots, r$. Throughout we let $x := (x_1, x_2, \ldots, x_r)$ denote an element of the Cartesian product $X := \prod_{i=1}^{r} X_i$. The dual space of X_i will be denoted by X_i'. The bounded multilinear operators from X into Y form a Banach space $ML_r(X, Y)$ with norm

$$\|M\| = \sup \{\|M(x_1, x_2, \ldots, x_r)\| : x_i \in U_i, i = 1, \ldots, r\} . \tag{5.50}$$

In the case $r = 1$ we use the standard notation $L(Y, Z) = ML_1(Y, Z)$. The definition of a multi-ideal given next is a natural generalization of an ideal of linear operators.

DEFINITION 5.8   We set $\mathcal{ML}_r := \bigcup_{X,Y} ML_r(X, Y)$. A subset $\mathcal{MA}_r$ is said to be a multi-ideal, or r-ideal, if it satisfies the following conditions.

1. $\mathcal{MA}_r(X, Y) = \mathcal{MA}_r \cap ML_r(X, Y)$ is a vector space.

2. *If $a_i \in X_i'$ for $i = 1, \ldots, r$ and $y \in Y$ then the "tensor-finite" map $a_1 \otimes \cdots \otimes a_r \otimes y$ is an element of $\mathcal{M}A_r(X, Y)$ where*

$$a_1 \otimes \cdots \otimes a_r \otimes y(x) = \langle a_1, x_1 \rangle \cdots \langle a_r, x_r \rangle y. \tag{5.51}$$

3. *If $T \in L(Y, Z)$ and $M \in \mathcal{M}A_r(X, Y)$ then $T \circ M \in \mathcal{M}A_r(X, Z)$.*

4. *If $T_i \in L(W_i, X_i)$ for $i = 1, \ldots, r$, and $M \in \mathcal{M}A_r(X, Y)$ then*

$$M \circ (T_1, \ldots, T_r) \in \mathcal{M}A_r(W, Y)$$

where W is the Cartesian product $\prod_{i=1}^{r} W_i$ and

$$M \circ (T_1, \ldots, T_r)(w) = M(T_1 w_1, \ldots, T_r w_r).$$

A multilinear operator is said to be compact if it carries each product of bounded sets to a conditionally convergent set. This class of operators forms a multi-ideal that is closed in the operator topology. Compact linear operators between Banach spaces form a subset of the completely linear operators, and the sets coincide when the domain space is reflexive. In general, the relationship between compact and completely continuous operators is not so clear in the multilinear case. To illustrate this, consider the following operator from elasticity theory.

Example 5.5
Let $B: L^2[a, b] \times L^2[a, b] \to L^2[a, b]$ be defined by

$$B(f, g)(t) = \int_a^b k(s, t) f(s) g(s) ds \tag{5.52}$$

where the kernel $k(s, t)$ is continuous. As we shall see later, B may not be completely continuous. However, a straightforward argument based on the Arzela–Ascoli Theorem shows that B is compact.

Linear completely continuous operators are those that carry weakly convergent sequences to norm convergent sequences. Naturally, any reasonable definition for r-linear operators should reduce to this one in the case $r = 1$. The following definition satisfies this criterion. We write $x_{in} \rightharpoonup x_{i0}$ whenever a sequence (x_{in}) converges to x_{i0} in the *weak* topology of X_i. Convergence in the norm topology will be denoted by $x_{in} \to x_{i0}$.

DEFINITION 5.9 *An r-linear operator is said to be completely continuous if $x_{ik} \rightharpoonup x_{i0}$ for $i = 1, \ldots, r$ implies that $M(x_{1k}, \ldots, x_{rk}) \to M(x_{10}, \ldots, x_{r0})$.*

The linear completely continuous operators form a closed operator ideal to be denoted by $\mathcal{V}$. This result generalizes to the multilinear setting.

THEOREM 5.28

The r-linear completely continuous operators $\mathcal{MV}_r$ form a closed multi-ideal.

PROOF   It is easy to verify that properties 1–4 of Definition 5.8 are satisfied. To see that $\mathcal{MV}_r$ is closed, let Y and X_i, $i = 1, \ldots, r$ be Banach spaces and suppose that M is in the closure of $\mathcal{MV}_r(X, Y)$. Fix $\varepsilon > 0$ and choose completely continuous N such that $\|M - N\| < \varepsilon$. Let $x_{ik} \hookrightarrow_k x_{i0}$ for $i = 1, \ldots, r$, so that $\|N(x_k) - N(x_0)\| < \varepsilon$ for sufficiently large k. By the weak convergence, we may assume that $x_{ik}, x_{i0} \in U_i$ for all i, k. Then

$$\|M(x_k) - M(x_0)\| \leq \|M(x_k) - N(x_k)\| + \|N(x_k) - N(x_0)\|$$

$$+ \|N(x_0) - M(x_0)\| < 3\varepsilon \qquad (5.53)$$

for sufficiently large k. Thus, M is completely continuous.   ∎

Observe that our definition of a multi-ideal does not require all finite rank multilinear maps to be elements of a multi-ideal. Only "tensor-finite" maps must be elements. As the next example will show, this is essential for viewing the classes of completely continuous and Hilbert–Schmidt operators as multi-ideals.

Example 5.6

Let H be a Hilbert space with orthonormal sequence (e_i) and inner product $\langle \cdot, \cdot \rangle$. Then define the finite rank operator $B \in ML_2(H \times H, H)$ by $B(x, y) = \langle x, y \rangle e_1$. Now $e_i \hookrightarrow 0$ but $\|B(e_i, e_i)\| = 1$ for all i, so B is not completely continuous. The map B is clearly compact, so compact r-linear maps need not be completely continuous for $r \geq 2$.

REMARK 5.7   For the operator given in Example 5.5, if the kernel $k(s, t)$ is taken to be 1, then the same argument as given in Example 5.6 shows that B is not completely continuous.

We also observe that an r-linear completely continuous map is a completely continuous linear operator in each variable. This follows directly from Definition 5.9. On the other hand, the example just given shows that the converse fails.   ∎

The following algebraic identity for r-linear mappings will be used frequently.

$$M(x_1, x_2, \ldots, x_r) - M(y_1, y_2, \ldots, y_r) = M(x_1 - y_1, x_2, \ldots, x_r)$$

$$+ M(y_1, x_2 - y_2, \ldots, x_r) + \cdots + M(y_1, y_2, \ldots, x_r - y_r) . \quad (5.54)$$

Recall that a multilinear map M induces linear maps

$$\phi_M^j : X_j \to ML_{r-1}(X_1 \times \ldots \times X_{j-1} \times X_{j+1} \times \ldots \times X_r, Y)$$

where

$$\phi_M^j(x_j)(x_1, \ldots, x_{j-1}, x_{j+1}, \ldots, x_r) = M(x_1, \ldots, x_r) . \quad (5.55)$$

It is well known that the map $M \mapsto \phi_M^j$ is an isometric isomorphism under the operator topology. However, not all operator properties are passed between M and ϕ_M^j under this map. Given linear operator ideal $\mathcal{A}$ and $r - 1$-ideal $\mathcal{B}$, define the r-ideal $[\mathcal{A}, \mathcal{B}]^j$ by its components

$$[\mathcal{A}, \mathcal{B}]^j(X_1 \times \ldots \times X_r, Y) = [\mathcal{A}, \mathcal{B}]^j \cap ML_r(X_1 \times \ldots \times X_r, Y) \quad (5.56)$$

where $[\mathcal{A}, \mathcal{B}]^j(X_1 \times \ldots \times X_r, Y)$ denotes

$$\left\{ M \in ML_r(X_1 \times \ldots \times X_r, Y) : \right.$$

$$\left. \phi_M^j \in \mathcal{A}(X_j, \mathcal{B}(X_1 \times \ldots \times X_{j-1} \times X_{j+1} \times \ldots \times X_r, Y)) \right\} . \quad (5.57)$$

It is clear that $\bigcap_{j=1}^r [\mathcal{A}, \mathcal{B}]^j$ is a multi-ideal, and for notational convenience we define

$$[\mathcal{A}, \mathcal{B}] = \bigcap_{j=1}^r [\mathcal{A}, \mathcal{B}]^j . \quad (5.58)$$

We now investigate the relationship between multi-ideals of this form and $\mathcal{MV}_r$.

THEOREM 5.29
The multi-ideal $[\mathcal{V}, ML_{r-1}]$ is contained in $\mathcal{MV}_r$.

PROOF   Suppose that $M \in [\mathcal{V}, ML_{r-1}]^j$ for $j = 1, \ldots, r$ and that $x_{in} \hookrightarrow x_{i0}$ for $i = 1, \ldots, r$. Applying Equation (5.54), we have

$$M(x_{1n}, x_{2n}, \ldots, x_{rn}) - M(x_{10}, x_{20}, \ldots, x_{r0})$$

$$= \phi_M^1 (x_{1n} - x_{10}) (x_{2n}, \ldots, x_{rn})$$

$$+ \cdots + \phi_M^r (x_{rn} - x_{r0}) (x_{1n}, \ldots, x_{r-1n}) . \tag{5.59}$$

The (x_{jn}) are uniformly bounded because of the weak convergence, and

$$\phi_M^j (x_{jn} - x_{j0}) \to 0 \tag{5.60}$$

for each j by hypothesis, so Equation (5.59) yields

$$M (x_{1n}, x_{2n}, \ldots, x_{rn}) \to M (x_{10}, x_{20}, \ldots, x_{r0}) . \quad \blacksquare \tag{5.61}$$

The converse to Theorem 5.29 is false, as the next example demonstrates.

Example 5.7

Define $W: \ell_2 \times \ell_1 \to \ell_2$ by $W((x_i), (y_i)) = (x_i y_i)$. Using Schur's lemma [11] and Equation (5.54) we see that W is completely continuous. However, ϕ_M^1 is not completely continuous since (e_n) converges weakly to 0 in ℓ_2 but $\|\phi_M^1(e_n)\| = 1$ for all $n \in \mathbf{N}$. Also note that $W(e_n, e_n) = e_n$ for all $n \in \mathbf{N}$, so multilinear completely continuous operators need not be compact.

Note that ϕ_M^1 is completely continuous by Schur's lemma. Indeed, we have the following partial converse to Theorem 5.29.

THEOREM 5.30

Suppose that X_i is a reflexive Banach space for $i \neq j$, and $M: X \to Y$ is completely continuous. Then

$$M \in [\mathcal{V}, \mathcal{M}\mathcal{V}_{r-1}]^j (X_1 \times \ldots \times X_r, Y) .$$

PROOF   For ease of notation, we give the proof in the case $j = 1$. For each $x \in X_1$, the map $\phi_M^1(x)$ is clearly completely continuous. Now assume for the sake of contradiction that ϕ_M^1 is not completely continuous. Then we can find a sequence (x_{1n}) and an $x_{10} \in X_1$ such that $x_{1n} \hookrightarrow x_{10}$ but $\phi_M^1(x_{1n})$ does not converge in norm to x_{10}. We can then choose an $\varepsilon_0 > 0$, a subsequence (x_{1n_k}) and sequences (x_{in_k}) in X_i, $i \neq 1$, for which

$$\left\| \phi_M^1 (x_{1n_k} - x_{10}) (x_{2n_k}, \ldots, x_{rn_k}) \right\| \geq 2\varepsilon_0 \tag{5.62}$$

for all n_k. Now X_i is a reflexive Banach space for $i \neq 1$ so, passing to subsequences if necessary, each (x_{in_k}) converges weakly to some x_{i0}. Furthermore, M

is completely continuous, so

$$M\left(x_{1n_k}, \ldots, x_{rn_k}\right) \to M\left(x_{10}, \ldots, x_{r0}\right) . \tag{5.63}$$

Applying and rearranging Equation (5.54) yields

$$\left\| \phi_M^1 \left(x_{1n_k} - x_{10}\right)\left(x_{2n_k}, \ldots, x_{rn_k}\right) \right\| \tag{5.64}$$

$$\leq \left\| M\left(x_{1n_k}, \ldots, x_{rn_k}\right) - M\left(x_{10}, \ldots, x_{r0}\right) \right\|$$

$$+ \left\| M\left(x_{10}, x_{2n_k} - x_{20}, \ldots, x_{rn_k}\right) \right\|$$

$$+ \cdots + \left\| M\left(x_{10}, x_{20}, \ldots, x_{rn_k} - x_{r0}\right) \right\| . \tag{5.65}$$

Since M is completely continuous, (5.63) forces each term on the right side of (5.64) to converge to 0, so

$$\left\| \phi_M^1 \left(x_{1n_k} - x_{10}\right)\left(x_{2n_k}, \ldots, x_{rn_k}\right) \right\| \to 0 ,$$

which contradicts (5.62). This finishes the proof. ∎

COROLLARY 5.6

Suppose that X_i is a reflexive Banach space for $i = 1 \ldots r$. Then $M: X \to Y$ is completely continuous if and only if ϕ_M^i is completely continuous for $i = 1 \ldots r$. That is,

$$\left[\mathcal{V}, ML_{r-1}\right](X, Y) = \mathcal{MV}_r(X, Y) .$$

PROOF   If M is completely continuous, then by Theorem 5.30 each ϕ_M^i is completely continuous. The converse follows from Theorem 5.29 ∎

The Newton-Kantorovich method is effective for solving certain operator equations involving bilinear operators on Banach spaces (see Chapters 1 through 3). Since this method involves Fréchet derivatives, we mention the following result:

COROLLARY 5.7

Suppose that X is a reflexive Banach space. For bilinear $M: X \times X \to Y$, define the "multipower" operator $\hat{M}: X \to Y$ by $\hat{M}(x) = M(x, x)$. If M is completely continuous then the Fréchet derivative $\hat{M}'$ of M is completely continuous.

PROOF   Recall that the Fréchet derivative of $\hat{M}$ is defined by

$$\hat{M}'(x)(y) = M(x, y) + M(y, x)$$

so

$$\hat{M}' = \phi_M^1 + \phi_M^2.$$

The result follows from the previous corollary.   ∎

It is well known that if T is a compact linear operator between Banach spaces, then T is completely continuous. We have seen in Example 5.6 that this is not true in general for r-linear operators with $r \geq 2$. It is also known that completely continuous operators are compact if the domain space is reflexive. Happily, this does generalize to the multilinear setting.

THEOREM 5.31
Let X_i, $i = 1, \ldots, r$ be reflexive Banach spaces. If $M \in ML_r(X, Z)$ is completely continuous, then M is compact.

PROOF   Let (x_{in}) be a sequence in the unit ball of X_i, for $i = 1, \ldots, r$. By the reflexivity of each X_i, we can find a subsequence (x_{in_k}) that converges weakly to some $x_{i0} \in X_i$, $i = 1, \ldots, r$. By hypothesis, $M(x_{1n_k}, \ldots, x_{rn_k})$ converges in norm to $M(x_{10}, \ldots, x_{r0})$. Thus M is compact.   ∎

We next examine finite rank operators $\mathcal{MF}_r$ and their connections with the multi-ideals $\mathcal{MK}_r$ and $\mathcal{MV}_r$.

THEOREM 5.32
Let Z be a Banach space with the approximation property, and let X_i be Banach spaces, $i = 1, \ldots, r$. Then

$$\mathcal{MK}_r(X, Z) = \overline{\mathcal{MF}_r(X, Z)} \tag{5.66}$$

where closure is in the operator topology.

PROOF   Since $\mathcal{MK}_r(X, Z)$ is closed in the operator topology, we have

$$\overline{\mathcal{MF}_r(X, Z)} \subset \overline{\mathcal{MK}_r(X, Z)} = \mathcal{MK}_r(X, Z). \tag{5.67}$$

To prove the other inclusion, let $M \in \mathcal{MK}_r(X, Z)$ and fix $\varepsilon > 0$. Now $K = M(U_1, \ldots, U_r)$ is compact, so by hypothesis we can choose a finite rank $A \in$

$L(Z, Z)$ for which $\|(I_Z - A)z\| < \varepsilon$ for all $z \in K$. Note that $A \circ M \in \mathcal{MF}_r(X, Z)$ and

$$\|M - A \circ M\| = \sup_{z \in K} \|z - Az\| < \varepsilon . \tag{5.68}$$

Hence $M \in \overline{\mathcal{MF}_r(X, Z)}$. ∎

In the multilinear case we consider a special subspace of $\mathcal{MF}_r$ known as the "tensor-finite" rank operators. We write

$\mathcal{TF}_r(X, Y)$

$$= \operatorname{span} \left\{ a_1 \otimes \cdots \otimes a_r \otimes y : a_i \in X_i', i = 1, \ldots, r; y \in Y \right\} . \tag{5.69}$$

According to Definition 5.8, $\mathcal{TF}_r$ is an ideal and a subset of any multi-ideal. We are interested in characterizing $\overline{\mathcal{TF}_r}$, the closure of $\mathcal{TF}_r$ in the operator topology. From the previous theorem, we know that $\overline{\mathcal{TF}_r(X, Y)}$ is a subspace of $\mathcal{MK}_r(X, Y)$ if Y has the approximation property. A special class of spaces with the approximation property is the set of Banach spaces with a Schauder basis. Recall that if X has a Schauder basis with a regular biorthogonal system $\{(e_n, e_n')\}$, then the "partial sum" operators $\{S_n\}$, defined by

$$S_n \left(\sum_{i=1}^{\infty} a_i e_i \right) = \sum_{i=1}^{n} a_i e_i , \tag{5.70}$$

are uniformly bounded in norm (see [181]). For ease of notation we define the associated "tail" operator by

$$T_n(x) = (I - S_n)(x) . \tag{5.71}$$

THEOREM 5.33
Let X_i, $i = 1, \ldots, r$, be reflexive Banach spaces. Suppose that the Banach space Y and the spaces X_i all have Schauder bases. Then

$$\mathcal{MV}_r(X, Y) = \overline{\mathcal{TF}_r(X, Y)}$$

where closure is in the operator topology.

PROOF   Since $\mathcal{MV}_r(X, Y)$ is a multi-ideal, $\mathcal{TF}_r(X, Y)$ is clearly a subset of $\mathcal{MV}_r(X, Y)$. Now $\mathcal{MV}_r(X, Y)$ is closed in the operator topology by Theorem 5.28, so $\overline{\mathcal{TF}_r(X, Y)} \subset \overline{\mathcal{MV}_r(X, Y)} = \mathcal{MV}_r(X, Y)$.

To prove that $\mathcal{M}\mathcal{V}_r(X, Y) \subset \overline{\mathcal{T}\mathcal{F}_r(X, Y)}$, let $M \in \mathcal{M}\mathcal{V}_r(X, Y)$ and fix $\varepsilon > 0$. For the given Schauder bases of the X_i and Y spaces, let S_n^i, S_n^Y denote the partial sum operators for X_i and Y, respectively. We will show that for sufficiently large n, the map

$$S_n^N \left(M \left(S_n^1(\cdot), \ldots, S_n^r(\cdot) \right) \right) \in \mathcal{T}\mathcal{F}_r(X, Y) \tag{5.72}$$

is within ε of M. As noted above, there is some constant $C \geq 1$ for which $\|S_n^i\|$, $\|S_n^Y\| \leq C$ for all $n \in \mathbf{N}$ and $i = 1, \ldots, r$. By Theorem 5.31, M is compact. By the corollary and linear theory each ϕ_M^i is compact. Hence, there is some $N \in \mathbf{N}$ for which $n \geq N$ implies that

$$\left\| T_n^Y \left(M \left(x_1, \ldots, x_r \right) \right) \right\| \leq \frac{\varepsilon}{2} \tag{5.73}$$

and

$$\left\| \phi_M^i \left(T_n^i \left(x_i \right) \right) \right\| \leq \frac{\varepsilon}{2rC^r}, \tag{5.74}$$

for all $x_i \in X_i$, $\|x_i\| \leq 1$, $i = 1, \ldots, r$. We need the following identity, which is easy to verify:

$$M \left(x_1, \ldots, x_r \right) = T_n^Y \left(M \left(x_1, \ldots, x_r \right) \right)$$

$$+ S_n^Y \left(M \left(T_n^1 \left(x_1 \right), x_2, \ldots, x_r \right) \right)$$

$$+ S_n^Y \left(M \left(S_n^1 \left(x_1 \right), T_n^2 \left(x_2 \right), x_3, \ldots, x_r \right) \right) \tag{5.75}$$

$$+ \cdots + S_n^Y \left(M \left(S_n^1 \left(x_1 \right), \ldots, S_n^r \left(x_r \right) \right) \right).$$

Rearranging Equation (5.75) and using (5.73) and (5.74), we obtain

$$\left\| M \left(x_1, \ldots, x_r \right) - S_n^Y \left(M \left(S_n^1 \left(x_1 \right), \ldots, S_n^r \left(x_r \right) \right) \right) \right\|$$

$$\leq S_n^Y \left(S_M^1 \left(T_n^1 \left(x_1 \right) \right) \left(x_2, \ldots, x_r \right) \right)$$

$$+ S_n^Y \left(\phi_M^2 \left(T_n^2 \left(x_2 \right) \right) \left(S_n^1 \left(x_1 \right), x_3, \ldots, x_r \right) \right)$$

$$+ \cdots + S_n^Y \left(\phi_M^r \left(T_n^r \left(x_r \right) \right) \left(S_n^1 \left(x_1 \right), \ldots, S_n^{r-1} \left(x_{r-1} \right) \right) \right) + \frac{\varepsilon}{2}$$

$$< C\frac{\varepsilon}{2rC^r} + C^2\frac{\varepsilon}{2rC^r} + \cdots + C^r\frac{\varepsilon}{2rC^r} + \frac{\varepsilon}{2} \tag{5.76}$$

$$\leq \varepsilon .$$

This completes the proof. ∎

The following result is useful in the study of integral equations.

THEOREM 5.34

Suppose that (P_n) *is a sequence of orthogonal projections in a Hilbert space* H *and that* $\lim P_n f = f$ *for all* $f \in H$. *If linear* $K: H \to H$ *is compact, then* $\lim \|K P_n - K\| = 0$.

This does not generalize to r-linear compact operators, for consider the map $B(x, y) = \langle x, y\rangle e_1$ of Example 5.6 and observe that

$$\|B(P_n, P_n) - B\| \geq \|B(P_n e_{n+1}, P_n e_{n+1}) - B(e_{n+1}, e_{n+1})\| = 1$$

for all n. However, the linear theorem does have an r-linear generalization for completely continuous operators. We use the notation from Theorem 5.33.

THEOREM 5.35

Let X_i, $i = 1, \ldots, r$, *be reflexive Banach spaces. Suppose that the Banach space* Y *and the spaces* X_i *all have Schauder bases. If* $M \in \mathcal{ML}_r(X, Y)$ *is completely continuous, then*

$$\lim_{n\to\infty} \left\| M\left(S_n^1, \ldots, S_n^r\right) - M \right\| = 0 .$$

PROOF   We first note that

$$\left\| M\left(S_n^1, \ldots, S_n^r\right) - M \right\| \leq \left\| M\left(S_n^1, \ldots, S_n^r\right) - S_n^Y M\left(S_n^1, \ldots, S_n^r\right) \right\|$$

$$+ \left\| S_n^Y M\left(S_n^1, \ldots, S_n^r\right) - M \right\|$$

$$= \left\| T_n^Y M\left(S_n^1, \ldots, S_n^r\right) \right\| + \left\| S_n^Y M\left(S_n^1, \ldots, S_n^r\right) - M \right\| .$$

Now $\|S_n^Y M(S_n^1, \ldots, S_n^r) - M\| \to 0$ as $n \to \infty$ by Theorem 5.33. Since M is compact by Theorem 5.31, $\|T_n^Y M(S_n^1, \ldots, S_n^r)\| \to 0$ as $n \to \infty$. ∎

The definition of a multilinear Hilbert–Schmidt operator between separable Hilbert spaces appears in [64]. We restate it but do not require the separability condition. Throughout this section, H_i, K denote real or complex Hilbert spaces with inner product $\langle \cdot, \cdot \rangle$, and we let $H = \prod_{i=1}^{r} H_i$.

DEFINITION 5.10   *An operator $M \in \mathcal{ML}_r(Y, K)$ is Hilbert–Schmidt if for any complete orthonormal families (CONF) $\{e_{ik}\} \subset H_i$, $i = 1, \ldots, r$, and $\{u_j\} \subset K$ the following condition is satisfied.*

$$\sum_{i_1} \cdots \sum_{i_r} \left\| M\left(e_{1i_1}, \ldots, e_{ri_r}\right) \right\|^2 = \sum_{j} \sum_{i_1}$$

$$\cdots \sum_{i_r} \left| \left\langle M\left(e_{1i_1}, \ldots, e_{ri_r}\right), u_j \right\rangle \right|^2 < \infty . \tag{5.77}$$

The Hilbert–Schmidt norm of M is defined by

$$|||M||| = \left(\sum_{i_1} \cdots \sum_{i_r} \left\| M\left(e_{1i_1}, \ldots, e_{ri_r}\right) \right\|^2 \right)^{1/2} . \tag{5.78}$$

The class of r-linear Hilbert–Schmidt operators will be denoted by $\mathcal{MHS}_r$.

The following theorem is clear.

THEOREM 5.36
The class $\mathcal{MHS}_r$ forms a multi-ideal.

In fact, $\mathcal{MHS}_r$ is a *normed* multi-ideal under the norm given in Equation (5.78) where we use the standard definition of a normed multi-ideal (see [157]).

COROLLARY 5.8
$[\mathcal{HS}, \mathcal{MHS}_{r-1}] = \mathcal{MHS}_r.$

PROOF   Let $\{e_{ik}\} \subset H_i$, $i = 1, \ldots, r$, and $\{u_j\} \subset K$ by CONF. For each e_{1i_1},

$$\left\| \left| \phi_M^1\left(e_{1i_1}\right) \right| \right\|^2 = \sum_{i_2} \cdots \sum_{i_r} \left\| \phi_M^1\left(e_{1i_1}\right)\left(e_{2i_2}, \ldots, e_{ri_r}\right) \right\|^2 , \tag{5.79}$$

so

$$\left|\left|\left|\phi_M^1\right|\right|\right|^2 = \sum_{i_1} \left|\left|\left|\phi_M^1\left(e_{1i_1}\right)\right|\right|\right|^2$$

$$= \sum_{i_1} \cdots \sum_{i_r} \left\|\phi_M^1\left(e_{1i_1}\right)\left(e_{1i_2}, \ldots, e_{ri_r}\right)\right\|^2$$

$$= \sum_{i_1} \cdots \sum_{i_r} \left\|M\left(e_{1i_1}, \ldots, e_{ri_r}\right)\right\|^2$$

$$= \left|\left|\left|M\right|\right|\right|^2 . \tag{5.80}$$

implies that M will be Hilbert–Schmidt if and only if ϕ_M^1 is Hilbert–Schmidt. Replacing ϕ_M^1 by ϕ_M^j for $j = 2, \ldots, r$ yields the desired result. ∎

REMARK 5.8　　The proof of this corollary shows the map

$$M \mapsto \phi_M^j \tag{5.81}$$

between $\mathcal{MHS}_r$ and $[\mathcal{HS}, \mathcal{MHS}_{r-1}]^j$ is an isometry under the Hilbert–Schmidt norm. ∎

COROLLARY 5.9
　The r-linear Hilbert–Schmidt operators are completely continuous: $\mathcal{MHS}_r \subset \mathcal{MV}_r$.

PROOF　　The result follows by induction on r. The case $r = 1$ is well known from linear theory. If we assume the result for $r = k - 1$, then the previous corollary, Corollary 5.7 and linear theory together yield:

$$\mathcal{MHS}_k = [\mathcal{HS}, \mathcal{HS}_{k-1}] \subset [\mathcal{V}, \mathcal{MV}_{k-1}] = \mathcal{MV}_k . \quad ∎$$

Consider the r-linear operator

$$K : \underbrace{L^2([a, b]) \times L^2([a, b]) \times \cdots \times L^2([a, b])}_{r \text{ times}} \to L^2([a, b])$$

defined by

$$K\left(f_1, f_2, \ldots, f_r\right)(s) = \int g\left(s, t_1, t_2, \ldots, t_r\right) f_1\left(t_1\right) f_2\left(t_2\right) \ldots f_r\left(t_r\right) dt_1 dt_2 \ldots dt_r .$$

Operators of this type are of interest in systems theory [158], radiative transfer [122], and stochastic control theory [64].

It is known that a linear operator T from an L^2 space into itself is Hilbert–Schmidt if and only if it is an integral operator with an L^2 kernel. We now give the multilinear version of this result.

THEOREM 5.37

Let (X, μ) be an arbitrary measure space, and let $H = L^2(X)$. Suppose that $K \in L^2(X^r)$. Then K induces a Hilbert–Schmidt integral operator $K \colon H^r \to H$, where

$$K\left(f_1, f_2, \ldots, f_r\right)(s) = \int g\left(s, t_1, t_2, \ldots, t_r\right) f_1\left(t_1\right) f_2\left(t_2\right) \ldots f_r\left(t_r\right) dt_1 dt_2 \ldots dt_r .$$

Conversely, if r-linear $K \colon H^r \to H$ is Hilbert–Schmidt, then K is an integral operator with an $L^2(X^r)$ kernel g satisfying $|||K||| = \|g\|_2$.

PROOF For notational convenience only, we give the proof with $r = 2$.

($\to$) Let $\{\varphi_i\}$ be a CONF in H. Recall that $\{\bar{\varphi}_i \bar{\varphi}_j \varphi_k\}$ forms a CONF in $L^2(X \times X \times X)$. Since

$$\langle K\left(\varphi_i, \varphi_j\right), \varphi_k \rangle = \int \int \int g(x, y, z) \varphi_i(x) \varphi_j(y) \overline{\varphi_k(z)} dx\, dy\, dz ,$$

we have

$$|||K|||^2 = \sum_{i,j,k} |\langle K\left(\varphi_i, \varphi_j\right), \varphi_k \rangle|^2 = \sum_{i,j,k} |\langle g, \bar{\varphi}_i \bar{\varphi}_j \varphi_k \rangle|^2 = \|g\|_2^2 .$$

($\leftarrow$) By hypothesis, $\sum_{i,j,k} |\langle K(\varphi_i, \varphi_j), \varphi_k \rangle|^2 < \infty$, so the function

$$g = \sum_{i,j,k} \langle K\left(\varphi_i, \varphi_j\right), \varphi_k \rangle \bar{\varphi}_i \bar{\varphi}_j \varphi_k$$

is an element of $L^2(X \times X \times X)$. By the first part of the proof, g induces a Hilbert–Schmidt operator M. In fact, for each i, j, k,

$$\langle M\left(\varphi_i, \varphi_j\right), \varphi_k \rangle = \int \int \int g(x, y, z)\varphi_i(x)\varphi_j(x)\overline{\varphi_k(z)}\, dx dy dz$$

$$= \langle g, \bar\varphi_i \bar\varphi_j \varphi_k \rangle = \langle K\left(\varphi_i, \varphi_j\right), \varphi_k \rangle .$$

By the completeness of $\{\varphi_i\}$, this yields $M = K$, whence g is the kernel of K.
∎

REMARK 5.9   For the special case of *separable* $H = L^2(X)$, Balakrishnan proves this theorem in [64]. His proof depends on the separability of H and is thus somewhat different than the one presented here.   ∎

Finally we apply some of the theory developed above to find error bounds for approximate solutions x in a Banach space X of so-called multipower equations

$$x = y + \tilde M(x) \tag{5.82}$$

where $y \in X$ and $\tilde M : X \to X$ is defined by $\tilde M(x) = M(x, \ldots, x)$ for some $M \in ML_r(X \times \cdots \times X, X)$. It is easy to show that the norms of M and $\tilde M$ are equal.

When the space X is infinite dimensional, the implementation of some iterative routines becomes impractical and an auxiliary equation

$$x_n = y_n + \tilde M_n(x) \tag{5.83}$$

is solved in a more tractable setting, such as an n-dimensional subspace of X. We are interested in bounding the difference between a solution to the original equation (5.82) and the auxiliary equation (5.83). The following lemma is required.

LEMMA 5.9
  Suppose that M_1, M_2 are elements of $ML_r(X \times \cdots \times X, X)$ with associated r-power operators $\tilde M_1$ and $\tilde M_2$, respectively. Then for $a, b \in X$,

$$\left\| \tilde M_1(a) - \tilde M_2(b) \right\| \le \|a - b\| \cdot \|M_1\| \left(\sum_{i=0}^{r-1} \|a\|^i \|b\|^{r-1-i} \right)$$

$$+ \|M_1 - M_2\| \cdot \|b\|^r . \tag{5.84}$$

PROOF   The proof follows from the identity (5.54) and the inequality

$$\left\|\tilde{M}_1(a) - \tilde{M}_2(b)\right\| \leq \left\|\tilde{M}_1(a) - \tilde{M}_1(b)\right\| + \left\|(\tilde{M}_1 - \tilde{M}_2)(b)\right\| . \quad \blacksquare \quad (5.85)$$

THEOREM 5.38
Suppose that $x = y + \tilde{M}(x)$ and $x_n = y_n + \tilde{M}_n(x_n)$. If $\|x\| \leq B$ and

$$\|M\| \left(\sum_{i=0}^{r-1} B^i \|x_n\|^{r-1-i} \right) < 1 \qquad (5.86)$$

then

$$\|x - x_n\| \leq \frac{\left(\|y - y_n\| + \|M - M_n\| \cdot \|x_n\|^r\right)}{1 - \|M\|(\sum_{i=0}^{r-1} B^i \|x_n\|^{r-1-i})} . \qquad (5.87)$$

PROOF   By hypothesis and Lemma 5.9 we have

$$\|x - x_n\| \leq \|y - y_n\| + \|x - x_n\| \cdot \|M\| \cdot \left(\sum_{i=0}^{r-1} \|x\|^i \|x_n\|^{r-1-i} \right)$$

$$+ \|M - M_n\| \cdot \|x_n\|^r . \qquad (5.88)$$

Some simple algebra yields the bound (5.87). $\blacksquare$

We now illustrate how this theorem can be employed.

Example 5.8
Consider solving the following equation for $x(t)$ in $L^2[0, 1]$.

$$x(t) = y(t) + \lambda \int_0^1 \frac{1}{u + v + t} x(u)x(v)dudv . \qquad (5.89)$$

Letting $K(f, g)(t) = \lambda \int_0^1 \frac{1}{u+v+t} f(u)g(v)dudv$, we see by Theorem 5.37 that K is Hilbert–Schmidt. For illustrative purposes we choose $\lambda = 0.1$ and $y(t) = 1$. By results from Chapter 1, the bound (5.86) is satisfied and Equation (5.89) must have a unique solution x_s with $\|x_s\| \leq 1.12$. We now solve the "auxiliary" equation

$$x_n = y_n + \tilde{K}_n (x_n) \qquad (5.90)$$

in an n-dimensional subspace X_n of $L^2[0, 1]$, where

$$K_n(f, g)(t) = \lambda \int_0^1 k_n(u, v, t) f(u) g(v) du dv .$$

For simplicity we choose $n = 3$, and X_3 to be the span of the first three Legendre polynomials $P_n(t)$. The "auxiliary" kernel k_n is the projection of the original kernel

$$k(u, v, t) = \frac{1}{u + v + t}$$

into $X_3 \times X_3 \times X_3$. Observe that $y_3(t) = y(t) = 1$. Using the iterative method

$$f_{j+1} = y_3 + \tilde{K}_3 \left(f_j \right)$$

in X_3 with $f_0(t) = 1$, we find a solution $x_3(t)$ to (5.90). To bound the error $\|x_s - x_3\|$, we first use the fact that K is Hilbert–Schmidt to obtain

$$\|K - K_3\| \leq \||K - K_3\|| = \lambda \left(\|k\|^2 - \|k_3\|^2 \right)^{1/2} \approx 0.018301692 .$$

Then (5.87) yields

$$\|x_s - x_3\| \leq \frac{\|K - K_3\| \cdot \|x_3\|^2}{1 - \lambda \|k\| (\|x_s\| + \|x_3\|)} \leq 0.024649371 .$$

REMARK 5.10   The Legendre polynomials were used solely for simplicity; wavelet bases have been shown to be superior in many aspects (see Section 4.6). If a solution in $C[0, 1]$ is desired, the benefits of wavelets can be retained; Wang [196] has reported a wavelet basis for a dense subspace of $C[0, 1]$. ∎

Exercises

  1. Prove Theorem 5.1.

  2. Prove Theorem 5.4.

  3. Prove Theorem 5.8.

  4. Prove Theorem 5.9.

5. Prove Theorem 5.20.

6. Prove Theorem 5.22 completely.

7. Fill in the details in Example 5.2.

8. Prove estimates (5.26), (5.27), (5.28), (5.30), and (5.32).

9. Let X, Y denote Banach spaces and $\mathbf{C}$ the complex numbers. Show:

   (i) The following two statements are equivalent:

   (a) If $P(x)$ is an operator on $\mathbf{C}$ to X, then it will be called a $\mathbf{C}$ polynomial if it can be expressed in the form

$$P(x) = c_0 + c_1 x + \cdots + c_n x^n$$

where $c_0, c_1, \ldots, c_n$ are fixed elements in X. If $c_n \neq 0$ it will be said to be of degree n.

   (b) Let $P(x)$ be an operator on $\mathbf{C}$ to X. Then $P(x)$ will be said to be a $\mathbf{C}$ polynomial if:

   (b$_1$) $P(x)$ is continuous

   (b$_2$) for some integer n, $\Delta^{n+1} P(x) = 0$, where $\Delta^n P(x) = \Delta_n [\Delta^{n-1} P(x)]$, $\Delta^0 P(x) = P(x)$, $\Delta_i P(x) = P(x + \Delta_i x) - P(x)$ and $\Delta_i x$ are arbitrary increments

   (b$_3$) $P(x)$ possess a Gateaux differential everywhere

   It will be said to be of degree n, if $\Delta^n P(x) \neq 0$.

   (ii) The following two statements are equivalent:

   (c) Let $P(x)$ be an operator on X to Y. Then $P(x)$ will be said to be an X polynomial if:

   (c$_1$) $P(x)$ is continuous

   (c$_2$) for every pair x, y, $P(x + dy)$ is a $\mathbf{C}$ polynomial in d

   It will be said to be of degree n, if for some x, y, $P(x + dy)$ is a $\mathbf{C}$ polynomial of degree n and for all x, y is a $\mathbf{C}$ polynomial of degree $\leq n$.

   (d) Let $P(x)$ be an operator on X to Y. Then $P(x)$ will be said to be an X polynomial if:

   (d$_1$) $P(x)$ is continuous

   (d$_2$) for some integer n, $\Delta^{n+1} P(x) = 0$

   (d$_3$) $P(x)$ possess a Gateaux differential everywhere

   It will be said to be of degree n, if $\Delta^n P(x) \neq 0$.

10. Let $P: X \to Y$ be a polynomial of degree n defined on a real Banach space X with values in a real Banach space Y. Let $P^{(k)}(x)$ denote the kth Fréchet derivative of P at x.

Show

(i) If $\|P(x)\| \leq 1$ for $\|x\| \leq 1$, then $\|P'(x)\| \leq n^2$ for $\|x\| \leq 1$;

(ii) If $\|P(x)\| \leq 1$ for $\|x\| \leq 1$, then the kth Fréchet-derivative of P satisfies

$$\|P(x; y, \ldots, y)\| \leq \frac{n^2 \left(n^2 - 1^1\right)\left(n^2 - 2^2\right) \cdots \left(n^2 - (k-1)^2\right)}{1 \cdot 3 \cdot 5 \cdots (2k-1)} \|y\|^k$$

$$= A_{nk} \|y\|^k \ (k \geq 1)$$

for all $\|x\| \leq 1$ and all $y \in X$, and

$$\left\| P\left(x; y_1, y_2, \ldots, y_k\right) \right\| \leq \frac{2^{\frac{(k-1)(k-2)}{2}}}{(k-1)! [1 \cdot 3 \cdot 5 \cdots (2k-5)(2k-3)]} A_{nk} \left\| y_1 \right\| \cdots \|y_k\|$$

for all $\|x\| \leq 1$ and all $y_1, \ldots, y_k \in X$.

(iii) If $P_n(x)$ is an nth degree homogeneous polynomial with $T_n(x_1, \ldots, x_n)$ as polar, then

$$\|T_n(y_1, \ldots, y_n)\| \leq \frac{2^{\frac{n(n-1)}{2}}}{(n-1)! [1 \cdot 3 \cdot 5 \cdots (2n-3)]}$$

for all $\|y_i\| \leq 1$ $(i = 1, 2, \ldots, n)$ if $\|P_n(x)\| \leq 1$ for $\|x\| \leq 1$ $(n \geq 2)$.

(iv) If X, Y are complex Banach spaces the conclusion of (i) holds for n^2 replaced by n.

For the rest of the exercises we will need to recall the definition of the degree of an operator.

For given $\alpha > 0$, let W_α be the set of all real functions $\varphi : [0, \infty) \subseteq \mathbf{R} \to \mathbf{R}$ which are continuous on $[0, \infty)$ and for which there exists a $\delta \in (0, \alpha)$ such that $\varphi(t) = 0$ whenever $t \notin [\delta, \alpha]$. We call every $\varphi \in W_\alpha$ a weight function of index α. Clearly, if $y \in W_\alpha$, then $g: \mathbf{R}^n \to \mathbf{R}$, $g(x) = \varphi(\|x\|_2)$ is a continuous function of compact support on $\mathbf{R}^n$, that is, $g(x) = 0$ for all x outside of some compact set S, called the support of g. Hence, the Riemannian integral $\int_{\mathbf{R}^n} \varphi(\|x\|_2) dx$ and the set

$$W_\alpha^1 = \left\{ \varphi \in W_\alpha \mid \int_{\mathbf{R}^n} \varphi(\|x\|_2) dx = 1 \right\}$$

are well defined. Let $F: D \subseteq \mathbf{R}^n \to \mathbf{R}^n$ be continuously differentiable on the open set D, C an open, bounded set with $\bar{C} \subseteq D$, and $y \notin F(C)$ a given point.

For any weight function φ of index

$$\alpha < \dot{\gamma} = \min\{\|F(x) - y\|_2 \mid x \in \dot{C}\},$$

define the operator $\varphi: \mathbf{R}^n \to \mathbf{R}^n$,

$$\varphi(x) = \begin{cases} \varphi\left(\|F(x) - y\|_2\right) \det F'(x), & x \in C \\ 0, & \text{otherwise}. \end{cases}$$

Then the integral

$$d_\varphi(F, C, y) = \int_{\mathbf{R}^n} \varphi(x)dx$$

is called the degree integral F on C with respect to y and the weight function φ.

11. Let X be a Hilbert space. If $F: X \to X$ is continuous and monotone, and satisfies $(x, F(x)) \geq 0$ for all $x \notin \bar{U}(0, r)$ for some $r > 0$, then show that the equation $F(x) = 0$ has a solution in $\bar{U}(0, r)$.

12. Let C be a closed, bounded convex set in a Banach space X and G_1, $G_2: C \to X$ such that (a) $G_1(x) + G_2(y) \in C$ for all $x, y \in C$, (b) G_1 is contractive on C, (c) G_2 is continuous and compact. Then $G_1 + G_2$ has a fixed point in C.

13. Suppose $H: C \times C \to C$ satisfies the conditions (a) there is a $c < 1$ so that, for each fixed $y \in C$, $H(\cdot, y)$ is a c-contraction; (b) there is a continuous compact operator $H_1: C \to X_1$, where X_1 is another Banach space, such that $\|H(x, y) - H(x, z)\| \leq \|H_1(x) - H_1(z)\|$ for all $x, y, z \in C_1$. Then there exists an $x^* \in C$ such that $H(x^*, x^*) = x^*$.

14. Let $F: \mathbf{R}^n \to \mathbf{R}^n$ be continuous and satisfy $\|F(x)\| \leq c_1\|x\| + c_2$ for all $x \in \mathbf{R}^n$, where $0 \leq c_1 < 1, c_2 > 0$. Then F has a fixed point.

15. Let C be open and bounded, and $F: \bar{C} \subseteq \mathbf{R}^n \to \mathbf{R}^n$ continuous. Suppose further, that $G: \bar{C} \subseteq \mathbf{R}^n \to \mathbf{R}^n$ is any continuous operator such that $F(x) = G(x)$ for $x \in \dot{C}$. Then for any $y \notin F(\dot{C})$ show that

$$\deg(F, C, y) = \deg(G, C, y).$$

16. Let $F: D \subseteq \mathbf{R}^n \to \mathbf{R}^n$ be continuously differentiable on the open set D, and C an open, bounded set with $\bar{C} \subset D$. Suppose further that, for given $y \notin F(\dot{C})$, $F'(x)$ is nonsingular for all $x \in \Gamma = \{x \in C \mid F(x) = y\}$. Then show that Γ consists of at most finitely many points, and there exists a

$$0 < \hat{\alpha} \leq \dot{\gamma} = \min\{\|F(x) - F(y)\| \mid x \in \dot{C}\}$$

with $\alpha \in (0, \hat{\alpha})$

$$d_\varphi(F, C, y) = \begin{cases} \sum_{j=1}^m \text{sgn det } F'(x^j), & \text{if } \Gamma = \{x^1, \ldots, x^m\} \\ 0, & \text{if } \Gamma \text{ is empty .} \end{cases}$$

Moreover, show that the above conclusion holds for any real polynomial.

17. Let $C \subset \mathbf{R}^n$ be an open, bounded set containing the origin and $G: \bar{C} \to \mathbf{R}^n$ a continuous operator. If $G(x) \neq \lambda x$ whenever $\lambda < 1$ and $x \in \dot{C}$, then G has a fixed point.

18. Let $L \in L(\mathbf{R}^n)$ be nonsingular and assume $\varphi: \mathbf{R}^n \to \mathbf{R}^n$ is continuous. Suppose that for some $r > 0$ every solution of $L(x) - t\varphi(x) = 0$ lies in $\bar{U}(0, r)$ for all $t \in [0, 1]$. Show that $L(x) = \varphi(x)$ has a solution in $\bar{U}(0, r)$.

19. Let $C \subseteq \mathbf{R}^n$ be an open bounded set and assume that $G: \bar{C} \to \mathbf{R}^n$ is continuous. Suppose there is an $x_0 \in C$ such that $G(x) \neq \lambda x + (1 - \lambda)x_0$ whenever $x \in \dot{C}$ and $\lambda > 1$.

20. Let $C \subseteq \mathbf{R}^n$ be an open, bounded set containing the origin and assume that $G: \bar{C} \to \mathbf{R}^n$ is continuous. If $x^T G(x) \leq x^T x$ for all $x \in \dot{C}$, then G has a fixed point in $\bar{C}$.

21. Let $C \subseteq \mathbf{R}^n$ be a convex, open set containing the origin, and assume that $G: \bar{C} \to \mathbf{R}^n$ is continuous. If $G\dot{C} \subset \bar{C}$, then G has a fixed point in $\bar{C}$.

22. Show that operator B defined in (5.52) is compact.

23. Show that the linear completely continuous operators form a closed ideal.

24. Show that operator W in Example 5.7 is completely continuous.

25. Show that if T is a compact linear operator between Banach spaces, then T is completely continuous.

26. Show that completely continuous linear operators are compact if the domain space is reflexive.

27. Show that $\mathcal{MK}_r(X, Z)$ is closed in the operator topology.

28. Show that operators S_n given by (5.70) are uniformly bounded in norm.

29. Show that $W \in BL(X \times Y, Z)$ is precompact if and only if W^x is compact, where

$$W^x : Z' \to B(X \times Y), \quad (W^x z')(x, y) = z'W(x, y) .$$

30. Show that the precompact bilinear operators in $B(X \times Y, Z)$ form a closed bideal.

31. Show that if $B \in BL(X \times Y, Z)$ is precompact, then $\text{Rang}(B)$ is separable.

32. Let X, Y, Z be normed linear spaces, $\varepsilon > 0$. If $B \in BL(X \times Y, Z)$ is pre-compact, then show there exists a linear subspace K of c_0 and precompact operators $B_1 \in BL(X \times Y, K)$, $A \in L(K, Z)$ such that

    (1) $B = A \circ B_1$,

    (2) $\|A\| \leq 1$, $\|B_1\| < \|B\|(1 + \varepsilon)$.

33. Let $B \in BL(X \times Y, Z)$ be compact, Z a B-space. Show that B^x sends bounded weak-convergent nets to norm-convergent nets.

34. Let X and Y be normed linear spaces. Then show $B \in BL(X \times Y, c_0)$ has the form $B(x, y) = \{b_n(x, y)\}_n$ with $\|b_n\| \to 0$, $b_b \in B(X \times Y)$ if and only if B is compact.

35. Show that a precompact linear mapping $A \in L(X, Y)$ has a precompact factorization through c_0 if and only if A is infinite-nuclear.

36. Let Z be an $\mathcal{L}_\infty$ space, X and Y normed linear spaces. Show that the compact and infinite number bilinear maps in $BL(X \times Y, Z)$ coincide.

37. Show that c_0 is an $\mathcal{L}_\infty$-space.

Chapter 6

General Methods for Solving Nonlinear Equations

In this chapter we introduce Newton-like methods to solve nonlinear equations. In Section 6.1 we discuss the accessibility of solutions of nonlinear equations using Newton-like methods with applications to the solution of quadratic equations [51]. Section 6.2 deals with the Super-Halley method [52]. Convergence rates for inexact Newton-like methods at singular points are studied in Section 6.3 [53]. Inexact Newton-like methods and their discretizations are discussed in Section 6.4 [50]. Convergence domains for inexact Newton-like methods in Banach spaces using outer and generalized inverses are studied in Section 6.5 [54]. Finally, in Section 6.6 we deal with inexact Newton-like methods on Banach spaces with a convergence structure [47].

6.1 Accessibility of Solutions of Equations by Newton-Like Methods and Applications

In this section we are concerned with the problem of approximating a locally unique solution x^* of the equation

$$F(x) = 0 \qquad (6.1)$$

where F is an operator defined on a closed convex domain D of a Banach space X with values in a Banach space Y. We use Newton-like methods of the form

$$x_{n+1} = x_n - A(x_n)^{-1} F(x_n), \quad (n \geq 0) \ (x_0 \in D) \qquad (6.2)$$

to generate an iteration $\{x_n\}$ $(n \geq 0)$ converging to x^*. Here $A(x) \in L(X, Y)$ $(x \in D)$ which is the space of bounded linear operators from X into Y. For

$A(x) = F'(x)$ $(x \in D)$ we obtain Newton's method [60]. Several other choices for A are also possible. We define the operator $P : D \subseteq X \to Y$ by

$$P(x) = x - A(x)^{-1} F(x) , \tag{6.3}$$

in which case (6.2) can also be written as

$$x_{n+1} = P(x_n) \quad (n \geq 0) \, (x_0 \in D) . \tag{6.4}$$

Sufficient conditions for the convergence of iteration $\{x_n\}$ $(n \geq 0)$ to x^* have been given by several authors (see, for example, [60] and the references there).

A solution x^* of Equation (6.1) is said to be accessible from x_0 by Newton-like method (6.2) if

$$x^* = \lim_{n \to \infty} x_n = \lim_{n \to \infty} P^n(x_0) . \tag{6.5}$$

The region of accessibility of x^* by method (6.2) is defined to be the set of all x_0 such that (6.5) is true.

Let us define operator $L_F \in L(X, Y)$ by

$$L_F(x) = P'(x) \quad (x \in D) . \tag{6.6}$$

This operator is the degree of logarithmic convexity of F in x and is a measure of the convexity of the function. It was used in [101] in the special case when $A(x) = F'(x)$ $(x \in D)$. These convergence results were used to find starting points x_0 lying outside previously found convergence regions, for which (6.2) converges to x^* in this case. However, this was done only for scalar as well as systems of equations when $X = Y = \mathbf{R}$. This is because for convergence we need to show $\|L_F(x)\| \leq c < 1$, and this is a very difficult problem in general.

Here we provide sufficient convergent conditions for our method (6.2) to a locally unique solution x^* of Equation (6.1). Our results reduce to the corresponding ones in [6]. Moreover, we show how to compute c for quadratic equations on X. We also suggest how to compute c for polynomial equations on X of degree $k \in N$.

Finally we show how to apply our results to solve quadratic integral equations appearing in radiative transfer (see Chapters 1 through 4).

Using contraction mapping techniques, we obtain the semilocal convergence results:

THEOREM 6.1

Let $F : D \subseteq X \to Y$ be Fréchet-differentiable on a closed convex domain D, $A(x) \in L(X, Y)$ for all $x \in D$. Assume:

(a)   linear operator $\Gamma(x) = A(x)^{-1}$ exists and is differentiable for all $x \in D$;

(b) *linear operator $L_F(x)$ exists on D and*

$$\|L_F(x)\| \le c < 1 \quad \text{for all } x \in D ; \tag{6.7}$$

(c) *for $x_0 \in D$, $r^* \ge \frac{\|x_0 - P(x_0)\|}{1-c}$, $\bar{U}(x_0, r^*) \subseteq D$.*

Then Newton-like iteration $\{x_n\}$ ($n \ge 0$) generated by (6.2) is well defined, remains in $\bar{U}(x_0, r^*)$ for all $n \ge 0$ and converges to a fixed point x^* of P in $\bar{U}(x_0, r^*)$ which is unique in D. Moreover, the following error bounds are true for all $n \ge 0$

$$\|x_n - x^*\| \le c^n r^* . \tag{6.8}$$

PROOF   Newton-like iteration $\{x_n\}$ ($n \ge 0$) is well defined on D for all $x_0 \in D$ since linear operator $A(x)$ is invertible on D. Using induction on $n \ge 0$ we can show

$$x_n \in U\left(x_0, r^*\right) \quad \text{and} \quad \|x_n - x_0\| \le \left(1 - c^n\right) r^* < r^* . \tag{6.9}$$

For $n = 1$ and hypothesis (c) we have $\|x_1 - x_0\| \le (1 - c)r^* < r^*$, which shows (6.9) in this case. Assume that (6.9) is true for all positive integers smaller or equal to n. Then we must show

$$x_{n+1} \in \bar{U}\left(x_0, r^*\right) \quad \text{and} \quad \|x_{n+1} - x_0\| \le \left(1 - c^{n+1}\right) r^* < r^* .$$

By hypothesis (b) and (6.4) we get

$$\|x_{n+1} - x_n\| = \|P(x_n) - P(x_{n-1})\|$$

$$\le \sup_{z \in [x_{n-1}, x_n]} \|P'(z)\| \|x_n - x_{n-1}\| \le c \|x_n - x_{n-1}\|$$

$$\le \cdots \le c^n \|x_1 - x_0\| = c^n(1 - c)r^* , \tag{6.10}$$

and

$$\|x_{n+1} - x_0\| \le \|x_{n+1} - x_n\| + \|x_n - x_0\| \le c^n(1 - c)r^* + \left(1 - c^n\right) r^*$$

$$= \left(1 - c^{n+1}\right) r^* < r^* ,$$

which completes the induction. Moreover, by (6.10) we obtain, for $n, m \in N$

$$\|x_{n+m} - x_n\| \le \left(1 - c^m\right) c^n r^* . \tag{6.11}$$

Estimate (6.11) shows that $\{x_n\}$ $(n \geq 0)$ is a Cauchy sequence in a Banach space Y and as such it converges to a limit $x^* \in \bar{U}(x_0, r^*)$ [since $\bar{U}(x_0, r^*)$ is a closed set]. By taking the limit as $n \to \infty$ is (6.4) and using the continuity of F and $A(x)$, we deduce $P(x^*) = x^*$. To show uniqueness, let $y^* \in D$ with $P(y^*) = y^*$. Then we can get

$$\|x^* - y^*\| = \|P(x^*) - P(y^*)\| \leq \sup_{z \in [x^*, y^*]} \|P'(z)\| \|x^* - y^*\|$$

$$\leq c \|x^* - y^*\|$$

which implies that $x^* = y^*$ (since $c \in [0, 1)$).

Finally, letting $n \to \infty$ in (6.11) we obtain (6.8).

That completes the proof of the theorem. ∎

The region of accessibility to x^* is extended to a closed ball around x_0 as the following result indicates:

THEOREM 6.2

Consider the iteration $y_{n+1} = P(y_n)$ for $y_0 \in \bar{U}(x_0, r^*)$ under the hypotheses of Theorem 6.1. Then iteration $\{y_n\}$ $(n \geq 0)$ is well defined, remains in $\bar{U}(x_0, r^*)$ and converges to a unique fixed point x^* of P in $\bar{U}(x_0, r^*)$. Moreover, the following error bounds are true for all $n \geq 0$:

$$\|y_n - x^*\| \leq \frac{c^n}{1-c} \|y_1 - y_0\| \quad \text{and} \quad \|y_n - x^*\| \leq c^n \|x^* - y_0\| .$$

PROOF   The result follows immediately by the contraction mapping principle provided we show that operator P maps $\bar{U}(x_0, r^*)$ into itself. Indeed let $x \in U(x_0, r^*)$, then we obtain

$$\|x_1 - P(x)\| = \|P(x_0) - P(x)\| \leq c \|x_0 - x\| \leq cr^* .$$

That completes the proof of the theorem. ∎

For $D = \bar{U}(v_0, r_0)$ we can obtain immediately from Theorem 6.2:

COROLLARY 6.1

If $\|v_0 - x_0\| \leq r_0 - r^*$ under the hypotheses of Theorem 6.2 Newton-like iteration $\{x_n\}$ $(n \geq 0)$ converges to x^* for any starting point in $\bar{U}(x_0, r^*)$.

In terms of the degree of logarithmic convexity we have the following result concerning the convergence of Newton-like method (6.2).

THEOREM 6.3

Assume that hypothesis (b) of Theorem 6.1 holds on $D = \bar{U}(v_0, r_0)$. If $\|A(v_0)^{-1}F(v_0)\| \leq (1-c)r_0$, Newton-like method $\{x_n\}$ $(n \geq 0)$ generated by (6.2) converges to the unique solution x^* of equation $F(x) = 0$ in D for any $x_0 \in D$.

PROOF We note that a fixed point P is a solution of equation $F(x) = 0$. The result now follows immediately from the proofs of the previous theorems. ∎

REMARK 6.1 For $A(x) = F'(x)$ $(x \in D)$, Theorems 6.1, 6.2, and 6.3, and Corollary 6.1, reduce respectively to Theorems 2.1, 2.2, and 2.4 and Corollary in [101]. ∎

The verification of condition (b) of Theorem 6.1 is a very hard problem in general. In [101] and the references there the authors verified this condition for scalar as well as systems of real or complex equations. Here we suggest a possible extension of our results in the case of quadratic equations of the form

$$F(x) = y + B(x, x) - x \tag{6.12}$$

where B is a bounded symmetric bilinear operator on $D \subseteq X$ and $y \in X$ is fixed. Hence, in the case of F given by (6.12) we obtain from (6.6) for $A(x) = F'(x)$ $(x \in D)$

$$L_F(x)(z) = 2(2B(x) - I)^{-1}B(2B(x) - I)^{-1}(z)(-x + y + B(x, x)). \tag{6.13}$$

Let $x_0 \in D$ be such that $(F'(x_0))^{-1} = (2B(x_0) - I)^{-1}$ exists and set $b \geq \|(2B(x_0) - I)^{-1}B\| \neq 0$. Let $r \in [0, \frac{1}{2b})$, and assume $\bar{U}(x_0, r) \subseteq D$. Then for $x \in \bar{U}(x_0, r)$ we have

$$2B(x) - I = (2B(x) - I) - (2B(x_0) - I) + (2B(x_0) - I)$$

$$= (2B(x_0) - I)\left[I + 2(2B(x_0) - I)^{-1}B(x - x_0)\right] \tag{6.14}$$

and

$$\left\| 2\left(2B\left(x_0\right)-I\right)^{-1}B\left(x-x_0\right)\right\| \leq 2\left\|\left(2B\left(x_0\right)-I\right)^{-1}B\right\|\left\|x-x_0\right\|$$

$$\leq 2br < 1. \tag{6.15}$$

It follows from the Banch Lemma on invertible operators [114], that $F'(x) = 2B(x) - I$ is invertible on $\bar{U}(x_0, r)$ and

$$\left\|\left(2B(x)-I\right)^{-1}\left(2B\left(x_0\right)-I\right)\right\| \leq (1-2br)^{-1}. \tag{6.16}$$

Moreover we have by (6.12)

$$\left(2B\left(x_0\right)-I\right)^{-1}F(x) = \left(2B\left(x_0\right)-I\right)^{-1}\left[\left(x_0-x\right)+\left(y-x_0\right)\right.$$

$$\left. + B\left(\left(x-x_0\right)+x_0,\left(x-x_0\right)+x_0\right)\right]$$

$$= \left(2B\left(x_0\right)-I\right)^{-1}\left(2B\left(x_0\right)-I\right)\left(x-x_0\right)$$

$$+ \left[\left(2B\left(x_0\right)-I\right)^{-1}B\right]\left(x-x_0, x-x_0\right)$$

$$+ \left(2B\left(x_0\right)-I\right)^{-1}\left(y-x_0+B\left(x_0, x_0\right)\right).$$

By taking norms, using the triangle inequality and (6.4) we get

$$\left\|\left(2B\left(x_0\right)-I\right)^{-1}F(x)\right\| \leq \left\|x-x_0\right\| + \left\|\left(2B\left(x_0\right)-I\right)^{-1}B\right\|$$

$$\cdot\left\|x-x_0\right\|^2 + \left\|x_0-P(x_0)\right\| \leq r+br^2+\left\|x_0-P\left(x_0\right)\right\|. \tag{6.17}$$

It follows from (6.13), (6.16), and (6.17) that

$$\left\|L_F(x)\right\| \leq c(r), \quad r \in \left[0, \frac{1}{2b}\right), \tag{6.18}$$

where

$$c(r) = \frac{2b}{(1-2br)^2}\left(r+\delta+br^2\right), \quad \delta \geq \left\|x_0-P\left(x_0\right)\right\|. \tag{6.19}$$

Define the scalar function h on $[0, +\infty)$ by

$$h(r) = c_1 r^2 + c_2 r + c_3 \tag{6.20}$$

where

$$c_1 = -2b^2, \quad c_2 = 6br \quad \text{and} \quad c_3 = 2b\delta - 1.$$

It is simple algebra to show that $c(r) \in [0, 1)$ if

$$h(r) < 0 \quad \text{and} \quad r \in \left[0, \frac{1}{2b}\right). \tag{6.21}$$

It can easily be seen that (6.21) is true if

$$q = 2b\delta < 1 \tag{6.22}$$

and

$$r \in [0, a)$$

where

$$a = \frac{3 - \sqrt{7 + 2q}}{2b}. \tag{6.23}$$

Note that estimate (6.22) is the Newton–Kantorovich hypothesis for Equation (6.12) and a is the smallest zero of the scalar equation $h(r) = 0$ where h is given by (6.20).

Hence we arrive at:

THEOREM 6.4

Let $F : U(x_0, a) \subseteq X \to Y$ be given by (6.12), and $A(x) = F'(x)$ $(x \in D)$ in (6.2). Assume that the Newton–Kantorovich hypothesis (6.22) is true for some $x_0 \in U(x_0, a)$ at which $F'(x_0)$ is invertible. Then (6.7) is true for all $r \in [0, a)$. Moreover, if there exists a minimum nonnegative number $r^* \in [0, a)$ satisfying the inequality

$$r \geq \frac{\delta}{1 - c(r)}, \tag{6.24}$$

then the conclusions of Theorem 6.1 for Equation (6.12) and iteration (6.2) are true.

PROOF It follows immediately from the above discussion, the proof of Theorem 6.1 and the observation that (6.24) is true if $g(r^*) \geq 0$ where g is a function defined on $[0, +\infty)$ given by

$$g(r) = d_1 r^3 + d_2 r^2 + d_3 r + d_4, \tag{6.25}$$

$d_1 = 2b^2$, $d_2 = -2b(3+q)$, $d_3 = 1+q$ and $d_4 = -\delta$.   ∎

REMARK 6.2   By Descartes' rule of signs the equation $g(r) = 0$ has three positive zeros or one. Let s denote the smallest such zero in either case. We note that Equation (6.25) can have zeros in $[0, a)$ even if $g(a) < 0$. However, it is simple algebra to check that $g(a) < 0$. Hence we can set $r^* = s$ in this case.   ∎

REMARK 6.3   Another approach will be to define the function g_1 on $[0, +\infty)$ by $g_1(r) = d_2 r^2 + d_3 r + d_4$. We note that by (6.25) $g_1(r) \geq 0$ implies $g(r) \geq 0$ for all $r \in [0, +\infty)$. The discriminant of this quadratic polynomial is nonnegative if $q \in \left[0, \frac{2\sqrt{7}-5}{3}\right]$. Let $t_1 \leq t_2$ be the real zeros of the equation $g_1(r) = 0$. Then we easily deduce $t_1 < a$. Hence in this case we can set $r^* = t_1$ in Theorem 6.4.
∎

The Newton–Kantorovich Theorem [114] for Equation (6.12) asserts that if hypothesis (6.22) is satisfied, then $x^* \in U(x_0, r_k)$ where $r_k = \frac{1-\sqrt{1-q}}{b}$. We easily show:

(i)   if $q \in \left[0, \frac{2\sqrt{13}-5}{9}\right)$ then $r_k < a$;

(ii)   if $q \in \left[\frac{2\sqrt{13}-5}{9}, 1\right]$ then $r_k \geq a$;

(iii) $g(r_k) \leq 0$ and $g(r_k) = 0$ if $q = 0$ ($r_k = 0$ in this case).

At the end of this study we provide an example where $x^* \in \bar{U}(x_0, r_k) \subseteq \bar{U}(x_0, t_1)$.

REMARK 6.4   Theorem 6.4 is a crude application of Theorem 6.1. In practice one hopes that (6.7) will be satisfied in cases that do not imply the Newton–Kantorovich hypothesis (6.22). Examples where Newton's method converges but (6.22) is violated were given in [101] for scalar or systems of real equations and in [15, 57] for quadratic integral equations on various Banach spaces. See also the example that follows.   ∎

REMARK 6.5   In conclusion, we note that both Newton–Kantorovich and Theorem 6.4 apply if condition (6.22) is satisfied. However, the balls centered at the same point x_0 that contain the solution x^* are not of the same radius.
Let the ring $U = \bar{U}(x_0, r^*) - \bar{U}(x_0, r_k) \neq \emptyset$. Then there exists a starting point $w_0 \notin U(x_0, r_k)$ such that Newton's method (6.2) converges. However, the Newton–Kantorovich Theorem does not guarantee convergence in this case. Hence, there exists a region of accessibility for the convergence of Newton's

method that is missed by the Newton–Kantorovich Theorem. We confront such a case in the example that follows. ∎

Example 6.1
Special cases of (6.12) are quadratic integral equations of the form

$$F(x)(s) = y(s) + \lambda x(s) \int_0^1 k(s,t)x(t)dt - x(s) = 0$$

in the space $Y = X = C[0, 1]$.
To apply Theorem 6.4 we need to compute b and δ initially. For example, choose $k(s,t) = \frac{s}{s+t}$, $k(0, 0) = 1$, $s, t \in (0, 1]$, and define

$$B(x, y)(s) = \frac{1}{2}\lambda \left[x(s) \int_0^1 \frac{s}{s+t} y(t)dt + y(s) \int_0^1 \frac{s}{s+t} x(t)dt \right].$$

Choose as an example $x_0(s) = y(s) = 1$ for all $s \in [0, 1]$, $\lambda = .2$, then as in Example 1.4 we obtain

$$\| F'(1)^{-1} \| \leq 1.3836213, \quad a = .8513131,$$

$$b = \delta \leq .1918106, \quad q = .0735826, \quad r_k = .1954752, \quad t_1 = .242025 = r^*$$

and $c(r^*) = .2074761$. All hypotheses of Theorem 6.4 are satisfied for $r^* = t_1$ and $\bar{U}(x_0, r_k) \subseteq \bar{U}(x_0, t_1)$ (also see Remark 6.5).
Finally, the results of Theorem 6.4 can be extended to include polynomial operator equations (see Chapter 3).

6.2 The Super-Halley Method

In this section we are concerned with the problem of approximating a locally unique solution x^* of Equation (6.1).
We recommend the Super-Halley method $\{x_n\}$ $(n \geq 0)$ defined for all $n \geq 0$ by

$$L_F(x) = F'(x)^{-1} F''(x) F'(x)^{-1} F(x) \quad (x \in D), \tag{6.26}$$

$$M(x) = [I - L_F(x)]^{-1} \quad (x \in D), \tag{6.27}$$

$$M_n = M(x_n) \,, \tag{6.28}$$

$$x_{n+1} = x_n - \left[I + \frac{1}{2}L_F(x_n)M_n\right]F'(x_n)^{-1}F(x_n) \,, \tag{6.29}$$

for some initial guess $x_0 \in D$.

The Super-Halley method has a fairly elaborate background [39, 44, 46, 52, 57, 60, 102, 107, 108]. Using a quadratic majorizing sequence, we showed that this method is in general of order three, and for quadratic polynomial operators F it is of order four. The same hypotheses were used in [102] to show that this method is of order four if the second Fréchet derivative $F''(x)$ of F is bounded on D. These results were both given in a nonaffine invariant form.

Here we use a cubic instead of a quadratic majorizing polynomial, and under weaker assumptions we show that method (6.29) is at least of order four. In cases where it is exactly four the rate of convergence in [102] is improved. Another advantage of our results is that they are given in an affine invariant form. The advantages of results given in this form were explained in [94].

Finally, numerical examples are given to show that our earlier bounds are better than later ones.

For some $\alpha > 0$, β, $\gamma \geq 0$, it is convenient for us to define the real function f by

$$f(t) = \frac{\alpha}{6}t^3 + \frac{\beta}{2}t^2 - t + \gamma \,. \tag{6.30}$$

We need the following result on the existence of roots for function f.

PROPOSITION 6.1

The real function f given by (6.30) has two positive roots r_2, r_3 with $r_2 \leq r_3$ and a negative root r_1 if and only if the following estimate holds

$$\gamma \leq \frac{2(2d + \beta)}{3(d + \beta)^2}, \quad d = \sqrt{\beta^2 + 2\alpha} \,. \tag{6.31}$$

PROOF   By the intermediate value theorem function f given by (6.30) has negative root r_1, since $f(0) > 0$ and $f(t) < 0$ as $t \to -\infty$. Moreover, there exists a root denoted by r^* of f' in $(0, \infty)$, since $f'(0) = -1$ and $f'(t) > 0$ as $t \to +\infty$. Setting $f'(t) = 0$, we obtain

$$r^* = \frac{d - \beta}{\alpha} \,. \tag{6.32}$$

Function f has two positive roots r_2, r_3 with $r_2 \leq r_3$ if and only

$$f\left(r^*\right) \leq 0, \tag{6.33}$$

which is true by (6.31).

That completes the proof of Proposition 6.1. ∎

Let us define the real functions L_f and G_f by

$$L_f(t) = \frac{f(t) f''(t)}{f'(t)^2}, \tag{6.34}$$

$$G_f(t) = t - \left[1 + \frac{1}{2} L_f(t)(1 - L_f(t))^{-1} \right] \frac{f(t)}{f'(t)}, \tag{6.35}$$

and the iteration $\{t_{n+1}\}$ $(n \geq 0)$ by

$$t_{n+1} = G_f(t_n), \quad t_0 = 0 \quad (n \geq 0). \tag{6.36}$$

We can prove the following result on the monotonicity of iteration (6.36).

PROPOSITION 6.2
Assume that condition (6.31) is satisfied. Then the following hold:

(i) $f(t) > 0$ *for all* $t \in [0, r_2)$;

(ii) $f'(t) < 0$ *for* $t \in [0, r^*)$, *and* $f^{(k)}(t) > 0$, $t \geq 0$, $k = 2, 3$;

(iii) $0 \leq L_f(t) \leq \frac{1}{2}$ *for all* $t \in [0, r_2]$; *and*

(iv) *iteration* $\{t_n\}$ $(n \geq 0)$ *given by (6.36) is well defined for all* $n \geq 0$, *monotonically increasing and bounded above by its limit* r_2.

PROOF   Statements (i), (ii) follow immediately by Proposition 6.1. For (iii) let $t \in [0, r_2]$, then function f given by (6.30) can be rewritten as

$$f(r) = f(t) + f'(t)(r - t) + \frac{f''(t)}{2}(r - t)^2 + \frac{f'''(t)}{6}(r - t)^3.$$

Define function f_1 by

$$f_1(r) = f(r) - \frac{f''(t)}{6}(r - t)^3.$$

Then we have $f_1(t) = f(t) \geq 0$ and $f_1(r_2) = -\frac{f'''(t)}{6}(r_2 - t)^3 \leq 0$ for $t \leq r_2$. By the intermediate value theorem there exists a real root of f_1 in $(t, r_2]$. Hence the discriminant of f_1 must be greater or equal to 0, which shows (iii). Finally part (iv) follows easily using mathematical induction on the integer n, and simple calculus to show $G'_f(t) > 0$ for all $t \in [0, r_2]$.

That completes the proof of Proposition 6.2.   ∎

Under the hypothesis of Proposition 6.2, set

$$a_n = r_2 - t_n, \quad b_n = r_3 - t_n \quad \text{and} \quad c_n = t_n - r_1 \quad (n \geq 0). \tag{6.37}$$

Then using (6.30), (6.34), (6.35), (6.36), and (6.37) we deduce

$$\frac{a_{n+1}}{b_{n+1}} = h_n \left(\frac{a_n}{b_n} \right)^3 \tag{6.38}$$

where

$$h_n = p_n \cdot q_n^{-1} \quad (n \geq 0), \tag{6.39}$$

$$p_n = b_n^3 a_n + a_n b_n c_n^2 - c_n b_n^2 a_n - a_n c_n^3 - b_n^2 c_n^2 \quad (n \geq 0) \tag{6.40}$$

and

$$q_n = a_n^3 b_n + a_n b_n c_n^2 - a_n^2 b_n c_n - b_n c_n^3 - a_n^2 c_n^2 \quad (n \geq 0). \tag{6.41}$$

Moreover denote by $a(r)$, $b(r)$, $c(r)$, $h(r)$, $p(r)$, $q(r)$ the functions defined by relations (6.37), (6.39), (6.40), and (6.41) by simply replacing t_n by r ($n \geq 0$). Furthermore, assume that there exist nonnegative constants k_1, k_2, m_1, m_2 such that

$$k_1 \left(\frac{a(r)}{b(r)} \right)^{m_1} \leq h(r) \leq k_2 \left(\frac{a(r)}{b(r)} \right)^{m_2} \quad \text{for all } r \in [0, r_2], \tag{6.42}$$

then

$$\frac{1}{k_4} (k_4 \theta)^{m_4^n} \leq \frac{a_{n+1}}{b_{n+1}} \leq \frac{1}{k_3} (k_3 \theta)^{m_3^n} \quad (n \geq 0) \tag{6.43}$$

where

$$k_4 = k_1^{\frac{1}{m_1 - 1}}, \, k_3 = k_2^{\frac{1}{m_2 - 1}}, \, \theta = \frac{r_2}{r_3},$$

$$m_3 = 3 + m_2 \quad \text{and} \quad m_4 = 3 + m_1. \tag{6.44}$$

That is, we showed:

PROPOSITION 6.3

Under the hypothesis of Proposition 6.2, assume estimate (6.42) is true. Then error bounds (6.43) hold.

REMARK 6.6   We hope that we can find $m_2 \geq 1$ and k_2 small so that (6.42) is satisfied [especially the right-hand side inequality, which will be used to majorize $\|x_{n+1} - x^*\|$ $(n \geq 0)$]. It can easily be seen that for $k_2 = m_2 = 1$, the right-hand side of inequality (6.42) reduces to showing

$$p(t) \geq 0 \quad \text{for all } t \in [0, r_2] \ , \tag{6.45}$$

where

$$p(t) = \left[(r_2 - t)^2 + (r_3 - t)^2 \right]$$

$$(t - r_1)^2 + (r_2 - t)(r_3 - t)(r_2 + r_3 - 2t)(t - r_1) - (r_2 - t)$$

$$(r_3 - t) \left[(r_2 - t)^2 + (r_2 - t)(r_3 - t) + (r_3 - t)^2 \right] . \tag{6.46}$$

By expanding and rearranging (6.45), we can write

$$p(t) = d_0 t^4 + d_1 t^3 + d_2 t^2 + d_3 t + d_4 \ , \tag{6.47}$$

where

$$d_0 = -3, \ d_1 = 7(r_2 + r_3) - 2r_1 \ , \ d_2 = 2r_1^2 - 5r_3^2 + r_1 r_2 - r_1 r_3 - 14 r_2 r_3 \ ,$$

$$d_3 = 4r_1 r_2 r_3 - r_1 r_2^2 - 2r_2 r_1^2 - r_1 r_3^2 - 2r_1^2 r_3 + 6r_2^2 r_3 + 6r_2 r_3^2 + r_2^3 + r_3^3 \ ,$$

$$d_4 = r_1^2 r_2^2 + r_3^2 r_1^2 - r_1 r_3 r_2^2 - r_1 r_2 r_3^2 - r_3 r_2^3 - r_2^2 r_3^3 - r_2 r_3^3 \ . \quad \blacksquare$$

Note that $p(r_2) > 0$ (for $r_2 \neq r_3$), which means that there exists $\varepsilon > 0$, $\varepsilon \leq r_2$ such that $p(t) > 0$ for $t \in [0, r_2 - \varepsilon]$. Hence, even if (6.45) is not true for all $t \in [0, r_2]$, after a certain n_0 the right-hand side inequality in (6.42) will hold.

We can now state and prove the main semilocal convergence theorem for the Super-Halley method generated by (6.29).

THEOREM 6.5

Let $F: D \subseteq X \to Y$ be a twice differentiable operator on a closed convex domain D of a Banach space X, with values in a Banach space Y, $x_0 \in D$ being such that $F'(x_0)^{-1}$ exists. Moreover, assume:

(a)   there exists $\alpha > 0$, $\beta, \gamma \geq 0$ satisfying

$$\left\| F'(x_0)^{-1} F(x_0) \right\| \leq \gamma , \tag{6.48}$$

$$\left\| F'(x_0)^{-1} F''(x_0) \right\| \leq \beta , \tag{6.49}$$

$$\left\| F'(x_0)^{-1} \left(F''(x) - F''(y) \right) \right\| \leq \alpha \|x - y\| , \tag{6.50}$$

and

$$\gamma \leq \frac{2(2d + \beta)}{3(d + \beta)^2} , \quad d = \sqrt{\beta^2 + 2\alpha} ,$$

for all $x \in U(x_0, r_3)$, where r_3 is the largest root of equation $f(t) = 0$, with f given by (6.30).

(b) $U(x_0, r_3) \subseteq D$.

Then the following hold:

(i) $\|F'(x_0)^{-1} F''(x)\| \leq f''(\|x - x_0\|) \ (x \in U(x_0, r_3))$;

(ii) $F'(x)^{-1}$ exists in $U(x_0, r_2)$ and

$$\left\| F'(x)^{-1} F'(x_0) \right\| \leq -\frac{1}{f'(\|x - x_0\|)} ;$$

(iii) $F'(x_n)^{-1} \ (n \geq 0)$ exists;

(iv) $\|F'(x_0)^{-1} F''(x_n)\| \leq -\dfrac{f''(t_n)}{f'(t_0)}, \ (n \geq 0)$;

(v) $\|F'(x_n)^{-1} F'(x_0)\| \leq \dfrac{f'(t_0)}{f'(t_n)}, \ (n \geq 0)$;

(vi) $\|F'(x_0) F(x_n)\| \leq -\dfrac{f(t_n)}{f'(t_0)}, \ (n \geq 0)$;

(vii) $\|L_F(x_n)\| \leq L_f(t_n), \ (n \geq 0)$;

(viii) $M_n = [I - L_F(x_n)]^{-1}$, *exists* $(n \geq 0)$,

$$\|M_n\| \leq (1 - L_f(t_n))^{-1}, \quad (n \geq 0)$$

and

$$\|I - M_n\| \leq L_f(t_n)(1 - L_f(t_n))^{-1}, \quad (n \geq 0);$$

(ix) *the Super-Halley iteration* $\{x_n\}$ $(n \geq 0)$ *generated by (6.29) is well defined, remains in* $\bar{U}(x_0, r_2)$ *for all* $n \geq 0$ *and converges to a solution* x^* *of equation* $F(x) = 0$ *which is unique in* $U(x_0, r_3)$. *Moreover the error bound holds for all* $n \geq 0$

$$\|x_{n+1} - x^*\| \leq a_{n+1} \leq b_{n+1} h_n \left(\frac{a_n}{b_n}\right)^2. \tag{6.51}$$

Furthermore if condition (6.42) holds then estimate (6.43) holds and

$$\|x_{n+1} - x^*\| \leq a_{n+1} \leq \frac{b_{n+1}}{k_3} (k_3\theta)^{m_3^n} \quad (n \geq 0), \tag{6.52}$$

where a_{n+1}, b_{n+1}, h_n, k_3, θ, m_3 *are given by (6.37), (6.38), (6.39), and (6.44), respectively.*

PROOF   (i) Using (6.49), (6.50) and the triangle inequality, we get

$$\left\| F'(x_0)^{-1} F''(x) \right\| \leq \left\| F'(x_0)^{-1} (F''(x) - F''(x_0)) \right\| + \left\| F'(x_0)^{-1} F''(x_0) \right\|$$

$$\leq \alpha \|x - x_0\| + \beta = f''(\|x - x_0\|) \quad \text{for all } x \in U(x_0, r_3).$$

(ii) By Taylor's formula we can have

$$F'(x_0)^{-1} F'(x) = I + \int_0^1 F'(x_0)^{-1} (F''(x_0 + t(x - x_0))$$

$$- F''(x_0)) (x - x_0) dt$$

$$+ F'(x_0)^{-1} F''(x_0)(x - x_0),$$

and

$$\left\| F'(x_0)^{-1} F'(x) - I \right\| \leq \int_0^1 \left\| F'(x_0)^{-1} \left(F''(x_0 + t(x - x_0)) \right. \right.$$

$$\left. \left. - F''(x_0) \right) \right\| \cdot \| x - x_0 \| \, dt$$

$$+ \left\| F'(x_0)^{-1} F''(x_0) \right\| \cdot \| x - x_0 \|$$

$$\leq \int_0^1 \alpha t \, \| x - x_0 \|^2 \, dt + \beta \, \| x - x_0 \|$$

$$= 1 + f'(\| x - x_0 \|) < 1 \quad \text{(by (ii) of Proposition 6.2).}$$

The result now follows from the Banach lemma on invertible operators.

At this point we need to reintroduce the approximation found in [57, 60] (see also [102]), for $y_n = x_n - F'(x_n)^{-1} F(x_n)$

$$F(x_{n+1}) = \int_{y_n}^{x_{n+1}} F''(x)(x_{n+1} - x) \, dx + \int_{x_n}^{y_n} F''(x)(x_{n+1} - y_n) \, dx$$

$$+ \int_{x_n}^{y_n} F''(x)[I - M_n](y_n - x) \, dx \tag{6.53}$$

$$+ + \int_{x_n}^{y_n} \left[F''(x) - F''(x_n) \right] M_n (y_n - x_n) \, dx \ .$$

Since $f'(t_0) = -1$ (iii)–(viii) hold for $n = 0$. Assume that they hold for all integer values smaller or equal to n. Then by (6.29), (6.53), and the induction hypotheses, we have

$$\| x_{n+1} - x_n \| \leq \left\| \left[I + \frac{1}{2} L_F(x_n)[I - L_F(x_n)]^{-1} \right] F'(x_n)^{-1} F(x_n) \right\|$$

$$\leq \left[1 + \frac{1}{2} L_f(t_n) \left(1 - L_f(t_n) \right)^{-1} \right] \frac{f(t_n)}{f'(t_n)}$$

$$= t_{n+1} - t_n \ . \tag{6.54}$$

That is $\|x_{n+1} - x_0\| \leq t_{n+1} < r_2$. By (i)–(ii), (iii)–(v) hold for $n + 1$. By (6.53) and (6.54), (vi) also holds for $n + 1$ and the conclusions of (vii)–(viii) follow from (iv)–(vi). It now follows from (6.54) and (iv) of Proposition 6.2 that the Super-Halley iteration $\{x_n\}$ ($n \geq 0$) generated by (6.29) is a Cauchy sequence in a Banach space X, and as such it converges to some $x^* \in \bar{U}(x_0, r_2)$ (since $\bar{U}(x_0, r_2)$ is a closed set). By letting $n \to \infty$ in (6.29) we obtain $F(x^*) = 0$. Moreover, standard majorization techniques (see Chapter 3) applied to (6.54) give

$$\|x^* - x_n\| \leq a_n \quad (n \geq 0) . \tag{6.55}$$

Estimate (6.38) follows from the analysis after Proposition 6.2, whereas (6.52) follows from Proposition 6.3.

The proof of the uniqueness of x^* in $U(x_0, r_3)$ is omitted, since it follows along the lines of our corresponding proofs in Section 6.1.

That completes the proof of the theorem. ∎

REMARK 6.7 Our results hold under the hypothesis $\alpha > 0$. The case $\alpha = 0$ has been handled in [57, 60, 102], or can be taken as a limiting case of results using a cubic majorizing polynomial as α approaches 0. See [60] and the references there. ∎

We now provide a simple example to show that we can achieve a better rate of convergence than the one given in [102].

Example 6.2
Let $X = Y = \mathbf{R}$, $x_0 = .9$, $D = U(x_0, r_3)$, and consider the function $F : D \to \mathbf{R}$ given by

$$F(x) = x^3 - 1 . \tag{6.56}$$

Using the above and (6.31), (6.48)–(6.50) we obtain

$$\alpha = 2.4691358, \ \beta = 2.2222222, \ \gamma = .1115226, \ d = 3.1426968 .$$

Condition (6.31) now becomes $\gamma \leq .1970559$, which is true.

By (6.30) we can have $f(t) = .4115226t^3 + 1.1111111t^2 - t + .1115226$, with roots $r_1 = -3.4312212$, $r_2 = .13175054$ and $r_3 = .5994705$. Moreover, the right-hand side of condition (6.42) holds for $k_2 = m_2 = 1$, since (6.45) becomes

$$p(t) = -3t^4 + 11.980989t^3 + 22.248857t^2 - 16.743011t + 4.5884011 \geq 0 ,$$

which is true for all $t \in [0, r_2]$.

Therefore, according to our Theorem 6.5, the Super-Halley method is of order four with ratio $\theta = \frac{r_2}{r_3} = .2197781$, and by (6.56), $x^* = 1$. To compare our ratio θ with the corresponding one in [102], using their notation we get for $\Omega_0 = U(.9, .2)$ say, $\beta = .4115226$, $N = 6$, $\eta = .1115226$, $M = 6.6$, $k = 7.1309$, $k\eta\beta = .3272669 < .5$, $r_1 = .1404776$, $r_2 = .5410616$, and $\bar{\theta} = \frac{r_1}{r_2} = .2596333 > \theta$.

That is our Theorem 6.5 provides a smaller ratio than the corresponding one obtained in [102, Thm 2.3, p. 117]. Moreover we found $t_1 = .119409502$ and $x_1 = 1.000448015$. Therefore, we can construct the following table on the corresponding error bounds [102, p. 117], (6.52) and (6.51) for $\|x_n - x^*\|$ $(n \geq 0)$.

Table 6.1 Error bounds comparison

n	Error bounds by Gutierez et al. [102, p. 117]	Our error bounds (6.52)	Our error bounds (6.51)	Actual error bounds
0	.1404776	.1317505	.1317505	.1
1	$1.828576423 \times 10^{-3}$	$1.120041823 \times 10^{-3}$	4.511445×10^{-4}	4.480146×10^{-4}

That is, our error bounds (6.52) and (6.51) are better than the corresponding ones in [102, p. 117]. Finally, our uniqueness ball $U(x_0, r_3)$ contains the corresponding one $U(x_0, r_2)$ given in [102].

6.3   Convergence Rates for Inexact Newton-Like Methods at Singular Points

In this section we are concerned with the problem of approximating a solution x^* of Equation (6.1).

We propose inexact Newton-like methods of the form

$$x_{n+1} = x_n - T(x_n)^{-1} F(x_n) - v_n, \quad v_n = v(x_n) \tag{6.57}$$

where $T(x) \in L(X, Y)$ $(x \in X)$ is an approximation to the Fréchet derivative $F'(x)$ of operator F for all $x \in X$, and $v: X \to X$ is a continuous operator. The points v_n are determined in such a way that the iteration $\{x_n\}$ $(n \geq 0)$ converges to x^*. By setting $T(x) = F'(x)$ $(x \in X)$ and $v_n = 0$ $(n \geq 0)$ we obtain the ordinary Newton method, which is known to converge quadratically to x^* provided that the initial guess is close to x^*, and the operator $F'(x^*)$ is nonsingular [65, 114, 119]. The importance of studying inexact Newton-like methods comes from the fact that many commonly used variants of Newton's method can be considered procedures of this type. Indeed, approximation (6.57) characterizes any iterative process in which the corrections are taken as approximate solutions of the Newton equations. Concerning approximation (6.57) we note that if for example an equation on the real line is solved, $F(x_n) \geq 0$ and $T(x_n)$ ($n \geq 0$) overestimates the derivative, then $x_n - T(x_n)^{-1}F(x_n)$ ($n \geq 0$) is always larger than the corresponding Newton iterate. In such cases, a positive v_n correction term is appropriate. Sufficient convergence conditions for inexact Newton-like methods have been given in [24, 37, 39, 41, 43, 200, 207] provided that $T(x^*)$ is nonsingular. Here we investigate the case when $T(x^*)$ is a Fredholm operator of index zero. It is well known that the ordinary Newton's method converges linearly rather than quadratically when operator $F'(x^*)$ is singular. Several authors have attempted to improve the rate of convergence in this case [60, 90, 171]. The limitations of their approach is twofold. First, the residuals at each step are assumed to be zero which is not the case in general. Second, these works are limited to the ordinary Newton method.

That is why in this study, motivated mainly by the elegant paper [90], we extend and improve their results by considering inexact Newton-like methods of the form (6.57). Convergence rates are provided for our methods. Our results reduce to the corresponding ones in [90] if we simply set $T(x) = F'(x)$ ($x \in X$). The proofs of some of our results are omitted since they follow similarly as the corresponding ones in [90], by simply replacing $F'(x)$ by $T(x)$ there.

Let F be a nonlinear operator defined on a Banach space X with values in a Banach space Y, $T(x) \in L(X, Y)$ ($x \in X$) be a linear operator such that at a point $x^* \in X$, $T(x^*)$ is a Fredholm operator of index zero. That is, there exists a finite dimensional subspace $N_1 \subseteq X$ and a closed subspace $X_2 \subseteq Y$ such that

$$N(T(x^*)) = N_1, \quad R(T(x^*)) = X_2, \quad \text{codim}(X_2) = \dim(N_1) .$$

We select complementing subspaces X_1, N_2 such that

$$X = N_1 \oplus X_1, \quad Y = N_2 \oplus X_2$$

and define for $i = 1, 2$ the projections P_{N_i} onto N_i parallel to X_i and $P_{X_i} = I - P_{N_i}$. Moreover, we define the operators

$$A(x) = P_{X_2}T(x)P_{X_1}, \quad B(x) = P_{X_2}T(x)P_{N_1} ,$$

$$C(x) = P_{N_2}T(x)P_{X_1}, \quad \text{and } D(x) = P_{N_2}T(x)P_{N_1} . \tag{6.58}$$

It follows easily that

$$T(x) = A(x) + B(x) + C(x) + D(x) ,$$

$$B\left(x^*\right) = 0, \; C\left(x^*\right) = 0, \; D\left(x^*\right) = 0 \tag{6.59}$$

and

$$\hat{T} = A\left(x^*\right) = P_{X_2} T(x^*) P_{X_1} \tag{6.60}$$

has an inverse as an operator from X_1 into X_2.

The following theorem provides an expression for $T(x)^{-1}$ when it exists for points x close to x^*.

THEOREM 6.6
For r sufficiently small and $x \in \bar{U}(x^*, r)$, $T(x)$ is nonsingular if and only if

$$\tilde{D}(x) = D(x) - C(x)A^{-1}(x)B(x) \tag{6.61}$$

is nonsingular as an operator from N_1 onto N_2. Moreover, for r sufficiently small and $\tilde{D}(x)$ invertible we have

$$T(x) = P_{X_1} \left[A(x)^{-1} + A(x)^{-1} B(x) \tilde{D}(x)^{-1} C(x) A^{-1}(x) \right] P_{X_2}$$

$$- P_{X_1} A(x)^{-1} B(x) \tilde{D}(x)^{-1} P_{N_2}$$

$$- P_{N_1} \tilde{D}(x)^{-1} C(x) A(x)^{-1} P_{X_2} + P_{N_1} \tilde{D}(x)^{-1} P_{N_2} . \tag{6.62}$$

PROOF   We define the operator $Q: X_1 \oplus N_1 \to X_2 \oplus N_2$ by

$$Q(x) = \begin{pmatrix} A(x) & B(x) \\ C(x) & D(x) \end{pmatrix} . \tag{6.63}$$

It follows from (6.59) that $T(x)$ is invertible if and only if $Q(x)$ is invertible. Define also the operator $\hat{Q}(x): X_2 \oplus N_2 \to X_1 \oplus N_1$ by

$$\hat{Q}(x) = \begin{pmatrix} \hat{A}(x) & \hat{B}(x) \\ \hat{C}(x) & \hat{D}(x) \end{pmatrix} \tag{6.64}$$

for $\hat{A}(x)$, $\hat{B}(x)$, $\hat{C}(x)$ and $\hat{D}(x)$ given by

$$\hat{A}(x) = P_{X_1} \left[A(x)^{-1} + A(x)^{-1} B(x) \tilde{D}(x)^{-1} C(x) A(x)^{-1} \right] P_{X_2}, \tag{6.65}$$

$$\hat{B}(x) = -P_{X_1} A(x)^{-1} B(x) \tilde{D}(x)^{-1} P_{N_2} , \tag{6.66}$$

$$\hat{C}(x) = -P_{N_1} \tilde{D}(x)^{-1} C(x) A(x)^{-1} P_{X_2} , \tag{6.67}$$

and

$$\hat{D}(x) = P_{N_1} \tilde{D}(x)^{-1} P_{N_2} . \tag{6.68}$$

It easily follows from (6.63) through (6.68) that $\hat{Q}(x) = Q(x)^{-1}$. Moreover, using this we get

$$\begin{pmatrix} P_{X_1} \\ P_{N_1} \end{pmatrix} = \begin{pmatrix} \hat{A}(x) & \hat{B}(x) \\ \hat{C}(x) & \hat{D}(x) \end{pmatrix} \begin{pmatrix} A(x) & B(x) \\ C(x) & D(x) \end{pmatrix} \begin{pmatrix} P_{X_1} \\ P_{N_1} \end{pmatrix} . \tag{6.69}$$

Furthermore by adding the components on each side of (6.69) we get

$$I = (\hat{A}(x) + \hat{B}(x) + \hat{C}(x) + \hat{D}(x))T(x)$$

which shows (6.62) [by (6.65)–(6.68)].

That completes the proof of Theorem 6.6.   ∎

REMARK 6.8   Theorem 6.6 shifts the question of invertibility of $T(x)$ to that of $\tilde{D}(x)$.   ∎

REMARK 6.9   For $T(x) = F'(x)$ $(x \in X)$ our Theorem 6.6 reduces to Theorem 1 in [90, p. 298].   ∎

From now on we assume that F, $T(x)$ are sufficiently many times Fréchet-differentiable and set $X = Y = X$ to eliminate subscripts on projections and subspaces. The general case can then follow easily.

We set $\tilde{x} = x - x^*$ and define candidates for regions of invertibility as

$$M(r, \theta, m) = \left\{ x \in X \mid 0 < \|\tilde{x}\| \le r, \|P_X \tilde{x}\| \le \theta \|P_N \tilde{x}\|^m \right\} \tag{6.70}$$

for some $m \ge 1$ and θ small. We will later choose the smallest m still guaranteeing invertibility. Define $\alpha_k(x)$ as any element of X or any operator on X whose norm is at least $O(\|\tilde{x}\|^k)$. Moreover if we write, say $F_1(x) = F_2(x) + \alpha_k(x)$ and $F_2(x) = O(\|\tilde{x}\|^k)$, then $\|F_1(x)\|$ is at least of order k.

For some $p \ge 1$ and $n \ge p$ we can write

$$T(x) = T\left(x^*\right) + \sum_{k=p}^{n} \frac{1}{k!} T^{(k)}\left(x^*\right) \left(\tilde{x}^k, \cdot\right) + \alpha_{n+1}(x) , \tag{6.71}$$

where $T^{(k)}(x^*)(\cdots, \cdot, \cdots, \cdot)$ is a $(k+1)$-linear operator and $\tilde{x}^k$ indicates that the first k arguments are all $\tilde{x}$.

By (6.59) there exist some integers a, b, c, d such that

$$A(x) = P_X T\left(x^*\right) P_X + \sum_{k=a}^{n} \frac{1}{k!} P_X T^{(k)}\left(x^*\right)\left(\tilde{x}^k, P_X\cdot\right) + \alpha_{n+1}(x), \quad (6.72)$$

$$\equiv \hat{T} + \sum_{k=a}^{n} A_k(x) + \alpha_{n+1}(x), \quad (6.73)$$

$$B(x) = \sum_{k=b}^{n} \frac{1}{k!} P_X T^{(k)}\left(x^*\right)\left(\tilde{x}^k, P_N\cdot\right) + \alpha_{n+1}(x), \quad (6.74)$$

$$\equiv \sum_{k=b}^{n} B_k(x) + \alpha_{n+1}(x), \quad (6.75)$$

$$C(x) = \sum_{k=c}^{n} \frac{1}{k!} P_N T^{(k)}\left(x^*\right)\left(\tilde{x}^k, P_X\cdot\right) + \alpha_{n+1}(x), \quad (6.76)$$

$$\equiv \sum_{k=c}^{n} C_k(x) + \alpha_{n+1}(x), \quad (6.77)$$

and

$$D(x) = \sum_{k=d}^{n} \frac{1}{k!} P_N T^{(k)}\left(x^*\right)\left(\tilde{x}^k, P_N\cdot\right) + \alpha_{n+1}(x), \quad (6.78)$$

$$\equiv \sum_{k=d}^{n} D_k(x) + \alpha_{n+1}(x). \quad (6.79)$$

That is, $p = \min(a, b, c, d)$, and we choose $n \geq \max(a, b, c, d)$. Moreover, we define

$$\bar{A}_k(x) \equiv \frac{1}{k!} P_X T^{(k)}\left(x^*\right)\left((P_N\tilde{x})^k, P_X\cdot\right), \quad (6.80)$$

etc. and assume $\bar{a}, \bar{b}, \bar{c}, \bar{d}$ to be the smallest integers such that for some $P_N\tilde{x} \neq 0$:

$$\bar{A}_{\bar{a}}(x) = \frac{1}{\bar{a}!} P_X T^{(\bar{a})}\left(x^*\right)\left((P_N\tilde{x})^{\bar{a}}, P_X\cdot\right) \neq 0, \quad (6.81)$$

$$\bar{B}_{\bar{b}}(x) = \frac{1}{\bar{b}!} P_X T^{(\bar{b})}(x^*)\left((P_N\tilde{x})^{\bar{b}}, P_N\cdot\right) \not\equiv 0, \tag{6.82}$$

$$\bar{C}_{\bar{c}}(x) = \frac{1}{\bar{c}!} P_N T^{(\bar{c})}(x^*)\left((P_N\tilde{x})^{\bar{c}}, P_X\cdot\right) \not\equiv 0, \tag{6.83}$$

$$\bar{D}_{\bar{d}}(x) = \frac{1}{\bar{d}!} P_N T^{(\bar{d})}(x^*)\left((P_N\tilde{x})^{\bar{d}}, P_N\cdot\right) \not\equiv 0. \tag{6.84}$$

Obviously, we have $\bar{a} \geq a$, $\bar{b} \geq b$, $\bar{c} \geq c$ and $\bar{d} \geq d$. Now for any $k, m \geq 1$ and for $x \in M(r, \theta, m)$ we note, recalling (6.70):

$$T^{(k)}(x^*)\left(\tilde{x}^k, \cdot\right) = T^{(k)}(x^*)\left((P_N\tilde{x})^k, \cdot\right) + \theta\alpha_{m+k-1}(x). \tag{6.85}$$

With this in mind we see that

$$A(x) = \hat{T} + \bar{A}_{\bar{a}}(x) + \alpha_{\bar{a}+1}(x) + \theta\alpha_{a+m-1}(x), \tag{6.86}$$

$$B(x) = \bar{B}_{\bar{b}}(x) + \alpha_{\bar{b}+1}(x) + \theta\alpha_{b+m-1}(x), \tag{6.87}$$

$$C(x) = \bar{C}_{\bar{c}}(x) + \alpha_{\bar{c}+1}(x) + \theta\alpha_{c+m-1}(x), \tag{6.88}$$

$$D(x) = \bar{D}_{\bar{d}}(x) + \alpha_{\bar{d}+1}(x) + \theta\alpha_{d+m-1}(x). \tag{6.89}$$

It should be noted that m is not yet determined, and the choice of m to be made shortly may force the last order symbol to be the dominant term in one or more of (6.86)–(6.89).

Assuming r so small that $A^{-1}(x)$ exists we have

$$A^{-1}(x) = \hat{T}^{-1} + \alpha_a(x) + \theta\alpha_{a+m-1}(x) = \hat{T}^{-1} + \alpha_1(x). \tag{6.90}$$

Now (6.85)–(6.89) in (6.62) yields

$$\bar{D}(x) = \bar{D}_{\bar{d}}(x) - \bar{C}_{\bar{c}}(x)\hat{T}^{-1}\bar{B}_{\bar{b}}(x)$$

$$+ \theta\left(\alpha_{d+m-1}(x) + \alpha_{\bar{b}+c+m-1}(x)\right.$$

$$+ \alpha_{b+\bar{c}+m-1}(x) + \theta\alpha_{b+c+2m-2}(x)) \tag{6.91}$$

$$+ \alpha_{\bar{b}+\bar{c}+1}(x) + \alpha_{\bar{d}+1}(x).$$

The nonsingularity of the dominant term of (6.91) will be shown to be sufficient to guarantee the invertibility of $\tilde{D}(x)$. We isolate the term we shall later require to be dominant by defining the operator $Q(x)$ as:

$$Q(x) = \begin{cases} \tilde{D}_{\bar{d}}(x) & \text{if } \bar{d} < \bar{b} + \bar{c}, \\ \tilde{D}_{\bar{d}}(x) - \bar{C}_{\bar{c}}(x)\hat{T}^{-1}\bar{B}_{\bar{b}}(x) & \text{if } \bar{d} = \bar{b} + \bar{c}, \\ -\bar{C}_{\bar{c}}(x)\hat{T}^{-1}\bar{B}_{\bar{b}}(x) & \text{if } \bar{d} > \bar{b} + \bar{c}. \end{cases} \tag{6.92}$$

We can now state the following result on the invertibility of operator $T(x)$.

PROPOSITION 6.4
Assume:

(a)   *there exists $r > 0$ such that $A(x)$ is invertible on $\bar{U}(x^*, r)$;*

(b)   *operator $Q(x)$ is invertible as an operator on N for all x with*

$$P_N \tilde{x} \neq 0. \tag{6.93}$$

Then there exists $\bar{r} > 0, \bar{\theta} > 0$ such that $T(x)$ is invertible for all $x \in M(\bar{r}, \bar{\theta}, m)$, where $m = 1 + q$ and q is the minimum nonnegative number satisfying

$$\min\left(d + q, \bar{b} + c + q, b + \bar{c} + q, b + c + 2q\right) \geq \min\left(\bar{d}, \bar{b} + \bar{c}\right). \tag{6.94}$$

Moreover, $T(x)$ is invertible in a region of type $M(\bar{r}, \bar{\theta}, 1)$ if
(c)
$$d = \bar{d} \text{ when } d < b + c; \tag{6.95}$$

(d)
$$d = \bar{d} \text{ or } b = \bar{b} \text{ and } c = \bar{c} \text{ when } d = b + c; \tag{6.96}$$

(e)
$$b = \bar{b} \text{ and } c = \bar{c} \text{ when } d > b + c. \tag{6.97}$$

Furthermore, if $\bar{d} > \bar{b} + \bar{c}$, then $T(x)$ is invertible in $M(r, \theta, m)$, where

$$m = 1 + \bar{b} + \bar{c} - \min\left(d, \bar{b} + c, b + \bar{c}\right). \tag{6.98}$$

REMARK 6.10   Our Proposition 6.4 reduces to Lemma 3.11, Corollary 3.15, and Corollary 3.17 in [90, p. 301] if we set $T(x) = F'(x)$ $(x \in X)$.   ∎

Let $x \in X$, then by approximation (6.57) the next iterate is given by

$$y = x - T(x)^{-1}F(x) - v(x) \tag{6.99}$$

provided that $T(x)$ is invertible. Set $\tilde{y} \equiv y - x^*, \tilde{x} \equiv x - x^*$ to obtain using (6.99) that

$$\tilde{y} = \tilde{x} - T(x)^{-1}F(x) - v(x) . \tag{6.100}$$

There exists an integer p_0 such that for $n \geq p_0$

$$F(x) = F'(\tilde{x}) + \sum_{j=p_0}^{n} \frac{1}{(j+1)!} F^{(j+1)}(x^*)\left(\tilde{x}^{j+1}\right) + \alpha_{n+2}(x) . \tag{6.101}$$

Assume that

$$F'(\tilde{x}) + \sum_{j=p_0}^{n} \frac{1}{(j+1)!} F^{(j+1)}(x^*)\left(\tilde{x}^{j+1}\right) = \hat{T}(\tilde{x})$$

$$+ \sum_{j=p}^{n} \frac{1}{j+1} \left[A_j(x) + B_j(x) + C_j(x)\right.$$

$$\left. + D_j(x)\right]\tilde{x} + \alpha_{n+2}(x) \text{ for all } x \in \bar{U}(x^*, r), \text{ some } r > 0 . \tag{6.102}$$

Then approximation (6.101) becomes

$$F(x) = \hat{T}(\tilde{x}) + \sum_{j=p}^{n} \frac{1}{j+1} \left[A_j(x) + B_j(x) + C_j(x) + D_j(x)\right]\tilde{x}$$

$$+ \alpha_{n+2}(x) . \tag{6.103}$$

Moreover approximation (6.71) gives

$$T(x)(\tilde{x}) = \hat{T}(\tilde{x}) + \sum_{j=p}^{n} \frac{1}{j!} T^{(j)}(x^*)\left(\tilde{x}^{j+1}\right) + \alpha_{n+1}(x) . \tag{6.104}$$

By solving (6.104) for $\hat{T}(x)$, approximation (6.103) gives

$$F(x) = T(x)(\tilde{x}) - \sum_{j=p}^{n} \frac{j}{j+1} \left[A_j(x) + B_j(x) + C_j(x) + D_j(x)\right]\tilde{x}$$

$$+ \alpha_{n+2}(x) . \tag{6.105}$$

Furthermore, assume

$$T(x)v(x) = \alpha_{n+2}(x) \quad \text{for all } x \in U(x^*, r), \text{ some } r > 0. \tag{6.106}$$

Finally using (6.105) and (6.106) approximation (6.100) gives

$$\tilde{y} = T(x)^{-1}$$

$$\left\{ \sum_{j=p}^{n} \frac{j}{j+1} \left[A_j(x) + B_j(x) + C_j(x) + D_j(x) \right] \tilde{x} + \alpha_{n+2}(x) \right\}. \tag{6.107}$$

With the above modifications the results obtained in Sections 5 through 8 in [90] can generalize to inexact Newton-like methods of the form (6.57). To avoid repetitions we will only prove one of the results listed below, whereas for the proof of the rest simply replace $F'(x)$ by $T(x)$ in the corresponding proofs in [90].

DEFINITION 6.1   *Denote by G the set of all $(T(x), v(x))$ where $T(x) \in L(X)$ $(x \in X)$, and $v: X \to X$ is a continuous function such that conditions (6.102) and (6.106) are satisfied. Then note that $G \neq \emptyset$ since $(F'(x), 0) \in G$.*

DEFINITION 6.2   *Let $\beta_p^q(x)$ represent any term with order at least $O(\|\tilde{x}\|^p)$ and such that $P_X \beta_p^q(x) = \alpha_{p+q}(x)$.*

We state the following convergence results:

THEOREM 6.7
Assume:

   (a) $G \neq \emptyset$ and $(T(x), v(x)) \in G$;

   (b) $\bar{D}_1$ is nonsingular as an operator on N for all $P_N \tilde{x} \neq 0$.

Then
   (i) for r and θ sufficiently small, $T(x_0)^{-1}$ exists for all $x_0 \in M(r, \theta, 1)$;
   (ii) inexact Newton-like method $\{x_n\}$ $(n \geq 0)$ generated by (6.57) remains in $M(r, \theta, 1)$ for all $n \geq 0$ and converges to x^*;
   (iii) the following error bounds hold for all $n \geq 0$

$$\|P_N(x_n - x^*)\| \leq K_1 \|x_{n-1} - x^*\|, \quad \text{some } K_1 > 0 \tag{6.108}$$

and

$$\lim_{n \to \infty} \frac{\|P_N (x_n - x^*)\|}{\|P_N (x_{n-1} - x^*)\|} = \frac{1}{2} .$$

(6.109)

PROOF By Proposition 6.4 the operator $T(x)^{-1}$ exists in $M(r, \theta, 1)$ for sufficiently small r and θ, since in this case $d = \bar{d} = 1$, $b, c \geq 1$. Using (6.62) and (6.72)–(6.79) $\tilde{D}(x) = D_1(x) + \alpha_2(x)$ and $D_1(x)^{-1}$ exists as an operator on N. It follows that $\tilde{D}(x)^{-1} = D_1(x)^{-1} + \alpha_0(x)$ and by (6.61), $T(x)^{-1} = P_N D_1(x)^{-1} P_N + \alpha_0(x) = \beta_{-1}^1(x)$. By (6.107) for $p = 1$

$$\tilde{x}_1 = \frac{1}{2} \left[P_N D_1(x_0)^{-1} P_N + \alpha_0(x_0) \right] \{ [A_1(x_0) + B_1(x_0)$$

$$+ C_1(x_0) + D_1(x_0)] \tilde{x}_0 + \alpha_3(x_0) \} .$$

(6.110)

But we have $P_N A_1(x_0) = P_N B_1(x_0) = 0$, $C_1(x_0)\tilde{x}_0 = \theta\alpha_2(x_0)$. Hence, we get $T(x_0)^{-1}C_1(x_0)\tilde{x}_0 = \theta\beta_1^1(x_0)$. Approximation (6.110) becomes

$$\tilde{x}_1 = \frac{1}{2} P_N \tilde{x}_0 + \theta\beta_1^1(x_0) + \alpha_2(x_0) .$$

(6.111)

By (6.111) there exist $K_0, K_1 > 0$ such that

$$\left(\frac{1}{2} - K_0\theta \right) \|P_N \tilde{x}_0\| \leq \|P_N \tilde{x}_1\| \leq \left(\frac{1}{2} + K_0\theta \right) \|P_N \tilde{x}_0\|$$

(6.112)

and

$$\|P_X \tilde{x}_1\| \leq K_1 \|x_0 - x^*\|^2 .$$

(6.113)

Define the sequences $\{r_n\}, \{\theta_n\}$ ($n \geq 0$) by

$$r_n = \|x_n - x^*\| , \theta_0 = \theta,$$

$$\theta_n = K_1 (1 + \theta_{n-1}) \left(\frac{1}{2} - K_0\theta_{n-1} \right)^{-1} r_{n-1} \quad (n \geq 1) .$$

Then, since $x_0 \in M(r_0, \theta_0, 1)$, approximations (6.112) and (6.113) can give

$$\|P_X \tilde{x}_1\| \leq K_1 r_0 (1 + \theta_0) \|P_N \tilde{x}_0\|$$

$$\leq K_1 r_0 (1 + \theta_0) \left(\frac{1}{2} - K_0\theta_0 \right)^{-1} \|P_N \tilde{x}_1\| ,$$

(6.114)

$$r_1 \leq \|P_N \tilde{x}_1\| + \|P_X \tilde{x}_1\|$$

$$\leq \left[\left(\frac{1}{2} + K_0 \theta_0 \right) (1 - \theta_0)^{-1} + K_1 r_0 \right] r_0 . \tag{6.115}$$

Hence, we get $\|P_X \tilde{x}_1\| \leq \theta_1 \|P_N \tilde{x}_1\|$, and $x_1 \in M(r_1, \theta_1, 1)$. For some given $t \in (\frac{1}{2}, 1)$, it follows from (6.114), (6.115) and the definition of θ_1 that we can choose r_0, θ_0, so small that $r_1 \leq t r_0$, $\theta_1 \leq t \theta_0$. This shows $M(r_1, \theta_1, 1) \subseteq M(r_0, \theta_0, 1)$ and x_0, x_1, θ may be replaced by x_1, x_2, θ_1 in (6.112)–(6.115). This way we get $r_2 < t r_1$ and $\theta_2 < t \theta_1$. That is we get $\lim_{n \to \infty} r_n = \lim_{n \to \infty} \theta_n = 0$, which shows $x_n \to x^*$ as $n \to \infty$. Moreover, (6.113) becomes (6.108) and by letting $n \to \infty$ in

$$\left(\frac{1}{2} - K_0 \theta_{n-1} \right) \|P_N \tilde{x}_{n-1}\| \leq \|P_N \tilde{x}_i\| \leq \left(\frac{1}{2} + K_0 \theta_{n-1} \right) \|P_N \tilde{x}_{n-1}\|$$

we obtain (6.109).

That completes the proof of Theorem 6.7.   ∎

REMARK 6.11   For $T(x) = F'(x)$ $(x \in X)$ and $v(x) = 0$ $(x \in X)$ Theorem 6.7 reduces to Theorem 5.1 in [90, p. 303].   ∎

THEOREM 6.8
Assume:

(a) $G \neq \emptyset$ and $(T(x), v(x)) \in G$;

(b) $d = \bar{d} \leq c$ and $\bar{D}_d$ is nonsingular for all $P_N \tilde{x} \neq 0$.

Then
(i) $T(x_0)^{-1}$ exists for r and θ sufficiently small and all $x_0 \in M(r, \theta, 1)$;
(ii) inexact Newton-like method $\{x_n\}$ $(n \geq 0)$ generated by (6.57) is well defined remains in $M(r, \theta, 1)$ for all $n \geq 0$ and converges to x^*;
(iii) the following error bounds hold for all $n \geq 0$

$$\|P_X (x_n - x^*)\| \leq K_2 \|x_{n-1} - x^*\|^{p+1}, \quad \text{some } K_2 > 0 \tag{6.116}$$

and

$$\lim_{n \to \infty} \frac{\|P_N (x_n - x^*)\|}{\|P_N (x_{n-1} - x^*)\|} = \frac{d}{d+1} . \tag{6.117}$$

REMARK 6.12   For $T(x) = F'(x)$ $(x \in X)$ and $v(x) = 0$ $(x \in X)$ Theorem 6.8 reduces to Theorem 5.9 in [90, p. 304].   ∎

REMARK 6.13   A similar result was given in [171] but under the additional restrictions that $\dim(N) = 1$, and

$$\left\| F^{(p+1)} \left(x^*\right) \left(\tilde{x}^p, g\right) \right\| \geq c \|g\| \|\tilde{x}\|^p$$

for all $x \in X$, $g \in N$, and some $c > 0$. In addition, the convergence rate obtained corresponding to (6.116) had $(p + 1)$ reduced to 2.   ∎

The hypothesis $\bar{D}_d$ invertible on N for all $P_N \tilde{x} \neq 0$ is strong and, if $d = 1$, implies that $\dim N = 1$ or 2. In this case we can consider a one-dimensional subspace $N_0 \subseteq N$ and set $N = N_0 \oplus R$ for some R. Let P_{N_0} be the projection onto N_0 parallel to $X \oplus R$. Define the sets

$$M(r, \theta, \eta, m) = \left\{ x \in X \mid 0 < \|\tilde{x}\| \leq r, \|P_X \tilde{x}\| \leq \theta \|P_N \tilde{x}\|^m, \right.$$

$$\left. \left\| (P_N - P_{N_0}) \tilde{x} \right\| \leq \eta \|P_N \tilde{x}\| \right\} . \tag{6.118}$$

We then modify operator $Q(x)$ of (6.92) by replacing $P_N \tilde{x}$ by $P_{N_0} \tilde{x}$ in (6.81)–(6.84). Denote the new operator by $\hat{Q}(x)$.

We can now state the following result on regions of invertibility of the form (6.118).

PROPOSITION 6.5
Assume there exists N_0 such that for all $\tilde{x}$ with $P_{N_0} \tilde{x} \neq 0$, $\hat{Q}(x)$ is nonsingular as a map on N. Then the conclusions of Proposition 6.4 remain valid with $M(r, \theta, m)$ replaced by $M(\bar{r}, \bar{\theta}, \bar{\eta}, m)$ and $\bar{n}$ sufficiently small.

REMARK 6.14   For $T(x) = F'(x)$ $(x \in X)$ Proposition 6.5 reduces to Lemma 3.22 in [90, p. 302].   ∎

Based on Proposition 6.5 we state:

THEOREM 6.9
Assume:

(a) $G \neq 0$ and $(T(x), v(x)) \in G$;

(b)   there exists a one-dimensional subspace $N_0 \subseteq N$ such that $\hat{D}_d(x)$ is nonsingular as an operator on N for all $P_{N_0}\tilde{x} \neq 0$.

Then
(i) for $\bar{r}$, $\bar{\theta}$, $\bar{\eta}$ sufficiently small $T(x)^{-1}$ exists for all $x \in M(\bar{r}, \bar{\theta}, \bar{\eta}, 1)$;
(ii) there exists $\hat{\eta} \leq \bar{\eta}$ such that for any $x_0 \in M(\bar{r}, \bar{\theta}, \hat{\eta}, 1)$ inexact Newton-like iterates $\{x_n\}$ $(n \geq 0)$ generated by (6.57) remain in $M(\bar{r}, \bar{\theta}, \bar{\eta}, 1)$ and converge to x^*;
(iii) error bounds (6.116) and (6.117) hold for all $n \geq 0$.

REMARK 6.15   For $T(x) = F'(x)$ $(x \in X)$ and $v(x) = 0$ $(x \in X)$ Theorem 6.9 reduces to Theorem 5.21 in [90, p. 305]. ∎

We also have the following result on higher orders of invertibility:

THEOREM 6.10
Assume:

(a) $G \neq \emptyset$ and $(T(x), v(x)) \in G$;

(b) $p = d \geq 2$, $\bar{d} = p + 1$, and $\bar{D}_{p+1}(x)$ is nonsingular as an operator on N for all $\tilde{x}$ with $P_N \tilde{x} \neq 0$.

Then
(i) for r, θ sufficiently small $T(x_0)^{-1}$ exists for all $x_0 \in M(r, \theta, 2)$;
(ii) inexact Newton-like method $\{x_n\}$ $(n \geq 0)$ generated by (6.57) is well defined, remains in $M(r, \theta, 2)$ for all $n \geq 0$ and converges to x^*;
(iii) the following error bounds hold for all $n \geq 0$

$$\left\| P_X \left(x_n - x^*\right) \right\| \leq K_3 \left\| x_{n-1} - x^* \right\|^{p+1}, \quad \text{some } K_3 > 0$$

and

$$\lim_{n \to \infty} \frac{\left\| P_N \left(x_n - x^*\right) \right\|}{\left\| P_N \left(x_{n-1} - x^*\right) \right\|} = \frac{p+1}{p+2}.$$

REMARK 6.16   For $T(x) = F'(x)$ $(x \in X)$ and $v(x) = 0$ $(x \in X)$ Theorem 6.10 reduces to Theorem 6.1 in [90, p. 306]. ∎

The following result proves convergence when $m > 2$.

THEOREM 6.11
Assume:

(a) $G \neq 0$ and $(T(x), v(x)) \in G$;

(b) $p = d \geq 2$ and $p < \bar{d} < \bar{b} + p$;

(c) $\bar{D}_d(x)^{-1}$ exists as an operator on N for all $\tilde{x}$ with $P_N \tilde{x} \neq 0$.

Then
   (i) for r, θ sufficiently small $T(x_0)^{-1}$ exists for all $x_0 \in M(r, \theta, m)$ with $m = 1 + \bar{d} - d$;
   (ii) inexact Newton-like method $\{x_n\}$ $(n \geq 0)$ generated by (6.57) is well defined, remains in $M(r, \theta, m)$ for all $n \geq 0$ and converges to x^*;
   (iii) the following error bounds hold for all $n \geq 0$

$$\lim_{n \to \infty} \frac{\|P_N(x_n - x^*)\|}{\|P_N(x_{n-1} - x^*)\|} = \frac{\bar{d}}{\bar{d} + 1},$$

and

$$\left\| P_X \left(x_n - x^* \right) \right\| \leq K_4 \left\| x_{n-1} - x^* \right\|^{s+1}, \quad \text{some } K_4 > 0,$$

where

$$s = \begin{cases} \min \left(\bar{b}, \bar{d} - d + a \right), & \text{if } \bar{d} - d \geq \bar{b} - b, \\ \bar{d} - d + \min(a, b), & \text{if } \bar{b} - b \geq \bar{d} - d. \end{cases}$$

REMARK 6.17   For $T(x) = F'(x)$ $(x \in X)$ and $v(x) = 0$ $(x \in X)$ Theorem 6.11 reduces to Theorem 6.20 in [90, p. 308]. ∎

We finally list two results on acceleration of convergence corresponding to Theorems 7.4 and 7.27 in [90, pp. 308, 311].

THEOREM 6.12
Assume:

(a) $G \neq \emptyset$ and $(T(x), v^i(x)) \in G$, $i = 1, 2, 3$;

(b) $d = \bar{d} \leq c, b \geq \min(2, d)$;

(c) $\bar{D}_d(x)$ is invertible as an operator on N for all $P_N \tilde{x} \neq 0$;

(d)   for all $x \in M(r, \theta, 1)$

$$\left\| \bar{D}_{d+1}(x) P_N \tilde{x} \right\| \geq K_5 \| P_N \tilde{x} \|^{d+2}, \quad \text{some } K_5 > 0. \tag{6.119}$$

Then
(i) $T(x_0)^{-1}$ exists for all $x_0 \in M(r, \theta, 1)$;

(ii) inexact Newton-like method iterates given by

$$y_n = x_n - T(x_n)^{-1} F(x_n) - v^1(x_n) ,$$

$$z_n = y_n - T(y_n)^{-1} F(y_n) - v^2(y_n),$$

and

$$x_{n+1} = z_n - (d+1)T(z_n)^{-1} F(z_n) - v^3(z_n)$$

are well defined, remain in $M(r, \theta, 1)$ for all $n \geq 0$ and converge to x^*;
(iii) the following error bounds hold for all $n \geq 0$

$$\|x_n - x^*\| \leq K_6 \|x_{n-1} - x^*\|^2 , \quad some\ K_6 > 0 .$$

THEOREM 6.13
Assume:

(a) $G \neq \emptyset$ and $(T(x), v^i(x)) \in G, i = 1, 2;$

(b) $b \geq 2, c \geq d = \bar{d} \geq 2;$

(c)   *estimate (6.119) holds.*

Then
(i) for r and θ sufficiently small the inexact Newton-like method iterates given by

$$y_n = x_n - T(x_n)^{-1} F(x_n) - v^1(x_n)$$

and

$$x_{n+1} = y_n - (d+1)T(y_n)^{-1} F(y_n) - v^2(y_n)$$

are well defined, remain in $M(r, \theta, 1)$ and converge to x^*;
(ii) the following error bounds hold for all $n \geq 0$

$$\|P_N(x_n - x^*)\| \leq K_7 \|x_{n-1} - x^*\|^2$$

and

$$\|P_X(x_n - x^*)\| \leq K_8 \|x_{n-1} - x^*\|^{p+1}$$

for some $K_7, K_8 > 0$.

Example 6.3
Let $X = Y = (-\infty, +\infty)$, $x^* = 0$, and consider the function F on X given by

$$F(x) = x^2 . \tag{6.120}$$

Choose

$$v_n = (1 - 2x_n)^{-1} (x_n - x^*) \quad (n \geq 0) \tag{6.121}$$

and

$$T(x) = F'(x) - I \quad (x \in X) . \tag{6.122}$$

With the above choices inexact Newton-like method (6.57) becomes

$$x_{n+1} - x^* = (2x_n - 1)^{-1} (x_n - x^*)^2 \quad (n \geq 0) . \tag{6.123}$$

It can easily be seen from (6.121) and (6.122) that the crucial conditions (6.106) and (6.102) are satisfied. Moreover, iteration (6.123) converges quadratically to x^* which is a zero of equation $F(x) = 0$ of multiplicity 2 provided that $x_0 \in (-\frac{1}{3}, \frac{1}{3})$. Furthermore, the ordinary Newton's method $y_{n+1} = y_n - F'(y_n)^{-1} F(y_n) (n \geq 0)$, cannot be computed in any neighborhood of x^*, for any starting point y_0 since $F'(y_n)^{-1}$ must be nonzero for all $n \geq 0$, and this is not true. We finally note that the corrected Newton's method $\delta_{n+1} = \delta_n - 2F'(\delta_n)^{-1} F(\delta_n) (n \geq 0)$ converges quadratically to x^* provided that $v_n = 0 (n \geq 0)$ which is not true in general.

Example 6.4
Let $X = Y = \mathbf{R}^2$ and define $F: X \to Y$ by

$$F(x) = \begin{bmatrix} f_1(x, y) \\ f_2(x, y) \end{bmatrix} = \begin{bmatrix} x + y^2 \\ \frac{3}{2}xy + y^2 + y^3 \end{bmatrix} . \tag{6.124}$$

For $x^* = (0, 0)^T$, $T(x) = F'(x) (x \in X)$, we obtain from (6.124) that

$$N\left(T\left(x^*\right)\right) = \text{span}\left(\ell_1\right), \ \ell_1 = (0, 1)^T, \ R\left(T\left(x^*\right)\right) = \text{span}\left(\ell_2\right), \ \ell_2 = (1, 0)^T,$$

$$P_N = \begin{bmatrix} 0 & 0 \\ 0 & 1 \end{bmatrix} \text{ and } P_X = \begin{bmatrix} 1 & 0 \\ 0 & 1 \end{bmatrix}. \text{ Here } P_N T'(x^*)(\ell, \ell) \neq 0, \text{ since } \frac{\partial^2 f_2}{\partial y^2} = 2 \neq 0$$ at $(0, 0)$. Take $\bar{x}_0 = .1$, $\bar{y}_0 = 1$, $x_n = [\bar{x}_n, \bar{y}_n]^T$, $v_n = \varepsilon x_n$, $\varepsilon = .5 \cdot 10^{-4}$ $(n \geq 0)$. It can easily be seen that the hypotheses of Theorem 6.7 are satisfied. Hence, using method (6.57), we can obtain the iterates given in Table 6.2.

Moreover, with the same initial guess via Theorem 6.11 we can have the acceleration scheme for the same example given in Table 6.3, for $v^1(x_n) = \varepsilon_1 x_n$, $v^2(y_n) = \varepsilon_2 y_n$, $v^3(z_n) = \varepsilon_3 z_n$, $\varepsilon_1 = .5 \cdot 10^{-4}$, $\varepsilon_2 = .25 \cdot 10^{-4}$ and $\varepsilon_3 = 10^{-4}$.

Finally we note that the results obtained here in Tables 6.2 and 6.3 constitute an improvement over the corresponding Tables 1 and 3, respectively, in [90, p. 312].

Table 6.2 Accelaration scheme I for Newton's method

i	$\bar{x}_i$	$\bar{y}_i$
1	−0.5346D–00	0.7672D–00
2	0.1712D–00	0.2720D–00
3	−0.3055D–01	0.1920D–00
4	0.9740D–03	0.9355D–01
5	−0.4740D–03	0.4930D–01
10	−0.1606D–07	0.1601D–02
20	−0.1540D–16	0.1566D–05

Table 6.3 Accelaration schme II for Newton's method

i	$\bar{x}_i$	$\bar{y}_i$
1	−0.2325D–00	0.1122D–00
2	−0.2134D–02	0.5866D–02
3	0.1993D–07	0.4145D–06
4	0.2740D–20	0.5754D–14
5	0.8931D–44	0.1428D–14

6.4 A Newton–Mysovskii-Type Theorem with Applications to Inexact Newton-Like Methods and their Discretizations

The goal of this section is to extend the validity of the mesh independence principle to include inexact Newton-like methods. Let us consider the problem of approximating a locally unique solution x^* of the equation

$$F(x) + Q(x) = 0 \qquad (6.125)$$

where F, Q are nonlinear operators defined on some closed convex subset D of a Banach space X with values in a Banach space Y.

Let $x_0 \in D$ be fixed and define the inexact Newton-like method for all $n \geq 0$ by

$$y_n = x_n - A(x_n)^{-1}(F(x_n) + Q(x_n)) \qquad (6.126)$$

and

$$x_{n+1} = y_n - z_n . \qquad (6.127)$$

Here, $A(x_n)$ denotes a linear operator which is an approximation to the Fréchet derivative $F'(x_n)$ of F evaluated at $x = x_n$ for all $n \geq 0$. The points $z_n \in D$ for all

$n \geq 0$, and are determined in such a way that the iteration $\{x_n\}$ ($n \geq 0$) converges to a solution x^* of Equation (6.125). By setting $z_n = 0$ for all $n \geq 0$, we obtain the Newton-like method.

Deuflhard and Heindl in [94] and Deuflhard and Potra in [95] used affine invariant versions of the so-called Newton–Mysovskii-type hypotheses [146] of the form

$$\|F'(y)^{-1}(F'(x + t(y - x)) - F'(x))\| \leq wt\|y - x\| \tag{6.128}$$

for all $x, y \in D$, some $w \in \mathbf{R}$ for all $t \in [0, 1]$, and

$$\|F'(z)^{-1}(F'(u) - F'(x))(u - x)\| \leq w\|u - x\|^2 \tag{6.129}$$

for all $u, x, z \in D$, for some $w \in \mathbf{R}$. They provided sufficient conditions for the convergence of iteration (6.126) to a solution x^* of Equation (6.125), when $Q(x) = 0$, $A(x) = F'(x)$ for all $x \in D$ and $z_n = 0$ ($n \geq 0$). Error bounds on the distances $\|x_{n+1} - x_n\|$ and $\|x_n - x^*\|$ ($n \geq 0$) were also given. Here under weaker and more general conditions [see (6.148)–(6.151)] we provide sharper error bounds on the same distances. This is important because if the error tolerance on the distances is ε and n_1 is the smallest nonnegative integer for which $\|x_{n_1} - x^*\| < \varepsilon$, then under (6.148)–(6.151) $\|x_{n_2} - x^*\| < \varepsilon$ and $n_2 \leq n_1$.

Moreover, the convergence of Newton methods has been examined extensively in [24, 37, 39, 41, 43, 200, 207]. Since the iterates of the inexact Newton-like method (6.126), (6.127) (whether $z_n = 0$, $n \geq 0$ or not) can rarely be computed in infinite dimensional spaces, (6.126) and (6.127) can be replaced in practice by a family of discretized equations

$$P(a) + P_1(a) = 0 \tag{6.130}$$

indexed by some real number $h > 0$, where P is a nonlinear operator between finite dimensional spaces X^1 and Y^1. Let the discretization on X be defined by the bounded linear operators $L: X \to X^1$. Consider also the iteration $\{a_n\}$ ($n \geq 0$) given for all $n \geq 0$ by

$$b_n = a_n - S(a_n)^{-1} P(a_n), \quad a_0 = L(x_0) \tag{6.131}$$

and

$$a_{n+1} = b_n - d_n . \tag{6.132}$$

Here $S(a_n)$ denotes a linear operator which is an approximation to the Fréchet derivative $P'(a_n)$ of P evaluated at $a = a_n$ for all $n \geq 0$. The points $d_n \in X^1$ for all $n \geq 0$, and are determined in such a way that the iteration $\{a_n\}$ $n \geq 0$ converges to a solution a^* of Equation (6.129). Note that all symbols introduced in (6.130)–(6.132) really depend on h. That is, $P = P_h$, $L = L_h$, $S = S_h$, etc. But to simplify the notation we do not use the latter.

In practice, the iterates y_n or even b_n can rarely be computed exactly. That is why we need to "correct" at every step by introducing z_n or d_n, respectively, for the iterations under consideration. This is the factor that the results in the studies mentioned above have not taken into account when proving the mesh independence principle. The mesh independence principle (proved in the studies listed above) asserts that the number of steps required by the two processes to converge to within a given tolerance is essentially the same.

Here we show that this is true for our inexact Newton-like method (6.126), (6.127). Our results can be reduced to the ones obtained earlier for appropriate choices of the factors involved.

The importance of the formulation of an efficient mesh size strategy based upon the mesh-independence has been extensively discussed in [29, 42, 45, 50] and the references there. We also apply our results to solve a nonlinear integral equation that appears in radioactive transfer.

The norms in all spaces will be denoted by the same symbol $\| \ \|$. For any bounded linear operator from X to Y or from X^1 to Y^1, the induced norm will be used. S^k, $k \in N$ will denote the k-tuple Cartesian product of the set S.

We find it convenient to introduce the following:

(G_1) Let $R > 0$ and assume there exists $x_0 \in D$ and a function $\bar{\alpha} : \bar{U}^3(x_0, R) \to [0, +\infty)$ such that

$$\|A(y)^{-1}[F(y) - F(x) - A(x)(y - x) + F'(y)(z - y)]\|$$

$$\leq \bar{\alpha}(x, y, z) \tag{6.133}$$

for all $x, y, z \in \bar{U}(x_0, R) \subseteq D$.

(G_2) There exist continuous, nondecreasing functions w, w_1, and w_2 such that

$$w : D \to [0, +\infty), \quad w_1, w_2 : [0, R] \to [0, +\infty)$$

with $w_1(0) = w_2(0) = 0$ and a sequence $\{z_n\}$ $(n \geq 0)$ of points from D with

$$\|z_i\| \leq w(z_i) \leq w_1(r) \quad \text{for all } i \geq 0 \tag{6.134}$$

and for all $k \in N$

$$\sum_{i=0}^{k+1} \|z_i\| \leq \sum_{i=0}^{k+1} w(z_i) \leq w_2(r), \quad r \in [0, R] \tag{6.135}$$

for all $z_i \in \bar{U}(x_0, r) \subseteq \bar{U}(x_0, R)$. We note that (6.134) implies that the sequence $\{z_n\}$ $(n \geq 0)$ is null.

(G_3) Let $F, Q: D \subseteq X \to Y$ be nonlinear operators satisfying:

$$\left\| \int_0^1 A(y)^{-1}(F'(x + t(y - x)) - A(x))(y - x)dt \right\|$$

$$\leq [C\,(\|x - x_0\|, \|y - x_0\|, \|y - x\|) + b_1]\,\|y - x\|, \quad (6.136)$$

$$\|A(y)^{-1}(F'(x) - A(x))(y - x)\|$$

$$\leq [C_1\,(\|x - x_0\|, \|y - x_0\|) + b_2]\,\|y - x\|, \qquad (6.137)$$

$$\|A(y)^{-1}(Q(y) - Q(x))\|$$

$$\leq [C_2\,(\|x - x_0\|, \|y - x_0\|) + b_3]\,\|y - x\|, \qquad (6.138)$$

and

$$\|A(y)^{-1}(A(x) - A(y))(y - x)\| \leq [C_3\,(\|x - x_0\|, \|y - x_0\|,$$

$$\|y - x\|) + b_4]\,\|y - x\| \qquad (6.139)$$

for all $x, y \in \bar{U}(x_0, r) \subseteq \bar{U}(x_0, R) \subseteq D$. The real functions $C, C_3, C_1,$ and C_2 are assumed to be continuous and nondecreasing on $[0, R]^3$, $[0, R]^3$, $[0, R]^2$, and $[0, R]^2$, respectively. Moreover $C(0, 0, 0) = C_1(0, 0) = C_2(0, 0) = C_3(0, 0) = 0$ and b_1, b_2 b_3, b_4 are fixed nonnegative real numbers.

(G_4) There exist continuous, nondecreasing functions $w_3, w_4: [0, R] \to [0, +\infty)$ with $w_3(0) = w_4(0) = 0$ such that

$$\overline{\alpha_i} = \bar{\alpha}(x_i, y_i, x_{i+1}) \leq w_3(r) \quad \text{for all } i \geq 0 \qquad (6.140)$$

and for all $k \in N$

$$\sum_{i=0}^{k+1} \overline{\alpha_i} \leq w_4(r) \qquad (6.141)$$

for all $x_i, y_i, x_{i+1} \in \bar{U}(x_0, r) \subseteq \bar{U}(x_0, R)$.

Let us define the functions φ, φ_1, $\varphi_2 : [0, R] \to [0, +\infty)$ by

$$\varphi(r) = r - [w_2(r) + (C(r, r, r) + C_1(r, r)$$

$$+ \; C_2(r, r) + b_1 + b_2 + b_3)r + w_4(r)] \;, \tag{6.142}$$

$$\varphi_1(r) = r - \varphi(r) \;, \tag{6.143}$$

and

$$\varphi_2(r) = C(0, r, r) + C_2(0, r) + C_3(0, r, r) + b_1 + b_4 + w_5(r) \;,$$

$$r \in [0, R] \;. \tag{6.144}$$

By the hypotheses on the C and w functions above, there exist constants p, h_1, δ_0, δ_1, δ_2, δ_3, c_6, c_5, c_4, c_3 with $c_4 > \delta_3$ such that for

$$0 < \delta_0 \leq c_4 - c_5 h^p \leq r_0(h) = r(h) \leq c_3 h^p \leq \delta_1 \leq r_0 \leq R \tag{6.145}$$

and $r(h) = 0$ when $a_0 = a^*$ or $h = 0$, the following are true for sufficiently small r_0, $R > 0$ and $h \in (0, h_1]$

$$0 < \delta_2 \leq \varphi_1(r(h)) \leq \delta_3 \quad \text{and} \quad \theta(r(h)) < 1 \;, \tag{6.146}$$

where

$$\theta(r) = C(R, r, r + R) + C_2(R, r) + C_3(R, r, r + R) + b_1 + b_4 \tag{6.147}$$

provided that $b_1 + b_4 \in [0, 1)$.

We can now show that for all $h \in (0, h_2]$, where

$$h_2 = \min \left\{ h_1, \left(\frac{\delta_2}{c_3 - c_6} \right)^{1/p}, \left(\frac{c_4 - \delta_3}{c_2 + c_5} \right)^{1/p} \right\}$$

$$\text{with } c_3 > c_6, c_6 \geq c_2 \tag{6.148}$$

the following is true

$$0 < c_2 h^p \leq \varphi(r(h)) \leq c_6 h^p \;. \tag{6.149}$$

Indeed from (6.146) and (6.149), we get $r(h) - \delta_3 \leq \varphi(r(h)) \leq r(h) - \delta_2$. It is enough to show that $r(h) - \delta_3 \geq c_2 h^p$ and $r(h) - \delta_2 \leq c_6 h^p$, which will be true if

$c_4 - c_5 h^p - \delta_3 \geq c_2 h^p$ and $c_3 h^p - \delta_2 \leq c_6 h^p$ respectively. The last inequalities are true by the choice of h and (6.148).

Similar arguments can show that for sufficiently small $h \in (0, h_2]$ there exist $\delta_4, \delta_5, \delta_6, \delta_7, c_7, c_8, c_9$ such that for

$$0 < \delta_4 \leq c_7 - c_8 h^p \leq r^*(h) = \left\| L(x^*) - a^* \right\| \leq c_9 h^p \leq \delta_5 \leq r^*$$

$$= \left\| x_0 - x^* \right\|, \quad \delta_5 \leq \delta_1 \tag{6.150}$$

and $r^*(h) = 0$ when $L(x^*) = a^*$ or $h = 0$, the following are true for a sufficiently small $r^* < r_0$

$$0 < \delta_6 \leq \varphi_2(r^*(h)) \leq \delta_7 < 1. \tag{6.151}$$

We will need to introduce constants s_0, ℓ_0, such that

$$t_0 = 0, \quad s_0 \geq \|y_0 - x_0\|, \quad \ell_0 \geq \|z_0\|, \tag{6.152}$$

iterations for all $n \geq 0$

$$s_{n+1} = t_{n+1} + h_{n+1}, \tag{6.153}$$

$$t_{n+1} = s_n + \ell_n, \quad \ell_n = w(z_n), \tag{6.154}$$

$$h_{n+1} = \left[C(s_n, t_{n+1}, t_{n+1} - s_n) + b_1 \right](t_{n+1} - s_n)$$

$$+ \left[C_1(s_n, t_{n+1}) + b_2 \right](t_{n+1} - s_n)$$

$$+ \left[C_2(t_n, t_{n+1}) + b_3 \right](t_{n+1} - t_n) \tag{6.155}$$

for some given sequences $\{\alpha_n\}$ and $\{\ell_n\}$ $(n \geq 0)$ with

$$\alpha_n \geq 0, \quad \sum_{i=0}^{k+1} \alpha_i \leq w_4(r), \quad \text{and} \quad \sum_{i=0}^{k+1} \ell_i \leq w_2(r) \tag{6.156}$$

for all positive integers k, some fixed real constants (which may depend on r_0) γ_0, γ_1, γ_2 and some fixed $r_0 \in [0, R]$.

Moreover, we define the iterations for all $n \geq 0$

$$\bar{v}_{n+1} = \left[C(\|y_n - x_0\|, \|x_{n+1} - x_0\|, \|x_{n+1} - y_n\|) + b_1 \right] \|x_{n+1} - y_n\|$$

$$+ \left[C_1 \left(\|y_n - x_0\| , \|x_{n+1} - x_0\| \right) + b_2 \right] \|x_{n+1} - y_n\| \tag{6.157}$$

$$+ \left[C_2 \left(\|x_n - x_0\| , \|x_{n+1} - x_0\| \right) + b_3 \right] \|x_{n+1} - x_n\| + \bar{\alpha}_n$$

where $\bar{\alpha}_n = \bar{\alpha}(x_n, y_n, x_{n+1})$ with

$$\bar{\alpha}_n \leq \alpha_n \quad \text{for all } n \geq 0 \tag{6.158}$$

and the function

$$T(r) = s_0 + \varphi_1(r) \tag{6.159}$$

on $[0, R]$.

We can now state the main result on semilocal convergence:

THEOREM 6.14

Let F, Q be operators defined on a closed convex subset D of a Banach space X with values in a Banach space Y.

Assume:

(i)   *F is Fréchet-differentiable on D whereas Q is only continuous there;*

(ii) *conditions (G_1) and (G_3) are satisfied;*

(iii) *iterations $\{\alpha_n\}$, $\{\ell_n\}$, $\{\bar{\alpha}_n\}$, $\{z_n\}$ $(n \geq 0)$ satisfy conditions (6.156), (6.158), and*

$$\|z_n\| \leq \ell_n \quad \text{for all } n \geq 0 ; \tag{6.160}$$

(iv) *there exists a minimum nonnegative number r_0 such that*

$$T(r_0) \leq r_0 \quad \text{and} \quad r_0 \leq R ; \tag{6.161}$$

(v)   *moreover r_0, R satisfy the inequality*

$$\theta(r_0) < 1 \quad \text{for } r_0 \leq R \tag{6.162}$$

where the function θ is given by (6.147); and

(vi) *the ball $\bar{U}(x_0, R) \subseteq D$.*

Then

(a)   *scalar sequences $\{t_n\}$ $\{s_n\}$ $(n \geq 0)$ generated by relations (6.152)–(6.155) are monotonically increasing and bounded above by their limit, which is r_0;*

(b)   sequences $\{x_n\}$, $\{y_n\}$ $(n \geq 0)$ generated by relations (6.126) and (6.127) are well defined, remain in $\bar{U}(x_0, r_0)$ for all $n \geq 0$, and converge to a solution x^* of the equation $F(x) + Q(x) = 0$, which is unique in $\bar{U}(x_0, R)$.

Moreover, the following estimates hold for all $n \geq 0$

$$\|y_n - x_n\| \leq s_n - t_n ,$$ (6.163)

$$\|x_{n+1} - y_n\| \leq t_{n+1} - s_n ,$$ (6.164)

$$\|x_n - x^*\| \leq r_0 - t_n ,$$ (6.165)

$$\|y_n - x^*\| \leq r_0 - s_n ,$$ (6.166)

$$\|y_{n+1} - x_{n+1}\| = \left\| A\left(x_{n+1}\right)^{-1} \left(F\left(x_{n+1}\right) + Q\left(x_{n+1}\right)\right) \right\|$$

$$\leq \bar{v}_{n+1} ,$$ (6.167)

and

$$\|y_n - x_n\| \leq \|x^* - x_n\| + \left[C \left(\|x_n - x_0\|, \|x^* - x_0\|, \|x^* - x_n\|\right) \right.$$

$$+ C_2 \left(\|x_n - x_0\|, \|x^* - x_0\|\right)$$

$$\left. + b_1 + b_3\right] \|x_n - x^*\| .$$ (6.168)

(We will be concerned only with the case $r_0 > 0$, since when $r_0 = 0$, $x_0 = x^*$.)

PROOF   (a) Using relations (6.152), (6.153), (6.154), (6.156), and (6.161), we deduce that the scalar sequence $\{t_n\}$ $n \geq 0$ is monotonically increasing, nonnegative, and $t_0 \leq s_0 \leq t_1 \leq s_1 \leq r_0$. Let us assume that $t_k \leq s_k \leq t_{k+1} \leq s_{k+1} \leq r_0$ for $k = 0, 1, 2, \ldots, n$. Then by relations (6.153), (6.154), and (6.156) we can have in turn

$$t_{k+2} = s_{k+1} + \ell_{k+1} \leq t_{k+1} + C\left(s_k, t_{k+1}, t_{k+1} - s_k\right)\left(t_{k+1} - s_k\right)$$

$$+ C_1\left(s_k, t_{k+1}\right)\left(t_{k+1} - s_k\right) + \left(b_1 + b_2\right)\left(t_{k+1} - s_k\right)$$

$$+ \left[C_2\left(t_k, t_{k+1}\right) + b_3\right]\left(t_{k+1} - t_k\right) + \alpha_k + \ell_{k+1}$$

$$\leq \cdots \leq s_0 + [C\,(r_0, r_0, r_0) + C_1\,(r_0, r_0) + C_2\,(r_0, r_0) + b_1 + b_2 + b_3]\,t_{k+1}$$

$$+ \sum_{i=0}^{k} \alpha_i + \sum_{i=0}^{k+1} \ell_i \leq T\,(r_0) \leq r_0\,.$$

Hence, the scalar sequence $\{t_n\}$ $(n \geq 0)$ is bounded above by r_0. By hypothesis (6.161) the number r_0 is the minimum nonnegative zero of the equation $T(r) - r = 0$ on $[0, r_0]$, and from the above $r_0 = \lim_{n \to \infty} t_n$.

(b) Using relations (6.126), (6.127), (6.152), (6.153), and (6.154) we deduce that $x_1, x_0 \in \bar{U}(x_0, r_0)$ and that estimates (6.163) and (6.164) are true for $n = 0$. Let us assume that they are true for $k = 0, 1, 2, \ldots, n - 1$. Using the induction hypothesis, we can have in turn

$$\|x_{k+1} - x_0\| \leq \|x_{k+1} - y_0\| + \|y_0 - x_0\|$$

$$\leq \|x_{k+1} - y_k\| + \|y_k - y_0\| + \|y_0 - x_0\|$$

$$\leq \cdots \leq (t_{k+1} - s_k) + (s_k - s_0) + s_0 \leq t_{k+1} \leq r_0\,,$$

and

$$\|y_{k+1} - x_0\| \leq \|y_{k+1} - y_0\| + \|y_0 - x_0\| \leq \|y_{k+1} - x_{k+1}\|$$

$$+ \|x_{k+1} - y_k\| + \|y_k - y_0\| + \|y_0 - x_0\|$$

$$\leq \cdots \leq (s_{k+1} - t_{k+1}) + (t_{k+1} - s_k) + (s_k - s_0) + s_0 \leq s_{k+1} \leq r_0\,.$$

That is, $x_n, y_n \in U(x_0, r_0)$ for all $n \geq 0$.

We can now have from approximations (6.126) and (6.127)

$$F\,(x_{k+1}) + Q\,(x_{k+1}) = F\,(x_{k+1}) - F\,(y_k) - A\,(y_k)\,(x_{k+1} - y_k)$$

$$+ A\,(y_k)\,(x_{k+1} - y_k) + F\,(y_k) + Q\,(x_{k+1})$$

$$= \int_0^1 \left[F'\,(y_k + t\,(x_{k+1} - y_k)) - A\,(y_k)\right](x_{k+1} - y_k)\,dt$$

$$+ \left(A\,(y_k) - F'\,(y_k)\right)(x_{k+1} - y_k) + (Q\,(x_{k+1}) - Q\,(x_k))$$

$$+ \left(F \left(y_k \right) - F \left(x_k \right) - A \left(x_k \right) \left(y_k - x_k \right) + F' \left(y_k \right) \left(x_{k+1} - y_k \right) \right) .$$

Hence, by using hypotheses (6.136)–(6.139) we obtain in turn

$$\left\| A \left(x_{k+1} \right)^{-1} \left(F \left(x_{k+1} \right) + Q \left(x_{k+1} \right) \right) \right\|$$

$$\leq \left\| \int_0^1 A \left(x_{k+1} \right)^{-1} \left[F' \left(y_k + t \left(x_{k+1} - y_k \right) \right) \right. \right.$$

$$\left. - A \left(y_k \right) \right] \left(x_{k+1} - y_k \right) \left\| dt \right.$$

$$+ \left\| A \left(x_{k+1} \right)^{-1} \left(A \left(y_k \right) - F' \left(y_k \right) \right) \left(x_{k+1} - y_k \right) \right\|$$

$$+ \left\| A \left(x_{k+1} \right)^{-1} \left(Q \left(x_{k+1} \right) - Q \left(x_k \right) \right) \right\|$$

$$+ \left\| A \left(x_{k+1} \right)^{-1} \left[F \left(y_k \right) - F \left(x_k \right) - A \left(x_k \right) \left(y_k - x_k \right) \right. \right.$$

$$\left. + F' \left(y_k \right) \left(x_{k+1} - y_k \right) \right] \right\| = \bar{v}_{k+1} \leq v_{k+1} \qquad (6.169)$$

by hypotheses (6.158) and relations (6.155) and (6.157). Hence, we have shown estimate (6.167) for all $n \geq 0$.

Using relations (6.126), (6.153), and (6.167) we obtain

$$\| y_{k+1} - x_{k+1} \| = \left\| A \left(x_{k+1} \right)^{-1} \left(F(x_{k+1}) + Q \left(x_{k+1} \right) \right) \right\|$$

$$\leq \bar{v}_{k+1} \leq v_{k+1} = s_{k+1} - t_{k+1} ,$$

which shows estimate (6.163) for all $n \geq 0$.

Similarly, from relations (6.127), (6.154), and (6.160) we obtain

$$\| x_{k+1} - y_k \| = \| -z_k \| \leq w \left(z_k \right) = t_{k+1} - s_k ,$$

from which it follows that estimate (6.164) is true for all $n \geq 0$.

It now follows from the estimates (6.163) and (6.164) that the sequence $\{x_n\}$ $(n \geq 0)$ is Cauchy in a Banach space X and as such it converges to some $x^* \in U(x_0, r_0)$, which by taking the limit as $n \to \infty$ in (6.126), we obtain $F(x^*) + Q(x^*) = 0$.

Using (6.134) we deduce that $\lim_{n\to\infty} z_n = 0$. Moreover, from (6.127) we can get

$$\lim_{n\to\infty} y_n = \lim_{n\to\infty} (x_{n+1} + z_n) = \lim_{n\to\infty} x_{n+1} + \lim_{n\to\infty} z_n = x^* .$$

Estimates (6.165) and (6.166) follow from (6.158) and (6.164), respectively, by using standard arguments in majorant theory (see also Chapter 3).

To show uniqueness, we assume that there exists another solution y^* of Equation (6.125) in $\bar{U}(x_0, R)$ with $x^* \neq y^*$. From the approximation

$$y_n - y^* = -A\,(x_n)^{-1} \left\{ \int_0^1 \left[F'\,(y^* + t\,(x_n - y^*)) - A\,(y^*) \right] (x_n - y^*)\,dt \right.$$

$$\left. + \left(A\,(y^*) - A\,(x_n) \right) (x_n - y^*) + \left(Q\,(x_n) - Q\,(y^*) \right) \right\},$$

we can obtain

$$\| y_n - y^* \| \leq \left[C\left(\| y^* - x_0 \|, \| x_n - x_0 \|, \| x_n - y^* \| \right) + b_1 \right] \| x_n - y^* \|$$

$$+ \left[C_3\left(\| y^* - x_0 \|, \| x_n - x_0 \|, \| x_n - y^* \| \right) + b_4 \right] \| x_n - y^* \|$$

$$+ C_2\left(\| y^* - x_0 \|, \| x_n - x_0 \| \right) \| x_n - y^* \|$$

$$\leq \theta\,(r_0) \| x_n - y^* \| . \tag{6.170}$$

By letting $n \to \infty$ in (6.170) we get

$$\| x^* - y^* \| \leq \theta\,(r_0) \| x^* - y^* \| < \| x^* - y^* \| . \tag{6.171}$$

Hence, we conclude from (6.171) that $x^* = y^*$. Finally using the approximation

$$y_n - x_n = x^* - x_n + A\,(x_n)^{-1} \left[\int_0^1 \left(F'\,(x_n + t\,(x^* - x_n)) \right. \right.$$

$$\left. \left. - A\,(x_n)\,(x^* - x_n)\,dt + Q\,(x^*) - Q\,(x_n) \right) \right]$$

and from conditions (6.136)–(6.139), (6.169) we obtain (6.168). That completes the proof of the theorem. ∎

We will need the following result on local convergence.

THEOREM 6.15

Let $F, Q: D \subset X \to Y$ be nonlinear operators as in Theorem 6.14, and assume:

(i)   there exists a regular solution $x^* \in D$ of the equation $F(x) + Q(x) = 0$;

(ii)   condition (G_3) is satisfied on $\bar{U}(x^*, r^*)$ (for $x_0 = x^*$) and $\bar{U}(x^*, r^*) \subseteq D$;

(iii)   there exists a sequence $\{z_n\}$ $(n \geq 0)$ of points from D satisfying

$$\|z_n\| \leq g_n = g(z_n) \leq w_5(r) \tag{6.172}$$

where $g: \bar{U}(x^*, r^*) \to [0, +\infty)$ is continuous; and

(iv)   constants b_1, b_4 are such that $b_1 + b_4 \in [0, 1)$.

Then the following hold:

(a)   for sufficiently small $r^* \in (0, R]$

$$0 < \varphi_2(r^*) < 1. \tag{6.173}$$

(b)   Sequences $\{y_n\}, \{x_n\}$ $(n \geq 0)$ are well defined, remain in $\bar{U}(x^*, r^*)$ for all $n \geq 0$ and $\lim_{n \to \infty} x_n = \lim_{n \to \infty} y_n = x^*$. Moreover, the solution x^* of Equation (6.125) is unique in $\bar{U}(x^*, r^*)$.

Furthermore, the following estimates hold for all $n \geq 0$:

$$\|x_{n+1} - x^*\| \leq \gamma_n \|x_n - x^*\| \leq \gamma \|x_n - x^*\| \tag{6.174}$$

and

$$\|y_n - x^*\| \leq \delta_n \|x_n - x^*\| \leq \delta \|x_n - x^*\|, \tag{6.175}$$

where

$$\delta_n = C\left(0, \|x_n - x^*\|, \|x_n - x^*\|\right) + C_2\left(0, \|x_n - x^*\|\right)$$

$$+ C_3\left(0, \|x_n - x^*\|, \|x_n - x^*\|\right) + b_1 + b_4, \tag{6.176}$$

$$\delta = C\left(0, r^*, r^*\right) + C_2\left(0, r^*\right) + C_3\left(0, r^*, r^*\right) + b_1 + b_4, \tag{6.177}$$

$$\gamma_n = \delta_n + g_n \tag{6.178}$$

and

$$\gamma = \delta + w_5(r^*). \tag{6.179}$$

PROOF   (a) By hypotheses (ii), (iii), and (iv) we have $C_0(0, 0, 0) = C_1(0, 0) = C_2(0, 0) = w_5(0) = 0$, $b_1 + b_2 \in [0, 1)$ and that all these functions are continuous and nondecreasing on $[0, \|x^* - x_0\|]$. Hence, we can find a number $r^* \in (0, \|x^* - x_0\|$ such that the estimate (6.173) holds.

(b) Let us assume that

$$x_m \in \bar{U}\left(x^*, r^*\right) \quad \text{for } m = 0, 1, 2, \ldots, k \tag{6.180}$$

since $x_0 \in U(x^*, r^*)$. Using (6.126) we get

$$y_k - x^* = -A\left(x_k\right)^{-1}$$

$$\left[F\left(x_k\right) - F\left(x^*\right) - A\left(x_k\right)\left(x_k - x^*\right) + Q\left(x_k\right) - Q\left(x^*\right)\right]. \tag{6.181}$$

We also introduce the approximation

$$F\left(x_k\right) - F\left(x^*\right) - A\left(x_k\right)\left(x_k - x^*\right) + Q\left(x_k\right) - Q\left(x^*\right)$$

$$= \int_0^1 \left[F'\left(x^* + t\left(x_k - x^*\right)\right) - A\left(x_k\right)\right]\left(x_k - x^*\right) dt + Q\left(x_k\right) - Q\left(x^*\right)$$

$$= \int_0^1 \left[F'\left(x^* + t\left(x_k - x^*\right)\right) - A\left(x_k\right)\right]\left(x_k - x^*\right) dt$$

$$+ Q\left(x_k\right) - Q\left(x^*\right). \tag{6.182}$$

We now compose both sides of (6.182) by $A(x_k)^{-1}$, and then by taking norms and using (6.148)–(6.151), we obtain that the left-hand side of (6.182) is bounded above by

$$\left[C\left(0, \|x_k - x^*\|, \|x_k - x^*\|\right) + b_1 + C_3\left(0, \|x_k - x^*\|, \|x_k - x^*\|\right)\right.$$

$$\left. + b_4 + C_2\left(0, \|x_k - x^*\|\right)\right] \|x_k - x^*\|. \tag{6.183}$$

From (6.180, (6.181), and (6.183) we now have

$$\|y_k - x^*\| \leq \left\|A\left(x_k\right)^{-1} \int_0^1 \left[F'\left(x^* + t\left(x_k - x^*\right)\right) - A\left(x_k\right)\right]\right.$$

$$(x_k - x^*) \, dt + (Q(x_k) - Q(x^*)) \Bigg\|$$

$$\leq \delta_k \left\| x_k - x^* \right\| \leq \delta \left\| x_k - x^* \right\| . \tag{6.184}$$

The above estimate shows that (6.175) is true and that $y_k \in \bar{U}(x^*, r^*)$ since $\delta_k \leq \delta < 1$ [by (6.173)].

Moreover, from (6.127), (6.184), and (6.172), we get

$$\left\| x_{k+1} - x^* \right\| \leq \left\| y_k - x^* \right\| + \left\| z_k \right\| \leq \gamma_k \left\| x_k - x^* \right\| \leq \gamma \left\| x_k - x^* \right\| , \tag{6.185}$$

which shows (6.174) and that $x_{k+1} \in \bar{U}(x^*, r^*)$. Hence the sequences $\{x_n\}$, $\{y_n\}$ $(n \geq 0)$ are well defined, remain in $\bar{U}(x^*, r^*)$ and satisfy (6.174) and (6.175) for all $n \geq 0$.

Let $m \geq 0$. Then by (6.174) we get

$$\left\| x_{n+m} - x^* \right\| \leq \gamma_{n+m-1} \left\| x_{n+m-1} - x^* \right\|$$

$$\leq \gamma_{n+m-1} \gamma_{n+m-2} \left\| x_{n+m-2} - x^* \right\|$$

$$\leq \cdots \leq \gamma^m \left\| x_n - x^* \right\| . \tag{6.186}$$

Similarly by (6.174) and (6.175) we get

$$\left\| y_{n+m} - x^* \right\| \leq \delta \cdot \gamma^m \left\| x_n - x^* \right\| . \tag{6.187}$$

Finally by letting $m \to \infty$ in (6.186) and (6.187) we obtain $\lim_{n \to \infty} x_n = \lim_{n \to \infty} y_n = x^*$ (since $0 < \gamma < 1$). The proof of the uniqueness of the solution x^* of Equation (6.125) in $U(x^*, r^*)$ is omitted as it is identical to the corresponding one in Theorem 6.14.

That completes the proof of the theorem. ∎

The points z_n $(n \geq 0)$ appearing in (6.134) depend on x_n $(n \geq 0)$. The points $\{z_n\}$ $(n \geq 0)$ in (6.172) depend on x_n $(n \geq 0)$ and may be the point x^*. That is why we can choose the functions g, w_5 [see (6.172)] to be the same or different from the functions w, w_1 [see (6.134)], respectively.

In many applications it turns out that the solution x^* of Equation (6.125) as well as iterates x_n, y_n have "better smoothness" properties than the elements of X. This is a motivation for considering a subset $X_1 \subseteq X$ such that

$$x^* \in X_1, x_n, y_n \in X_1, x_n - x^*, y_n - x^*$$

$$\in X_1, \quad x_{n+1} - x_n, \, y_{n+1} - y_n \in X_1 \, (n \geq 0) \, . \tag{6.188}$$

We consider a family

$$\{P, P_1, L, L_0\} \, , \quad h > 0 \tag{6.189}$$

where

$$P, P_1 : X_1^1 \subseteq X^1 \to Y^1, \quad h > 0$$

are nonlinear operators and

$$L : X \to X^1, \quad L_0 : Y \to Y^1, \quad h > 0$$

are bounded linear discretization operators such that

$$L\left(X_1 \cap U\left(x^*, \delta_5\right)\right) \subseteq X_1^1, \quad h > 0 \, . \tag{6.190}$$

The operators P, P_1, L, L_0, S depend on h. That is, $P = P_h$ etc. To simplify the notation we assume that this is understood and hence we avoid the use of the subscript h.

The discretization (6.174) is called uniform if there exists a number δ_1 such that

$$\bar{U}\left(L\left(x^*\right), \delta_1\right) \subseteq X_1^1, \quad h > 0 \tag{6.191}$$

and the triplet $(P, Q, L(x^*))$ satisfies the "G" conditions that the triplet (F, A, x_0) satisfies for all $h > 0$ in the ball $\bar{U}(L(x^*), \delta_1)$.

Moreover, the discretization family (6.189) is called: *bounded* if there is a constant $q > 0$ such that

$$\|L(x)\| \leq q\|x\|, \quad x \in X_1, \, h > 0 \, , \tag{6.192}$$

consistent of order $p > 0$ if there are two constants $c_0, c_1 > 0$ such that

$$\|S(L(x))^{-1}\left(L_0(F + Q) - (P + P_1)\left(L(x)\right)\right)\| \leq c_0 h^p \, ,$$

$$x \in X_1 \cap \bar{U}\left(L\left(x^*\right), \delta_5\right), \quad h > 0 \tag{6.193}$$

and

$$\|S(L(x))^{-1}\left(L_0(A(x)(y) - S(L(x))L(y))\right)\| \leq c_1 h^p \, ,$$

$$x \in X_1 \cap \bar{U}\left(L\left(x^*\right), \delta_5\right), \, y \in X_1, \, h > 0 \, . \tag{6.194}$$

Concerning the function w_5, we can easily see from (6.135) and (6.172) that it can be identified with the function w_2. Choose for example $z_n = d_n$ for all $n \geq 0$, $h > 0$. However, we do not need this to prove our discretization results. By choosing $L(x_0) \in U(x_0, \frac{R}{3})$, $\delta_5 \leq \frac{R}{3}$ and $r^* \leq \frac{R}{3q}$, one can easily show that $\bar{U}(L(x^*), \delta_5) \subseteq \bar{U}(x_0, R)$. Hence, the C^h, w^h, $\bar{\alpha}^h$, b^h are just the C, w, $\bar{\alpha}$, b respectively restricted on $\bar{U}(L(x^*), \delta_5)$.

With the notation introduced above we can now formulate our main result.

THEOREM 6.16
Let $F, Q: D \subseteq X \to Y$ be nonlinear operators as in Theorem 6.14. Assume:

(i)   *hypotheses of Theorem 6.15 are satisfied;*

(ii) *discretization (6.189) is bounded, stable, and consistent of order p and $\bar{U}(L(x^*), \delta_5) \subseteq \bar{U}(x_0, R)$ for all $h \in (0, h_2]$;*

(iii) *the following estimate holds:*

$$\left\| S(v)^{-1}(S(w) - S(v)) \right\| \leq C_4 \left(\|v - L(x^*)\| \, , \right.$$

$$\left. \|w - L(x^*)\| \, , \|v - w\| \right) + b_5 \, , \tag{6.195}$$

for all $v, w \in \bar{U}(L(x^*), \delta_5)$, some $b_5 \in [0, 1)$ and a function C_4 satisfying the same properties as the C function.

Then

(a)   *Equation (6.130) has a locally unique solution*

$$a^*(h) = a^* = L(x^*) + O(h^p) \tag{6.196}$$

for all $h \in (0, \overline{h_0}]$ with $\overline{h_0}$ being a constant.

(b)   *There exist constants $\overline{h_1} \in (0, \overline{h_0}]$, $r_1 \in (0, \delta_5]$ such that the discrete iteration (6.131) and (6.132) converges to a^*.*

(c)   *If there exist constants c_{10}, c_{11} with $4(b_1 c_{11} + b_3 c_{11} + 2q c_3 b_5 + c_0 + c_1) \leq c_{10} \leq c_{11}$, such that for all $n \geq 0$,*

$$\|d_n - L(z_n)\| \leq (c_{11} - c_{10}) h^p \, ,$$

$$h \in (0, \overline{h_0}] \, , \quad r \in (0, \delta_5] \, , \tag{6.197}$$

then there exist constants $\overline{h_1} \in (0, \overline{h_0}]$, $r_3 \in (0, r_1]$ such that the following estimates hold for all $n \geq 0$

$$b_n = L(y_n) + O(h^p) , \qquad (6.198)$$

$$a_n = L(x_n) + O(h^p) , \qquad (6.199)$$

$$S(b_n)^{-1}(P + P_1)(b_n) = S(b_n)^{-1} L_0((F + Q)(y_n))$$

$$+ O(h^p) , \qquad (6.200)$$

(provided that $(P + P_1)$ is b-Lipschitz continuous on $\bar{U}(L(x^*), \delta_5)$)

$$S(a_n)^{-1}(P + P_1)(a_n) = S(a_n)^{-1} L_0((F + Q)(x_n))$$

$$+ O(h^p) , \qquad (6.201)$$

$$b_n - a^* = L(y_n - x^*) + O(h^p) \qquad (6.202)$$

and

$$a_n - a^* = L(x_n - x^*) + O(h^p) . \qquad (6.203)$$

PROOF   (a) The C, w functions are continuous, vanish at the origin and $b_1 + b_4 \in [0, 1)$. Hence we can find intervals $(0, h_0]$ and $(0, 1]$ such that conditions (6.161) and (6.162) are satisfied for all $h \in (0, h_0]$ and $r(h) \in (0, \delta_1]$. Set $h_3 = \min\{h_0, h_2, (\frac{\delta_1}{c_3})^{1/p}\}$. Then using (6.145) and (6.193) we obtain in turn

$$s_0(h) = \left\| S(L(x))^{-1}(P + P_1)(L(x^*)) \right\|$$

$$\leq \left\| (P + P_1)(L(x^*)) - L_0(F + Q)(x^*) \right\|$$

$$\leq c_0 h^p \leq c_2 h^p \leq \varphi(r(h)) , \qquad (6.204)$$

and

$$r(h) \leq c_3 h^p \leq \delta_1, \quad \text{for all } h \in (0, h_3] , \qquad (6.205)$$

which shows that (6.161) and (6.162) hold for all $h \in (0, h_3]$. Since all hypotheses of Theorem 6.14 are satisfied, Equation (6.130) has a solution $a^*(h) = a^* \in$

$\bar{U}(L(x^*), r(h))$ which is a unique solution in $\bar{U}(L(x^*), \delta_1)$. Thus, (6.196) follows from

$$\left\| a^* - L\left(x^* \right) \right\| \leq r(h) \leq c_3 h^p \leq \delta_1 \tag{6.206}$$

by setting $\overline{h_0} = h_3$.

(b) As in part (a) we can find intervals $(0, h_3]$ and $(0, \delta_5]$ such that conditions (6.142) and (6.173) are satisfied for all $h \in (0, h_3]$ and $r^*(h) \in (0, \delta_5]$. By applying Theorem 6.14 to (6.130) we see that the sequence (6.131)–(6.132) converges to a^* if

$$\left\| L\left(x_0 \right) - a^* \right\| < r^*(h) \tag{6.207}$$

and

$$\bar{U}\left(a^*, \left\| L\left(x_0 \right) - a^* \right\| \right) \subseteq \bar{U}\left(L\left(x^* \right), \delta_1 \right) . \tag{6.208}$$

But (6.208) certainly holds if

$$\left\| a^* - L\left(x^* \right) \right\| + \left\| L\left(x_0 \right) - a^* \right\| \leq \delta_1 . \tag{6.209}$$

By (6.192) and (6.205) we obtain

$$\left\| L\left(x_0 \right) - a^* \right\| \leq \left\| L\left(x_0 \right) - L\left(x^* \right) \right\| + \left\| L\left(x^* \right) - a^* \right\|$$

$$\leq q \left\| x_0 - x^* \right\| + c_3 h^p . \tag{6.210}$$

Thus (6.207), (6.208) hold if

$$q \left\| x_0 - x^* \right\| + 2c_3 h^p \leq \delta_1 , \tag{6.211}$$

and

$$q \left\| x_0 - x^* \right\| + c_3 h^p \leq c_7 - c_8 h^p \tag{6.212}$$

hold, respectively. Conditions (6.211) and (6.212) will certainly hold if

$$q \left\| x_0 - x^* \right\| \leq \frac{\delta_1}{2}, \quad 2c_3 h^p \leq \frac{\delta_1}{2}, \quad q \left\| x_0 - x^* \right\| \leq \frac{c_7}{2}$$

and

$$c_3 h^p \leq \frac{c_7}{2} - c_8 h^p .$$

We choose

$$\left\| x_0 - x^* \right\| \leq r_1 = \min \left\{ \frac{\delta_1}{2q}, \frac{c_7}{2q} \right\} ,$$

and

$$h_4 = \min\left\{h_2, h_3, \left(\frac{\delta_1}{4c_3}\right)^{1/p}, \left[\frac{c_7}{2(c_3 + c_8)}\right]^{1/p}\right\}.$$

It is now easily verified that (6.207) and (6.208) are satisfied for all $h \in (0, h_4]$ and $x_0 \in \bar{U}(x^*, r_1)$. Therefore, for these h and x_0, the iteration (6.131)–(6.132) converges to a^*.

(c) We will now show that there exist $\bar{h_1} \in (0, \bar{h_0}]$, $r_3 \in (0, r_1]$ such that

$$\|a_n - L(x_n)\| \le c_{11}h^p. \tag{6.213}$$

For $n = 0$ (6.213) is true since $a_0 = L(x_0)$. Suppose that (6.213) holds for $n = 0, 1, \ldots, i$. We note that if we show that

$$\|b_n - L(y_n)\| \le c_{10}h^p, \tag{6.214}$$

then from (6.131), (6.132), (6.214), (6.197) and the estimate

$$\|a_{i+1} - L(x_{i+1})\| = \|b_i - d_i - L(y_i - z_i)\| \le \|b_i - L(y_i)\| + \|d_i - L(z_i)\|$$

$$\le c_{10}h^p + (c_{11} - c_{10})h^p = c_{11}h^p, \tag{6.215}$$

we can complete the induction for (6.213). But (6.214) is true for $n = 0$ by (6.197). We now suppose that (6.199) is true for $n = 0, 1, \ldots, i$. Using (6.126), (6.127), (6.131), and (6.132) we can obtain the approximation

$$b_i - L(y_i) = S(a_i)^{-1}\{[S(a_i)(a_i - L(x_i)) - (P + P_1)(a_i)$$

$$+ (P + P_1)(L(x_i))]$$

$$+ \left[(S(a_i) - S(L(x_i)))L\left(A(x_i)^{-1}(F + Q)(x_i)\right)\right]$$

$$+ \left[S(L(x_i))L\left(A(x_i)^{-1}(F + Q)(x_i)\right) - L_0((F + Q)(x_i))\right]$$

$$+ [L_0((F + Q)(x_i)) - ((P + P_1)(x_i))]\}. \tag{6.216}$$

From (6.192) and (6.213) we obtain

$$\|a_i - L(x^*)\| \le \|a_i - L(x_i)\| + \|L(x_i) - L(x^*)\| \le c_{11}h^p + qr_1. \tag{6.217}$$

By composing the first bracket in (6.216) by $S(a_i)^{-1}$ and taking norms, using (6.136), (6.139), and (6.217), we obtain that this term is bounded above by

$$\left\| \int_0^1 S(a_i)^{-1} \left\{ \left[P'(L(x_i) + t(a_i - L(x_i)) - S(a_i) \right] \right. \right.$$

$$\left. \left. \cdot (a_i - L(x_i)) \, dt + (P_1(L(x_i)) - P_1(a_i)) \right\} \right\|$$

$$\leq \left[C \left(\|L(x_i) - L(x^*)\|, \|a_i - L(x^*)\|, \|a_i - L(x_i)\| + b_1 \right] \|a_i - L(x_i)\|$$

$$+ \left[C_2 \left(\|L(x_i) - L(x^*)\|, \|a_i - L(x^*)\| \right) + b_3 \right] \|a_i - L(x_i)\| \quad (6.218)$$

$$\leq \left[C \left(qr_1, c_{11} h^p + qr_1, c_{11} h^p \right) + b_1 \right] c_{11} h^p$$

$$+ \left[C_2 \left(qr_1, c_{11} h^p + qr_1 \right) + b_3 \right] c_{11} h^p . \quad (6.219)$$

Moreover, by adding and subtracting $S(a_i)^{-1}$ inside the parenthesis of the second bracket, composing by $S(L(a_i)^{-1}$ and using (6.217) and (6.192), we obtain that this term is bounded above by

$$\left[C_4 \left(\|L(x_i) - L(x^*)\|, \|a_i - L(x^*)\|, \right. \right.$$

$$\left. \|a_i - L(x_i)\| + b_5 \right] q \left(\|y_i - x^*\| + \|x_i - x^*\| \right)$$

$$\leq 2qc_3 \left[C_4 \left(qr_1, c_{11} h^p + qr_1, c_{11} h^p \right) + b_5 \right] h^p \quad \text{[by (6.141)]} \quad (6.220)$$

Furthermore using (6.194) and (6.196), we obtain that the third and fourth brackets in (6.216), after being composed by $S(a_i)^{-1}$, are bounded above by $c_1 h^p$ and $c_0 h^p$, respectively.

Finally, by collecting all the above majorizations, we obtain that estimate (6.214) will be true if

$$C \left(qr_1, c_{11} h^p + qr_1, c_{11} h^p \right) c_{11} + C_2 \left(qr_1, c_{11} h^p + qr_1 \right) c_{11}$$

$$+ 2q C_4 \left(qr_1, c_{11} h^p + qr_1, c_{11} h^p \right) c_3$$

$$+ (b_1 c_{11} + b_3 c_{11} + 2q c_3 b_5 + c_1 + c_1) \leq c_{10} . \quad (6.221)$$

Inequality (6.221) will certainly be true if each term on the left is bounded above by $\frac{1}{4}$. Since the functions C and C_0 vanish at the origin, we can find $h_5, r_2 > 0$ such that this will happen for the first three terms. By the hypothesis of (c), the last term is also bounded above by $\frac{1}{4}$. Finally, set $\overline{h_1} = \min\{h_4, h_5, (\frac{1}{4c_{11}})^{1/p}\}$ and $r_3 = \min\{r_1, r_2\}$. With the above choices of $\overline{h_1}$ and r_3, estimates (6.198) and (6.199) follow.

Using (6.193) and (6.214) we obtain

$$\left\| S(b_n)^{-1}((P+P_1)(b_n) - L_0((F+Q)(y_n))) \right\|$$

$$\leq \left\| S(b_n)^{-1}((P+P_1)(b_n) - (P+P_1)(L(y_n))) \right\|$$

$$+ \left\| S(b_n)^{-1}((P+P_1)(L(y_n)) - L_0((F+Q)(y_n))) \right\|$$

$$\leq b \|b_n - L(y_n)\| + c_0 h^p \leq (bc_{10} + c_0) h^p . \tag{6.222}$$

Estimate (6.222) shows (6.200). Estimate (6.201) is obtained similarly by replacing b_n, y_n, and c_{10} by a_n, x_n, and c_{11}.

Moreover, from (6.214) and (6.205) we obtain

$$\left\| (b_n - a^*) - L(y_n - x^*) \right\| \leq \|b_n - L(y_n)\| + \|a^* - L(x^*)\|$$

$$\leq c_{10} h^p + c_3 h^p = (c_{10} + c_3) h^p , \tag{6.223}$$

which shows (6.202).

Furthermore, from (6.213) and (6.205), we finally obtain

$$\left\| (a_n - a^*) - L(x_n - x^*) \right\| \leq \|a_n - L(x_n)\| + \|a^* - L(x^*)\|$$

$$\leq c_{11} h^p + c_3 h^p = (c_{11} + c_3) h^p , \tag{6.224}$$

from which (6.203) follows.

That completes the proof of the theorem. ∎

We note that conditions $\delta_1 \leq r_0$ and $\delta_5 \leq r^*$ [see (6.145) and (6.150)] are not used in the proof. However, in many practical applications we may want these conditions to be true.

We can now prove the mesh-independence principle for inexact Newton-like methods.

THEOREM 6.17
Assume:

(i) *hypotheses of Theorem 6.16 hold;*

(ii) *there exists a constant $\delta > 0$ such that*

$$\liminf_{h \to 0} \|L(u)\| \geq \delta \|u\| \quad \text{for each } u \in X_1 . \tag{6.225}$$

Then for some $r_6 \in (0, r_3]$, and for any fixed $\varepsilon > 0$ and $x_0 \in \bar{U}(x^*, r_6)$ there exists a constant $\bar{h} = \bar{h}(\varepsilon, x_0) \in (0, \bar{h}_1]$ such that

$$|\min\{n \geq 0, \|x_n - x^*\| < \varepsilon\} - \min\{n \geq 0, \|a_n - a^*\| < \varepsilon\} \leq 1 \tag{6.226}$$

for all $h \in \mathbb{Q}, \bar{h}$.

PROOF   By hypotheses there exists a unique integer $i > 0$ such that

$$\|x_{i+1} - x^*\| < \varepsilon \leq \|y_i - x^*\| \tag{6.227}$$

and $h_6 = h_6(x_0)$ such that

$$\|L(y_i - x^*)\| \geq \delta \|y_i - x^*\| \quad \text{for all } h \in (0, h_6] . \tag{6.228}$$

We will prove that the theorem holds for

$$r_6 = \min\left\{r_3, \frac{r_4}{q}\right\}, \beta = \min\{\delta, 2q, [(C(0, r_5, r_5)$$

$$+ C_3(0, r_5, r_5) + C_2(0, r_5) + b_1 + b_3 + b_4) 2q]^{-1}\}, \tag{6.229}$$

$$\bar{h} = \min\left\{\bar{h}_1, h_6, \left(\frac{\beta \varepsilon}{2c_{12}}\right)^{1/p}, \left(\frac{\delta \varepsilon}{2c_{13}}\right)^{1/p}\right\}, \tag{6.230}$$

where $c_{12} = c_3 + c_{11}, c_{13} = c_{10} + c_3$ and $r_5 = \delta_1 + r_4$.
From (6.224) and (6.230) it follows that

$$\|a_{i+1} - a^*\| \leq \|L(x_{i+1} - x^*)\| + c_{12}h^p \leq q\varepsilon + \frac{\beta \varepsilon}{2} < 2q\varepsilon . \tag{6.231}$$

Using (6.229), (6.230), and Theorem 6.14 we obtain in turn that

$$
\begin{aligned}
\left\| b_{i+1} - a^* \right\| &\leq \left[C \left(0, \left\| a_{i+1} - a^* \right\|, \left\| a_{i+1} - a^* \right\| \right) + b_1 \right. \\
&\quad + C_3 \left(0, \left\| a_{i+1} - a^* \right\|, \left\| a_{i+1} - a^* \right\| \right) \\
&\quad \left. + b_4 + C_2 \left(0, \left\| a_{i+1} - a^* \right\| \right) + b_3 \right] \left\| a_{i+1} - a^* \right\| \qquad (6.232) \\
&\leq \left[C \left(0, r_5, r_5 \right) + b_1 + C_3 \left(0, r_5, r_5 \right) + b_4 + C_2 \left(0, r_5 \right) \right] 2\beta q \varepsilon < \varepsilon
\end{aligned}
$$

(since $\left\| a_{i+1} - a^* \right\| \leq \left\| a_0 - a^* \right\| = \left\| L(x_0) - a^* \right\| \leq \left\| L(x_0) - L(x^*) \right\| + \left\| L(x^*) - a^* \right\| \leq qr_3 + c_3 h^p \leq r_4 + \delta_1 = r_5$).

Moreover, from (6.228) and (6.223) we obtain

$$
\varepsilon \leq \left\| y_i - x^* \right\| \leq \frac{1}{\delta} \left\| L \left(y_i - x^* \right) \right\| \leq \frac{1}{\delta} \left(\left\| b_i - a^* \right\| + c_{13} h^p \right)
$$

or

$$
\left\| b_i - a^* \right\| \geq \delta \varepsilon - c_{13} h^p \geq \delta \varepsilon - \frac{\delta \varepsilon}{2} = \frac{\delta \varepsilon}{2}. \qquad (6.233)
$$

Furthermore, if $\left\| a_i - a^* \right\| < \varepsilon$ then as in (6.232) we get

$$
\left\| b_i - a^* \right\| < \frac{1}{2} \beta \varepsilon \leq \frac{\delta \varepsilon}{2}
$$

which contradicts (6.233).

Hence we must have

$$
\left\| a_i - a^* \right\| \geq \varepsilon. \qquad (6.234)
$$

The result now follows from (6.227), (6.232), and (6.234).

That completes the proof of the theorem. ∎

As it was observed in [29] condition (6.225) follows from the condition

$$
\lim_{h \to 0} \left\| L(u) \right\| = \left\| u \right\| \quad \text{for each } u \in X_1, \qquad (6.235)
$$

which is standard in most discretization studies. In fact, for some discretization studies, we have

$$
\lim_{h \to 0} \left\| L(u) \right\| = \left\| u \right\| \quad \text{uniformly for } u \in X_1. \qquad (6.236)
$$

If this is the case, we can have a stronger version of the mesh independence principle.

COROLLARY 6.2
Assume:

(i) *hypotheses of Theorem 6.16 hold;*

(ii) *condition (6.236) holds uniformly for $u \in X_1$.*

Then there exists a constant $r_7 \in (0, r_3]$ and, for any fixed $\varepsilon > 0$, some $\overline{h_2} = \overline{h_2}(\varepsilon) \in (0, \overline{h_1}]$ such that (6.226) holds for all $h \in (0, \overline{h_2}]$ and all starting points $x_0 \in \overline{U}(x^*, r_7)$.

REMARK 6.18 (1) In all our previous results we assumed that $A(y)$ is invertible for all $y \in D$. It turns out that our results hold under the weaker condition that $A(x_0)$ is invertible only. For Theorem 6.14 replace $A(y)^{-1}$ by $A(x_0)^{-1}$ in (6.136), (6.137), (6.138), and (6.139), and add the hypotheses

$$\| A(x_0)^{-1} (A(x) - A(x_0)) \| \le C_0 (\|x - x_0\|) + b_0, \quad b_0 \in [0, 1) \quad (6.237)$$

and

$$C_0(r_0) + b_0 < 1 \quad (6.238)$$

for all $x \in \overline{U}(x_0, r) \subseteq \overline{U}(x_0, R)$, $r \in [0, R]$, where the function C_0 is continuous and nondecreasing on $[0, R]$ with $C_0(0) = 0$. Moreover, define the function C_5 on $[0, R]$ by

$$C_5(r) = [1 - (C_0(r) + b_0)]^{-1} . \quad (6.239)$$

Then using (6.237), (6.239), and the Banach lemma on invertible operators, we get that $A(x_n)$ is invertible and

$$\left\| A(x_n)^{-1} A(x_0) \right\| \le C_5 (\|x_n - x_0\|) . \quad (6.240)$$

Furthermore multiply the "C", w_4, $\bar{\alpha}$ functions by $C_5(r)$ (or $C_5(\|x_n - x_0\|)$) and the "b" constants by $C_5(r)$ (or $C_5(\|x_n - x_0\|)$).

With the above modifications one can easily see that the conclusions of Theorem 6.14 can now follow.

(2) Similarly, for Theorem 6.15, we can argue as in Remark 6.18 (1) but with the following modifications x_0, C_0, b_0, C_5, w_4 are x^*, C_0^*, b_0^*, C_5^*, and w_5, respectively. The "C^*" functions and the point b_0^* have properties similar to the ones without the stars. We note that they can even be taken to be equal to each other, and if this is true the rest of the results in this study can follow.

(3) The results obtained in this study will also hold if the left-hand side of (6.136) is replaced by the conditions

$$\|A(y)^{-1}(F'(x + t(y - x)) - A(x))(y - x)\| \quad \text{for all } t \in [0, 1] \,,$$

or

$$\int_0^1 \|A(y)^{-1}(F'(x + t(y - x)) - A(x))(y - x)\| dt$$

or

$$\|A(y)^{-1}(F'(x + t(y - x)) - A(x))\| \|y - x\| \quad \text{for all } t \in [0, 1] \,,$$

or

$$\int_0^1 \|A(y)^{-1}(F'(x + t(y - x)) - A(x))\| \|y - x\| dt$$

or any combination of the above conditions in the nonaffine form whether $A(y) = A(x_0)$ (or not). (See the proof of Theorem 6.14). In particular see relation (6.169). In all cases the function C and the point b_1 appearing at the right-hand side of (6.136) may become larger which will result in larger upper bounds on the distances

$$\|y_n - x_n\| \,, \|x_{n+1} - y_n\| \,, \|x_n - x^*\| \quad \text{and} \quad \|y_n - x^*\| \quad (n \geq 0) \,.$$

Similar remarks can be made for the left-hand sides of conditions (6.137), (6.138), and (6.139).

(4) Our results can be further generalized if we can find a continuous nondecreasing and vanishing at the origin function C_6 on $[0, R]^6$ and a null sequence $\{\varepsilon_n\}$ $(n \geq 0)$ such that

$$\|y_{n+1} - x_{n+1}\| \leq C_6 (x_n - x_0, y_n - x_0, x_{n+1} - x_0, x_{n+1} - y_n, y_n - x_n, z_n)$$

$$= \varepsilon_{n+1} \,.$$

It suffices to show that the iteration $\{x_n\}$ $(n \geq 0)$ is Cauchy, etc. A choice for ε_n is given by

$$\varepsilon_n = \bar{v}_n \quad (n \geq 1) \quad [\text{see } (6.167)] \,.$$

Many other choices are possible.

(5) Our results can be extended to include more general iterations of the forms

$$y_n = x_n - A(x_n)^{-1}(F(x_n) + Q(x_n)), \quad x_{n+1} = e_n y_n - z_n \ (n \geq 0) \ \text{for } e_n \in R \,.$$

We note that for $e_n = 1$ $(n \geq 0)$ the above iteration reduces to (6.126)–(6.127). Moreover, it can easily be seen from the proof of Theorem 6.14 that if we just replace z_i by $(e_i - 1)y_i - z_i$ $(i \geq 0)$ in (6.135), all the results obtained here will hold for this more general iteration.

(6) In [43] we show how to choose the functions $C, C_1, C_2, C_3, \bar{\alpha}, w, w_1, w_2,$ $w_3, w_4, w_5,$ and the sequence $\{z_n\}$ $(n \geq 0)$. We also showed that special choices of the above can reduce our results to earlier ones involving single step methods (Newton's method, Secant method, the method of tangent parabolas, the method of tangent hyperbolas and other) as well as two step methods, and by making use of Theorem 6.15.

(7) As an application of Theorem 6.14, we note that this theorem can be realized for operators F which satisfy an autonomous differential equation of the form

$$F'(x) = B(F(x)), \quad \text{for some given operator } B .$$

Assume for simplicity that $A(x) = F'(x)$ for all $x \in D$. As $F'(x^*) = B(0)$, the inverse $F'(x^*)^{-1}$ can be evaluated without knowing the actual solution x^*. Consider, for example, the scalar equation

$$F(x) = 0 \tag{6.241}$$

where F is given by $F(x) = e^x - s, s > 0$. Note that $F'(x) = F(x) + s$. That is $F'(x^*) = s$. Under the hypotheses of Theorem 6.14 and provided that $x_0 \in U(x^*, r^*)$, the iterations (6.126)–(6.127) converges to the solution $x^* = \ln(s)$ of Equation (6.241).

(8) It can easily be seen that our results can be reduced to the ones in [3] for $A(x) = F'(x)$, $Q(x) = 0$ $(x \in D)$ and $z_n = 0$ for all $n \geq 0$. With the notation used in [3] we can define the crucial "c" constants appearing in (6.145), (6.150), and (6.197) as follows:

$$c_3 = 2\sigma c_0, \quad c_4 = \frac{1}{\sigma L}, \quad c_5 = \sqrt{\eta} \ (0 < \eta < 1) \text{ for } h \leq \left(\frac{1-\eta}{2\sigma^2 L c_0}\right)^{1/p},$$

$$c_7 = \frac{2}{3L\sigma}, \quad c_8 = \frac{4\sigma c_0}{3}, \quad c_{11} = c_{12} = 8\sigma \max\{c_0, c_1\} \text{ and } \delta_1 = \rho$$

(see the proof of Theorem 2 in [3, pp. 163–164]). Moreover, they can be reduced to the ones in [29] for $z_n = 0$ for all $n \geq 0$. Furthermore, our condition (6.226) and the corresponding ones in [3] state that if

$$\min\{n \geq 0, \|x_n - x^*\| < \varepsilon\} = i + 1, \ i > 0 \tag{6.242}$$

then

$$\min\{n \geq 0, \|a_n - a^*\| < \varepsilon\} = i + 1, \text{ or } i, \text{ or } i + 2. \tag{6.243}$$

However, we can actually show that if (6.242) is true, then

$$\min\{n \geq 0, \|a_n - a^*\| < \varepsilon\} = i + 1 \text{ or } i, \tag{6.244}$$

which improves (6.243).

Let us assume that $q \in (0, \gamma^*)$, for some $\gamma^* \in (0, 1)$, and under the hypotheses of Theorem 6.17, set

$$\overline{h_3} = \min\left\{\bar{h}, \left(\frac{(\gamma^* - q)\varepsilon}{c_{12}}\right)^{1/p}\right\}. \tag{6.245}$$

The estimate (6.231) can also be written as

$$\|a_{i+1} - a^*\| \leq q\varepsilon + c_{12}h^p \leq \gamma^*\varepsilon < \varepsilon, \text{ for } h \in (0, \overline{h_3}], \tag{6.246}$$

which shows (6.244).

If $q \geq \gamma^*$ in (6.192), we can consider the linear operators M, M_0 instead of L, L_0 given by $M = \lambda L$ and $M_0 = \lambda M_0$, where λ is such that $|\lambda|q < \gamma^*$ and $\lambda \neq 0$. Then conditions (6.192), (6.193), and (6.194) will still be true with $\bar{q}$, $\bar{\sigma}$, $\overline{c_0}$, $\overline{c_1}$ replacing q, σ, c_0, c_1 and given by $\bar{q} = |\lambda|q$, $\bar{\sigma} = \frac{\sigma}{|\lambda|}$, $\overline{c_0} = |\lambda|c_0$ and $\overline{c_1} = |\lambda|c_1$.

(9) Concerning the choices of the "corrector" sequences $\{z_n\}$ and $\{d_n\}$ $(n \geq 0)$ appearing in (6.127) and (6.132), respectively, we state the following. Once the z_ns $n \geq 0$ are chosen (see Remark 6.18), then the d_ns will be chosen in such a way that condition (6.197) is satisfied. Note that condition (6.197) will certainly be satisfied if we simply set $d_n = L(z_n)$ for all $n \geq 0$, which is a logical choice but not the only one.

(10) As we showed in [29, 42] the discretization method $\{P, P_1, L, L_0\}$ can be used to solve boundary value problems involving operators F of the form

$$F(y) = \{y'' - F(x, y, y'); 0 \leq x \leq 1, y(0) - v, y(1) - w\}$$

or

$$F(y) = \{y' - f(x, y), 0 \leq x \leq 1, sy(0) + ty(1) - v\},$$

or integral operators of the form

$$(F(y))(x) = y(x) - \int_0^1 f(x, b, y(t))dt + g(x), \quad 0 \leq x \leq 1$$

or operators of the form

$$F(y) = \left\{ -y_{x_1 x_1} - y_{x_2 x_2} + f\left(x_1, x_2, y, y_{x_1}, y_{x_2}\right) \text{ in } \Omega, y = 0 \text{ on } \partial\Omega \right\}$$

involving partial differential equation boundary value problems. We leave the details to the motivated reader.

(11) Theorem 6.14 can be reduced to Theorem 1 in [94]. Indeed, set $Q(x) = 0$, $A(x) = F'(x)$ $(x \in D)$, $z_n = 0$ $(n \geq 0)$, $C_1 = C_2 = 0$, $b_2 = b_3 = 0$. Assume that condition (6.128) is satisfied. Then (6.136) is satisfied also if we set

$$2C_3 = C\left(\|x - x_0\|, \|y - x_0\|, \|y - x\|\right) = \frac{1}{2}\omega\|y - x\| \text{ and } b_1 = b_4 = 0 .$$

(12) Our conditions (6.136)–(6.139) are more general than the ones by Chen-Yamamoto [89], which in turn are more general than Potra's-Ptak's [164, 165], Kanno's [112], Yamamoto's [201], and Zabrejko's-Nguen's [207]. Hence, they can be used to solve a wider range of problems. In particular our conditions can be reduced to the ones obtained by Chen-Yamamoto if by using their versions and the last choice in Remark 6.18(3), we choose special cases of our functions given by $C = w(\|x - x_0\| + t\|y - x\|) - w_0(\|x - x_0\|)$, $(A(y) = A(x_0))$, $C_2 = e(\|x - x_0\|)\|x - y\|$, $C_3 = w_0(\|x - x_0\|)$, $C_1 = C_3 = 0$, $b_2 = b_4 = 0$, $b_1 = c$, $b_4 = b$ and $A(y)^{-1} = A(x_0)^{-1}$ for all $t \in [0, 1]$, $x, y \in \bar{U}(x_0, R)$. ∎

Example 6.5

We will provide an application for Theorem 6.14. Let us set $A(x) = F'(x)$, $Q(x) = 0$ $(x \in D)$ and choose

$$z_n = \frac{1}{2}F'(x_n)^{-1} F''(x_n)(y_n - x_n)^2 \quad (n \geq 0) . \tag{6.247}$$

Then the inexact Newton-like iteration $\{x_n\}$ $(n \geq 0)$, generated by (6.126)–(6.127) becomes (see also Section 6.2)

$$y_n = x_n - F'(x_n)^{-1} F(x_n) \tag{6.248}$$

$$x_{n+1} = y_n - \frac{1}{2}F'(x_n)^{-1} F''(x_n)(y_n - x_n)^2 \quad (n \geq 0) . \tag{6.249}$$

Moreover by eliminating y_n from relations (6.248) and (6.249), we obtain the method of tangent parabolas (or Euler-Chebysheff method) given by

$$x_{n+1} = x_n - \left[I + \frac{1}{2} F'(x_n)^{-1} F''(x_n) F'(x_n)^{-1} F(x_n) \right]$$

$$F'(x_n)^{-1} F(x_n) \quad (n \geq 0) . \tag{6.250}$$

This method has been examined extensively (see Section 6.2 and the references there) under the hypotheses

$$\left\| F'(x_0)^{-1} (F'(x) - F'(y)) \right\| \leq L \|x - y\| , \tag{6.251}$$

$$\left\| F'(x_0)^{-1} F''(x) \right\| \leq M \tag{6.252}$$

and

$$\left\| F'(x_0)^{-1} (F''(x) - F''(y)) \right\| \leq N \|x - y\| \tag{6.253}$$

for some $L, M, N > 0$ and all $x, y \in \bar{U}(x_0, R)$. Under the above hypotheses we can set

$$\bar{\alpha}_n = \left[\frac{N}{6} \|y_n - x_n\|^3 + M \|y_n - x_n\| \cdot \|x_{n+1} - y_n\| \right] \cdot \frac{1}{1 - L \|y_n - x_0\|}$$

$$w_4(r) = \left(\frac{N}{6} r^3 + M r^2 - M s_0 r \right) \frac{1}{1 - Lr}$$

$$w(z_n) = \frac{M}{2(1 - L \|y_n - x_0\|)} \|y_n - x_n\|^2 ,$$

$$w_2(r) = \frac{M r^2}{2(1 - Lr)} \quad \text{for all } r \in [0, R]$$

[for the computational details see [43, 60] and Remark 6.18(1)]. Moreover from conditions (G_3) we can set

$$C_1 = C_2 = 0, \quad b_1 = b_2 = b_3 = b_4 = 0 ,$$

$$\tfrac{1}{2} C_3 (\|x - x_0\|, \|y - x_0\|, \|y - x\|) = C (\|x - x_0\|, \|y - x_0\|, \|y - x\|)$$

$$= \frac{L}{2(1 - L\|y - x_0\|)} \|y - x\|$$

for all $x, y \in \bar{U}(x_0, R)$,

$$C(r, r, r) = \frac{L}{2(1 - Lr)} r \, ,$$

$$T(r) = s_0 + w_2(r) + w_4(r) + C(r, r, r)r$$

and

$$\theta(r) = (C + C_3)(R, r, r, +R)$$

for all $r \in [0, R]$.

With the above choices conditions (6.161) and (6.162) of Theorem 6.14 will be satisfied if the pair (r_0, R) satisfies

$$Nr^3 + 9(M + L)r^2 - 6(s_0 L + 1 + M s_0)r + 6s_0 \le 0, \quad Lr < 1 \, ,$$

and

$$0 < R < \frac{2(1 - Lr)}{3L} - r \, .$$

Finally we note that in all but our references it is assumed that $N > 0$ (see [147, 177, 207]), which means that their results cannot apply to solve quadratic operator equations of the form

$$P_2(x) = B(x, x) + L(x) + q$$

where B, L are bounded, symmetric, bilinear, and linear operators, respectively, with q fixed in X. We have that $P_2'(x) = 2B(x) + L$ and $P_2''(x) = 2B$. Hence, we get $M = 2\|B\|$ and $N = 0$.

We can now apply the conditions above to solve a nonlinear integral equation appearing in radioactive transfer with $N = 0$.

Example 6.6

Let us consider the quadratic integral equation of the form (1.22).

Define the operator F on X by

$$F(x) = \lambda x(s) \int_0^1 \frac{s}{s+t} x(t) dt - x(s) + y(s) \, .$$

Note that every zero x^* of the equation $F(x) = 0$ satisfies equation (1.22).

Set $y(s) = x_0(s) = 1$, $\lambda = .24$ and by using the definition of the first and second Fréchet-derivative of the operator F we obtain in turn

$$b = \left\| F'(1)^{-1} \right\| = 1.53039421, \quad N = 0 ,$$

$$L = M = 2|\lambda| \max_{0 \le s \le 1} \left| \int_0^1 \frac{s}{s+t} dt \right| b = 2b|\lambda| \ln 2 = .509178447 ,$$

$$s_0 = .254589224, \quad r_1 = .355336454 \quad \text{and} \quad r_2 = .469039018$$

where r_1 and r_2 are the solutions of the equation

$$3Lr^2 - (2s_0 L + 1) r + s_0 = 0 .$$

The hypotheses of Theorem 6.14 are satisfied if we set $r_1 = r_0$ and choose $R \in [r_0, R_0)$ with $R_0 = .717071253$. Therefore, the iteration of tangent parabolas $\{x_n\}$ $(n \ge 0)$ generated by (6.248)–(6.249) converges to a solution x^* of equation $F(x) = 0$ in $\bar{U}(x_0, r_0)$ which is unique in $\bar{U}(x_0, R)$.

6.5 Convergence Domains for some Iterative Processes in Banach Spaces Using Outer and Generalized Inverses

In this section we are concerned with the problem of approximating a locally unique solution x^* of the equation

$$\Gamma(F(x) + G(x)) = 0 , \tag{6.254}$$

where F is a Fréchet-differentiable operator defined on a convex subset D of a Banach space X with values in a Banach space Y, $G: D \to Y$ is a continuous operator, and $\Gamma \in L(X, Y)$.

We propose inexact Newton-like methods of the form

$$y_{n+1} = y_n - A(y_n)^{\#} (F(y_n) + G(y_n)) - z_n \quad (n \ge 0) , \tag{6.255}$$

to generate a sequence $\{y_n\}$ $(n \ge 0)$ converging to x^*. Here $A_1(x)$, $A_2(x)$, $A(x) \in L(X, Y)$ $(x \in D)$ with $A_1(x) + A_2(x) = A(x)$ $(x \in D)$ and $A(x)^{\#}$ denotes an outer inverse of $A(x)$ $(x \in D)$. Operator A_1 approximates the Fréchet-derivative $F'(x)$ of F $(x \in D)$, whereas $A_2(x)(y - x)$ approximates $G(y) - G(y)$ for all $x, y \in D$. The points $z_n \in X$ are chosen so that iteration $\{y_n\}$ $(n \geq 0)$ converges to x^*.

Outer inverses as well as generalized inverses have been used in the context of Newton's method by several authors. Deuflhard and Heindl [94], Häubler [105], Argyros [42, 43], and Yamamoto [201] gave Mysovskii-type and Kantorovich-type theorems for Gauss–Newton methods. They all assume the very strong condition on either an outer inverse or the Moore–Penrose inverse:

$$\|F'(y)^{\#}(I - F'(x)F'(x)^{\#})F(x)\| \leq \delta(x)\|x - y\|, \quad \delta(x) \leq \delta < 1 \quad (6.256)$$

for all $x, y \in D$. Condition (6.256) can hardly be satisfied in concrete cases. Ben-Israel [66, 67] used the more stringent conditions that there exist positive constants M and N such that for all x and y in a neighborhood of x_0,

$$\|F(x) - F(y) - F'(y)(x - y)\| \leq M\|x - y\|$$

$$\left\|(F'(x)^{\dagger} - F'(y)^{\dagger})F(y)\right\| \leq N\|x - y\|$$

and

$$M\left\|F'(y)^{\dagger}\right\| + N < 1 \quad \text{for all } x, y \in D,$$

where $A^{\dagger}$ denotes the generalized inverse of A. He also used the same conditions with $F'(x)^{\dagger}$ replaced by an outer inverse $F'(x)^{\#}$. His results are not semilocal, since they require information about $F'(y)^{\#}$ for all $y \in D$. Thus, when specialized to the case when $F'(x_0)$ is invertible, they impose conditions that are not needed in the Kantorovich theory. In the elegant paper by Chen and Nashed [80] sharp generalizations were given of the Kantorovich and Mysovskii theory for operator equations in Banach spaces where the derivative is not necessarily invertible. They used iteration (6.255) for $G(x) = 0$, $A_2(x) = 0$ $(x \in D)$, $z_n = 0$ $(n \geq 0)$, and $A(x) = A_1(x)$ $(x \in D)$ approximating the Fréchet-derivative of $F'(x)$ $(x \in D)$.

Based on Banach-type lemmas and perturbation bounds for outer as well as generalized inverses, we determine a domain $\Omega \subseteq D$ such that starting from any point of Ω, approximation (6.255) converges to a locally unique solution x^* of Equation (6.254). The advantages of our results over earlier ones are the following:

(a)   They cover a wider range of problems (see Remark 6.20);

(b)   They reduce to earlier ones for special choices of the functions involved (see Remark 6.20);

(c)   They provide a sharper error bounds than before (see Theorem 6.19, Example 6.7, and Remark 6.20);

(d)   They can be used to solve problems that cannot be handled with earlier results. (See Example 6.8, where we solve a nonlinear integral equation of Uryson type.)

In this section we restate some of the definitions and lemmas given in the elegant paper [80].

Let $A \in L(X, Y)$. A linear operator $B: Y \rightarrow X$ is called an inner inverse of A if $ABA = A$. A linear operator B is an outer inverse of A if $BAB = B$. If B is both an inner and an outer inverse of A, then B is called a generalized inverse of A. There exists a unique generalized inverse $B = A^{\dagger}_{P,Q}$ satisfying $ABA = A$, $BAB = B$, $BA = I - P$, and $AB = Q$, where P is a given projector on X onto $N(A)$ (the null set of A) and Q is a given projector of Y onto $R(A)$ (the range of A). In particular, if X and Y are Hilbert spaces, and P, Q are orthogonal projectors, then $A^{\dagger}_{P,Q}$ is called the Moore–Penrose inverse of A.

We will need five lemmas of Banach-type and perturbation bounds for outer inverses and for generalized inverses in Banach spaces. Lemmas 6.1–6.5 stated here correspond to Lemmas 2.2–2.6 in [80], respectively. See also [146] for a comprehensive study of inner, outer, and generalized inverses.

LEMMA 6.1

Let $A \in L(X, Y)$ and $A^{\#} \in L(Y, X)$ be an outer inverse of A. Let $B \in L(X, Y)$ be such that $\|A^{\#}(B - A)\| < 1$. Then $B^{\#} = (I + A^{\#}(B - A))^{-1} A^{\#}$ is a bounded outer inverse of B with $N(B^{\#}) = N(A^{\#})$ and $R(B^{\#}) = R(A^{\#})$. Moreover, the following perturbation bounds hold:

$$\left\| B^{\#} - A^{\#} \right\| \leq \frac{\left\| A^{\#}(B - A)A^{\#} \right\|}{1 - \left\| A^{\#}(B - A) \right\|} \leq \frac{\left\| A^{\#}(B - A) \right\| \left\| A^{\#} \right\|}{1 - \left\| A^{\#}(B - A) \right\|}$$

and

$$\left\| B^{\#}A \right\| \leq \left(1 - \left\| A^{\#}(B - A) \right\| \right)^{-1}.$$

LEMMA 6.2

Let $A, B \in L(X, Y)$ and $A^{\#}, B^{\#} \in L(Y, X)$ be outer inverses of A and B, respectively. Then $B^{\#}(I - AA^{\#}) = 0$ if and only if $N(A^{\#}) \subseteq N(B^{\#})$.

LEMMA 6.3

Let $A \in L(X, Y)$ and suppose X and Y admit the topological decompositions $X = N(A) \oplus M$, $Y = R(A) \oplus S$. Let $A^{\dagger} (= A^{\dagger}_{M,S})$ denote the generalized inverse of A relative to these decompositions. Let $B \in L(X, Y)$ satisfy

$$\left\| A^{\dagger}(B - A) \right\| \leq 1$$

and

$$(I + (B - A)A^\dagger)^{-1}B \quad maps \ N(A) \ into \ R(A) \ .$$

Then $B^\dagger = B^\dagger_{R(A^\dagger), N(A^\dagger)}$ *exists and is equal to*

$$B^\dagger = A^\dagger \left(I + TA^\dagger \right)^{-1} = \left(I + A^\dagger T \right)^{-1} A^\dagger \ ,$$

where $T = B - A$. *Moreover,* $R(B^\dagger) = R(A^\dagger)$, $N(B^\dagger) = N(A^\dagger)$ *and* $\|B^\dagger A\| \le (1 - \|A^\dagger(B - A)\|)^{-1}$.

LEMMA 6.4

Let $A \in L(X, Y)$ *and* $A^\dagger$ *be the generalized inverse of Lemma 6.3. Let* $B \in L(X, Y)$ *satisfy the conditions* $\|A^\dagger(B - A)\| < 1$ *and* $R(B) \subseteq R(A)$. *Then the conclusion of Lemma 6.3 holds and* $R(B) = R(A)$.

LEMMA 6.5

Let $A \in L(X, Y)$ *and* $A^\dagger$ *be a bounded generalized inverse of* A. *Let* $B \in L(X, Y)$ *satisfy the condition* $\|A^\dagger(B - A)\| < 1$. *Define* $B^\# = (I + A^\dagger(B - A))^{-1}A^\dagger$. *Then* $B^\#$ *is a generalized inverse of* B *if and only if* $\dim N(B) = \dim N(A)$ *and* $codim \ R(B) = codim R(A)$.

THEOREM 6.18

Let $F: D \subseteq X \to Y$ *be Fréchet-differentiable,* $G: D \subseteq X \to Y$ *be continuous,* $A_1(x)$, $A_2(x)$, $A(x) \in L(X, Y)$ ($x \in D$) *with* $A(x) = A_1(x) + A_2(x)$ ($x \in D$), $A^\#$ *be a bounded outer inverse of* A (= $A(x^0)$), $x^0 \in D$, $\bar{U}(x^0, R)$ *be a closed convex ball centered at* $x^0 \in D$ *and of radius* $R > 0$, $w_i: [0, R] \to [0, +\infty)$ *be nondecreasing, nonnegative functions that vanish at the origin,* $i = 0, 1, 2, 3$, η, α, β *be numbers with* $\eta > 0$, $\alpha \ge 0$, $\beta \ge 0$, *and* $\alpha + \beta < 1$, $\{z_n\}$ ($n \ge 0$) *be a null sequence in* X *with* $z_n = z(y_n)$ ($n \ge 0$), $z: D \to Y$, *be a continuous function such that for all* $x, y \in \bar{U}(x^0, R) \subseteq D$, $t \in [0, 1]$ *the following conditions hold:*

$$0 \le w_1(r + q) - w_0(r), \ q \ge 0 \ is \ a \ nondecreasing \ function \ , \tag{6.257}$$

w_0, w_2, w_3 *are continuously differentiable functions on* $[0, R]$ *with*

$$w_0'(r) > 0, \quad r \in [0, R] \ , \tag{6.258}$$

$$0 \le w_i(s_1 + s_2) - w_i(s_1) \le w_i(s_3 + s_4) - w_i(s_3) \ , $$

$$i = 2, 3 \tag{6.259}$$

for $0 \leq s_1 \leq s_3$, $0 \leq s_2 \leq s_4$, $0 < \|A^{\#}(F(y_0) + G(y_0)) + z_0\| \leq \eta$,

$$\left\| A^{\#} \left[F'(x + t(y - x)) - A_1(x) \right] \right\|$$

$$\leq w_1 \left(\left\| x - x^0 \right\| + t \| y - x \| \right) - w_0 \left(\left\| x - x^0 \right\| \right) + \alpha , \quad (6.260)$$

$$\left\| A^{\#} \left[G(y) - G(x) - A_2(x)(y - x) \right] \right\|$$

$$\leq w_2 \left(\left\| x - x^0 \right\| + \| y - x \| \right) - w_2 \left(\left\| x - x^0 \right\| \right) , \quad (6.261)$$

$$\left\| A^{\#}(A(x) - A) \right\| \leq w_0 \left(\left\| x - x^0 \right\| \right) + \beta , \quad (6.262)$$

$$\left\| A^{\#} \left(A(y_{n+1})(z_{n+1}) - A(y_n)(z_n) \right) \right\|$$

$$\leq w_3 \left(\| y_{n+1} - y_n \| + \left\| y_n - x^0 \right\| \right) - w_3 \left(\left\| y_n - x^0 \right\| \right) . \quad (6.263)$$

Define functions f_1, f_2 *and* h *on* $[0, R]$ *by*

$$f_1(r) = \eta - r + \int_0^r w_1(t)dt , \quad (6.264)$$

$$f_2(r) = w_2(r) + w_3(r) \quad (6.265)$$

and

$$h(r) = f_1(r) + f_2(r) + (\alpha + \beta)r . \quad (6.266)$$

Denote the minimal value of $h(r)$ *in* $[0, R]$ *by* f^*, *and the minimal point by* r^*.
Define also functions $f(r)$, $g(r)$ *by*

$$f(r) = h(r) - f^* , \quad (6.267)$$

$$g(r) = 1 - \beta - w_0(r) \quad (6.268)$$

and the scalar iteration $\{r_n\}$ $(n \geq 0)$ *by*

$$r_0 \in [0, R], \quad r_{n+1} = r_n + \frac{f(r_n)}{g(r_n)} \quad (n \geq 0) . \quad (6.269)$$

Moreover suppose that

$$h(R) \leq 0. \qquad (6.270)$$

Then

(i) *function $h(r)$ has a unique zero t^* in $(0, r^*]$.*

(ii) *Scalar iteration $\{r_n\}$ $(n \geq 0)$ generated by (6.255) is monotonically increasing and converges to r^* for any $r_0 \in [0, R]$ which is the unique zero of f in $[0, R]$.*

(iii) *Equation $A^{\#}(F(x) + G(x)) = 0$ has a unique solution x^* in $\tilde{U}(x^0, t^*) \cap \{R(A^{\#}) + x^0\}$, where*

$$\tilde{U} = \begin{cases} \tilde{U}(x^0, R) \text{ (if } h(R) < 0 \text{ or } h(R) = 0 \text{ and } t^* = R), \\[2mm] U(x^0, R) \text{ (if } h(R) = 0 \text{ and } t^* < R), \end{cases} \qquad (6.271)$$

and

$$R\left(A^{\#}\right) + x^0 := \left\{x + x^0 : x \in R\left(A^{\#}\right)\right\}.$$

Define the set

$$\Omega = \bigcup_{r \in [0, r^*)}$$

$$\left\{y \in \tilde{U}\left(x^0, r\right) \mid \|A^{\#}(y)(F(y) + G(y)) + z(y)\| \leq \frac{f(r)}{g(r)}\right\}. \qquad (6.272)$$

Then, for any $y \in \Omega$, iteration $\{y_n\}$ $(n \geq 0)$ generated by (6.255) with $A(y_k)^{\#} = [I + A^{\#}(A(x_k) - A)]^{-1}A^{\#}$ is well defined, remains in $U(x^0, r^*)$ for all $n \geq 0$ and converges to x^*. Moreover the following error bounds hold for all $n \geq 0$

$$\|y_{n+1} - y_n\| \leq r_{n+1} - r_n \qquad (6.273)$$

and

$$\|y_n - x^*\| \leq r^* - r_n \qquad (6.274)$$

provided that r_0 is chosen in (6.269) such that $r_0 \in R_{y_0}$, where for $y \in \Omega$

$$R_y = \left\{r \in [0, r^*] \mid \|A^{\#}(y)(F(y) + G(y))\right.$$

$$\left. + z(y)\| \leq \frac{f(r)}{g(r)}, \|y - x^0\| \leq r\right\}. \qquad (6.275)$$

(iv) $\bar{U}\left(x^0, \frac{|f^*|}{2-\beta}\right) \subseteq \Omega.$

PROOF   (i) By hypothesis (6.270) function h has a unique zero $t^* \in (0, r^*]$, since h is strictly convex.

(ii) We have $f^* = h(r^*) \leq h(R) \leq 0$. That is $f(0) \geq \eta > 0$. Moreover, since $f(r^*) = 0$ and f is strictly convex, we get that r^* is the unique zero of f in $[0, r^*]$. We also have

$$w_0(r) + \beta \leq f'(r) + 1 < f'(r^*) + 1 \leq 1, \quad r \in [0, r^*) \qquad (6.276)$$

and $w_0'(r) > 0$ $(r \in [0, r^*])$. The result now follows exactly as in Proposition 3 in [10, p. 677].

(iii) By induction on $n \geq 0$ we will first show (6.273) and $\|y_n - x^0\| \leq r_n$. Choose $y_0 \in \Omega$. Then there exists $r_0 \in R_{y_0}$ such that

$$\left\|y_0 - x^0\right\| \leq r_0 < r^*, \qquad (6.277)$$

and by (6.255) and (6.275),

$$\|y_1 - y_0\| = \left\|A(y_0)^{\#}(F(y_0) + G(y_0)) + z(y_0)\right\|$$

$$\leq \frac{f(r_0)}{g(r_0)} = r_1 - r_0. \qquad (6.278)$$

By (6.277) and (6.278) we get

$$\left\|y_1 - x^0\right\| \leq \|y_1 - y_0\| + \left\|y_0 - x^0\right\| \leq r_1 - r_0 + r_0 = r_1.$$

For $n = 0$, (6.273) is (6.278) and $\|y_0 - x^0\| \leq r_0$ is (6.277). By (6.262) we get

$$\left\|A^{\#}(A(y_1) - A)\right\| \leq w_0\left(\left\|y_1 - x^0\right\|\right) + \beta \leq w_0(r_1 - r_0) + \beta$$

$$\leq w_0(r^*) + \beta < 1. \qquad (6.279)$$

Hence, from Lemma 6.1 we have that $A(y_1)^{\#} := [I + A^{\#}(A(y_1) - A)]^{-1}A^{\#}$ is an outer inverse of $A(y_1)$;

$$\left\| A\left(y_1\right)^{\#} A \right\| \leq \left(1 - \beta - w_0\left(\left\| y_1 - x^0 \right\|\right)\right)^{-1}$$

$$\leq \left(1 - \beta - w_0\left(r_1 - r_0\right)\right)^{-1}, \tag{6.280}$$

and $N(A(x_1)^{\#}) = N(A^{\#})$. Assume that for $1 \leq n \leq k$

$$\| y_n - y_{n-1} \| \leq r_n - r_{n-1} \quad \text{and} \quad N\left(A\left(y_{n-1}\right)^{\#}\right) = N\left(A^{\#}\right).$$

Then $\| y_k - x^0 \| \leq r_k - r_0 \leq r_k$ and $N(A(y_k)^{\#}) = N(A(y_{k-1})^{\#}) = N(A^{\#})$. Hence, by Lemma 6.2 we have

$$A\left(y_k\right)^{\#}\left(I - A\left(y_{k-1}\right) A\left(y_{k-1}\right)^{\#}\right) = 0.$$

Using (6.255) we can get the approximation

$$F\left(y_k\right) + G\left(y_k\right) + A\left(y_k\right) z_k = \left[F\left(y_k\right) - F\left(y_{k-1}\right) - A_1\left(y_{k-1}\right)\left(y_k - y_{k-1}\right)\right]$$

$$+ \left[G\left(y_k\right) - G\left(y_{k-1}\right) - A_2\left(y_{k-1}\right)\left(y_k - y_{k-1}\right)\right]$$

$$+ \left[A\left(y_k\right)\left(z_k\right) - A\left(y_{k-1}\right)\left(z_{k-1}\right)\right]. \tag{6.281}$$

By (6.255), (6.257)–(6.263), (6.269), and (6.280) we obtain in turn

$$\| y_{k+1} - y_k \| \leq \left\| A\left(y_k\right)^{\#} A \right\| \left[\int_0^1 \left\| A^{\#}\left[F'\left(y_{k-1} + t\left(y_k - y_{k-1}\right)\right)\right.\right.$$

$$\left. - A_1\left(y_{k-1}\right)\right] \| \, \| y_k - y_{k-1} \| \, dt$$

$$+ \left\| A^{\#}\left[G\left(y_k\right) - G\left(y_{k-1}\right) - A_2\left(y_{k-1}\right)\left(y_k - y_{k-1}\right)\right]\right\|$$

$$\left. + \left\| A^{\#}\left[A\left(y_k\right)\left(z_k\right) - A\left(y_{k-1}\right)\left(z_{k-1}\right)\right]\right\|\right]$$

$$\leq g\left(\left\| y_k - x^0 \right\|\right)^{-1}\left[\int_0^1 \left(w_1\left(\left\| y_{k-1} - x^0 \right\| + t\, \| y_k - y_{k-1} \|\right)\right.\right.$$

$$\left. - w_0\left(\left\| y_{k-1} - x^0 \right\|\right)\right) \| y_k - y_{k-1} \| \, dt$$

$$+ \alpha \, \|y_k - y_{k-1}\| + w_2 \left(\left\| y_{k-1} - x^0 \right\| + \|y_k - y_{k-1}\| \right)$$

$$- w_2 \left(\left\| y_{k-1} - x^0 \right\| \right) \tag{6.282}$$

$$+ w_3 \left(\|y_k - y_{k-1}\| + \left\| y_{k-1} - x^0 \right\| \right) - w_3 \left(\left\| y_{k-1} - x^0 \right\| \right) \Bigg]$$

$$\leq g \, (r_k)^{-1} \left[\int_0^1 (w_1 \, (r_{k-1} - r_0 + t \, (r_k - r_{k-1})) \right.$$

$$- w_0 \, (r_{k-1} - r_0)) \, (r_k - r_{k-1}) \, dt + \alpha \, (r_k - r_{k-1})$$

$$+ w_2 \, (r_{k-1} - r_0 + r_k - r_{k-1}) - w_2 \, (r_{k-1} - r_0)$$

$$\left. + w_3 \, (r_k - r_{k-1} + r_{k-1} - r_0) - w_3 \, (r_{k-1} - r_0) \right]$$

$$= g \, (r_k)^{-1} \, f \, (r_k) = r_{k+1} - r_k \, ,$$

which completes the induction for (6.273). Moreover, we have

$$\left\| y_{k+1} - x^0 \right\| \leq \|y_{k+1} - y_k\| + \left\| y_k - x^0 \right\| \leq r_{k+1} \, ,$$

which shows $y_n \in U(x^0, r^*)$ since $r_n < r^*$ by part (ii). Hence, we have for any $k \geq 0$

$$\|y_{k+1} - y_k\| \leq r_{k+1} - r_k, \quad \left\| y_k - x^0 \right\| \leq r_k - r_0 \leq r^* \, ,$$

$$\left\| A^{\#} \, (A \, (y_{k+1}) - A) \right\| \leq w_0 \left(\left\| y_{k+1} - x^0 \right\| \right) + \beta \leq w_0 \, (r_{k+1} - r_0) + \beta$$

$$\leq w_0 \, (r_{k+1}) + \beta \leq w_0 \, (r^*) + \beta < 1 \, ,$$

and

$$A \, (y_{k+1})^{\#} := \left(I + A^{\#} \, (A \, (y_{k+1}) - A) \right)^{-1} A^{\#}$$

is an outer inverse of $A(x)$. It now follows from (6.273) and part (ii) that iteration $\{y_n\}$ $(n \geq 0)$ is Cauchy and as such it converges to some point $x^* \in \bar{U}(x^0, r^*)$.

The point x^* is a solution of $A^\#(F(x) + G(x)) = 0$, since by definition $A(y_k)^\# = (I + A^\#(A(y_k) - A))^{-1}A^\#$ for all k, and

$$0 = \lim_{k \to \infty} \left(I + A^\# (A(y_k) - A)\right)(y_k - y_{k-1})$$

$$= \lim_{k \to \infty} A^\# (F(y_k) + G(y_k) + A(y_k)(z_k)) = A^\#(F(x^*) + G(x^*)) .$$

Estimate (6.274) follows from (6.273) by using standard majorization techniques. The uniqueness part follows exactly as in [81, p. 42] and [80, p. 244]. Indeed, for $x^0 \in \Omega$ the iteration $\{x_n\}$ $(n \geq 0)$ defined by

$$x^0 = x_0, \ x_{n+1} = x_n - A(x_n)^\# (F(x_n) + G(x_n))$$

$$- \tilde{z}_n, \tilde{z}_n = z(x_n) \quad (n \geq 0) \tag{6.283}$$

converges to a solution x^* of $A^\#(F(x) + G(x)) = 0$ and satisfies

$$\|x_{n+1} - x_n\| \leq t_{n+1} - t_n, \ \|x^* - x_n\| \leq r^* - t_n \quad (n \geq 0)$$

where $\{t_n\}$ $(n \geq 0)$ is defined by (6.269) with

$$t_0 = 0, \ t_{n+1} = t_n + \frac{f(t_n)}{g(t_n)} \quad (n \geq 0) . \tag{6.284}$$

To prove the uniqueness of x^* in $\tilde{U}$, let y^* be any solution in $\tilde{U}$. Then as in [81, p. 42] we show $h(\|y^* - x^0\|) \geq 0$, $\|y^* - x^0\| \leq t^*$ and

$$\|y^* - x_n\| \leq r^* - t_n \quad (n \geq 0) . \tag{6.285}$$

By letting $n \to \infty$ in (6.285) we obtain $y^* = \lim_{n \to \infty} x_n$. But $x^* = \lim_{n \to \infty} x_n$. That is $x^* = y^*$ and $x^* \in \bar{U}(x^0, t^*)$.

(iv) For $r \in [0, r^*]$ and $0 < t^* \leq r^*$ we have $-1 \leq h'(r) \leq 0$, which implies

$$|f^*| = |h(r^*) - h(t^*)| \leq |r^* - t^*| < r^* .$$

Choose $y_0 \in \bar{U}\left(x^0, \frac{|f^*|}{2-\beta}\right)$, then $\|y_0 - x^0\| \leq \frac{|f^*|}{2-\beta} < r^*$. Let $\|y_0 - x^0\| = r_0$. Then from (6.262), (6.268) and (6.280) we get $\|A(y_0)^\# A\| \leq g(r_0)^{-1}$. Moreover,

(6.281) and (6.282) for y_k replaced by y_0 give

$$\left\| A(y_0)^\# (F(y_0) + G(y_0)) + z_0 \right\| \le g(r_0)^{-1} \left\{ \int_0^{r_0} w_1(t)dt + f_2(r_0) \right.$$

$$\left. + (1+\alpha)r_0 + \eta + |f^*| - (2-\beta)r_0 \right\} = g(r_0)^{-1} (h(r_0) - f^*) .$$

That is $y_0 \in \Omega$.

That completes the proof of Theorem 6.18.   ∎

The function $h(r)$ can be generalized. Indeed for any $y \in \Omega$, fix $r_y \in R_y$ and set

$$\eta_y = \left\| A(y)^\# (F(y) + G(y)) + z_y \right\|, \quad z_y = z(y) ,$$

$$\gamma_y = \begin{cases} 1 & \text{(if } y = x^0 \text{ and } r_y = 0) \\ g(r_y)^{-1} & \text{(otherwise)} , \end{cases}$$

and

$$h_y(r) = \eta_y + \gamma_y \left(\int_0^r w_1(r_y + t)dt + f_2(r_y + r) + (\alpha + \beta - 1)r \right) .$$

Define also the scalar sequence $\{v_n\}$ $(n \ge 0)$ by

$$v_0 = 0, \quad v_{n+1} = v_n + \frac{h_y(v_n)}{\gamma_y g(r_y + v_n)} \quad (n \ge 0) .$$

Then we can show:

THEOREM 6.19
Assume that the hypotheses of Theorem 6.18 are satisfied. Then the following hold:

(i) $h_y(r^* - r_y) \le 0$ *and* $h_y(r)$ *has a unique zero* $v^* \in [0, r^* - r_y]$;

(ii)   *iteration* $\{y_n\}$ $(n \ge 0)$ *generated by* (6.255) *with* $y_0 = y$ *satisfies*

$$\|y_{n+1} - y_n\| \le v_{n+1} - v_n \tag{6.286}$$

and

$$\left\| x^* - y_n \right\| \leq v^* - v_n \leq r^* - r_y \quad (n \geq 0) . \tag{6.287}$$

PROOF As in [81, p. 45] we first have

$$r^* - r_y = \eta_y + \gamma_y \left\{ \int_{r_y}^{r^*} w_1(t)dt + f_2\left(r^*\right) - f_2\left(r_y\right) + (\alpha + \beta - 1)\left(r^* - r_y\right) \right\}$$

$$\leq \frac{h\left(r_y\right) - f^*}{g\left(r_y\right)} + \frac{h\left(r^*\right) - h\left(r_y\right)}{g\left(r_y\right)} = 0 ,$$

which shows that function $h_y(r)$ has a unique zero $v^* \in [0, r^* - r_y]$, and sequence $\{v_n\}$ $(n \geq 0)$ monotonically increases and converges to v^*. We only need to show (6.286), since estimate (6.287) follows from (6.286). Estimate (6.286) holds for $n = 0$, since

$$\| y_1 - y_0 \| = \left\| A\left(y_0\right)^{\#}\left(F\left(y_0\right) + G\left(y_0\right)\right) + z_0 \right\| = \eta_y \leq v_1 - v_0 .$$

Assume (6.286) holds for all $n < k$. Then as in (6.280) and (6.281) for $\| y_k - x^0 \| \leq r_y + v_k < r^*$ we get $\| A(y_k)^{\#} A(x^0) \| \leq g(r_y + v_k)^{-1}$, and

$$\| y_{k+1} - y_k \| \leq g\left(r_y + v_k\right)^{-1} \left\{ \int_{r_y + v_{k-1}}^{r_y + v_k} w_1(t)dt + f_2\left(v_k\right) - f_2\left(v_{k-1}\right) \right.$$

$$\left. + (\alpha + \beta - 1)\left(v_k - v_{k-1}\right) + g\left(r_y + v_{k-1}\right)\left(v_k - v_{k-1}\right) \right\}$$

$$= \frac{1}{\gamma_y g\left(r_y + v_k\right)} \left\{ h_y\left(v_k\right) - h_y\left(v_{k-1}\right) + \gamma_y g\left(r_y + v_{k-1}\right)\left(v_k - v_{k-1}\right) \right\}$$

$$= \frac{h_y\left(v_k\right)}{\gamma_y g\left(r_y + v_k\right)} = v_{k+1} - v_k .$$

That completes the induction for (6.286) and the proof of Theorem 6.19. ∎

REMARK 6.19 Under the hypotheses of Theorem 6.18, assume that

$$(I + (A(x) - A)A^{\dagger})^{-1} A(x) \text{ maps } N(A) \text{ into } R(A) \tag{6.288}$$

where $I + (A(x) - A)A^\dagger$ is invertible for some $x \in D$. Then by Lemma 6.3 $A(x_k)^\# = (I + A^\dagger(A(x_k) - A))^{-1}A^\dagger$ is a generalized inverse of $A(x_k)$. Therefore, we can restate Theorems 6.18 and 6.19 for generalized inverses. Call these Theorems 6.18′ and 6.19′, respectively. Moreover, by Lemma 6.5 hypothesis (6.288) can be replaced by $\operatorname{rank}(A(x)) \leq \operatorname{rank}(A)$ $(x \in D)$ provided that X and Y are finite dimensional spaces. Furthermore, hypothesis (6.288) in general Banach spaces can be replaced by $R(A(x)) \subseteq R(A)$ $(x \in D)$ which is a stronger condition. This follows from Lemma 6.4. Finally (6.288) can be replaced by the necessary and sufficient conditions of Lemma 6.5. ∎

REMARK 6.20   (a) Our Theorems 6.18 and 6.19 can be reduced to Theorems 1 and 2, respectively, in Chen and Yamamoto [81, p. 40]. Indeed, set $A^\#(x) = A(x)^{-1}$, $A_2(x) = 0$ $(x \in D)$, $z_n = 0$ $(n \geq 0)$, $w_3(r) = 0$, $w_2(t) = e(r)t$ $r, t \in [0, R]$, $\alpha = b$ and $\beta = c$.

(b) Moreover, the results by Chen and Yamamoto can be reduced to Zabrejko and Nguen in [207, p. 677, Proposition 4] for $x^0 = y_0$, $A(x) = F'(x)$ $(x \in D)$, $t = 1$, $w_0(s) = w_1(s) = \|F'(x_0)^{-1}\|k(r)s$ $s \in [0, R]$, and $b = c = 0$.

(c) Furthermore, the results by Zabrejko and Nguen can be reduced to the Newton-Kantorovich [114] theorem for $G(x) = 0$ $(x \in D)$, $e(r) = 0$, $k(r) = \ell$, $r \in [0, R]$.

(d) Our Theorem 6.18 also reduces to Theorem 3.1 given in [80, p. 241] by Chen and Nashed. Just choose $A_2(x) = 0$, $G(x) = 0$ $(x \in D)$, $w_2(r) = w_3(r) = 0$, $w_0(r) = Lr$ $r \in [0, R]$, $\beta = \ell$, $\alpha = \mu$, $y_0 = x^0$, $z_n = 0$ $(n \geq 0)$, $L + m = K$, and $w_1(r) = Kr$, $r \in [0, R]$. Note that in this special case if $L + M > K$ the error bounds on the distances $\|x_{n+1} - x_n\|$ and $\|x_n - x^*\|$ given in [11, p. 242] are sharper than our (6.273) and (6.274), respectively. However, the converse is true if $L + M < K$. Finally similar remarks can be made for Theorems 6.18′ and 6.19′ (see Remark 6.19). ∎

We can give a Mysovskii-type theorem corresponding to Theorem 6.18 (similarly for Theorem 6.19) for inexact Newton-like methods using outer or generalized inverses. We now state such a theorem using outer inverses. A similar theorem can be stated for generalized inverses. The proof is identical to Theorem 6.18 and is omitted.

THEOREM 6.20

Let $F: D \subseteq X \to Y$ be Fréchet-differentiable, $G: D \subseteq X \to Y$ be continuous, $A_1(x)$, $A_2(x)$, $A(x) \in L(X, Y)$ $(x \in D)$ with $A(x) = A_1(x) + A_2(x)$ $(x \in D)$, $A^\#$ be a bounded outer inverse of A $(= A(x^0))$, $x^0 \in D$, $w_i: [0, R] \to [0, +\infty)$ $(R > 0)$ be nondecreasing, nonnegative functions that vanish at the origin, $i = 1, 2, 3$, η, α be numbers with $\eta > 0$, $0 \leq \alpha < 1$, $\{z_n\}$ $(n \geq 0)$ be a null sequence in X with $z_n = z(y_n)$ $(n \geq 0)$, $z: D \to Y$ be a continuous function, such that for *all $x, y \in \bar{U}(x^0, R) \subseteq D$, $t \in [0, 1]$ the following conditions hold:*

$$0 \leq w_i (s_1 + s_2) - w_i (s_1) \leq w_i (s_3 + s_4) - w_i (s_3), \quad i = 2, 3$$

for $0 \leq s_1 \leq s_3$, $0 \leq s_2 \leq s_4$, $0 < \|A^{\#}(F(y_0) + G(y_0)) + z_0\| \leq \eta$,

$$\left\| A(y)^{\#} \left[F'(x + t(y - x)) - A_1(x) \right] \right\|$$

$$\leq w_1 \left(\left\| x - x^0 \right\| + t \|y - x\| \right) + \alpha,$$

$$\left\| A(y)^{\#} \left[G(y) - G(x) - A_2(x)(y - x) \right] \right\|$$

$$\leq w_2 \left(\left\| x - x^0 \right\| + \|y - x\| \right) - w_2 \left(\left\| x - x^0 \right\| \right),$$

$$\left\| A(y)^{\#} \left(A(y_{n+1})(z_{n+1}) - A(y_n)(z_n) \right) \right\|$$

$$\leq w_3 \left(\|y_{n+1} - y_n\| + \left\| y_n - x^0 \right\| \right) - w_3 \left(\left\| y_n - x^0 \right\| \right).$$

Moreover assume that for all $x \in \bar{U}(x^0, R)$ there exists an outer inverse $A(x)^{\#}$ of $A(x)$ satisfying $N(A(x)^{\#}) = N(A^{\#})$. Then the conclusions of Theorem 6.18 (except the uniqueness part) hold with $A(A_k)^{\#}$ satisfying $N(A(x_k)^{\#}) = N(A^{\#})$ [where we have set $w_0(r) = 0$ for $r \in [0, R]$ and $\beta = 0$ in (6.266) and (6.268)].

Remarks similar to 6.19 and 6.20 can now follow for Theorem 6.20.

Results concerning modified Newton-like methods were not given in [80] or [81]. However, they were given in Proposition 2 in [207]. Consider the modified inexact Newton-like methods of the form

$$y_{n+1} = y_n - A^{\#} (F(y_n) + G(y_n)) - z_n \quad (n \geq 0). \tag{6.289}$$

It follows immediately that under the hypotheses of Theorem 6.18 [excluding (6.262), for $w_0(r) = 0$ $r \in [0, R]$ and $\beta = 0$] the conditions of Theorems 6.18 and 6.19 hold for iteration (6.289). Call such Theorems 6.18″ and 6.19″, respectively. Moreover, it follows from the proof of Theorem 6.18 that the conclusions of Theorems 6.18 and 6.19 still hold for iteration (6.289) if (6.260) is replaced by the weaker condition

$$\left\| A^{\#} \left[F'(x + t(y - x)) - A_1 \left(x^0 \right) \right] \right\|$$

$$\leq w_1 \left(\left\| x - x^0 \right\| + t \| y - x \| \right) + \alpha . \tag{6.290}$$

Call the corresponding result Theorem 6.18'''.

This is an important observation because one can construct examples where the results of Theorem 6.18''' applies where the results of Proposition 2 in [207] or the results that can be derived from [80] and [81] for iteration (6.289) cannot apply (see Example 6.8 that follows).

We complete this section with two examples. For simplicity we take $A(x)^{\#} = A(x)^{-1}$ ($x \in D$), $z_n = 0$ ($n \geq 0$).

In Example 6.7 we justify claim (c) made at the introduction.

Example 6.7

Let $X = Y = (\mathbf{R}^2, \| \cdot \|_{\infty})$. Consider the system

$$3x^2 y + y^2 - 1 + |x - 1| = 0$$

$$x^4 + xy^3 - 1 + |y| = 0$$

and the methods

$$x_{n+1} = x_n - F'(x_n)^{-1} (F(x_n) + G(x_n)) \quad (n \geq 0) \tag{6.291}$$

$$w_{n+1} = w_n - [w_{n-1}, w_n; G]^{-1} (F(x_n) + G(x_n)) \quad (n \geq 0) \tag{6.292}$$

$$v_{n+1} = v_n - [F'(v_n) + [v_{n-1}, w_n; G]]^{-1} (F(x_n) + G(x_n))$$

$$(n \geq 0) , \tag{6.293}$$

where $[x, y; G]$ denotes the first order divided difference of G at the points $x, y \in X$. Methods (6.291)–(6.293) are obviously special cases of (6.255). Set $\|x\|_{\infty} = \|(x', x'')\|_{\infty} = \max\{|x'|, |x''|\}$, $F = (F_1, F_2)$, $G = (G_1, G_2)$. For $x = (x', x'') \in \mathbf{R}^2$ we take $F_1(x', x'') = 3(x')^2 x'' + (x'')^2 - 1$, $F_2(x', x'') = (x')^4 + x'(x'')^3 - 1$, $G_1(x', x'') = |x' - 1|$, $G_2(x', x'') = |x''|$. We shall take $[x, y; G] \in M_{2 \times 2}(\mathbf{R})$ as $[x, y; G]_{i,1} = \frac{G_1(y', y'') - G_1(x', y'')}{y' - x'}$, $[x, y; G]_{i,2} = \frac{G_1(x', y'') - G_1(x', x'')}{y'' - x''}$, $i = 1, 2$.

Using method (6.291) with $x_0 = (1, 0)$ we obtain Table 6.4

Table 6.4 Newton's method

n	$x_n^{(1)}$	$x_n^{(2)}$	$\|x_n - x_{n-1}\|$
0	1	0	
1	1	0.333333333333333	3.333E–1
2	0.906550218340611	0.354002911208151	9.344E–2
3	0.885328400663412	0.338027276361322	2.122E–2
4	0.891329556832800	0.326613976593566	1.141E–2
5	0.895238815463844	0.326406852843625	3.909E–3
6	0.895154671372635	0.327730334045043	1.323E–3
7	0.894673743471137	0.327979154372032	4.809E–4
8	0.894598908977448	0.327865059348755	1.140E–4
9	0.894643228355865	0.327815039208286	5.002E–5
10	0.894659993615645	0.327819889264891	1.676E–5
11	0.894657640195329	0.327826728208560	6.838E–6
12	0.894655219565091	0.327827351826856	2.420E–6
13	0.894655074977661	0.327826643198819	7.086E–7
...			
39	0.894655373334687	0.327826511746298	5.149E–19

Using the method of chord (6.292) with $w_0 = (5, 5)$, $w_1 = (1, 0)$, we obtain Table 6.5

Table 6.5 Secant method

n	$w_n^{(1)}$	$w_n^{(2)}$	$\|w_n - w_{n-1}\|$
0	5	5	
1	1	0	5.000E+00
2	0.989800874210782	0.012627489072365	1.262E–02
3	0.921814765493287	0.307939916152262	2.953E–01
4	0.900073765669214	0.325927010697792	2.174E–02
5	0.894939851625105	0.327725437396226	5.133E–03
6	0.894658420586013	0.327825363500783	2.814E–04
7	0.894655375077418	0.327826521051833	3.045E–06
8	0.894655373334698	0.327826521746293	1.742E–09
9	0.894655373334687	0.327826521746298	1.076E–14
10	0.894655373334687	0.327826521746298	5.421E–20

Using our method (6.293) with $v_0 = (5, 5)$, $v_1 = (1, 0)$, we obtain Table 6.6

Hence, method (6.293) (i.e., method (6.255) in this case) converges faster than (6.291) suggested by [80, 81, 207] in this case and the method of chord (6.292) [21].

In Example 6.8 we justify claim (d) made at the introduction.

Table 6.6   Newton-like method

n	$v_n^{(1)}$	$v_n^{(2)}$	$\|v_n - v_{n-1}\|$
0	5	5	
1	1	0	5
2	0.909090909090909	0.363636363636364	3.636E–01
3	0.894886945874111	0.329098638203090	3.453E–02
4	0.894655531991499	0.327827544745569	1.271E–03
5	0.894655373334793	0.327826521746906	1.022E–06
6	0.894655373334687	0.327826521746298	6.089E–13
7	0.894655373334687	0.327826421746298	2.710E–20

Example 6.8

Let $X = Y = D = C = C[0, 1]$, the space of continuous functions on $[0, 1]$ equipped with the sup-norm. For simplicity set $G(x) = 0$, $z(x) = 0$ ($x \in X$) in (6.254) and (6.255). Consider the nonlinear integral equation of Uryson type given by

$$F(x)(t) = x(t) - \int_0^1 p(t, s, x(s))ds \ . \tag{6.294}$$

Equation (6.294) is from nonlinear elasticity theory [9, 60]. Let $x_0 = 0$, $A = F'(0)$, and consider the choices of the function involved as in Remark 6.20(b). Suppose $p(t, s, u) = p_1(t)p_2(s)p_3(u)$ with two continuous functions p_1, p_2 and $p_3 \in C^2$. Setting

$$\gamma = \int_0^1 p_2(s)ds, \quad \delta = \int_0^1 p_1(s)p_2(s)ds \ , \tag{6.295}$$

using [14, p. 278], we get

$$k(r) = \|p_1\|_C \cdot \gamma \cdot \sup_{\|u\| \le r} \|p_3''(u)\| \ , \tag{6.296}$$

$$a = \left\| F'(x_0)^{-1} F(x_0) \right\| = \frac{\gamma p_3(0)}{1 - \delta p_3'(0)} \|p_1\|_C \tag{6.297}$$

and

$$b = \left\| F'(x_0)^{-1} \right\| = 1 + \frac{\gamma p_3'(0)}{1 - \delta p_3'(0)} \|p_1\|_C \ , \tag{6.298}$$

provided that

$$\delta p_3'(0) < 1 \ . \tag{6.299}$$

Choose: $p_1(s) = \frac{1}{(s+1)\sqrt{s+1}}$, $p_2(s) = \beta\sqrt{s+1}$, $\beta = \frac{3}{20(2\sqrt{2}-1)}$ and $p_3(s) = e^s$.
Then by (6.295) we get

$$\gamma = \frac{2}{3}\beta\left(2\sqrt{2} - 1\right) \quad \text{and} \quad \delta = \beta \ln 2 .$$

Condition (6.299) becomes

$$\delta p_3'(0) = \beta \ln 2 = .5686421 < 1$$

which is true. Moreover by (6.296)–(6.298), and the above values we get

$$a = \frac{2\beta(2\sqrt{2} - 1)}{3(1 - \beta \ln 2)}, \quad b = 1 + \frac{2\beta(2\sqrt{2} - 1)}{3(1 - \beta \ln 2)}$$

and

$$k(r) = \frac{b}{10}e^r .$$

By Equation (12) in [207, p. 673] we get $\chi(r) = a - r + b\int_0^r (r - t)k(t)dt$, or using the above values $\chi(r) = .1106029e^r - 1.1106029r - .0045736$. It is simple calculus to show that $\min_{r\in[0,+\infty)} \chi(r) = \chi(0) - .1060293 > 0$, χ is increasing on $[2.306712, +\infty)$ and decreasing on $[0, 2.306712]$. Proposition 2 in [207, p. 674] fails to apply because $\chi(r)$ has no zero r_0 in the interval $[0, R]$ and $\chi(R) > 0$ for all $R \in [0, +\infty)$. However, our Theorem 6.18''' with the choices $A_2 = 0$, $A_1 = F'(x_0)$, $w_0 = 0$, $\alpha = \beta = 0$, $w_1(r) = b \cdot \bar{k}(r)r$, $t = 0$, $z_n = 0$ $(n \geq 0)$, $w_2 = w_3 = 0$ can apply.

Indeed, using (6.270) and the above values we get $\bar{k}(r) = \frac{1}{10}$, and $h(r) = .1106029r^2 - 2r + .2120586$. Hypothesis (6.270) of our Theorem 6.18''' is now satisfied if we set $t^* = .1066586$ and $R = 17.97605$. Hence, according to our Theorem 6.18''' iteration (6.289) converges to a solution $x^* \in U(0, r^*)$ of equation $F(x)(t) = 0$, where F is given by (6.294). Moreover, x^* is the unique solution of the same equation in $U(0, R)$. Furthermore, estimates (6.273) and (6.274) hold in this case for all $n \geq 0$.

6.6 Convergence of Inexact Newton Methods on Banach Spaces with a Convergence Structure

In this section we are concerned with approximating a solution x^* of the non-linear operator equation

$$F(x) + Q(x) = 0 , \tag{6.300}$$

where F is a Fréchet-differentiable operator defined on a convex subset D of a Banach space X with values in X, and Q is a nondifferentiable nonlinear operator with the same domain and values in X.

We introduce the inexact Newton-method

$$x_{n+1} = x_n + F'(x_n)^* [-(F(x_n) + Q(x_n))] - z_n, \quad z_0 = 0 \ (n \geq 0) \quad (6.301)$$

to approximate a solution x^* of Equation (6.300). Here $F'(x_n)^*$ $(n \geq 0)$ denotes a linear operator which is an approximation for $F'(x_n)^{-1}$ $(n \geq 0)$.

The notion of a Banach space with a convergence structure was used in the elegant paper in [134] (see also [28, 47]) to solve Equation (6.300), when $Q(x) = 0$ for all $x \in D$ and $z_n = 0$ for all $n \geq 0$. However, there are many interesting real life applications already in the literature, where Equation (6.300) contains a nondifferentiable term. See, for example, the applications at the end of this study. The case when $Q(x) = 0$ for all $x \in D$ and $F'(x_n)^* = F'(x_n)^{-1}$ $(n \geq 0)$ has already been considered but on a Banach space without generalized structure. (See previous sections of this chapter.)

By imposing very general Lipschitz-like conditions on the operators involved, on the one hand, we cover a wider range of problems and, on the other hand, by choosing our operators appropriately, we can find sharper error bounds on the distances involved than before. As in [135], we provide semi-local results of Kantorovich-type and global results based on monotonicity considerations from the same general theorem. Moreover, we show that our results can be reduced to the one obtained in [135], when $Q(x) = 0$ for all $x \in D$ and $z_n = 0$ $(n \geq 0)$, and furthermore to the ones obtained in [28, 134] by further reading the requirements on X.

Finally, our results apply to solve a nonlinear integral equation involving a nondifferentiable term that cannot be solved with existing methods.

We will need the definitions:

DEFINITION 6.3   *The triple* (X, V, X) *is a Banach space with a convergence structure if*

(C_1) $(X, \| \cdot \|)$ *is a real Banach space;*

(C_2) $(V, C, \|\|\|_V)$ *is a real Banach space which is partially ordered by the closed convex cone C; the norm $\|\|\|_V$ is assumed to be monotone on C;*

(C_3) X *is a closed convex cone in $X \times V$ satisfying $\{0\} \times C \subseteq X \subseteq X \times C$;*

(C_4)   *the operator $\| : \Delta \to C$ is well defined:*

$$|x| = \inf\{q \in C \mid (x, q) \in X\}$$

for

$$x \in \Delta = \{x \in X \mid \exists q \in C : (x, q) \in X\} \, ;$$

and

(C5) *for all $x \in \Delta$ $\|x\| \le \|\|x\|\|_V$.*

The set

$$U(a) = \{x \in X \mid (x, a) \in X\}$$

defines a sort of generalized neighborhood of zero.
  Let us give some motivational examples for $X =: \mathbf{R}^m$ with the maximum-norm

  (a) $V =: \mathbf{R}$, $X := \{(x, e) \in \mathbf{R}^m \times \mathbf{R}^m \mid \|x\|_\infty \le e\}$.

  (b) $V =: \mathbf{R}^m$, $X := \{(x, e) \in \mathbf{R}^m \times \mathbf{R}^m \mid |x| \le e\}$
       (componentwise absolute value).

  (c) $V := \mathbf{R}^m$, $X := \{(x, e) \in \mathbf{R}^m \times \mathbf{R}^m \mid 0 \le x \le e\}$.

Case (a) involves classical convergence analysis in a Banach space, (b) concerns componentwise analysis and error estimates, and (c) is used for monotone convergence analysis.

DEFINITION 6.4   *An operator $L \in C^1(V_1 \to V)$ defined on an open subset V_1 of an ordered Banach space V is order convex on $[a, b] \subseteq V_1$ if*

$$c, d \in [a, b], \ c \le d \Rightarrow L'(d) - L'(c) \in L_+(V) \, ,$$

where for $m \ge 0$

$$L_+(V^m) = \left\{ L \in L\left(V^m\right) \mid 0 \le x_i \Rightarrow 0 \le L\left(x_1, x_2, \ldots, x_m\right) \right\}$$

and $L(V^m)$ denotes the space of m-linear, symmetric, bounded operators on V.

DEFINITION 6.5   *The set of bounds for an operator $H \in L(X^m)$ is defined to be*

$$B(H) = \left\{ L \in L_+ \left(V^m\right) \mid (x_i, q_i) \in X \Rightarrow \left(H\left(x_1, \ldots, x_m\right), L\left(q_1, \ldots, q_m\right)\right) \in X \right\} \, .$$

DEFINITION 6.6   *Let $H \in L(X)$ and $y \in X$ be given, then*

$$H^*(y) = z \Leftrightarrow z = T^\infty(0) = \lim_{n \to \infty} T^n(0) \, ,$$

$$T(x) = (I - H)(x) + y \Leftrightarrow z = \sum_{i=0}^{\infty}(I - H)^i y \,,$$

if this limit exists.

We will also need the Lemmas:

LEMMA 6.6
Let $L \in L_+(V)$ and $a, q \in C$ be given such that:

$$L(q) + a \leq q \quad and \quad L^n(q) \rightarrow \quad as \; n \rightarrow \infty \,.$$

Then the operator
$$(I - L)^* : [0, a] \rightarrow [0, a]$$

is well defined and continuous.

The following is a generalization of Banach's lemma.

LEMMA 6.7
Let $H \in L(X)$, $L \in B(H)$, $y \in \Delta$ and $q \in C$ be such that

$$L(q) + |y| \leq q \quad and \quad L^n(q) \rightarrow 0 \; as \; n \rightarrow \infty \,.$$

Then the point $x = (I - H)^*(y)$ is well defined, $x \in S$ and

$$|x| \leq (I - L)^*|y| \leq q \,.$$

Moreover, the sequence

$$b_{n+1} = L(b_n) + |y|, \quad b_0 = 0$$

is well defined and

$$b_{n+1} \leq q, \quad \lim_{n \to \infty} b_n = b = (I - L)^*|y| \leq q \,.$$

LEMMA 6.8
Let $H_1 : [0, 1] \rightarrow L(X^m)$ and $H_2 : [0, 1] \rightarrow L_+(V^m)$ be continuous operators, then for all $t \in [0, 1]$: $H_2(t) \in B(H_1(t)) \Rightarrow \int_0^1 H_2(t)dt \in B\left(\int_0^1 H_1(t)dt\right)$ which will be used in the remainder of Taylor's formula [134].

The convergence analysis will be based on monotonicity considerations in the space $X \times V$. Let (x_n, e_n) be an increasing sequence in X^N, then

$$(x_n, e_n) \leq (x_{n+k}, e_{n+k}) \Rightarrow 0 \leq (x_{n+k} - x_n, e_{n+k} - e_n) .$$

If $e_n \to e$, we obtain: $0 \leq (x_{n+k} - x_n, e - e_n)$ and hence by (C_5)

$$\|x_{n+k} - x_n\| \leq \|e - e_n\|_V \to 0 \quad \text{as } n \to \infty .$$

Hence $\{x_n\}$ $(n \geq 0)$ is a Cauchy sequence. When deriving error estimates, we shall as well use sequences $e_n = w_0 - w_n$ with a decreasing sequence $\{w_n\}$ $(n \geq 0)$ in C^N to obtain the estimate

$$0 \leq (x_{n+k} - x_n, w_n - w_{n+k}) \leq (x_{n+k} - x_n, w_n) .$$

If $x_n \to x^*$ as $n \to \infty$ this implies the estimate $|x^* - x_n| \leq w_n$. Note also that $(x, e) \in X$, $x \in \Delta$ and by (C_4) we get $|x| \leq e$.

Let $a \in C$, operators $K_1, K_2, M, M_1, K_3(w) \in C(V_1 \to C)$ $V_1 \subseteq V$, $w \in [0, a]$, and points $x_n \in D$ $(n \geq 0)$. It is convenient to define the sequences c_n, d_n, a_n, b_n $(n \geq 0)$ by

$$c_{n+1} = |x_{n+1} - x_n| \quad (n \geq 0) , \tag{6.302}$$

$$d_{n+1} = (K_1 + K_2 + M + M_1)(d_n) + K_3(|x_n|) c_n,$$

$$d_0 = 0 (n \geq 0) , \tag{6.303}$$

$$a_n = (K_1 + K_2 + M + M_1)^n (a) , \tag{6.304}$$

$$b_n = (K_1 + K_2 + M + M_1)^n (0) , \tag{6.305}$$

and the point b by

$$b = (K_1 + K_2 + M + M_1)^\infty (0) . \tag{6.306}$$

We can now state and prove the main result of this section:

THEOREM 6.21

Let X be a Banach space with convergence structure (X, V, X) with $V = (V, C, \|\|\|_V)$, an operator $F \in C^1(D \to X)$ $(D \subseteq X)$, an operator $Q \in C(D \to X)$, a continuous operator A_t such that for each $v, w \in V_1 \subseteq V$,

$A_t(v, w) : [0, 1] \to L_+(V)$, *a point* $a \in C$, *operators* K_1, $K(w)$, $K_3(w)$, $M_1 \in L_+(V)$ ($w \in [0, a]$), *an operator* $M_0 = M_0(v, w) \in C(V_1 \times V_1 \to V)$, *operators* M, $K_2 \in C(V_1 \to C)$, *and a null sequence* $\{z_n\} \in D$ ($n \geq 0$) *such that the following conditions are satisfied:*

(C_6) $U(a) \subseteq D$, $[0, a] \subseteq V_1$, $K_3(0) \in B(I - F'(0))$, $(-(F(0) + Q(0) + F'(0)(z_0))$, $((K_1 + K_2 + M + M_1)(0))) \in X$;

(C_7) $K_1 + K(|x| + t|y|) - K(|x|) \geq A_t(|x| + t|y|, |x|) \in B(F'(x) - F'(x + ty))$ *for all* $t \in [0, 1]$, x, $y \in U(a)$ *with* $|x| + |y| \leq a$;

(C_8) $0 \leq (Q(x) - Q(x + y), M_0(|x|, |y|)) \in X$ *and* $M_0(v, w) \leq M(v + w) - M(v)$ *for all* x, $y \in U(a)$ *with* $|x| + |y| \leq a$, *and* v, $w \in [0, a]$;

(C_9) $0 \leq (F'(x_n)(z_n) - F'(x_{n-1})(z_{n-1}), M_1(c_{n-1})) \in X$ ($n \geq 1$);

(C_{10})

$$R(a) := (K_1 + K_2 + M + M_1)(a) \leq a ; \qquad (6.307)$$

(C_{11}) $(K_1 + K_2 + M + M_1)^n a \to 0$ *as* $n \to \infty$;

(C_{12}) $K_3(|x|) - K_3(0) \in B(F'(0) - F'(x))$ *and* $K_3(|x|) \leq K_1 + K_2$ ($x \in U(a)$);

(C_{13}) $\int_0^1 K(w + t(v - w))(v - w)dt \leq K_2(v) - K_2(w)$ *for all* v, $w \in [0, a]$ *with* $w \leq v$;

(C_{14}) $M_2(w) \geq 0$ *and* $0 \leq M_1(w_1 + w_2) - M_2(w_1) \leq M_2(w_3 + w_4) - M_2(w_3)$ *for all* w, w_1, w_2, w_3, $w_4 \in [0, a]$ *with* $w_1 \leq w_3$ *and* $w_2 \leq w_4$, *where* M_2 *is* M *or* K_2.

Then

(i)   *the sequences* (x_n, d_n), $(x_n, b_n) \in (X \times V)^N$ *are: well defined, remain in* X^N, *monotone and satisfy* $b_n \leq d_n \leq b$, $b_n \leq a_n$ *and* $\lim_{n \to \infty} b_n = \lim_{n \to \infty} d_n = b$.

(ii)   *Iteration* $\{x_n\}$ ($n \geq 0$) *generated by* (6.301) *is: well defined, remains in* $U(b)$ *and converges to a solution* $x^* \in U(b)$ *of equation* $F(x) + Q(x) = 0$, *where* b *is the smallest fixed point of* R *on* $[0, a]$. *Moreover if* $z_n = 0$ ($n \geq 0$) x^* *is unique in* $U(a)$.

(iii)   *Furthermore the following error bounds are true:*

$$|x_{n+1} - x_n| \leq d_{n+1} - d_n ,$$

$$|x_n - x^*| \leq b - d_n ,$$

and

$$\left|x_n - x^*\right| \le a_n - b_n \quad if \, z_n = 0 \quad (n \ge 0)$$

where d_n, a_n and b_n are given by (6.303), (6.304), and (6.305), respectively.

PROOF We first note that b replacing a also satisfies the conditions of the theorem. Using conditions (C_6) and (C_{12}) we obtain

$$\left|I - F'(0)|(b) + \right| - (F(0) + Q(0) + A(0) \, (z_0))| \le K_3(0)(b)$$

$$+ (K_1 + K_2 + M + M_1) \, (0)$$

$$\le (K_1 + K_2) \, (b - 0) + (K_1 + K_2 + M + M_1) \, (0)$$

$$\le (K_1 + K_2 + M + M_1) \, (b - 0) + (K_1 + K_2 + M + M_1) \, (0)$$

$$= (K_1 + K_2 + M + M_1) \, (b) = R(b) \le b \quad \text{[by (6.312)]}.$$

Hence, by Lemma 6.7, x_1 is well defined and $(x_1, b) \in X$. We also get

$$x_1 = (I - F'(0)) \, (x_2) + (-(F(0) + Q(0) + A(0) \, (z_0)))$$

$$\Rightarrow$$

$$|x_1| \le K_3(0) \, |x_1| + (K_1 + K_2 + M + M_1) \, (0)$$

$$\le (K_1 + K_2) \, |x_1| + (K_1 + K_2 + M + M_1) \, (0) = d_1 \, ,$$

and by the order convexity of L

$$d_1 = (K_1 + K_2) \, |x_1| + (K_1 + K_2 + M + M_1) \, (0)$$

$$\le (K_1 + K_2) \, |x_1| + (K_1 + K_2 + M + M_1) \, (0)$$

$$\le (K_1 + K_2) \, (b - 0) + (K_1 + K_2 + M + M_1) \, (0)$$

$$\le (K_1 + K_2 + M + M_1) \, (b - 0) + (K_1 + K_2 + M + M_1) \, (0) = R(b) = b \, .$$

That is we get $|x_1 - x_0| \leq d_1 - d_0$ or $0 \leq (x_0, d_0) \leq (x_1, d_1)$. We assume that

$$0 \leq (x_{k-1}, d_{k-1}) \leq (x_k, d_k) \quad \text{and} \quad d_k \leq b \text{ for } k = 1, 2, \ldots, n.$$

We need to find a bound for $I - F'(x_n)$ $(n \geq 0)$. We will show that $K_3(|x_n|) \in B(I - F'(x_n))$ $(n \geq 0)$.

This fact follows from (C$_6$), (C$_{12}$) and the estimate

$$\left| I - F'(x_n) \right| \leq \left| I - F'(0) \right| + \left| F'(0) - F'(x_n) \right|$$

$$\leq K_3(0) + K_3(|x_n|) - K_3(0) = K_3(|x_n|) \quad (n \geq 0).$$

Using (6.301) we obtain the approximation

$$-\left[F(x_n) + Q(x_n) + F'(x_n)(z_n) \right] = -F(x_n) - Q(x_n)$$

$$- F'(x_n)(z_n) + F'(x_{n-1})(x_n - x_{n-1})$$

$$+ F(x_{n-1}) + Q(x_{n-1}) + F'(x_{n-1})(z_{n-1}).$$

Hence, by (C$_7$), (C$_8$), (C$_9$), (C$_{10}$), (C$_{14}$), (C$_{15}$), (C$_{16}$), Lemma 6.8, and the induction hypotheses, we obtain in turn

$$\left| -F(x_n) + F(x_{n-1}) + F'(x_{n-1})(x_n - x_{n-1}) \right| + \left| Q(x_{n-1}) - Q(x_n) \right|$$

$$+ \left| F'(x_n)(z_n) - F'(x_{n-1})(z_{n-1}) \right|$$

$$\leq \int_0^1 A_t \left(|x_{n-1}| + tc_{n-1}, |x_{n-1}| \right) c_{n-1} dt + M_0 \left(|x_{n-1}|, |x_n| \right) + M_1 c_{n-1}$$

$$\leq \int_0^1 \left[K \left(|x_{n-1}| + tc_{n-1} \right) - K \left(|x_{n-1}| \right) + K_1 \right] c_{n-1} dt$$

$$+ M \left(|x_{n-1}| + c_{n-1} \right) - M \left(|x_{n-1}| \right) + M_1 c_{n-1}$$

$$\leq \int_0^1 \left[K \left(d_{n-1} + t(d_n - d_{n-1}) \right) (d_n - d_{n-1}) dt - K \left(|x_{n-1}| \right) c_{n-1} \right.$$

$$+ K_1 c_{n-1} + M \left(d_{n-1} + d_n - d_n \right) - M \left(d_{n-1} \right) + M_1 \left(d_n \right) - M_1 \left(d_{n-1} \right) \Big]$$

$$\leq K_2 (d_n) - K_2 (d_{n-1}) - K (|x_{n-1}|) c_{n-1} + K_1 c_{n-1} + M (d_n)$$

$$-M (d_{n-1}) + M_1 (d_n) - M_1 (d_{n-1})$$

$$\leq (K_1 + K_2 + M + M_1) (d_n) - d_n . \tag{6.308}$$

We can now obtain that

$$K_3 (|x_n|) (b - d_n) + |- (F (x_n) + Q (x_n) + A (x_n) (z_n))|$$

$$+d_n \leq (K_1 + K_2 + M + M_1) (b - d_n)$$

$$+ (K_1 + K_2 + M + M_1) (d_n) = R(b) = b \tag{6.309}$$

(where we have used that $\sum_{k=1}^{n-1} c_k \leq d_n - d_1$ and $|x_k| \leq d_k$ for $k = 0, 1, 2, \ldots, n$).
That is, x_{n+1} is also well defined by Lemma 6.7 and $c_n \leq b - d_n$. Hence d_{n+1} is well defined too and as in (6.309), we obtain that:

$$d_{n+1} \leq R(b) \leq b .$$

The monotonicity $(x_n, d_n) \leq (x_{n+1}, d_{n+1})$ can be derived from

$$c_n + d_n \leq K_3 (|x_n|) c_n + |- (F (x_n) + Q (x_n) + A (x_n) (z_n))| + d_n$$

$$\leq K_3 (|x_n|) c_n + (M + M_1 + K_1 + K_2) d_n \leq d_{n+1} .$$

The induction has now been completed. We need to show that

$$b_n \leq d_n \quad \text{for all } n \geq 1 .$$

For $n = 1$ and from the definitions of b_n d_n

$$b_1 = (K_1 + K_2 + M + M_1)^1 (0) \leq d_1 .$$

Assume that

$$b_k \leq d_k \quad \text{for } k = 1, 2, \ldots, n .$$

Then, we obtain in turn

$$b_{n+1} \leq (K_1 + K_2 + M + M_1)^{n+1} (0)$$

$$= (K_1 + K_2 + M + M_1) (K_1 + K_2 + M + M_1)^n (0)$$

$$\leq (K_1 + K_2 + M + M_1) (d_n) \leq d_n \leq d_{n+1} .$$

Since $d_n \leq b$, we have $b_n \leq d_n \leq b$. By (6.304) and (6.305) it follows that

$$0 \leq a_n - b_n \leq (K_1 + K_2 + M + M_1)^n (a) \quad (n \geq 1) \, [28, 134].$$

By condition (C_{13}) and the above, we deduce that the sequence $\{b_n\}$ $(n \geq 0)$ is Cauchy in a Banach space C, and as such it converges to some $b = (K_1 + K_2 + M + M_1)^\infty (0)$. From $(K_1 + K_2 + M + M_1)(b) = (K_1 + K_2 + M + M_1)(\lim_{n \to \infty} (K_1 + K_2 + M + M_1)^n (0)) = \lim_{n \to \infty} (K_1 + K_2 + M + M_1)^{n+1} (0) = b$, we obtain

$$(K_1 + K_2 + M + M_1) (b) = b \leq a ,$$

which makes b smaller than any solution of the inequality

$$(K_1 + K + M + M_1) (p) \leq p \quad \text{(see also [28, 134])} .$$

It also follows that the sequence $\{x_n\}$ $(n \geq 0)$ is Cauchy in X, and as such, it converges to some $x^* \in U(b)$. By letting $n \to \infty$ in (6.308) and using the hypotheses that $\lim_{n \to \infty} z_n = 0$, we deduce that x^* is a solution of the equation $F(x) + Q(x) = 0$.

To show uniqueness, let us assume that there exists another solution y^* of the equation $F(x) + Q(x) = 0$ in $\bar{U}(a)$. Then, exactly as in [28, 134], by considering the modified Newton-process

$$x_{n+1} = x_n - (F(x_n) + Q(x_n)) ,$$

we can show that this sequence converges, under the hypotheses of the theorem. Moreover, as above, we can easily show (see also [28, 134]) that

$$|y^* - x_n| \leq a_n - b_n \quad \text{for } z_n = 0 \quad (n \geq 0) ,$$

from which follows that $x_n \to y^*$ as $n \to \infty$. Finally, the estimates in (iii) are obtained by using standard majorization techniques (see also Chapter 3).

That completes the proof of the theorem. ∎

We will now introduce results on *a posteriori* estimates. It is convenient to define the operator

$$R_n(q) = \left(I - K_3 (|x_n|)\right)^* S_n(q) + c_n$$

where

$$S_n(q) = (K_1 + K_2 + M + M_1)(|x_n| + q)$$

$$- (K_1 + K_2 + M + M_1)(|x_n|) - K_3(|x_n|)(q)$$

and the interval

$$I_n = [0, a - |x_n|] .$$

It can easily be seen that the operators S_n are monotone on I_n. Moreover, the operators $R_n : [0, a - d_n] \to [0, a - d_n]$ are well defined and monotone. This fact follows from Lemma 6.6 and the scheme

$$d_n + c_n \le d_{n+1} \Rightarrow R(a) - d_{n+1} \le a - d_n - c_n$$

$$\Rightarrow S_n(a - d_n) + K_3(|x_n|)(a - d_n - c_n)$$

$$\le a - d_n - c_n (n \ge 0) .$$

Then, exactly as in [135], we can show:

PROPOSITION 6.6
The following implications are true:

(i)   *if $q \in I_n$ satisfy $R_n(q) \le q$, then*

$$c_n \le R_n(q) = p \le q$$

   and
$$R_{n+1}(p - c_n) \le p - c_n \quad \text{for all } n \ge 0 ;$$

(ii)   *under the hypotheses of Theorem 6.21, let $q_n \in I_n$ be a solution of $R_n(q) \le q$, then*
$$|x^* - x_m| \le a_m \quad (m \ge n)$$

   where
$$a_n = q_n \quad \text{and} \quad a_{m+1} = R_m(a_m) - c_m .$$

REMARK 6.21   (a) The results obtained in Theorem 6.21 and the Proposition 6.6 reduce immediately to the corresponding ones in [135, Theorem 5 and Lemmas 10–12] when $Q(x) = 0$ $(x \in D)$, $z_n = 0$ $(n \ge 0)$, $t = 1$, $K_1 = 0$, $K_2 = L$, where

L is order convex on $[0, a]$, $K = L'$ and $K_3(0) = L'(0)$. On the one hand, using our conditions we cover a wider range of problems and, on the other hand, it is because it may be possible to choose K_t, K, K_1 so that $K_t(p + tq, p) \leq K_1 + K(p + tq, p) \leq L'(p + tq) - L'(p)$ for all $p, q \in V_1$, $t \in [0, 1]$. Then it can easily be seen that our estimates on the distances $|x_{n+1} - x_n|$ and $|x^* - x_n|$ ($n \geq 0$) will be sharper. One such choice for K_t could be

$$K_t(p + tq, p) = \sup_{\substack{|x|+|y|\leq a, t\in[0,1] \\ |x|\leq p, |y|\leq q}} |F'(x) - F'(x + ty)|$$

for all $x, y \in U(a)$, $p, q \in [0, a]$.

(b) As in [47, 134, 135], we can show that if conditions (C_6)–(C_{10}), (C_{13})–(C_{15}) are satisfied and there exists $t \in (0, 1)$ such that $(K_1 + K_2 + M + M_1)(a) \leq ta$, then there exists $a_1 \in [0, ta]$ satisfying conditions (C_6)–(C_{15}). The solution $x^* \in U(a_1)$ is unique in $U(a)$ (when $z_n = 0$ ($n \geq 0$)).

(c) From the approximation

$$F'(x_n)(z_n) - F'(x_{n-1})(z_{n-1})$$

$$= \left(F'(x_n)(z_n) - z_n\right) + \left(F'(x_{n-1})(z_{n-1}) - z_{n-1}\right) + I(z_n - z_{n-1})$$

we observe that (C_9) will be true if $M_1 = 2K_3(b) + I$ and points z_n ($n \geq 0$) are such that $|z_n| + |z_{n-1}| + |z_n - z_{n-1}| \leq c_{n-1}$ ($n \geq 0$).

(d) Another choice for M_1, z_n can be $M_1 = |\varepsilon|I$, $z_n = z_{n-1} + \varepsilon_n(x_n - x_{n-1})$ ($n \geq 1$) with $|\varepsilon_n| \leq |\varepsilon|$ ($n \geq 0$), where e, e_n ($n \geq 0$) are numbers or operators in $L_+(V)$ and provided that $F'(x) = I$ ($x \in D$). It can then easily be seen that (C_9) is satisfied. The sequence $\{z_n\}$ ($n \geq 0$) must still be chosen to be null. At the end of this paper, in part V, we have given examples for this case. Several other choices are also possible.

(e) Define the residuals $r_n = -A(x_n)(z_n)$ ($n \geq 0$) and set $\delta_n = x_{n+1} - x_n$ ($n \geq 0$). Then from the approximation

$$r_n = \left[(I - F'(x_n)) - I\right](z_n)$$

we obtain

$$|r_n| \leq (K_3(|x_n|) + I)(z_n)$$

which shows that $r_n \to 0$ as $n \to \infty$ if $z_n \to 0$ as $n \to \infty$. Consequently the results obtained in Theorem 6.21 and Proposition 6.6 remain true for the system

$$x_{n+1} - x_n = \delta_n, \quad F'(x_n)\delta_n = -(F(x_n) + Q(x_n)) + r_n \quad (n \geq 0).$$

(f) It can easily be seen from the proof of Theorem 6.21 that the results obtained in Theorem 6.21 remain valid if (C₉) is replaced by the condition

$(C_9)'$ $(F'(x_n)(z_n) - F'(x_{n-1})(z_{n-1}), M_2(d_n^* - d_{n-1}^*)) \in X$
    with $d_n^* = d_n$ or $d_n^* = b_n$ $(n \geq 0)$, for some $M_2 \in L_+(V)$.

This is equivalent to the condition $(F'(x_n)(z_n), M_2(d_n^*)) \in X^N$ $(n \geq 1)$ is an increasing sequence.

We now examine the monotone case. Let $J \in L(X \to X)$ be a given operator. Define the operators P, T $(D \to X)$ by

$$P(x) = JT(x + u), \quad T(x) = G(x) + G_1(x), \quad P(x) = F(x) + Q(x),$$

$$F(x) = JG(x + u) \quad \text{and} \quad Q(x) = JG_1(x + u),$$

where G, G_1 are as F, Q, respectively. We deduce immediately that under the hypotheses of Theorem 6.21, the zero x^* of P is a zero of JT also, if $u = 0$.

We will now provide a monotonicity result to find a zero x^* of JT. The space X is assumed to be partially ordered and satisfies the conditions for V given in (C₁)–(C₅). Moreover, we set $X = V$, $D = C^2$ so that $| \, |$ turns out to be I. ∎

THEOREM 6.22

Let V be a partially ordered Banach space satisfying conditions (C₁)–(C₅), Y be a Banach space, G, G₁ as F, Q D ⊆ V, J ∈ L (V → V), A_t, M, M₁; K, K₁, K₂, K₃ as in Theorem 6.21 and u, v ∈ V such that

(C₁₆) *[u, v] ⊆ D;*

(C₁₇) *sequence {zₙ} (n ≥ 0) and iteration*

$$y_0 = u, \quad y_{n+1} = y_n + \left[JG'(y_n)\right]^* (-JT(y_n)) - z_n \quad (n \geq 0) \quad (6.310)$$

are such that

$$y_n + \left[JG'(y_n)\right]^* (-JT(y_n)) - v \leq z_n \leq \left[JG'(y_n)\right]^*$$

$$(-JT(y_n)), \, z_n \in [u, v] \quad (n \geq 0).$$

(C₁₈) *conditions (C₆)–(C₁₅) are satisfied for $a = v - u$.*

Then iteration (6.310) is well defined for all $n \geq 0$, monotone and converges to a zero x^* of JT in $[u, v]$. Moreover, x^* is unique in $[u, v]$ if $z_n = 0$ $(n \geq 0)$.

PROOF   It follows immediately from Theorem 6.21 by setting $a = v - u$.   ∎

We will complete this section with two examples that show how to choose the terms introduced in Theorem 6.21, in practical applications. From now on we choose $t = 1$, $K_1 = 0$, $K = L'$ (order convex), $K_2 = L$ and $K_3(0) = L'(0)$. It can then easily be seen from the proof of Theorem 6.21 that conditions (C$_{12}$) and (C$_{13}$) can be replaced by $(L + M + M_1)(a) \leq a$ and $(L'(a) + M + M_1)^n(a) \to 0$ as $n \to \infty$ respectively [see also Remark 6.21(a)].

Example 6.9
We discuss the case of a real Banach space with norm $\| \ \|$. Assume that $F'(0) = I$ and there exists a monotone operator

$$f : [0, a] \to IR$$

such that

$$\|F''(x)\| \leq f(\|x\|) \quad \text{for all } x \in U(a)$$

and a continuous nondecreasing function g on $[0, r]$, $r \leq a$ such that

$$\|Q(x) - Q(y)\| \leq g(r)\|x - y\| \tag{6.311}$$

for all $x, y \in U\left(\frac{r}{2}\right)$.
We showed in [60] that (6.311) implies that

$$\|Q(x + \ell) - Q(x)\| \leq h(r + \|\ell\|) - h(r), \quad x \in U(a), \quad \|\ell\| \leq a - r, \tag{6.312}$$

where

$$h(r) = \int_0^r g(t)dt \ .$$

Conversely, it is not hard to see that we may assume, without loss of generality, that the function h and all functions $h(r + t) - h(r)$ are monotone in r. Hence, we may assume that $h(r)$ is convex and hence differentiable from the right. Then, as in [60], we show that (6.312) implies (6.311) and $g(r) = h'(r + 0)$. Hence,

$$L(q) = \|F(0) + Q(0)\| + \int_0^q ds \int_0^s f(t)dt \tag{6.313}$$

and

$$M(q) = \int_0^q g(t)dt \ . \tag{6.314}$$

In Remarks 6.21 (c) and (d) we have already provided some choices for M_1, z_n. Here, however, for simplicity let us choose $z_n = 0$ $(n \geq 0)$ and $M_1 = 0$.

Then condition (C_{10}) will be true if

$$\frac{1}{2} f(a) a^2 - (1 - g(a))a + \|F(0) + Q(0)\| \leq 0 . \tag{6.315}$$

If we set $Q = 0$ and $g = 0$, (6.315) is true if $\|F(0)\| f(a) \leq \frac{1}{2}$, which is a well-known condition due to Kantorovich [114, Ch. 18]. If $Q \neq 0$, condition (6.315) is the same condition with the one found in [206, 210] for the Zincenko iteration.

In the example that follows, we show that our results can apply to solve nonlinear integral equations involving a nondifferentiable term, whereas the results obtained in [135] cannot apply.

Example 6.10
Let $X = V = C[0, 1]$, and consider the integral equation

$$x(t) = \int_0^1 k(t, s, x(s)) ds \text{ on } X , \tag{6.316}$$

where the kernel $k(t, s, x(s))$ with $(t, s) \in [0, 1] \times [0, 1]$ is a nondifferentiable operator on X. Consider (6.316) in the form (6.300) where $F, Q : X \rightarrow X$ are given by

$$F(x)(t) = Ix(t) \quad \text{and} \quad Q(x)(t) = -\int_0^1 k(t, s, x(s)) ds .$$

The operator $|\ |$ is defined by considering the sup-norm. We assume that V is equipped with natural partial ordering, and there exists α, $a \in [0, +\infty)$, and a real function $\alpha(t, s)$ such that

$$\|k(t, s, x) - k(t, s, y)\| \leq \alpha(t, s) \|x - y\|$$

for all $t, s \in [0, 1]$, $x, y \in U\left(\frac{\alpha}{2}\right)$, and

$$\alpha \geq \sup_{t \in [0,1]} \int_0^1 \alpha(t, s) ds .$$

Define the real functions h, f, g on $[0, a]$ by $h(r) = \alpha r$, $f(r) = 0$ and $g(r) = \alpha$ for all $r \in [0, a]$. By choosing L, M, M_1 as in (6.313), (6.314), and Remarks 6.21 (c), respectively, we can easily see that the conditions (C_1)–(C_9), (C_{12})–(C_{15}) of Theorem 6.21 are satisfied. In particular, condition (C_{10}) becomes

$$(1 - \alpha - |\varepsilon|)a - \|Q(0)\| \geq 0 \qquad (6.317)$$

which is true in the following cases: if $0 \leq \alpha < 1 - |\varepsilon|$, choose $a \geq \beta = \frac{\|Q(0)\|}{1-\alpha-|\varepsilon|}$; if $\alpha = 1 - |\varepsilon|$ and $Q(0) = 0$, choose $a \geq 0$; if $\alpha > 1 - |\varepsilon|$ and $Q(0) = 0$, choose $a = 0$. If in (6.317) strict inequality is valid, then there exists a solution a^* of Equation (6.317) satisfying condition (C_{13}). Note that if we choose $\alpha \in (0, 1-|\varepsilon|)$, $a \in (\beta, +\infty)$ and $\varepsilon \in (-1, 1)$ condition (6.317) is valid as a strict inequality. Finally, we remark that the results obtained in [132, 135] cannot apply here to solve Equation (6.316), since Q is nondifferentiable on X and the z_ns are not necessarily zero. This example is useful, especially when the z_ns are not necessarily all zero. Otherwise, results on (6.301) with general convergence structure have already been found (see, e.g., [114] and the references there).

Exercises

1. Using either the secant method or Newton's method, find the indicated roots of the following equations.

    (a) The positive root of $x^3 - x^2 - x - 1 = 0$.

    (b) All roots of $x = 1 + .3\cos(x)$.

    (c) The smallest positive root of

    $$\cos(x) = \frac{1}{2} + \sin(x) .$$

2. (a) Use Newton's method to calculate the smallest positive root of $x - \tan x = 0$.

    (b) Calculate the root of $x - \tan x = 0$ closest to $x = 100$. Explain the difference in the behavior of Newton's method as compared with that in part (a).

3. Using Newton's method, solve the nonlinear system $x^2+y^2 = 1, x^2-y^2 = 1$. The true solutions are easily determined to be $x = \pm\sqrt{2.5}, y = \pm\sqrt{1.5}$. As an initial guess, use $(x_0, y_0) = (1.6, 1.2)$.

4. Using Newton's method, find all roots of the system

    $$x^2 + y^2 - 2x - 2y + 1 = 0, \quad x + y - 2xy = 0 .$$

5. Assume that the operator F defined on a subset of a Banach space X with values in a Banach space Y satisfies a Hölder condition

$$\|F'(x) - G'(y)\| \leq L\|x - y\|^\alpha$$

with $0 < \alpha < 1$, on some convex set. Show that for any x, y in this set

$$\|F(x) - F(y) - F'(y)(x - y)\| \leq \frac{L}{1+\alpha}\|x - y\|^{1+\alpha} .$$

6. Let $x_0, x_1, \ldots, x_n$ be distinct real numbers, and let f be a given real-valued function. Show that the following divided difference satisfies

$$[x_0, x_1, \ldots, x_n] = \sum_{j=0}^{n} \frac{f(x_j)}{g_n'(x_j)}$$

and

$$[x_0, x_1, \ldots, x_n](x_n - x_0) = [x_1, \ldots, x_n] - \big[x_0, \ldots, x_{n-1}\big]$$

where

$$g_n(x) = (x - x_0) \cdots (x - x_n) .$$

7. Let $x_0, x_1, \ldots, x_n$ be distinct real numbers, and let f be n times continuously differentiable function on the interval $\{x_0, x_1, \ldots x_n\}$. Then show that

$$[x_0, x_1, \ldots, x_n] = \int_{\tau_n} \cdots \int f^{(n)}(t_0 x_0 + \cdots + t_n x_n)\, dt_1 \cdots dt_n$$

in which

$$\tau_n = \left\{ (t_1, \ldots, t_n) \mid t_1 \geq 0, \ldots, t_n \geq 0, \sum_{i=1}^{n} t_i \leq 1 \right\}$$

$$\tau_n = 1 - \sum_{i=1}^{n} t_i .$$

8. If f is a real polynomial of degree m, then show:

$$[x_0, x_1, \ldots, x_n, x] = \begin{cases} \text{polynomial of degree } m - n - 1, & n \leq m - 1 \\ a_m, & n = m - 1 \\ 0, & n > m - 1 \end{cases}$$

where $f(x) = a_m x^n + $ lower-degree terms.

9. The tensor product of two matrices $M, N \in L(\mathbf{R}^n)$ is defined as the $n^2 \times n^2$ matrix $M \times N = (m_j N \mid i, j = 1, \ldots, n)$, where $M = (m_{ij})$. Consider two F-differentiable operators $H, K : L(\mathbf{R}^n) \to L(\mathbf{R}^n)$ and set $F(X) = H(X)K(X)$ for all $X \in L(\mathbf{R}^n)$. Show that $F'(X) = [H(X) \times I]K'(X) + [I \times K(X)^T]H'(X)$ for all $X \in L(\mathbf{R}^n)$.

10. Let $F: \mathbf{R}^2 \to \mathbf{R}^2$ be defined by $f_1(x) = x_1^3$, $f_2(x) = x_2^2$. Set $x = 0$ and $y = (1, 1)^T$. Show that there is no $z \in [x, y]$ such that

$$F(y) - F(x) = F'(z)Y(y - x) .$$

11. Let $F: D \subset \mathbf{R}^n \to \mathbf{R}^m$ and assume that F is continuously differentiable on a convex set $D_0 \subset D$. For $x, y \in D_0$, show that

$$\|F(y) - F(x) - F'(x)(y - x)\| \leq \|y - x\| w(\|y - x\|) ,$$

where w is the modulus of continuity of F' on $[x, y]$. That is

$$w(t) = \sup\{\|F'(x) - F'(y)\| \mid x, y \in D_0, \|x - y\| \leq t\} .$$

12. Let $F: D \subset \mathbf{R}^n \to \mathbf{R}^m$. Show that F'' is continuous at $z \in D$ if and only if all second partial derivatives of the components $f_1, \ldots, f_m$ of F are continuous at z.

13. Let $F: D \subset \mathbf{R}^n \to \mathbf{R}^m$. Show that $F''(z)$ is symmetric if and only if each Hessian matrix $H_1(z), \ldots, H_m(z)$ is symmetric.

14. Let $M \in L(\mathbf{R}^n)$ be symmetric, and define $f: \mathbf{R}^n \to \mathbf{R}$ by $f(x) = x^T M x$. Show, directly from the definition that f is convex if and only if M is positive semidefinite.

15. Show that $f: D \subset \mathbf{R}^n \to \mathbf{R}$ is convex on the set D if and only if, for any $x, y \in D$, the function $g: [0, 1] \to \mathbf{R}$, $g(t) = g(tx + (1 - t)y)$ is convex on $[0, 1]$.

16. Show that if $g_i: \mathbf{R}^n \to \mathbf{R}$ is convex and $c_i \geq 0$, $i = 1, 2, \ldots, m$, then $g = \sum_{i=1}^{m} c_i g_i$ is convex.

17. Suppose that $g: D \subset \mathbf{R}^n \to \mathbf{R}$ is continuous on a convex set $D_0 \subset D$ and satisfies

$$\frac{1}{2}g(x) + \frac{1}{2}g(y) - g\left(\frac{1}{2}(x+y)\right) \geq \gamma \|x - y\|^2$$

for all $x, y \in D_0$. Show that g is convex on D_0 if $\gamma = 0$.

18. Let $M \in L(\mathbf{R}^n)$. Show that M is a nonnegative matrix if and only if it is an isotone operator.

19. Let $M \in L(\mathbf{R}^n)$ be diagonal, nonsingular, and nonnegative. Show that $\|x\| = \|D(x)\|_\infty$ is a monotonic norm on $\mathbf{R}^n$.

20. Let $M \in L(\mathbf{R}^n)$. Show that M is invertible and $M^{-1} \geq 0$ if and only if there exist nonsingular, nonnegative matrices $M_1, M_2 \in L(\mathbf{R}^n)$ such that $M_1 M M_2 = I$.

21. Show that the integral representation of $[x_0, \ldots, x_k]$ is indeed a divided difference of kth order of F. Let us assume that all divided differences have such an integral representation. In this case for $x_0 = x_1 = \cdots = x_k = x$ we shall have

$$\underbrace{[x, x, \ldots, x]}_{k+1 \text{ times}} = \frac{1}{k} f^{(k)} x .$$

Suppose now that the nth Fréchet-derivative of F is Lipschitz continuous on D, i.e., there exists a constant c_{n+1} such that

$$\left\| F^{(n)}(u) - F^{(n)}(v) \right\| \leq c_{n+1} \|u - v\|$$

for all $u, v \in D$. In this case, set

$$R_n(y) = \left([x_0, \ldots, x_{n-1}, y] - [x_0, \ldots, x_{n-1}, x_n] \right) (y - x_{n-1}), \ldots, (y - x_0)$$

and show that

$$\|R_n(y)\| \leq \frac{c_{n+1}}{(n+1)!} \|y - x_n\| \cdot \|y - x_{n-1}\| \cdots \|y - x_0\|$$

and

$$\left\| F(x+h) - \left(F(x) + F'(x)h + \frac{1}{2}F''(x)h^2 + \cdots + \frac{1}{n!}F^{(n)}(x)h^n \right) \right\|$$

$$\leq \frac{c_{n+1}}{(n+1)!} \|h\|^{n+1} .$$

22.   We recall the definitions:

(a)   An operator $F: D \subset \mathbf{R}^n \to \mathbf{R}^m$ is Gateaux- (or G-) differentiable at an interior point x of D if there exists a linear operator $L \in L(\mathbf{R}^n, \mathbf{R}^m)$ such that, for any $h \in \mathbf{R}^n$

$$\lim_{t \to 0} \frac{1}{t} \|F(x + th) - F(x) - tL(h)\| = 0.$$

L is denoted by $F'(x)$ and called the G-derivative of F at x.

(b)   An operator $F: D \subset \mathbf{R}^n \to \mathbf{R}^m$ is semicontinuous at $x \in D$ if, for any $h \in \mathbf{R}^n$ and $\varepsilon > 0$, there is a $\delta = \delta(\varepsilon, h)$ so that whenever $|t| < \delta$ and $x + th \in D$, then $\|F(x + th) - F(x)\| < \varepsilon$.

(c)   If $F: D \subset \mathbf{R}^n \to \mathbf{R}^m$ and if for some interior point x of D, and $h \in \mathbf{R}^n$, the limit

$$\lim_{t \to 0} \frac{1}{t}[F(x + th) - F(x)] = A(x, h)$$

exists, then F is said to have a Gateaux-differential at x in the direction h.

(d)   If the G-differential exists at x for all h and if, in addition

$$\lim_{h \to 0} \frac{1}{\|h\|} \|F(x + H) - F(x) - A(x, h)\| = 0,$$

then F has a Fréchet differential at x.

Show:

(i)   The linear operator L is unique;

(ii)   If $F: D \subset \mathbf{R}^n \to \mathbf{R}^m$ is G-differentiable at $x \in D$, then F is hemicontinuous at x

(iii) G-differential and "uniform in h" implies F-differential;

(iv) F-differential and "linear in h" implies F-derivative;

(v)   G-differential and "linear in h" implies G-derivative;

(vi)   G-derivative and "uniform in h" implies F-derivative. Here "uniform in h" indicated the validity of (d). Linear in h means that $A(x, h)$ exists for all $h \in \mathbf{R}^n$ and $A(x, h) = M(x)h$, where $M(x) \in L(\mathbf{R}^n, \mathbf{R}^m)$.

(vii) Define $F: \mathbf{R}^2 \to \mathbf{R}$ by $F(x) = \operatorname{sgn}(x_2) \min(|x_1|, |x_2|)$. Show that, for any $h \in \mathbf{R}^2$, $A(0, h) = F(h)$, but F does not have a G-derivative at 0.

(viii)   Define $F: \mathbf{R}^2 \to \mathbf{R}$ by $F(0) = 0$ if $x = 0$ and

$$F(x) = x_2 \left(x_1^2 + x_2^2 \right)^{3/2} \Big/ \left[\left(x_1^2 + x_2^2 \right)^2 + x_2^2 \right], \quad \text{if } x \neq 0 .$$

Show that F has a G-derivative at 0, but not an F-derivative. Show, moreover, that the G-derivative is hemicontinuous at 0.

(ix)   If the G-differential $A(x, h)$ exists for all x in an open neighborhood of an interior point x_0 of D and for all $h \in \mathbf{R}^n$, then F has an F-derivative at x_0 provided that for each fixed h, $A(x, h)$ is continuous in x at x_0.

(e)   Assume that $F: D \subset \mathbf{R}^n \to \mathbf{R}^m$ has a G-derivative at each point of an open set $D_0 \subset D$. If the operator $F': D_0 \subset \mathbf{R}^n \to L(\mathbf{R}^n, \mathbf{R}^m)$ has a G-derivative at $x \in D_0$, then $(F')'(x)$ is denoted by $F''(x)$ and called the second G-derivative of F at x.

Show:

(i)   If $F: \mathbf{R}^n \to \mathbf{R}^m$ has a G-derivative at each point of an open neighborhood of x, then F' is continuous at x if and only if all partial derivatives $\partial_i F_i$ are continuous at x.

(ii) F'' is continuous at $x_0 \in D$ if and only if all second partial derivatives of the components $f_1, \ldots, f_m$ of F are continuous at x_0. $F''(x_0)$ is symmetric if and only if each Hessian matrix $H_1(x_0), \ldots, H_m(x_0)$ is symmetric.

Assume:

(a)   The Fréchet-derivative $F'(x)$ of the nonlinear operator $F: D \subseteq X \to Y$ satisfies conditions

$$\left\| F'(x) - F'(y) \right\| \leq k(r) \|x - y\|, \quad x, y \in U(x_0, r) ,$$

$$a = \left\| F'(x_0)^{-1} F(x_0) \right\|, \quad b = \left\| F'(x_0)^{-1} \right\|, \quad (6.318)$$

for k a nondecreasing function on $[0, R]$.

(b)   The real function $\varphi(r) = a + b \int_0^r (r - t) k(t) dt$ has a unique zero r^* in the interval $[0, R]$, and $\varphi(R) \leq 0$.

Then show in turn the following:

23.   Equation (6.1) has a solution x^* in $U(x_0, r^*)$, this solution is unique in $U(x_0, R)$, and the Newton iterates are well defined, remain in $U(x_0, r^*)$ and satisfy the estimates

$$\|x_{n+1} - x_n\| \leq t_{n+1} - t_n$$

and

$$\|x^* - x_n\| \leq r^* - t_n \text{ for all } n \geq 0,$$

where the sequence $\{t_n\}$ $(n \geq 0)$ which is monotonically increasing and converges to r^*, is defined by the recursive formula

$$t_{n+1} = t_n - \frac{\varphi(t_n)}{\varphi'(t_n)} \quad (n \geq 0) \text{ with } t_0 = 0.$$

24. The real function $u(r) = -\frac{\varphi(r)}{\varphi'(r)}$ is strictly monotonically decreasing on $[0, r^*]$.

25. The function u is a bijection between $[0, r^*]$ and $[0, a]$.

26. Define $\Delta(r) = u(r + v(r))$, $v(r) = u^{-1}(r)$ on $[0, a]$, $\Delta^0(r) = r$, $\Delta^{n+1}(r) = \Delta(\Delta^{(n)}(r))$ $(n \geq 0)$ and $w(r) = \sum_{n=0}^{\infty} \Delta^{(n)}(r)$. Then show that:

$$t_{n+1} - t_n = \Delta^{(n)}(a)$$

and

$$r^* - t_n = w\left(\Delta^{(n)}(a)\right) \quad \text{for all } n \geq 0.$$

27. Consider the case when $k(r) = k$ on $[0, r]$. Then show that hypothesis (b) above is satisfied if

$$2abk \leq 1$$

and

$$1 - \sqrt{1 - 2abk} \leq bkR \leq 1 + \sqrt{1 - 2abk}$$

where φ becomes

$$\varphi(r) = a + \frac{bkr^2}{2} - r$$

and r^* is the small zero of the equation $\varphi(r) = 0$.

28. Moreover under the hypothesis in (27) show that:

$$\Delta(r) = \frac{r^2}{2\sqrt{r^2 + d}},$$

$$w(r) = r + \sqrt{r^2 + d} - \sqrt{d},$$

$$\Delta^{(n)}(r) = \frac{2\sqrt{d}g^{2^n}(r)}{1 - g^{2^{n+1}}(r)} \quad (n \geq 0),$$

$$w\left(\Delta^{(n)}(r)\right) = \frac{2\sqrt{d}g^{2^{n+1}}(r)}{1 - g^{2^n}(r)} \quad (n \geq 0),$$

where

$$d = (bk)^{-2}(1 - 2abk)$$

and

$$g(r) = \frac{\sqrt{r^2 + d} - \sqrt{d}}{r} .$$

29. Define the sequences $r_n = \|x_n - x_0\|$, $k_n(r) = k(r_n + r)$ for $r \in [0, R - t_n]$ and set $a_n = \|x_{n+1} - x_n\|$, $b_n = (1 - h(r_n))^{-1}$ for all $n \geq 0$, where $h(r) = \int_0^r k(t)dt$. Assume that $a_n > 0$ for all $n \geq 0$. Show that the equation

$$r = a_n + b_n \int_0^r (r - t)k_n(t)dt$$

has a unique positive zero r_n^* in the interval $[0, R - t_n]$,

$$\|x^* - x_n\| \leq r_n^* \quad (n \geq 0)$$

$$\leq (r^* - t_n) a_n / (t_{n+1} - t_n) \quad (n \geq 0)$$

$$\leq (r^* - t_n) a_{n-1} / (t_n - t_{n-1}) \quad (n \geq 1)$$

$$\leq r^* - t_n \quad (n \geq 0) ,$$

and

$$\|x^* - x^*\| \geq s_n^* \quad (n \geq 0)$$

where s_n^* is the positive zero of the equation

$$a_n = r + b_n \int_0^r (r - t)k_n(t)dt \quad \text{for all } n \geq 0 .$$

30. Consider the Cauchy problem from the theory of ordinary differential equations in the form

$$\frac{dx}{dt} = f(t, x), \quad x(0) = x_0 .$$

Find conditions on f defined on an appropriately chosen space of your choice so that the operator

$$F(x(t)) = x_0 + \int_0^t f(s, x(s))ds - x(t)$$

satisfies condition (6.318) for some real function $k(r)$ to be specified accordingly. Then under hypotheses (a) and (b) above apply the results of Exercises 23–29 to the Cauchy problem.

31. Consider the Goursat problem from the theory of partial differential equations in the form

$$\frac{\partial^2 x}{2t \partial s} = f\left(t, s, s, \frac{\partial x}{\partial t}, \frac{\partial x}{\partial s}\right),$$

$$x(0, s) = x_1(s), \quad x(t, 0) = x_2(t),$$

where $x_1(s)$ and $x_2(t)$ are given functions with $x_1(0) = x_2(0) = x_0$. Find conditions on f defined on an appropriately chosen space of your choice so that the operator

$$F(v(t, s)) = f(t, s, v(t), s), \quad x_1(s) + x_2(t) - x_0$$

$$+ \int_0^t \int_0^s v(p, q) dp dq, \quad x_2(t) + \int_0^s v(t, q) dq,$$

$$x_1(s) + \int_0^t v((p, s) dp) - v(t, s))$$

$$\left(v = \frac{\partial^2 x}{\partial t \partial s}\right)$$

satisfies condition (6.318) for some real function $k(r)$ to be specified accordingly. Then under hypotheses (a) and (b) above apply the results of Exercises 23–29 to the Goursat problem.

32. In addition to (a) above consider the hypothesis

(c) The nonlinear operator $G: D \subset X \to Y$ satisfies

$$\|G(x) - G(y)\| \le e(r)\|x - y\|$$

on $\bar{U}(x_0, r) \subseteq D$ for $r \in [0, R]$ and some fixed $R > 0$. Define also the iteration

$$x_{n+1} = x_n - F'(x_n)^{-1}(F(x_n) + G(x_n)) \quad (n \ge 0), \quad (6.319)$$

the function

$$q(r) = \varphi(r) + \int_0^r e(t) dt$$

and the equation

$$F(x) + G(x) = 0. \qquad (6.320)$$

Moreover, assume that (b) is satisfied with φ replaced by q and call the new hypothesis (b'). Under hypotheses (a), (b') and (c), which of the results in Exercises 23–29 hold for the new iteration (6.319) and Equation (6.320)?

33. Consider the Cauchy problem of Exercise 30 and replace the function $f(t, x)$ with $f(t, x) + g(t, x)$ where the latter is chosen in such a way that (c) above is satisfied. Then answer to the same questions posed in Exercise 30.

34. Do as in Exercise 33 for the Goursat problem of Exercise 31.

35. Assume that $F: D \subset \mathbf{R}^n \to \mathbf{R}^m$ has a G-derivative which satisfies $\|F'(x)\| \le c < 1$ for all x in a convex set $D_0 \subset D$. Show that F is contractive on D_0.

36. Assume that $F: \mathbf{R}^n \to \mathbf{R}^n$ is G-differentiable on an open set D and, for some $c > 0$, satisfies $\|F(x) - F(y)\| \ge c\|x - y\|$, $x, y \in D$. Show that, for any $x \in D$, $F'(x)$ is invertible and $\|F'(x)^{-1}\| \le c^{-1}$.

37. Let $K: [0, 1] \times [0, 1] \times \mathbf{R} \to \mathbf{R}$ satisfy

$$|K(t, s, y) - K(s, t, v)| \le \eta|u - v| \quad \text{for all } s, t \in [0, 1], \ u, v \in \mathbf{R},$$

where $\eta < 1$, and let $\gamma_1, \gamma_2, \ldots, \gamma_n$ be positive constants such that $\sum_{j=1}^{n} \gamma_j = 1$. Show that the equation

$$x_i = \psi(t_i) + \sum_{j=1}^{n} \gamma_j k\left(t_i, t_j, x_j\right), \quad i = 1, 2, \ldots, n,$$

where $t_1, \ldots, t_n \in [0, 1]$ and $\psi: [0, 1] \to \mathbf{R}$ is a given function has a unique solution x^*, and that the iterates

$$x_i^{k+1} = \psi(t_i) + \sum_{j=1}^{n} \gamma_j K\left(t_i, t_j, x_j^k\right) \quad i = 1, 2, \ldots, n, \ k = 0, 1, \ldots,$$

converge to x^* for any x^0.

38. Let x^* be a simple zero of Equation (6.1) in X and set $\Gamma = F'(x^*)^{-1}$. Define the iteration

$$x_{n+1} = x_n - \Gamma F(x_n) \quad (n \ge 0)$$

This method can be realized for operators F which satisfy an autonomous differential equation

$$F'(x) = P(F(x))$$

as $F(x^*) = P(0)$ can be evaluated without knowing the value of x^*. Show:

(i) If Γ exists, $\|\Gamma\| \leq b$, the Fréchet derivative F' of F satisfies a Lipschitz condition with constant K in X, and x_0 is such that $h = \frac{1}{2}Kb\|x_0 - x^*\| < 1$, then the sequence $\{x_n\}$ $(n \geq 0)$ defined above converges to x^* with

$$\|x_n - x^*\| \leq h^{2^n - 1} \|x_0 - x^*\| \quad (n \geq 0).$$

(ii) If P is Lipschitz with constant c, $\|F(x)\| \leq d$, and

$$e = cdb < 1,$$

then the equation $F(x) = 0$ has a unique solution x^* to which the sequence $\{x_n\}$ $(n \geq 0)$ converges with

$$\|x_n - x^*\| \leq \frac{e^n}{1 - n} \|x_1 - x_0\| \quad (n \geq 0).$$

(iii) If P and F are Lipschitz continuous with constants c and β and $q = (1 - cb\|F(x_0)\|)^2 - 4bK_0\|x_1 - x_0\| \geq 0$, $K_0 = c\beta$, then a solution x^* of the equation $F(x) = 0$ exists in $\bar{U}(x_0, r_1)$ which is unique in $U(x_0, r_2)$, and

$$\|x_n - x^*\| \leq d_1^n r_1 \quad (n \geq 0)$$

where

$$r_1 = \frac{1 - cb\|F(x_0)\| - \sqrt{d}}{2bK_0}, \quad r_2 = \frac{1 - cb\|F(x_0)\|}{bK_0}$$

and

$$d_1 = \frac{1}{2}\left(1 + cb\|F(x_0)\| - \sqrt{q}\right).$$

39. Let $F: D \subset X \to Y$ be a nonlinear operator with Lipschitz continuous Fréchet derivative on the open domain D with constant $c > 0$. Assume Equation (6.1) has a solution x^* which is singular in the sense that $F'(x^*)$ has a bounded inverse with norm $b = \|F'(x^*)^{-1}\|$. Set $r^* = \frac{2}{3cd}$ and

suppose that $U = \bar{U}(x^*, r^*) \subset D$. Then show that for any $x_0 \in U$, Newton's method converges to x^* and

$$\|x_{n+1} - x^*\| \leq \frac{cd \|x_n - x^*\|^2}{2(1 - cd\|x_n - x^*\|)} \quad (n \geq 0).$$

40. Let $F: D \subseteq X \to Y$ be Fréchet differentiable in an open convex set $D_0 \subset D$ and, for some $x_0 \in D_0$, $F'(x_0)^{-1}$ exists. Assume that $F(x_0) \neq 0$ without loss of generality and that

$$\left\| F'(x_0)^{-1} \left(F'(x) - F'(y) \right) \right\| \leq K \|x - y\| \quad \text{for all } x, y \in D_0,$$

$$\eta = \left\| F'(x_0)^{-1} F(x_0) \right\|, \quad h = K\eta \leq \tfrac{1}{2}, \quad t^* = \frac{2\eta}{1 + \sqrt{1 - 2h}}$$

$$\bar{U} = \bar{U}(x_1, t^* - \eta) \subseteq D_0.$$

Then show:

(i) Newton's method is well defined, remains in U for all $n \geq 1$ and converges to a solution $x^* \in \bar{U}$ of (6.1).

(ii) The solution x^* is unique in $U(x_0, t^{**}) \cap D_0$, $t^{**} = \frac{1 + \sqrt{1 - 2h}}{k}$ if $2h < 1$, and in $\bar{U}(x_0, t^{**}) \cap D_0$ if $2h = 1$.

(iii) The following estimates are true:

$$\|x_n - x^*\| \leq t^* - t_n = \frac{2\eta_n}{1 + \sqrt{1 - 2h_n}} \leq 2^{1-n}(2h)^{2^n - 1}\eta \quad (n \geq 0)$$

where

$$t_0 = 0, \quad t_{n+1} = t_n - \frac{f(t_n)}{f'(t_n)} \quad (n \geq 0),$$

$$f(t) = \frac{1}{2}kt^2 - t + \eta,$$

$$B_0 = 1, \quad \eta_0 = \eta, \quad h_0 = h = k\eta, \quad B_n = \frac{B_{n-1}}{1 - h_{n-1}}, \quad \eta_n = \frac{h_{n-1}\eta_{n-1}}{2(1 - h_{n-1})},$$

and

$$h_n = K B_n h_n \quad (n \geq 1).$$

41.   Consider the Stirling method

$$z_{n+1} = z_n - \left[I - F'\left(F\left(z_n\right)\right)\right]^{-1}\left[z_n - F\left(z_n\right)\right]$$

for approximating a fixed point x^* of the equation

$$x = F(x)$$

in a Banach space X.
Show:

(i)   If $\|F'(x)\| \le \alpha < \frac{1}{3}$, then the sequence $\{x_n\}$ $(n \ge 0)$ converges to the unique fixed point x^* of equation $x = F(x)$ for any $z_0 \in X$. Moreover, show that:

$$\left\|x^* - z_n\right\| \le \left(\frac{2\alpha}{1-\alpha}\right)^n \frac{\|z_0 - F(z_0)\|}{1-\alpha} \quad (n \ge 0) .$$

(ii)   If F' is Lipschitz continuous with constant K and $\|F'(x)\| \le \alpha < 1$, then Newton's method converges to x^* for any $x_0 \in X$ such that

$$h_N = \frac{1}{2}K\frac{\|x_0 - F(x_0)\|}{(1-\alpha)^2} < 1$$

and

$$\left\|x_n - x^*\right\| \le (h_N)^{2^n-1} \frac{\|x_0 - F(x_0)\|}{1-\alpha} \quad (n \ge 0) .$$

(iii)   If F' is Lipschitz continuous with constant K, and $\|F'(x)\| \le \alpha < 1$, then $\{z_n\}$ $(n \ge 0)$ converges to x^* for any $z_0 \in X$ such that

$$h_s = \frac{K}{2}\frac{1+2\alpha}{1-\alpha}\frac{\|z_0 - F(z_0)\|}{1-\alpha} < 1$$

and

$$\left\|z_n - x^*\right\| \le (h_s)^{2^n-1} \frac{\|z_0 - F(z_0)\|}{1-\alpha} \quad (n \ge 0) .$$

42.   Let H be a real Hilbert space and consider the nonlinear operator equation $P(x) = 0$ where $P: U(x_0, r) \subseteq H \to H$. Let P be differentiable in $U(x_0, r)$ and set $F(x) = \|P(x)\|^2$. Then $P(x) = 0$ reduces to $F(x) = 0$. Define the iteration

$$x_{n+1} = x_n - \frac{\|P(x_n)\|^2}{2\|Q(x_n)\|^2}Q(x_n) \quad (n \ge 0)$$

where $Q(x) = P'(x)P(x)$, and the linear operator $P'(x)$ is the adjoint of $P'(x)$. Show that if:

(a)   there exist two positive constants B and K such that $B^2 K < 4$;

(b) $\|P'(x)y\| \geq B^{-1}\|y\|$ for all $y \in H$, $x \in U(x_0, r)$;

(c) $\|Q'(x)\| \leq K$ for all $x \in U(x_0, r)$;

(d) $\|x_1 - x_0\| < \eta_0$ and $r = \dfrac{2\eta_0}{2 - B\sqrt{K}}$.

Then equation $P(x) = 0$ has a solution $x^* \in U(x_0, r)$ and the sequence $\{x_n\}$ $(n \geq 0)$ converges to x^* with

$$\|x_n - x^*\| \leq \eta_0 \frac{\alpha^n}{1 - \alpha}$$

where

$$\alpha = \frac{1}{2} B\sqrt{K} .$$

43.   Consider the equation

$$x = T(x)$$

in a Banach space X, where $T: D \subset X \rightarrow X$ and D is convex. Let $T_1(x)$ be another nonlinear continuous operator acting from X into X, and let P be a projection operator in X. Then the operator $PT_1(x)$ will be assumed to be Fréchet differentiable on D. Consider the iteration

$$x_{n+1} = T(x_n) + PT_1'(x_n)(x_{n+1} - x_n) \quad (n \geq 0).$$

Assume:

(a) $\|[I - PT_1'(x_0)]^{-1}(x_0 - T(x_0))\| \leq \eta$,

(b) $\Gamma(x) = \Gamma = [I - PT_1'(x)]^{-1}$ exists for all $x \in D$ and $\|\Gamma\| \leq b$,

(c) $PT_1'(x)$, $QT_1(x)(Q = I - P)$ and $T(x) - T_1(x)$ satisfy a Lipschitz condition on D with respective constants M, q, and f,

(d) $\bar{U}(x_0, H\eta) \subseteq D$, where

$$H = 1 + \sum_{j=1}^{\infty} \prod_{i=1}^{j} J_i, \quad J_1 = b + \frac{h}{2},$$

$$J_i = b + \frac{h}{2} J_1 \cdots J_{i-1}, \quad i \geq 2, \quad J_0 = \eta,$$

(e) $h = BM\eta < 2(1 - b)$, $b = B(q + f) < 1$. Then show that the equation $x = T(x)$ has a solution $x^* \in \bar{U}(x_0, H\eta)$ and the sequence $\{x_n\}$ $(n \geq 0)$ converges to x^* with

$$\|x_n - x^*\| \subseteq H\eta \prod_{i=1}^{n} J_i \ .$$

44. Let H be a real separable Hilbert space. An operator F on H is said to be weakly closed if

(a) x_n converges weakly to x

and

(b) $F(x_n)$ converges weakly to y to imply that

$$F(x) = y \ .$$

Let F be a weakly closed operator defined on $\bar{U}(x_0, r)$ with values in H. Suppose that F maps $\bar{U}(x_0, r)$ into a bounded set in H provided the following condition is satisfied:

$$(F(x), x) \leq (x, x) \quad \text{for all } x \in S$$

where $S = \{x \in H \mid \|x\| = r\}$. Then show that there exists $x^* \in U(x_0, r)$ such that
$$F(x^*) = x^* \ .$$

45. Let X be a Banach space, $LB(X)$ the Banach space of continuous linear operators on X equipped with the uniform norm, B_1 the unit ball. Recall that a nonlinear operator K on X is compact if it maps every bounded set into a set with compact closure. We shall say a family H of operators on X is collectively compact if and only if every bounded set $B \subset X$, $\bigcup_{P \in H} H(B)$ has compact closure.
Show:

(i) If

(a) H is a collectively compact family of operators on X,

(b) K is in the pointwise closure of H,

then K is compact.

(ii) If

(a) H is a collectively compact family on X,

(b) H is equidifferentiable on $D \subset X$,

then for every $x \in D$, the family $\{P'(x) \mid P \in H\}$ is collectively compact.

46. Consider the equations

$$x - K(x) = 0$$

and

$$x - K_n(x) = 0,$$

where K is a compact operator from a domain D of a Banach space X into X, $\{K_n\}$ $(n \geq 1)$ are collectively compact operators.
Moreover, assume:

(a) The family $\{K_n\}$ $(n \geq 1)$ is pointwise convergent to K on D, i.e.,

$$K_n(x) \to K(x) \text{ as } n \to (x) \text{ as } n \to \infty, \ x \in D.$$

(b) The family $\{K_n\}$ $(n \geq 1)$ has continuous first and second derivatives on $\bar{U}(x_0, r)$.

(c) The linear operator $I - K'(x^*)$ is nonsingular.

Then show there exists a constant r^*, $0 < r^* \leq r$ such that for all sufficiently large n, equation $x - K_n(x) = 0$ has a unique solution $x_n \in \bar{U}(x^*, r^*)$ and $\lim x_n = x^*$ as $n \to \infty$.

47. Let X be a regular partially ordered Banach space. Denote the order by $\leq$ and consider the iteration

$$x_{n+1} = x_n - \frac{F(x_n)}{c_1} \quad c_1 > 0$$

for approximating a solution x^* of equation $F(x) = 0$ in X. Assume that there exist positive numbers c_1 and c_2 such that

$$c_2(x - y) \leq F(x) - F(y) \leq c_1(x - y) \quad \text{for all } x \leq y$$

and

$$\|F(x_0)\| \leq \frac{c_2 r}{2} \quad \text{for some fixed } r > 0.$$

Then show that:

(i) The sequence $\{x_n\}$ $(n \geq 0)$ converges to a solution x^* of the equation $F(x) = 0$.

(ii) The following estimates are true:

$$\|x_n - x^*\| \leq \frac{c}{c_2} \|F(x_0)\| c_3^n,$$

and

$$\|x_n - x^*\| \le cc_3 \|x_n - x_{n-1}\| \,,$$

where $c_3 = \frac{c_1 - c_2}{c_1}$ and c is such that $\|x\| \le c\|y\|$ whenever $0 \le x \le y$.

(iii)   The sequence $\{x_n\}$ $(n \ge 0)$ belongs to the set $\{x \in X \mid x < x_0, \|x - x_0\| \le r\}$ if $0 < F(x_0)$, or the set $\{x \in X \mid x_0 < x, \|x - x_0\| \le r\}$ if $F(x_0) < 0$.

48.   Let $F: D \subset \mathbf{R}^n \to \mathbf{R}^n$ be continuous and G-differentiable on D. We consider $\mathbf{R}^n$ endowed with the natural component-wise partial ordering. Suppose that there exist $v, w \in D$ such that

$$v \le w, \quad \langle v, w \rangle \subset D, \quad R(v) \le 0 \le R(w)$$

and

$$F(y) - F(x) \ge F'(x)(y - x) \text{ for all comparable } x, y \in \langle v, w \rangle \,,$$

$F'(x)^{-1}$ exists and is nonnegative for all $x \in \langle v, w \rangle$. Then show that

(i)   The Newton iterates

$$y_{n+1} = y_n - F'(y_n)^{-1} F(y_n) \quad (n \ge 0)$$

are monotonically decreasing and converge to $y^* \in \langle v, w \rangle$ as $n \to \infty$.

(ii)   If F' is either continuous at y^* or isotone on $\langle v, w \rangle$ then y^* is the unique solution of the equation $F(x) = 0$ in $\langle v, w \rangle$.

(iii)   If F' is isotone on $\langle v, w \rangle$ then the sequence

$$x_{n+1} = x_n - F'(y_n)^{-1} F(x_n) \quad (n \ge 0)$$

is monotonically increasing and converges to y^* as $n \to \infty$.

49.   As in Exercise 48, let $F: D \subseteq \mathbf{R}^n \to \mathbf{R}^n$ and suppose that the G-derivative $F'(x)$ exists and is nonsingular at each point of a compact subset D_0 of D. Set $A = \{x \in D_0 \mid F'(x)^{-1} F(x) \ge 0\}$ and assume that

$$F'(x)^{-1}[F(x) - F(y) - F'(y)(x - y)] \ge 0$$

$$x - F'(x)^{-1} F(x) \in D_0$$

whenever $x \in D_0$, $y \in A$ and $x \le y$. Then show that:

(i) For any $y_0 \in A$ the Newton iteration

$$y_{n+1} = y_n - F'(y_n)^{-1} F(y_n) \quad (n \geq 0)$$

is monotonically decreasing and converges to $y^* \in D_0$ as $n \to \infty$.

(ii) If F and F' are continuous at y^* then

$$F(y^*) = 0.$$

50. Show the claims made in Remark 6.3.

51. Show part (iv) of Proposition 6.2.

52. Show estimate (6.55) in Theorem 6.5.

53. Prove Theorem 6.8.

54. Prove Theorem 6.7.

55. Prove Proposition 6.5.

56. Prove Theorem 6.10.

57. Prove Theorem 6.11.

58. Prove Theorem 6.12.

59. Prove Theorem 6.13.

60. Fill the computational details in Example 6.5.

61. Prove Lemma 6.1.

62. Prove Lemma 6.2.

63. Prove Lemma 6.3.

64. Prove Lemma 6.4.

65. Prove Lemma 6.5.

66. Fill the details in the proof of uniqueness for Theorem 6.18.

67. Prove the statements made in Remark 6.19.

68. Prove the statements made in Remark 6.20.

69. Prove Theorem 6.20.

70. Prove estimates (6.296)–(6.299).

71. Prove Lemma 6.6.

72. Prove Lemma 6.7.

73.   Prove Lemma 6.8.

74.   Show the uniqueness part in Theorem 6.21.

75.   Prove Proposition 6.6.

76.   Prove the statements made in Remarks 6.21.

77.   Show (6.312).

78.   Let $F: D \subseteq X \to Y$ and let D be an open set. Assume:

(a)   the divided difference $[x, y]$ of F satisfies

$$[x, y](y - x) = F(y) - F(x) \quad \text{for all } x, y \in D .$$

$$\|[x, y] - [y, u]\| \leq I_1 \|x - y\|^p + I_2 \|x - y\|^p + I_2 \|y - u\|^p$$

for all $x, y, u \in D$ where $I_1 \geq 0$, $I_2 \geq 0$ are constants that do not depend on x, y and u, while $p \in (0, 1]$;

(b) $x^* \in D$ is a simple solution of Equation (6.300);

(c)   there exists $\varepsilon > 0$, $b > 0$ such that $\|[x, y]^{-1}\| \leq b$ for every $x, y \in U(x^*, \varepsilon)$;

(d)   there exists a convex set $D_0 \subset D$ such that $x^* \in D_0$, and there exists $\varepsilon_1 > 0$, with $0 < \varepsilon_1 < \varepsilon$ such that $F'(\bullet) \in H_{D_0}(c, p)$ for every $x, y \in D_0$ and $U(x^*, \varepsilon_1) \subset D_0$.

Let $r > 0$ be such that:

$$0 < r < \min \left\{ \varepsilon_1, (q(p))^{-1/p} \right\}$$

where

$$q(p) = \frac{b}{p + 1} \left[2^p (I_1 + I_2)(1 + p) + c \right] .$$

Then

(i)   if $x_0, x_1 \in \bar{U}(x^*, r)$, the secant iterates are well defined, remain in $\bar{U}(x^*, r)$ for all $n \geq 0$, and converge to the unique solution x^* of Equation (6.1) in $\bar{U}(x^*, r)$. Moreover, the following estimation:

$$\|x_{n+1} - x^*\| \leq \gamma_1 \|x_{n-1} - x^*\|^p \|x_n - x^*\| + \gamma_2 \|x_n - x^*\|^{1+p}$$

holds for sufficiently large n, where

$$\gamma_1 = b(I_1 + I_2) 2^p \quad \text{and} \quad \gamma_2 = \frac{bc}{1 + p} .$$

(ii) If the above conditions hold with the difference that x_0 and x_1 are chosen such that:

$$\|x^* - x_0\| \le ad_0 ;$$

$$\|x^* - x_1\| \le \min \left\{ ad_0^{t_1}, \|x^* - x_0\| \right\} ,$$

where $0 < d_0 < 1$, $a = (q(p))$, while t_1 is the positive root of the equation:

$$t^2 - t - p = 0 ,$$

then show that for every $n \in N$, $x_n \in U = \{ x \in X \mid \|x - x^*\| < a \}$ and

$$\|x_{n+1} - x^*\| \le ad_0^{t_1^{n+1}} \quad (n \ge 0) .$$

79. Let $F: D \subseteq X \to Y$ and let D be an open set. Assume:

(a) $x_0 \in X$ is fixed, and consider the nonnegative real numbers: $B, v, w, p \in (0, 1]$, $\alpha, \beta, q \ge 1$, I_1, I_2, and I_3, where

$$w = B\alpha \left(I_1 B^p + I_2 \beta^p + I_3 B^p \alpha^p \|F(x_0)\|^{p(q-1)} \right)$$

and

$$v = w^{\frac{1}{p+q-1}} \|F(x_0)\| .$$

Denote $r = \max\{B, \beta\}$ and suppose $\bar{U}(x_0, r^*) \subseteq D$, where

$$r^* = \frac{rv}{w^{\frac{1}{p+q-1}}(1 - v^{p+q-1})} .$$

(b) Condition (a) of the previous exercise holds with the last I_2 replaced by I_3;

(c) for every $x, y \in \bar{U}(x_0, r^*)$, $[x, y]^{-1}$ exists, and $\|[x, y]^{-1}\| \le B$;

(d) for every $x \in \bar{U}(x_0, r^*)$, $\|F(g(x))\| \le \alpha \|F(x)\|^q$ where $g: X \to Y$ is an operator having at least one fixed point which coincides with the solution x^* of Equation (6.1);

(e) for every $x \in \bar{U}(x_0, r^*)$, $\|x - g(x)\| \le \beta \|F(x)\|$;

(f) the number v is such that: $0 < v < 1$.

Then show that the Steffensen-type method

$$x_{n+1} = x_n - [x_n, g(x_n)]^{-1} F(x_n) \quad (n \ge 0)$$

is well defined, remains in $\bar{U}(x_0, r^*)$ for all $n \geq 0$ and converges to a solution x^* of Equation (6.1) with

$$\|x^* - x_n\| \leq \frac{rw^{(p+q)^n}}{w^{\frac{1}{p+q-1}} \left(1 - v^{p+q-1}\right)} \quad (n \geq 0).$$

80. Let us consider the iteration

$$x_{n+1} = x_n - F'(x_n)^{-1} (F(x_n) + G(x_n)) \quad (n \geq 0).$$

Assume:

(a) the following conditions are satisfied:

$$\left\| F'(x_0)^{-1} (F'(x_1) - F'(x_2)) \right\| \leq k(r) \|x_1 - x_2\|$$

and

$$\left\| F'(x_0)^{-1} (G(x_1) - G(x_2)) \right\| \leq \varepsilon(r) \|x_1 - x_2\| .$$

for all $x_1, x_2 \in \bar{U}(x_0, r), 0 < r \leq R$ and $R > 0$ fixed.

(b) the function $\chi(r) = \varphi(r) + \psi(r)$ has a unique zero r^* in the interval $[0, R]$ and $\chi(R) \leq 0$, where $\varphi(r) = a + \int_0^r w(t)dt - r$, $\psi(r) = \int_0^r \varepsilon(t)dt$, $w(r) = \int_0^r k(t)dt$, and $a = \|x_1 - x_0\| > 0$.

Then show:

(i)   The iteration $\{x_n\}$ $(n \geq 0)$ is well defined, remains in $\bar{U}(x_0, r^*)$ for all $n \geq 0$, and converges to a solution x^* of the equation $F(x) + G(x) = 0$, which is unique in $\bar{U}(x_0, R)$.

(ii) The following statements are true:

$$\|x_{n+1} - x_n\| \leq r_{n+1} - r_n, \quad \|x^* - x_n\| \leq r^* - r_n \quad (n \geq 0)$$

where the sequence $\{r_n\}$ $(n \geq 0)$ is monotonically increasing, converges to r^* and is given by

$$r_{n+1} = r_n - \frac{\chi(r_n)}{\varphi'(r_n)} r_0 = 0 \quad (n \geq 0) .$$

81. Consider the equation $F(x) + G(x) = 0$ and the iteration $x_{n+1} = x_n - A(x_n)^{-1}(F(x_n) + G(x_n))$ $(n \geq 0)$

Assume:

(a) $\|A(x_n)^{-1}(A(x) - A(x_n))\| \leq v_n(r) + b_n,$

$$\left\| A(x_n)^{-1} \left(F'(x + t(y - x)) - A(x) \right) \right\|$$

$$\leq w_n(r + t\|y - x\|) - v_n(r) + c_n$$

and

$$\left\| A(x_n)^{-1} (G(x) - G(y)) \right\| \leq e_n(r)\|x - y\|$$

for all $x_n, x, y \in \bar{U}(x_0, r) \subseteq \bar{U}(x_0, R)$, $t \in [0, 1]$, where $w_n(r + t) - v_n(r) \, t \geq 0$ and $e_n(r)$ $(n \geq 0)$ are nondecreasing nonnegative functions with $w_n(0) = v_n(0) = e_n(0) = 0$ $(n \geq 0)$, $v_n(r)$ are differentiable, $v'_n(r) > 0$ $(n \geq 0)$ for all $r \in [0, R]$, and the constants b_n, c_n satisfy $b_n \geq 0, c_n \geq 0$ and $b_n + c_n < 1$ for all $n \geq 0$. Introduce for all $n, i \geq 0$, $a_n = \|A(x_n)^{-1}(F(x_n) + G(x_n))\|$ $\varphi_{n,i}(r) = a_i - r + c_{n,i} \int_0^r w_n(t)dt$, $z_n(r) = 1 - v_n(r) - b_n$, $\psi_n(r) = c_{n,i} \int_0^r e_n(t)dt$, $c_{n,i} = z_n(r_i)^{-1}$, $h_{n,i}(r) = \varphi_{n,i}(r) + \psi_{n,i}(r)$, $r_n = \|x_n - x_0\|$, $a_n = \|x_{n+1} - x_n\|$, the equations

$$r = a_n + c_{0,n}$$

$$\left(\int_0^r (w_0(r_n + t) + e_n(r_n + t))\, dt + (b_n + c_n - 1)r \right) \quad (6.321)$$

$$r = a_n + c_{n,n}$$

$$\left(\int_0^r (w_n(r_n + t) + e_n(r_n + t))\, dt + (b_n + c_n - 1)r \right) \quad (6.322)$$

$$a_n = r + c_{0,n}$$

$$\left(\int_0^r (w_0(r_n + t) + e_n(r_n + t))\, dt + (b_n + c_n - 1)r \right) \quad (6.323)$$

$$a_n = r + c_{n,n}$$

$$\left(\int_0^r (w_n(r + t) + e_n(r_n + t))\, dt + (b_n + c_n - 1)r \right) \quad (6.324)$$

and the scalar iterations

$$s_{0,n} = s_{n,n}^0 = 0, \quad s_{k+1,n}^0 = s_{k,n}^0 + \frac{h_{0,n}\left(s_{k,n}^0 + r_n\right)}{c_{0,n} z_0 \left(s_{k,n}^0 + r_n\right)} \quad (k \geq 0)$$

$$s_{k+1,n} = s_{k,n} + \frac{h_{n,n}\left(s_{k,n} + r_n\right)}{c_{n,n} z_{n,n}\left(s_{k,n} + r_n\right)} \quad (k \geq n) .$$

(b) The function $h_{0,0}(r)$ has a unique zero s_0^* in the interval $[0, R]$ and $h_{0,0}(R) \leq 0$;

(c) The following estimates are true:

$$\frac{h_{n,n}\left(r + r_n\right)}{c_{n,n} z_{n,n}\left(r + r_n\right)} \leq \frac{h_{0,n}\left(r + r_n\right)}{c_{0,n} z_n\left(r + r_n\right)}$$

for all $r \in [0, R - r_n]$ and for each fixed $n \geq 0$.

Then show:

(i) the scalar iterations $\{s_{k+1,n}^0\}$ and $\{s_{k+1,n}\}$ for $k \geq 0$ are monotonically increasing and converge to s_n^* and s_n^{**} for each fixed $n \geq 0$, which are the unique solutions of Equations (6.321) and (6.322) in $[0, R - s_n]$, respectively, with $s_n^{**} \leq s_n^*$ ($n \geq 0$);

(ii) the iteration $\{x_n\}$ is well defined, remains in $\bar{U}(x_0, s^*)$ for all $n \geq 0$ and converges to a solution x^* of the equation $F(x) + G(x) = 0$, which is unique in $\bar{U}(x_0, R)$;

(iii) the following estimates are true:

$$\|x_{n+1} - x_n\| \leq s_{n+1,n+1} - s_{n,n} \leq s_{n+1,n+1}^0 - s_{n,n}^0 ,$$

$$\|x^* - x_n\| \leq s_n^{**} - s_{n,n} \leq s_n^* - s_{n,n}^0 \leq s_0^* - s_{n,0}^0$$

$$\|x^* - x_n\| \geq I_n^*, \quad \|x^* - x_n\| \geq I_n^{**}$$

and

$$I_n^{**} \leq I_n^* \quad (n \geq 0)$$

where I_n^* and I_n^{**} are the solutions of the equations (6.323) and (6.324) respectively for all $n \geq 0$;

(iv)   Define the sequence $\{s_n^1\}$ $(n \geq 0)$ by

$$s_0^1 = 0, \quad s_{n+1}^1 = s_n^1 + \frac{h_{n,n}\left(s_n^1 + r_n\right)}{c_{n,n}} \quad (n \geq 0).$$

Then under the hypotheses (a), (b), and (c) above show that:

$$\|x_{n+1} - x_n\| \leq s_{n+1}^1 - s_n^1 \leq s_{n+1,n+1} - s_{n,n} \leq s_{n+1,n+1}^0 - s_{n,n}^0$$

and

$$\left\|x^* - x_n\right\| \leq t^* - s_n^1 \leq s_n^{**} - s_{n,n} \leq s_n^* - s_{n,n}^0 \leq s_0^* - s_{n,0}^0 \quad (n \geq 0),$$

where

$$t^* = \lim_{n \to \infty} s_n^1.$$

82.   Consider the Newton-like method (6.2). Let $A: D \to L(X, Y)$, $x_0 \in D$, $M_{-1} \in L(X, Y)$, $X \subset Y$, $L_{-1} \in L(X, X)$. For $n \geq 0$ choose $N_n \in L(X, X)$ and define $M_n = M_{n-1} N_n + A(x_n) L_{n-1}$, $L_n = L_{n-1} + L_{n-1} N_n$, $x_{n+1} = x_n + L_n(y_n)$, y_n being a solution of $M_n(y_n) = -[F(x_n) + z_n]$ for a suitable $z_n \in y$.
Assume:

(a) F is Fréchet-differentiable on D;

(b)   there exist nonnegative numbers α, α_0 and nondecreasing functions $w, w_0: \mathbf{R}^+ \to \mathbf{R}^+$ with $w(0) = w_0(0) = 0$ such that

$$\|F(x_0)\| \leq \alpha_0, \quad \|R_0(y_0)\| \leq \alpha, \quad \|A(x) - A(x_0)\| \leq w_0(\|x - x_0\|)$$

and

$$\|F'(x + t(y - x)) - A(x)\| \leq w(\|x - x_0\| + t\|x - y\|)$$

for all $x, y \in \bar{U}(x_0, R)$ and $t \in [0, 1]$.

(c)   Let M_{-1} and L_{-1} be such that M_{-1} is invertible,

$$\left\|M_{-1}^{-1}\right\| \leq \beta, \quad \|L_{-1}\| \leq \gamma \quad \text{and} \quad \|M_{-1} - A(x_0)L_{-1}\| \leq \delta.$$

(d)   There exist nonnegative sequences $\{a_n\}, \{\bar{a}_n\}, \{b_n\}$ and $\{c_n\}$ such that for all $n \geq 0$

$$\|N_n\| \leq a_n, \quad \|I + N_n\| \leq \bar{a}_n, \quad \left\|M_{-1}^{-1}\right\| \cdot \|M_{-1} - M_n\| \leq b_n < 1$$

and

$$\|z_n\| \le c_n \|F(x_n)\| .$$

(e)   The scalar sequence $\{t_n\}$ $(n \ge 0)$ given by

$$t_{n+1} = t_{n+1} + e_{n+1}d_{n+1}(1 + c_{n+1})\left[I_n + \sum_{i=1}^{n} h_i w(t_i)(t_i - t_{i-1}) \right.$$

$$\left. + w(t_{n+1})(t_{n+1} - t_n) \right] \cdot (n \ge 0), \ t_0 = 0, \ t_1 = \alpha$$

is bounded above by a t_0^* with $0 < t_0^* \le R$, where

$$e_0 = \gamma \bar{a}_0, \ e_n = I_{n-1}\bar{a}_n \ (n \ge 1), \ d_n = \frac{\beta}{1 - d_n} \ (n \ge 0)$$

$$I_n = \varepsilon_n \varepsilon_{n-1} \ldots \varepsilon_0 \alpha_0 \ (n \ge 0) \quad \varepsilon_n = p_n d_n (1 + c_n) + c_n ,$$

$$p_n = q_{n-1} a_n \ (n \ge 1), \ p_0 = \delta a_0, \ q_n = p_n + w_0 (t_{n+1}) e_n \ (n \ge 1)$$

and

$$h_i = \prod_{m=i}^{n} \varepsilon_m \quad (i \le n) .$$

(f)   The following estimate is true $\varepsilon_n \le \varepsilon < 1 \ (n \ge 0)$.

Then show:

(i)   The scalar sequence $\{t_n\}$ $(n \ge 0)$ is nondecreasing and converges to a t^* with $0 < t^* \le t_0^*$ as $n \to \infty$.

(ii)   The Newton-like method (6.2) is well defined, remains in $\bar{U}(x_0, t^*)$ and converges to a solution x^* of Equation (6.1).

(iii)   The following estimates are true:

$$\|x_{n+1} - x_n\| \le t_{n+1} - t_n$$

and

$$\|x_n - x^*\| \le t^* - t_n \ (n \ge 0) .$$

For Exercises 83–91 we will need the definitions.

Let $\{x_n\}$ $(n \geq 0)$ be a convergent sequence with limit x^* in a metric space (X, d). Then the quantities

$$Q_\tau \{x_n\} = \begin{cases} 0, & \text{if } x_n = x^*, \text{ for all but finitely many } n, \\ \lim \sup_{n \to \infty} \frac{\|x_{n+1} - x^*\|}{\|x_n - x^*\|^\tau}, & \text{if } x_n \neq x^*, \text{ for all but finitely many } n, \\ +\infty, & \text{otherwise}, \end{cases}$$

defined for all $\tau \in [1, \infty)$, are the quotient convergence factors, or Q-factors for short, of $\{x_n\}$ with respect to the norm $\| \cdot \|$ on X. Let $C(P, x^*)$ denote the set of all sequences with limit x^* generated by an iterative process P. Then

$$Q_\tau (P, x^*) = \sup \{ Q_\tau \{x_n\} \mid \{x_n\} \in C(P, x^*) \}, \quad 1 \leq \tau < \infty$$

are the Q-factors of P at x^* with respect to the norm in which the $Q_\tau\{x_n\}$ are computed.

Moreover, the numbers

$$R_\tau \{x_n\} = \begin{cases} \lim \sup_{n \to \infty} \|x_n - x^*\|, & \text{if } \tau = 1 \\ \lim \sup_{n \to \infty} \|x_n - x^*\|^{1/\tau_n} & \text{if } \tau > 1, \end{cases}$$

are the root convergence factors, or R-factors, for short, of the sequence. Furthermore,

$$R_\tau (P, x^*) = \sup \{ R_\tau (x_n) \mid \{x_n\} \in C(P, x^*) \}, \quad 1 \leq \tau < \infty$$

are the R-factors of P at x^* with respect to the norm in which the $R_\tau\{x_n\}$ are computed.

83. Let $Q_\tau(P, x^*)$, $\tau \in [1, \infty)$ denote the Q-factors of an iterative process at x^* in some fixed norm on $\mathbf{R}^n$. Then exactly one of the following conditions holds:

   (i) $Q_\tau(P, x^*) = 0$, for all $\tau \in [1, \infty)$;

   (ii) $Q_\tau(P, x^*) = \infty$, for all $\tau \in [1, \infty)$;

   (iii) There exists a $\tau_0 \in [1, \infty)$ such that $Q_\tau(P, x^*) = 0$, for all $\tau \in [1, \tau_0)$, and $Q_\tau(P, x^*) = \infty$, for all $\tau \in (\tau_0, \infty)$.

84. Let $\{x_n\}$, $\{y_n\}$ $(n \geq 0)$ be real sequences which converge to some x^*, and let $\| \cdot \|$ be any norm on $\mathbf{R}^n$. Define the "scaled" norm $\| \cdot \|' = c\| \cdot \|$, where $c > 0$, and denote by Q_τ, Q_τ^1 the Q-factors with respect to $\| \cdot \|$ and $\| \cdot \|'$. Prove

   (i)   If $0 < Q_\tau\{x_n\} < \infty$ for $z > 1$, then $Q_\tau^1\{x_n\} = Q_\tau\{x_n\}$ if and only if $c = 1$.

(ii)   There is a τ such that $Q_\tau\{x_n\} < Q_\tau\{y_n\}$ if and only if $Q_\tau^1\{x_n\} < Q_\tau^1\{y_n\}$, that is, the Q-factor relation is invariant under scaling of the norm.

85.   Compute the Q-factors and the corresponding orders of the following real sequences:

(a) $x_n = 2^{-\tau^n}$;

(b) $x_n = 22^{-t^n}$;

(c) $x_n = c^{-n^2}$, with $c > 1$.

86.   Let $\{x_n\}$ ($n \geq 0$) be a real convergent sequence, and, for some $n' \geq 1$, define the sequence $y_n = x^{n+n'}$ ($n \geq 0$). Show that, in any norm, $Q_\tau\{x_n\} = Q_\tau\{y_n\}$ for all $\tau \in [1, \infty)$.

87.   Compute the R-orders for the sequences in Exercise 85.

88.   Let $\{x_n\}$, $\{y_n\}$ ($n \geq 0$) be as defined in Exercise 86. Show that

(i) $R_1\{x_n\} = R_1\{y_n\}$;

(ii) $R_\tau\{y_n\} = [R_\tau\{x_n\}]^{\tau^{n'}}$, for $\tau > 1$.

89.   Let $\{x_n\}$ ($n \geq 0$) be a real convergent sequence with limit x^*, and suppose that

$$\lim_{n\to\infty} \sup \left(\|x_{n+r} - x^*\| / \|x_n - x^*\|\right) = b_r < 1$$

for some integer $r \geq 1$. If $\|\cdot\|'$ is another norm on $\mathbf{R}^n$ such that

$$c\|x\| \leq \|x\|' \leq d\|x\|, \quad \text{for all } x \in \mathbf{R}^n, \ 0 < c \leq d.$$

Show that

$$\lim_{n\to\infty} \sup \left(\|x_{n+mr} - x^*\| / \|x_n - x^*\|\right) \leq \frac{d}{c}b^m \quad (m \geq 1)$$

for some $b < 1$.

90.   Let b_r be as defined in Exercise 89. Show that $R_1\{x_n\} < 1$, $\tau = 1$.

91.   Suppose that the real sequence $\{x_n\}$ ($n \geq 0$) satisfies

$$\|x_{n+1} - x^*\| = c_n \|x_n - x^*\| \quad (n \geq 0)$$

where $\lim_{n\to\infty} c_n = c \in (0, 1)$. Show that $\lim_{n\to\infty} x_n = x^*$ and that $Q_1\{x_n\} = R_1\{x_n\} = c$.

92. Let F be an operator whose domain and range are subsets of a Banach space X. Assume:

   (a) $x^* \in X$ is a fixed point of F;

   (b) the operator F is Fréchet-differentiable at x^*; and

   (c) the spectral radius of the derivative of F at x^* is less than one.

   Then show that there exists a neighborhood U of x^* such that

   $$\lim_{n \to \infty} F(x_n) = x^* ,$$

   for each $x_0 \in U$.

93. Consider the problem of approximating a solution x^* of the equation $f(x) = 0$ in the complex plane by the iteration defined by

   $$y_n = x_n + \alpha f(x_n)$$

   $$z_n = y_n + \frac{\alpha f(y_n)}{1 - f(y_n)/f(x_n)}$$

   and

   $$x_{n+1} = z_n + \frac{\alpha f(z_n)}{1 - f(y_n)/f(x_n)}, \quad \alpha > 0 \ (n \geq 0) \ x_0 \in X \text{ given.}$$

   Show: Let $f: D \subset C \to C$ where C is the complex space and D is a convex open domain. Assume that f has 2nd order continuous derivatives on D, and satisfies:

   $$|f''(x)| \leq k \text{ for all } x \in D, \quad \left| \frac{1}{\alpha} + f'(x_0) \right| \leq \frac{1}{\alpha} - \frac{1}{\beta} ,$$

   $$|y_0 - x_0| \leq \eta, \quad h = k\beta\eta \leq \frac{1}{2} ,$$

   $$\bar{U}(t^* - \eta) \subset D, \quad t^* = \frac{1 - \sqrt{1 - 2h}}{h} \eta .$$

   Define also the real function

   $$g(t) = \frac{K}{2} t^2 - \frac{1}{\beta} t + \frac{\eta}{\beta} ,$$

and the iterations

$$s_n = t_n + \alpha f(t_n)$$

$$u_n = s_n + \frac{\alpha f(s_n)}{1 - f(s_n)/f(t_n)}$$

and

$$t_{n+1} = u_n + \frac{\alpha f(u_n)}{1 - f(s_n)/f(t_n)}, \quad t_0 = 0.$$

Then the sequence $\{x_n\}$ $(n \geq 0)$ is well defined, remains in $\tilde{U}(z_0, t^* - \eta)$ and converges to a solution x^* of equation $f(x) = 0$. Moreover the following are true:

$$|x_n - x^*| \leq t^* - t_n$$

$$|y_n - x^*| \leq t^* - s_n$$

$$|z_n - x^*| \leq t^* - u_n$$

$$|f(x_{n+1})| \leq g(t_{n+1})$$

and

$$|x_n - x^*| \leq t^* - t_n < \frac{(1 - \theta^2)\,\eta}{1 - \theta^{3n}} \theta^{3^n - 1}$$

for all $n \geq 0$, where

$$\theta = \frac{1 - \sqrt{1 - 2h}}{1 + \sqrt{1 - 2h}}.$$

94. Consider the problem of approximating a multiple root x^* with multiplicity $m \geq 1$ of the real equation $f(x) = 0$. Assume that f has derivatives as high as we desire. Show that

(i) the iterative function

$$h(x) = g(x) \tag{6.325}$$

$$- \frac{f(x)(g(x) - x)}{a_1 f'(g(x))(g(x) - x) + a_2 f(x) + a_3 f'(x)(g(x) - x)}$$

is convergent with order three, where

$$g(x) = x - a\frac{f(x)}{f'(x)} ,$$

$$b = g'(x^*) = 1 - \frac{a}{m} ,$$

$$a_1 = -1/mb^m \left[(m+1)b^2 - 2mb + m - 1\right]$$

$$a_2 = (b-1)\left[(m+1)b^2 - 3mb - b + 2m\right]/b$$

$$\left[(m+1)b^2 - 2mb + m - 1\right]$$

and

$$a_3 = [(m+1)b - m]/mb\left[(m+1)b^2 - 2mb + m - 1\right] .$$

(ii) Choose $a = 1$ or $a = \frac{m}{m+1}$ and consider comparing the resulting iterations with Newton's or the modified Newton method on the example

$$f(x) = (x^*x - 1)(x^*x - 1), \quad m = 2, \quad x^* = 1 .$$

Then verify the results:

Table 6.7 Modified Newton method

$a = 1$	$a = \frac{m}{m+1}$ (6.325)	(N)	(NM)
$x_0 = .8$	$x_0 = .8$	$x_0 = .8$	$x_0 = .8$
$x_1 = .9995$	$x_1 = 1.0031$	$x_1 = .9125$	$x_1 = 1.025$

95. Let $F: D \subseteq \mathbf{R} \to \mathbf{R}$ be a function with continuous derivatives of second order on an open interval D, and let $x_0 \in D$ be fixed. Assume:

$$\|F''(x)\| \leq K \text{ for all } x \in D, \quad -\frac{1}{\alpha} \leq F'(x_0) \leq -\frac{1}{\beta} ,$$

$$|y_0 - x_0| \le \frac{\eta}{\beta}, \ \bar{U}(y_0, r_1 - \eta) \subseteq D, \ h = K\beta\eta \le \frac{1}{2}.$$

Consider the iterations

$$y_n = x_n + \alpha F(x_n), \ x_{n+1} = y_n + \frac{\alpha F(y_n)}{1 - \frac{F(y_n)}{F(x_n)}} \quad (n \ge 0)$$

$$s_n = t_n + \alpha\varphi(t_n), \ t_{n+1} = s_n + \frac{\alpha\varphi(s_n)}{1 - \frac{\varphi(s_n)}{\varphi(t_n)}}, \ t_0 = 0 \quad (n \ge 0)$$

where

$$\varphi(t) = \frac{1}{2}kt^2 - \frac{1}{\beta}t + \frac{\eta}{\beta}.$$

Then show:

(i) There exists a unique solution x^* of the equation $F(x) = 0$ in $\bar{U}(y_0, r_1 - \eta)$, where r_1 is the small solution of the equation $\varphi(t) = 0$ (r_2 is the large solution).

(ii) Moreover, we have

$$\left|x_n - x^*\right| \le r_1 - t_n < \frac{(1+\theta)\eta}{b_n}\theta^{2^n - 1}, \ b_n = \sum_{k=0}^{2^n - 1} \theta^k, \ \theta = \frac{r_1}{r_2},$$

$$\left|y_n - x^*\right| \le r_1 - s_n,$$

$$0 = t_0 < s_0 < t_1 < s_1 < \cdots < t_{n-1} < s_{n-1} < \cdots < r_1,$$

and

$$\lim_{n \to \infty} t_n = \lim_{n \to \infty} s_n = r_1, \ \lim_{n \to \infty} x_n = x^*.$$

96. Let $F: D \subseteq \mathbf{R} \to \mathbf{R}$ be a function with continuous derivatives of third order on an open interval D, and $x_0 \in D$ be fixed.
Assume:

$$\left|F'(x_0)^{-1}\right| \le \beta, \ |y_0 - x_0| \le \eta, \ \left|F''(x)\right|$$

$$\le M, \ \left|F'''(x)\right| \le N \text{ for all } x \in D,$$

$$\left(M^3 + \frac{2}{3}\frac{NM}{\beta}\right)^{1/3} \le K, \ h = K\beta\eta \le \frac{1}{2}, \text{ and } 0 \le \alpha \le 2.$$

Consider the iterations

$$y_n = x_n - \frac{F(x_n)}{F'(x_n)}, \quad x_{n+1} = y_n - \frac{1 + \alpha \frac{F(y_n)}{F'(x_n)}}{1 - (a - 2)\frac{F(y_n)}{F'(x_n)}} \quad (n \geq 0)$$

$$s_n = t_n - \frac{\varphi(t_n)}{\varphi'(t_n)}, \quad t_0 = 0$$

$$t_{n+1} = s_n - \frac{\varphi(s_n)}{\varphi'(t_n)} 1 + \alpha \frac{\varphi(s_n)}{\varphi'(t_n)} 1 + (\alpha - 2)\frac{\varphi(s_n)}{\varphi'(t_n)} \quad (n \geq 0)$$

where

$$\varphi(t) = \frac{1}{2}Kt^2 - \frac{1}{\beta}t + \frac{\eta}{\beta}.$$

Then show:

(i)   There exists a unique solution x^* of the equation $F(x) = 0$ in $\bar{U}(x_0, r_1)$, provided that $\bar{U}(x_0, r_1) \subseteq D$, where r_1 is the small solution of the equation $\varphi(t) = 0$ (r_2 is the large solution).

(ii) Moreover, we have

$$\left| y_n - x^* \right| \leq r_1 - s_n, \quad \left| x_{n+1} - x^* \right| \leq r_1 - t_{n+1} \quad (n \geq 0)$$

$$0 = t_0 < s_0 < t_1 < s_1 < \cdots < t_{n-1} < s_{n-1} < \cdots < r_1$$

and

$$\lim_{n \to \infty} t_n = \lim_{n \to \infty} s_n = r_1, \quad \lim_{n \to \infty} x_n = x^*.$$

Furthermore, we have

$$r_1 - t_n(\alpha) \geq \frac{(1 + \theta)}{b_n}\theta^{4^n - 1} = r_1 - t_n(0)$$

where

$$b_n = \sum_{i=0}^{4^n - 1} \theta^i \quad \text{and} \quad \theta = \frac{r_1}{r_2}.$$

(iii) Finally, let us assume that $h = k\beta\eta \leq \frac{2\sqrt[3]{t}}{(1 + \sqrt[3]{5})^2}$. Then we have:

$$\frac{(1 - \theta^2)\,\eta}{1 - \theta^{4^n}}\theta^{4^n - 1} \leq r_1 - t_n(\alpha)$$

$$\leq \frac{(1-\theta^2)\eta}{1-\frac{1}{\sqrt[3]{5}}(\sqrt[3]{5\theta})^{4^n}}(\sqrt[3]{5\theta})^{4^n-1} \qquad (n \geq 0).$$

97. Assume $F \in C^4[a,b]$, $F'(x) \neq 0$, $x^* \in (a,b)$, $F(x^*) = 0$, $H = H(x,y) = F'(x - \frac{2}{3}u) - F'(x)$. Consider also the iterative functions

$$\Phi_\theta(x) = x - u + \frac{3}{4}u\frac{H}{F'(x)}\frac{1 + \frac{\theta H}{F'(x)}}{1 + \frac{\left(\frac{3+\theta}{2}\right)H}{F'(x)}}, \qquad (6.326)$$

$$\Phi(x) = x + \alpha u + \beta u \frac{F'(\varphi(x)) - F'(x)}{F'(x)}$$

$$\frac{1 + \theta \frac{F'(\varphi(x)) - F'(x)}{F'(x)}}{1 + \delta \frac{F'(\varphi(x)) - F'(x)}{F'(x)}},$$

where

$$u = u(x) = \frac{F(x)}{F'(x)} \quad \text{and} \quad \varphi(x) = x - \lambda u.$$

Then show that (6.326) converges to the solution x^* of the equation $F(x) = 0$ with order four if the parameters α, β, λ and θ are chosen to satisfy the equations

$$1 + \alpha = 0, \qquad \frac{\alpha}{2} + \beta\lambda = 0,$$

$$\frac{\beta}{2}(\lambda - 2)\lambda - \frac{\alpha}{3} = 0,$$

and

$$\frac{\alpha}{2} + \beta\lambda(2 - \delta\lambda) + \beta\theta\lambda^2 = 0.$$

98. Let a and b be real numbers such that $b \geq 0$, $0 \leq a < \frac{2}{3}$. We set $a_0 = 1$, $c_0 = 1$, $b_0 = \frac{a}{2}$, $d_0 = \frac{2}{2-a}$, and, for $n \geq 0$, $a_{n+1} = \frac{a_n}{1-aa_nd_n}$, $c_{n+1} = a_{n+1}\left[\frac{b}{6} + (1 - b_n)a_n\frac{a^2}{4}\right]d_n^3$, $b_{n+1} = \frac{a}{2}a_{n+1}c_{n+1}$, $d_{n+1} = \frac{c_{n+1}}{1-b_{n+1}}$, $r_n = d_0 + d_1 + \cdots + d_n$ and $r = \lim_{n\to\infty} r_n$ (if the limit exists!). We denote by $R(a,b) = \{a_n, c_n, b_n, d_n\}$, we say $R(a,b)$ is positive if $a_n \geq 1$ ($n \geq 0$), stable if there exists a constant $M \geq 1$ such that $a_N \leq M$ for all $n \geq 0$, and convergent if there exists $\lim_{n\to\infty} r_n = r$. Prove the result:

Let X and Y be Banach spaces. Let D be an open convex subset of X. Let F be from D into Y, an operator which is twice Fréchet-differentiable on D. Assume:

(a) there exists a constant k_2 such that

$$\|F''(x)\| \leq k_2 \quad \text{for all } x \in D \,;$$

(b) there exists a constant k_3 such that

$$\|F''(x) - F'(y)\| < k_3 \|x - y\| \quad \text{for all } x, y \in D \,.$$

(c) Let $x_0 \in D$ be fixed with $\|F'(x_0)^{-1}\| \leq B$, $\|F'(x_0)^{-1}F(x_0)\| \leq \eta$ and $F(x_0) \neq 0$, such that if we set $a = k_2 B \eta$ and $b = k_3 B \eta^2$, $R(a, b)$ is positive, convergent and $U(x_0, r\eta) \subseteq D$, where $r = \lim_{n \to \infty}(d_0 + \cdots + d_n)$. Then

(i) the Halley iteration

$$x_{n+1} = x_n - (I - T(x_n))^{-1} F'(x_n)^{-1} F(x_n) \,,$$

where

$$T(x) = \frac{1}{2} F'(x)^{-1} F''(x) F'(x)^{-1} F(x) \,,$$

is well defined and lies in $U(x_0, r\eta)$ for all $n \geq 0$. Moreover, the sequence $\{x_n\}$ $(n \geq 0)$ converges to a point $x^* \in \bar{U}(x_0, r\eta)$, such that $F(x^*) = 0$.

(ii) The following estimates are true:

$$\left\| F'(x_n)^{-1} \right\| \leq a_n B, \quad \left\| F'(x_n)^{-1} F(x_n) \right\|$$

$$\leq c_n \eta \, \|T(x_n)\| \leq b_n \,,$$

$$\|x_{n+1} - x_n\| \leq d_n \eta, \quad \left\| x^* - x_{n+1} \right\|$$

$$\leq (r - r_n) \eta = \sum_{k=n+1}^{\infty} d_k \eta \qquad (n \geq 0) \,,$$

and

$$\|x_{n+1} - x_n\| \leq \frac{4M}{3\eta^2} \left(\frac{b}{6} + \frac{5a^2 M}{16} \right) \|x_n - x_{n-1}\|^3 \quad (n \geq 1) \,.$$

(iii)   For all $0 < a \leq \frac{1}{2}$ there exists a second degree polynomial F and a point x_0 such that $R(a, 0)$ produces optimal estimates for the Halley method, i.e., for such F and x_0 the above estimates (excluding the last one) hold as equalities.

(iv)   For all $0 < a \leq \frac{1}{2}$, $R(a, 0)$ is convergent with

$$\sum_{k=0}^{\infty} d_k = \frac{1 - \sqrt{1 - 2a}}{a} .$$

Glossary of Symbols

$\mathbf{R}^n$	real n-dimensional space
$\mathbf{C}^n$	complex n-dimensional space
$X \times Y, X \times X = X^2$	Cartesian product space of X and Y
$e^1, \ldots, e^n$	the coordinate vectors of $\mathbf{R}^n$
$x = (x_1, \ldots, x_n)^T$	column vector with component x_i
x^T	transpose of x
$\{x_n\}_{n \geq 0}$	sequence of points from X
$\| \cdot \|$	norm on X
$\| \|_p$	L_p norm
$\|$	absolute value symbol
$/\!/$	norm symbol of a generalized Banach space X
$\langle x, y \rangle$	set $\{z \in X \mid z = tx + (1 - t)y, t \in [0, 1]\}$
$U(x_0, R)$	open ball $\{z \in X \mid \|x_0 - z\| < R\}$
$\bar{U}(x_0, R)$	closed ball $\{z \in X \mid \|x_0 - z\| \leq R\}$
$U(R) = U(0, R)$	ball centered at the zero element in X and of radius R
$U, \bar{U}$	open, closed balls, respectively no particular reference to X, x_0 or R
$M = \{m_{ij}\}$	matrix $1 \leq i, j \leq n$
M^{-1}	inverse of M
M^+	generalized inverse of M
det M or $|M|$	determinant of M
M^k	kth power of M
rank M	rank of M
I	identity matrix (operator)
L	linear operator
L^{-1}	inverse
null L	null set of L
rad L	radical set of L
$F: D \subseteq X \to Y$	an operator with domain D included in X, and values in Y

$F'(x), F''(x)$	first, second Fréchet-derivatives of F evaluated at x
δ_{ij}	Kronecker delta
$\sum$	summation symbol
$\prod$	product of factors symbol
$\int$	integration symbol
$\in$	element inclusion
$\subset, \subseteq$	set inclusion (strict, non-strict, respectively)
$\forall$	for all
$\Rightarrow$	implies
$\cap, \cup$	union, intersection
$A - B$	difference between sets A and B
$\bar{B}$	mean of a bilinear operator B
$Q(X, Z)$	set of all quadratic operators from X to Z
$Q_F^*(X)$	set of all bounded quadratic operator Q in X such that Q has finite rank
X^{2*}	set of all bounded quadratic functionals
$\oplus$	direct sum
$A^{\#}$	outer inverse of A
$\dim A$	dimension of A
$\text{codim } A$	codimension of A

References

[1] Alefeld, G. "Monotone Regular-Falsi-ähnliche verfahren bei nichtkonvexen operator gleichungen," *Beitrage zür Numerischen Mathematik*, 8 (1979), 39–48.

[2] Allgower, E.L. and Prenter, P.M. "On the branching of solutions of quadratic differential equations," *Aequationes Mathematicae*, 10, 1 (1974), 81–96.

[3] Allgower, E.L., Böhmer, K., Potra, F.A. and Rheinboldt, W.C. "A mesh independence principle for operator equations and their discretizations," *S.I.A.M. J. Num. Anal.* 23, (1) (1986), 160–169.

[4] Altman, M. "Iterative methods of higher order," *Bull. Acad. Polon. Sci. Ser. Math. Astr. Phys.* 9 (1961), 63–68.

[5] Altman, M. "Concerning the method of tangent hyperbolas for operator equations," *Bull. Acad. Polon. Sci. Ser. Math. Astr. Phys.* 9 (1961), 633–637.

[6] Ames, W. *Nonlinear Ordinary Differential Equations in Transport Processes*, Academic Press, New York, 1968.

[7] Ambartsumian, V.A. *Theoretical Astrophysics*, Transl. by J.B. Sykes, Pergamon Press, New York, NY, 1948.

[8] Amson, J.C. "Representations of multilinear and polynomial operators on vector spaces," *J. London Math. Soc.* 2, 4 (1972), 394–400.

[9] Anselone, P.M. and Moore, R.H. "An extension of the Newton-Kantorovich method for solving nonlinear equations with application to elasticity," *J. Math. Anal. Appl.* 13 (1966), 476–501.

[10] Anselone, P.M. *Collectively Compact Operator Approximation Theory*, Prentice-Hall, Englewood Cliffs, NJ, 1971.

[11] Antosik, P. and Schwarz, C. *Matrix Methods in Analysis*, Lecture notes in Mathematics, 1113, Springer-Verlag, 1985.

[12] Appel, J., Massabó, I., Vignoli, A., and Zabrejko, P.P. "Lipschitz and Darbo conditions for the superposition operators in ideal spaces," *Annali Mat. Pura Applic.* 152 (1988), 123–137.

[13] Appel, J. and Zabrejko, P.P. *Nonlinear Superposition Operators*, Cambridge University Press, Cambridge, England, 1990.

[14] Appel, J., Pascale, E., and Zabrejko, P.P. "On the application of the Newton-Kantorovich method to nonlinear equations of Uryson type," *Num. Funct. Anal. Optimiz.* 12 (3 and 4) (1991), 271–283.

[15] Argyros, I.K. "Quadratic equations and applications to Chandrasekhar's and related equations," *Bull. Austral. Math. Soc.* 32 (1985), 275–297.

[16] Argyros, I.K. "On the cardinality of solutions of multilinear differential equations and applications," *Intern. J. Math. Math. Sci.* 9, 4 (1986), 757–766.

[17] Argyros, I.K. "On the approximation of some nonlinear equations," *Aequationes Mathematicae*, 32 (1987), 87–95.

[18] Argyros, I.K. "On polynomial equations in Banach space, perturbation techniques and applications," *Intern. J. Math. Math. Sci.* 10, 1 (1987), 69–78.

[19] Argyros, I.K. "Newton-like methods under mild differentiability conditions with error analysis," *Bull. Austral. Math. Soc.* 37 (1987), 131–147.

[20] Argyros, I.K. "On Newton's method and nondiscrete mathematical induction," *Bull. Austral. Math. Soc.* 38 (1988), 131–140.

[21] Argyros, I.K. "On the secant method and fixed points of nonlinear equations," *Monatschefte für Mathematik*, 106 (1988), 85–94.

[22] Argyros, I.K. "On a class of nonlinear integral equations arising in neutron transport," *Aequationes Mathematicae*, 36 (1988), 99–111.

[23] Argyros, I.K. "On the number of solutions of some integral equations arising in radiative transfer," *Intern. J. Math. Math. Sci.* 12, 2 (1989), 297–304.

[24] Argyros, I.K. "Improved error bounds for a certain class of Newton-like methods," *J. Approx. Th. Applic.* (6:1) (1990), 80–98.

[25] Argyros, I.K. "Error bounds for the modified secant method," *BIT*, 20 (1990), 92–200.

[26] Argyros, I.K. "On some projection methods for the approximation of implicit functions," *Appl. Math. Letters*, 32 (1990), 5–7.

[27] Argyros, I.K. "The Newton-Kantorovich method under mild differentiability conditions and the Pták error estimates," *Monatschefte für Mathematik*, 109, 3 (1990), 110–128.

[28] Argyros, I.K. "The secant method in generalized Banach spaces," *Appl. Math. Comp.* 39 (1990), 111–121.

[29] Argyros, I.K. "A mesh independence principle for operator equations and their discretizations under mild differentiability conditions," *Computing*, 45 (1990), 265–268.

[30] Argyros, I.K. "On the convergence of some projection methods with perturbation," *J. Comp. Appl. Math.*, 36 (1991), 255–258.

[31] Argyros, I.K. "On an application of the Zincenko method to the approximation of implicit functions," *Public. Math. Debrecen*, 39 (3-4) (1991), 1–7.

[32] Argyros, I.K. "On an iterative algorithm for solving nonlinear equations," *Beitrage zür Numerischen Math.* 10 1 (1991), 83–92.

[33] Argyros, I.K. "On a class of quadratic equations with perturbation," *Funct. et Approx. Comm. Math.* XX (1992), 51–63.

[34] Argyros, I.K. "Improved error bounds for the modified secant method," *Intern. J. Computer Math.* 43, (1 + 2) (1992), 99–109.

[35] Argyros, I.K. "Some generalized projection methods for solving operator equations," *J. Comp. Appl. Math.* 39, 1 (1992), 1–6.

[36] Argyros, I.K. "On the convergence of generalized Newton-methods and implicit functions," *J. Comp. Appl. Math.* 43 (1992), 335–342.

[37] Argyros, I.K. "On the convergence of inexact Newton-like methods," *Publ. Math. Debrecen* 42, (1–2) (1992), 1–7.

[38] Argyros, I.K. "On the convergence of a Chebysheff-Halley-type method under Newton-Kantorovich hypothesis," *Appl. Math. Letters* 6, 5 (1993), 71–74.

[39] Argyros, I.K. "Newton-like methods in partially ordered linear spaces," *J. Approx. Th. Applic.* 9, 1 (1993), 1–10.

[40] Argyros, I.K. "On the solution of underdetermined systems of nonlinear equations in Euclidean space," *Pure Math. Appl.* 4, 3 (1993), 199–209.

[41] Argyros, I.K. "A convergence theorem for Newton-like methods under generalized Chen-Yamamoto type assumptions," *Appl. Math. Comput.* 61, 1 (1994), 25–37.

[42] Argyros, I.K. "On the discretization of Newton-like methods," *Internat. J. Computer Math.* 52 (1994), 161–170.

[43] Argyros, I.K. "A unified approach for constructing fast two-step Newton-like methods," *Mh. Math.* 119 (1995), 1–22.

[44] Argyros, I.K. "On the method of tangent hyperbolas," *J. Appr. Th. Appl.* 12, 1 (1996), 78–96.

[45] Argyros, I.K. "On an extension of the mesh-independence principle for operator equations in Banach space," *Appl. Math. Lett.* 9, 3 (1996), 1–7.

[46] Argyros, I.K. "Chebysheff-Halley-like methods in Banach spaces," *Korean J. Comp. Appl. Math.* 4, 1 (1997), 83–107.

[47] Argyros, I.K. "Concerning the convergence of inexact Newton methods," *J. Comp. Appl. Math.* 79 (1997), 235–247.

[48] Argyros, I.K. "A generalization of Edelstein's theorem on fixed points and applications," *Southwest J. Pure Appl. Math.* 2 (1996), 60–64.

[49] Argyros, I.K. "General ways of constructing accelerating Newton-like iterations on partially ordered topological spaces," *Southwest J. Pure Appl. Math.* 2 (1997), 1–12.

[50] Argyros, I.K. "On a new Newton-Mysovskii-type theorem with applications to inexact Newton-like methods and their discretizations," *IMA J. Num. Anal.* (to appear).

[51] Argyros, I.K. "Accessibility of solutions of equations on Banach spaces by Newton-like methods and applications," submitted for publication.

[52] Argyros, I.K. "Improving the order and rates of convergence for the Super-Halley method in Banach spaces," submitted for publication.

[53] Argyros, I.K. "Convergence rates for inexact Newton-like methods of singular points and applications," *Appl. Math. Comp.* (to appear).

[54] Argyros, I.K. "Convergence domains for some iterative processes in Banach spaces using outer and generalized inverses," submitted for publication.

[55] Argyros, I.K. and Chen, D. "On the midpoint method for solving nonlinear operator equations in Banach spaces," *Appl. Math. Letters* 5, 4 (1992), 7–9.

[56] Argyros, I.K. and Chen, D. "A fourth order iterative method in Banach spaces," *Appl. Math. Letters* 6, 4 (1993), 97–98.

[57] Argyros, I.K. and Chen, D. "A note on the Halley method in Banach spaces," *Appl. Math. Comp.* 58 (1993), 215–224.

[58] Argyros, I.K. and Szidarovsky, F. "On the monotone convergence of general Newton-like methods," *Bull. Austral. Math. Soc.* 45 (1992), 489–502.

[59] Argyros, I.K. and Szidarovsky, F. "Convergence of general iteration schemes," *J. Math. Anal. Applic.* 168 (1992), 42–62.

[60] Argyros, I.K. and Szidarovsky, F. *The Theory and Applications of Iteration Methods*, CRC Press, Boca Raton, Florida, 1993.

[61] Argyros, I.K. and Szidarovsky, F. "On the convergence of modified contractions," *J. Comput. Appl. Math.*, 55, 2 (1994), 97–108.

[62] Atkinson, K.E. "The numerical evaluation of fixed points for completely continuous operators," *S.I.A.M. J. Num. Anal.* 10 (1973), 799–807.

[63] Atkinson, K.E. *A Survey of Numerical Methods for the Solution of Fredholm Integral Equations of the Second Kind*, S.I.A.M., Philadelphia, 1976.

[64] Balakrishnan, A.V. *Applied Functional Analysis*, Springer-Verlag, 1976.

[65] Baluev, A.N. "On the abstract theory of Chaplygin's method," *Dokl. Akad. Nauk. SSSR*, 83 (1952), 781–784.

[66] Ben-Israel, A. "A Newton-Raphson method for the solution of systems of operators," *J. Math. Anal. Applic.* 15 (1966), 243–252.

[67] Ben-Israel, A. and Greville, T.N.E. *Generalized Inverses: Theory and Applications*, John Wiley and Sons, 1974.

[68] Boh, E. Monotonie: *Lösbarkeit and Numerik bei Operatorgleichun-gen*, Springer-Verlag, New York, 1974.

[69] Brent, R.P. *Algorithms for Minimization Without Derivatives*, Prentice-Hall, Englewood Cliffs, New Jersey, 1973.

[70] Browder, F.E. and Petryshyn, W.V. "The solution by iteration of linear functional equations in Banach spaces," *Bull. Amer. Math. Soc.* 72 (1966), 566–570.

[71] Brown, P.T. "A local convergence theory for combined inexact-Newton/finite-difference projection methods," *S.I.A.M. J. Numer. Anal.* 24 (1987), 407–434.

[72] Busbridge, L.W. *The Mathematics of Radiative Transfer*, Cambridge University Publ., Cambridge, England, 1960.

[73] Cahlon, B. "Numerical solution of nonlinear Volterra integral equations," *J. Comp. Appl. Math.* 7, 2 (1981), 121–128.

[74] Cahlon, B. and Eskin, M. "Existence theorems for an integral equation of the Chandrasekhar *H*-equation with perturbation," *J. Math. Anal. Applic.* 83 (1981), 159–171.

[75] Candela, V. and Marquina, A. "Recurrence relations for rational cubic methods I: The Halley method," *Computing* 44 (1990), 169–184.

[76] Case, K.M. and Zweifel, P.F. *Linear Transport Theory*, Addison Wesley Publ., Reading, MA, 1967.

[77] Chandrasekhar, S. *Radiative Transfer*, Dover Publ., New York, 1960.

[78] Chen, D. "On the convergence and optimal error estimates of King's iteration for solving nonlinear equations," *Intern. J. Comp. Math.* 26, (3–4) (1989), 229–237.

[79] Chen, D. "Kantorovich-Ostrowski convergence theorems and optimal error bounds for Jarratt's iterative method," *Intern. J. Computer. Math.* 31, (3–4) (1990), 221–235.

[80] Chen, X. and Nashed, M.Z. "Convergence of Newton-like methods for singular operator equations using outer inverses," *Numer. Math.* 66 (1993), 235–257.

[81] Chen, X. and Yamamoto, T. "Convergence domains of certain iterative methods for solving nonlinear equations," *Numer. Funct. Anal. Optimiz.* 10 (1–2) (1989), 37–48.

[82] Chui, C.K. and Quak, F. "Wavelets on a bounded interval." In: *Numerical Methods of Approximation Theory*, Vol. 9 (eds: D. Braess and Larry L. Schumaker, Intern. Ser. Num. Math., Vol. 105), Basel: Birkhäuser Verlag, 1992, 53–75.

[83] Chow, S.N.and Hale, J.K. *Methods of Bifurcation Theory*, Springer-Verlag, New York, 1962.

[84] Collatz, L. *Functional Analysis and Numerisch Mathematik*, Springer-Verlag, New York, 1964.

[85] Cronin, J. "Upper and lower bounds for the number of solutions of nonlinear equations," *J. Math. Anal. Applic.* 29 (1973), 50–61.

[86] Danes, J. "Fixed point theorems, Nemytskii and Uryson operators, and continuity of nonlinear mappings," *Comm. Math. Univ. Carolinae*, 11 (1970), 481–500.

[87] Darbo, G. "Punti uniti in transformationa codominio non compatto," *Rend Sem. Mat. Univ. Padova*, 24 (1955), 84–92.

[88] Daubechies, I. *Ten Lectures in Wavelets*, (Conf. Board Math. Sci. (CBMS): Vol. 61), Philadelphia, SIAM, 1992.

[89] Davis, H.T. *Introduction to Nonlinear Differential and Integral Equations*, Dover Publ., New York, 1962.

[90] Decker, D.W., Keller, H.B. and Kelley, C.T. "Convergence rates of Newton's method at singular points," *SIAM J. Numer. Anal.* 20, 2 (1983), 296–314.

[91] Dembo, R.S., Eisenstat, S.C. and Steinhaug, T. "Inexact Newton methods," *S.I.A.M. J. Num. Anal.* 19 (1982), 400–408.

[92] Dennis, J.E. "Towards a Unified Convergence Theory for Newton-like Methods," in: *Nonlinear Functional Analysis and Applications* (L.B. Rall, Ed.), Academic Press, New York, 1971.

[93] Dennis, J.E. and Schnabel, R.B. *Numerical Methods for Unconstrained Optimization and Nonlinear Equations*, Prentice-Hall, Englewood Cliffs, New Jersey, 1983.

[94] Deuflhard, P. and Heindl, G.A. "Affine invariant convergence theorems for Newton's method, and extensions to related methods," *SIAM J. Numer. Anal.* 16, 1 (1979), 1–10.

[95] Deuflhard, P. and Potra, F.A. "Asymptotic mesh independence of Newton-Galerkin methods and a refined Mysovskii theorem," *SIAM J. Numer. Anal.* 29, 5 (1992), 1395–1412.

[96] Diallo, O.W. *On the Theory of Linear Integro-differential Equations of Barbashin Type in Lebesgue Spaces*, (Russian), VINITI, 1013, 88, Minsk, 1988.

[97] Doring, B. "Iterative lösung gewisser randwertprobleme and integralgleichungen," *Apl. Mat.* 24 (1976), 1–31.

[98] Dunford, N. and Schwartz, J.T. *Linear Operators I*, Int. Publ. Leyden, 1963.

[99] Fox, C. "A solution of Chandrasekhar's integral equation," *Trans. Amer. Math. Soc.* 99 (1961), 285–291.

[100] Friedman, *Partial Differential Equations of Parabolic Type*, Prentice-Hall Publ., Englewood Cliffs, New Jersey, 1964.

[101] Gutierez, J.M., Hernandez, M.A. and Salanova, M.A. "Accessibility of solutions by Newton's method," *Intern. J. Computer. Math.* 57 (1995), 239–247.

[102] Gutierez, J.M., Hernandez, M.A. and Salanova, M.A. "Resolution of quadratic equations in Banach spaces," *Numer. Funct. Anal. Optimiz.* 17 (1 and 2), (1996), 113-121.

[103] Hanson, R.J. and Lawson, C.L. *Solving Least Squares Problems*, Prentice-Hall, Englewood Cliffs, New Jersey, 1974.

[104] Hartman, P. *Ordinary Differential Equations*, Wiley, New York, 1964.

[105] Häubler, W.M. "A Kantorovich-type convergence analysis for the Gauss-Newton method," *Numer. Math.* 48 (1986), 119–125.

[106] Hellinger, E. and Toeplitz, O. *Integralgleichungen and Gleichungen Mit Unendlich Vielen Unbekannten*, Chelsea, New York, 1953.

[107] Hernandez, M.A. "A note on Halley's method," *Num. Math.* 59, 3 (1991), 273–276.

[108] Hernandez, M.A. and Salanova, M.A. "A family of Chebyshev-Halley type methods," *Intern. J. Comp. Math.* 47 (1993), 59–63.

[109] Hille, E. and Phillips, R.S. *Functional Analysis and Semigroups*, Amer. Math. Soc. Coll. Publ., New York, 1957.

[110] Jarrat, P. "Some efficient fourth order multipoint methods for solving equations," *BIT*, 9 (1969), 119–124.

[111] Jerri, A.J. *Introduction to Integral Equations with Applications*, Dekker Publ., New York, 1985.

[112] Kanno, S. "Convergence theorems for the method of tangent hyperbolas," *Math. Japonica* 37, 4 (1992), 711–722.

[113] Kantorovich, L.V. "The method of successive approximation for functional equations," *Acta Math.* 71 (1939), 63–97.

[114] Kantorovich, L.V. and Akilov, G.P. *Functional Analysis in Normed Spaces*, Pergamon Press, New York, 1964.

[115] Kelley, C.T. "Solution of the Chandrasekhar H-equation by Newton's method," *J. Math. Phys.* 21, 7 (1980), 1625–1628.

[116] Kelley, C.T. "Approximation of solutions of some quadratic integral equations in transport theory," *J. Integr. Eq.* 4 (1982), 221–237.

[117] Krasnosel'skii, M.A. *Positive Solution of Operator Equations*, Goz. Izdat. Fiz. Mat. Moscow 1962; Transl. by R. Flaherty and L. Boron, P. Noordhoff, Groningen, 1964.

[118] Krasnosel'skii, M.A. *Topological Methods in the Theory of Nonlinear Integral Equations*, Pergamon Press, London, 1966.

[119] Krasnosel'skii, M.A. *Approximate Solutions of Operator Equations*, Walter Noordhoff Publ. Groningen, 1972.

[120] Krasnosel'skii, M.A., Zabrejko, P.P., Pustyl'nik, E.I., and Sobolevskii, P.E. *Integral Operators in Spaces of Summable Functions* (Russian), Nauka, Moscow, 1966.

[121] Krein, S.F. *Linear Equations in Banach Space*, Birkhauser Publ., Boston, MA, 1982.

[122] Kupsch, J. "Estimates of the unitary integral," *Commun. Math. Phys.* 19 (1960), 65–82.

[123] Kuratowski, C. "Sur les espaces complets," *Fund. Math.* 15 (1930), 301–309.

[124] Kwon, U.K. and Redheffer, R.M. "Remarks on linear equations in Banach space," *Arch. Rational Mech. Anal.* 32 (1969), 247–254.

[125] Lancaster, P. "Error analysis for the Newton-Raphson method," *Num. Math.* 9, 55 (1968), 55–68.

[126] Langer, R.E. "On the zeros of exponential sums and integrals," *Bull. Amer. Math. Soc.* 37 (1931), 213–239.

[127] Leggett, R.W. "On certain nonlinear integral equations," *J. Math. Anal. Applic.* 57 (1977), 462–468.

[128] Liusternik, L.A. and Sobolev, V.I. *Elements of Functional Analysis*, Ungar Publ., 1961.

[129] MacFarland, J.E. "An iterative solution of the quadratic equation in Banach space," *Trans. Amer. Math. Soc.* (1958), 824–830.

[130] Martinez, J.M. "Quasi-Newton methods for solving undetermined nonlinear simultaneous equations," *J. Comp. Appl. Math.* 34 (1991), 171–190.

[131] McCormick, S.F. *A Revised Mesh Refinement Strategy for Newton's Method Applied to Two-point Boundary Value Problems.* Lecture Notes in Mathematics 674, Springer-Verlag, Berlin, 1978, 15–23.

[132] Meyer, P.W. *Die Anwendung Verallgemeinerter Normen zer Fehlerab-schätzung Bei Iteration Sverfahren*, Dissertation, Düsseldort, 1980.

[133] Meyer, P.W. "Das modifizierte Newton-verfahren in verallgemeinerten Banach-Räumen," *Num. Math.* 43 (1984), 91–104.

[134] Meyer, P.W. "Newton's method in generalized Banach spaces," *Numer. Funct. Anal. Optimiz.* 9, (3–4) (1987), 244–259.

[135] Meyer, P.W. "A unifying theorem for Newton's method," *Numer. Funct. Anal. Optimiz.* 13 (5 and 6), (1992), 463–473.

[136] Miel, G.J. "Majorizing sequences and error bounds for iterative methods," *Math. Comp.* 34 (1980), 185–202.

[137] Migovich, F.M. "On the convergence of projection-iterative methods for solving nonlinear operator equations," *Dopov. Akad. Nauk. Ukr. RSR*, Ser. A, 1 (1970), 20–23.

[138] Mirsky, L. *An Introduction to Linear Algebra*, Clarendon Press, Oxford, England, 1955.

[139] Moore, R.H. "Approximation to nonlinear operator equations and Newton's method," *Numer. Math.* 12 (1968), 23–34.

[140] Moret, I. "On the behaviour of approximate Newton-methods," *Computing* 37 (1986), 185–193.

[141] Moret, I. "On a general iterative scheme for Newton-type methods," *Numer. Funct. Anal. Optimiz.* 9, (10–12) (1987–88), 1115–1137.

[142] Moret, I. "A Kantorovich-type theorem for inexact Newton methods," *Numer. Funct. Anal. Optimiz.* 10, (3–4) (1989), 351–365.

[143] Mullikin, T.W. "Some probability distributions for neutron transport in a half-space," *J. Appl. Prob.* 5 (1968), 357–374.

[144] Muroya, Y. "Practical monotonous iterations for nonlinear equations," *Memoirs of the Faculty of Science Kyushu University*, Ser. A, 22, 1 (1968), 56–73.

[145] Muroya, Y. "Left subinverses of matrices and monotonous iterations for nonlinear equations," *Memoirs of the Faculty of Science and Engineering*, Waseda University, 34 (1970), 157–171.

[146] Mysovskii, I. "On the convergence of Newton's method," *Trudy Mat. Inst. Steklov*, 28 (1949), 145–147 (in Russian).

[147] Natanson, I.P. *The Theory of Functions of a Real Variable* (Russian), Goste-hizdat, Moscow, 1957.

[148] Necepurenko, M.T. "On Chebysheff's method for functional equations" (Russian), Usephi, *Mat. Nauk* 9 (1954), 163–170.

[149] Nerekenov, T.K. *Necessary and Sufficient Conditions for Uryson and Ne-mytskii Operators to Satisfy a Lipschitz Condition* (Russian), VINITI 1459, 81, Alma-Ata, 1981.

[150] Nguen, D.F. and Zabrejko, P.P. "The majorant method in the theory of the Newton-Kantorovich approximations and the Pták error estimates," *Numer. Funct. Anal. Optimiz.* 9, (5–6) (1987), 671–686.

[151] Noble, B. *The Numerical Solution of Nonlinear Integral Equations and Related Topics*, University Press, Madison, WI, 1964.

[152] Ojnarov, R. and Otel'baev, M. "A criterion for a Uryson operator to be a contraction" (Russian), *Dokl. Akad. Nauk. SSSR*, 255 (1980), 1316–1318.

[153] Ortega, J.M. and Rheinboldt, W.C. *Iterative Solution of Nonlinear Equations in Several Variables*, Academic Press, New York, 1970.

[154] Ostrowski, A.M. *Solution of Equations in Euclidean and Banach Spaces*, Academic Press, New York, 1973.

[155] Pandian, M.C. "A convergence test and componentwise error estimates for Newton-type methods," *S.I.A.M. J. Num. Anal.* 22 (1985), 779–791.

[156] Pavaloiu, I. "Rezolvarea equatiilor prin interplare," *Dacia Publ. Cluj-Napoca*, Romania, 1981.

[157] Pietsch, A. "Ideals of multilinear functionals," Proc. of the II Intern. Conf. on Operator Ideals and Their Applic. In *Theoretical Physics*, Leipzig, 1983.

[158] Porter, W. "Synthesis of polynomic systems," *SIAM J. Math. Anal.* 11, 2 (1980), 308–315.

[159] Prenter, P. "On polynomial operators and equations," in *Nonlinear Functional Analysis and Applications* (Rall, L. ed.), Academic Press, New York 361–398, 1971.

[160] Prisyazhnyuk, N.M. "The convergence of some approximate methods for solving nonlinear operator equations with nondifferentiable operators," *Ukrainskii Mathematicheskii Zhürnal*, 30, 2 (1978), 261–265.

[161] Potra, F.A. "On the convergence of a class of Newton-like methods," Iterative solution of nonlinear systems of equations, Lecture Notes in Mathematics 953, Springer-Verlag, New York, 1982.

[162] Potra, F.A. "On an iterative algorithm of order 1.839 . . . for solving nonlinear operator equations," *Numer. Funct. Anal. Optimiz.* 7, (1) (1984–85), 75–106.

[163] Potra, F.A. "Newton-like methods with monotone convergence for solving nonlinear operator equations," *Nonlin. Anal. Th. Meth. Appl.*, 11, 6 (1987), 697–717.

[164] Potra, F.A. and Pták, V. "Sharp error bounds for Newton's method," *Numer. Math.* 34 (1980), 63–72.

[165] Potra, F.A. and Pták, V. *Nondiscrete Induction and Iterative Processes*, Pitman Publ., London, 1984.

[166] Pták, V. "The rate of convergence of Newton's process," *Numer. Math.* 25 (1976), 279–285.

[167] Rall, L.B. "Quadratic equations in Banach space," *Rend. Circ. Math. Palermo*, 10 (1961), 314–332.

[168] Rall, L.B. *Computational Solution of Nonlinear Operator Equations*, Wiley, New York, 1968.

[169] Rall, L.B. *Nonlinear Functional Analysis and Applications*, Academic Press, New York, 1971.

[170] Redheffer, R.M. "Remarks on a paper of Taussky," *J. Alg.* 2 (1965), 42–47.

[171] Reddien, G.W. "On Newton's method for singular problems," *SIAM J. Numer. Anal.* 15 (1978), 993–996.

[172] Rockne, J. "Newton's method under mild differentiability conditions with error analysis," *Numer. Math.* 18 (1972), 401–412.

[173] Ruch, D. "On uniformly contractive systems and quadratic equations in Banach space," *Bull. Austral. Math. Soc.* 95 (1995), 441–455.

[174] Ruch, D. "Completely continuous and related multilinear equations," *Rend. Circolo Matematico di Palermo*, 45 (1996).

[175] Ruch, D. "Solving polynomial operator equations in ordered Banach spaces," submitted for publication.

[176] Ruch, D. and Van Fleet, P. "On multipower equations: Some iterative solutions and applications," *Zeitschrift für Analysis und ihre Anwendungen*, 15, 1 (1996), 201–222.

[177] Safiev, R.A. "The method of tangent hyperbolas," *Sov. Math. Dokl.* 4 (1963), 482–485.

[178] Schmidt, J.W. "Monotone einschliessung mit Regula-Falsi bei konvexen functionen," *ZAMM*, 50 (1970), 640–643.

[179] Schmidt, J.W. and Leonhardt, H. "Eingrenzung von losungen mit hilfe der Regular-Falsi," *Computing*, 6 (1970), 318–329.

[180] Schomber, H. "Monotonically convergent iterative methods for nonlinear systems of equations," *Numer. Math.* 32 (1979), 97–104.

[181] Singer, I. *Bases in Banach Spaces I*, Springer-Verlag, New York, 1970.

[182] Slugin, S. "Approximate solution of operator equations by the method of Chaplygin" (Russian), *Dokl. Nauk. SSSR*, 103 (1955), 565–568.

[183] Slugin, S. "Monotonic processes of bilateral approximation in a partially ordered convergence group," *Soviet Math.* 3 (1962), 1547–1551.

[184] Stuart, C.A. "Existence theorems for a class of nonlinear integral equations," *Math. Z.* 137 (1974), 49–66.

[185] Szidarovszky, F. and Yakowitz, S. *Principles and Procedures in Numerical Analysis*, Plenum, New York, 1978.

[186] Szidarovsky, F. and Bahill, T. *Linear Systems Theory*, CRC Press, Boca Raton, FL, 1992.

[187] Taylor, A.E. *Introduction to Functional Analysis*, Wiley, New York, 1957.

[188] Törnig, W. "Monoton konvergente iterationsverfahren zür Lösung michtlinearer differenzen-randwertprobleme," *Beitrage zür Numerischen Mathematik*, 4 (1975), 245–257.

[189] Traub, J.F. *Iterative Methods for the Solution of Equations*, Prentice-Hall, Englewood Cliffs, NJ, 1964.

[190] Triconi, F.G. *Integral Equations*, Interscience Publ., 1957.

[191] Ul'm, S. "Iteration methods with divided differences of the second order" (Russian), *Dokl. Akad. Nauk. SSSR*, 158 (1964), 55–58.

[192] Urabe, M. "Convergence of numerical iteration in solution of equations," *J. Sci. Hiroshima University*, Ser. A, 19 (1976), 479–489.

[193] Vandergraft, J.S. "Newton's method for convex operators in partially ordered spaces," *S.I.A.M. J. Numer. Anal.* 4 (1967), 406–432.

[194] Varga, R.S. *Matrix Iterative Analysis*, Prentice-Hall, Englewood Cliffs, NJ, 1962.

[195] Walker, H.F. and Watson, L.T. "Large change secant update methods for undetermined systems," *S.I.A.M. J. Numer. Anal.* 27 (1990), 1227–1262.

[196] Wang, J.Z. "A cubic sphere wavelet basis of Sobolev spaces and multilevel interpolations," preprint.

[197] Weiss, R. "On the approximation of fixed points of nonlinear compact operators," *S.I.A.M. J. Num. Anal.* 11 (1974), 550–553.

[198] Yamamoto, T. "Error bounds for computed eigenvalues and eigenvectors," *Numer. Math.* 39 (1980), 189–199.

[199] Yamamoto, T. "A method for finding chart error bounds for Newton's method under the Kantorovich assumptions," *Numer. Math.* 44 (1986), 203–220.

[200] Yamamoto, T. "A convergence theorem for Newton-like methods in Banach spaces," *Numer. Math.* 51 (1987), 545–557.

[201] Yamamoto, T. "On the method of tangent hyperbolas in Banach spaces," *J. Comp. Appl. Math.* 21 (1988), 75–86.

[202] Yamamoto, T. and Chen, Z. "Convergence domains of certain iterative methods for solving nonlinear equations," *Numer. Funct. Anal. Optimiz.* 10 (1989), 34–48.

[203] Ypma, T.J. "Affine invariant convergence results for Newton's methods," *BIT*, 22 (1982), 108–118.

[204] Ypma, T.J. "Local convergence of inexact Newton methods," *S.I.A.M. J. Numer. Anal.* 21 (1984), 583–590.

[205] Zaanen, A.C. *Linear Analysis*, North-Holland Publ., Amsterdam, 1953.

[206] Zabrejko, P.P. and Majorova, N.L. "On the solvability of nonlinear Uryson integral equations" (Russian), *Kach. Pribl. Metody Issled. Oper. Uravn.* 3 (1978), 61–73.

[207] Zabrejko, P.P. and Nguen, D.F. "The majorant method in the theory of Newton-Kantorovich approximations and the Ptak error estimates," *Numer. Funct. Anal. Optimiz.* 9 (5 and 6), (1987), 671–684.

[208] Zabrejko, P.P. and Zlepko, P.P. "On majorants of Uryson integral operators" (Russian), *Kach. Pribl. Metody Issled. Oper. Uravn.* 8 (1983), 67–76.

[209] Zincenko, A.I. "A class of approximate methods for solving operation equations with nondifferentiable operators," *Dopovidi Akad. Nauk Ukrain. RSR* (1963), 156–161.

[210] Zlepko, P.P. and Migovich, F.M. "An application of a modification of the Newton-Kantorovich method to the approximate construction of Implicit functions" (Ukrainian), *Ukrainskii Mathematischeskii Zhürnal*, 30, 2 (1978), 222–226.

Index

569

Printed and bound by CPI Group (UK) Ltd, Croydon, CR0 4YY

23/10/2024

01778227-0020